# Current Topics in Microbiology and Immunology

## Volume 323

Series Editors

Richard W. Compans
Emory University School of Medicine, Department of Microbiology and
Immunology, 3001 Rollins Research Center, Atlanta, GA 30322, USA

Max D. Cooper
Howard Hughes Medical Institute, 378 Wallace Tumor Institute, 1824 Sixth
Avenue South, Birmingham, AL 35294-3300, USA

Tasuku Honjo
Department of Medical Chemistry, Kyoto University, Faculty of Medicine,
Yoshida, Sakyo-ku, Kyoto 606-8501, Japan

Hilary Koprowski
Thomas Jefferson University, Department of Cancer Biology, Biotechnology
Foundation Laboratories, 1020 Locust Street, Suite M85 JAH, Philadelphia,
PA 19107-6799, USA

Fritz Melchers
Biozentrum, Department of Cell Biology, University of Basel, Klingelbergstr.
50–70, 4056 Basel Switzerland

Michael B.A. Oldstone
Department of Neuropharmacology, Division of Virology, The Scripps Research
Institute, 10550 N. Torrey Pines, La Jolla, CA 92037, USA

Sjur Olsnes
Department of Biochemistry, Institute for Cancer Research, The Norwegian
Radium Hospital, Montebello 0310 Oslo, Norway

Peter K. Vogt
The Scripps Research Institute, Dept. of Molecular & Exp. Medicine, Division
of Oncovirology, 10550 N. Torrey Pines. BCC-239, La Jolla, CA 92037, USA

Steven Tracy • M. Steven Oberste
Kristen M. Drescher
Editors

# Group B Coxsackieviruses

Dr. Steven Tracy
University of Nebraska
   Medical Center
Department of Pathology & Microbiology
Omaha, NE
USA
e-mail: stracy@unmc.edu

Dr. M. Steven Oberste
Centers for Disease Control and Prevention
National Center for Immunization &
   Respiratory Diseases
Division of Viral Diseases
Atlanta, GA, USA
e-mail: soberste@cdc.gov

Dr. Kristen M. Drescher
Creighton University School of Medicine
Department of Medical Microbiology and Immunology
2500 California Plaza
Omaha, NE
USA
email: kdresche@creighton.edu

*Cover Illustration:* (clockwise from top right): 1) The image of the coxsackievirus B3 (CVB) capsid is color-coded to reflect distinct structural proteins (VP1, blue; VP2, green; VP3, red). The five, three and two-fold axes of symmetry are clearly visible. The canyons where the CVB receptor, CAR, binds upon infection are also evident. 2) The sequence shows the 5′ cloverleaf secondary structure observed in CVB and other human enteroviruses. The colored letters indicate various deletions that can naturally occur during CVB replication in human or mouse heart tissue or in primary cell cultures. These mutations severely attenuate viral replication, permitting long-term persistence in the immunocompetent host. 3) A one week-old infant with hepatic and cardiac failure due to perinatally-acquired coxsaxckievirus B4 infection experienced a complicated clinical course because of severe hepatitis with disseminated intravascular coagulopathy (hemorrhage-hepatitis syndrome), myocarditis and seizures (encephalomyocarditis syndrome). The infant survived after prolonged neonatal intensive care. 4) Dual-immunofluorescent staining of a pancreatic Islet of Langerhans from a NOD mouse previously inoculated with CVB3 using antibodies for enterovirus capsid proteins and insulin. Pancreatic beta cells are susceptible to CVB3 infection, as both enterovirus capsid protein (red) and insulin (green), a marker of beta cells, co-localize.

   The viral image is courtesy of and was created by, Dr. Jean-Yves Sgro, Institute for Molecular Virology, University of Wisconsin-Madison, Madison WI USA (http://virology.wisc.edu/virusworld). It was created using PyMol [DeLano, W.L. "The PyMOL Molecular Graphics System" DeLano Scientific LLC, San Carlos CA, USA. http://www.pymol.org] from PDB coordinates 1COV [Muckelbauer, J.K. et al. (1995) Structure Determination of Coxsackievirus B3 (CVB3) to 3.5 Angstroms Resolution. Acta Crystallogr., Sect.D v51 p.871]. The 5′ terminal sequence figure is adapted from Kim et al., J Virol (2005) 79:7024-7041, "5′-Terminal deletions occur in coxsackievirus B3 during replication in murine hearts and cardiac myocyte cultures and correlate with encapsidation of negative-strand viral RNA". The image of the child has been altered to disguise identity and is courtesy of Dr. Jose R. Romero, Department of Pediatrics, University of Nebraska Medical Center, Omaha, NE USA. The islet image is courtesy of Dr. K. M. Drescher, Department of Medical Microbiology and Immunology, Creighton University School of Medicine, Omaha, NE USA.

ISBN 978-3-540-75545-6       e-ISBN 978-3-540-75546-3
DOI: 10.1007/978-3-540-75546-3

Current Topics in Microbiology and Immunology ISSN 007-0217x

Library of Congress Catalog Number: 72-152360

© 2008 Springer-Verlag Berlin Heidelberg

This work is subject to copyright. All rights reserved, whether the whole or part of the material is concerned, specifically the rights of translation, reprinting, reuse of illustrations, recitation, broadcasting, reproduction on microfilm or in any other way, and storage in data banks. Duplication of this publication or parts thereof is permitted only under the provisions of the German Copyright Law of September, 9, 1965, in its current version, and permission for use must always be obtained from Springer-Verlag. Violations are liable for prosecution under the German Copyright Law.

The use of general descriptive names, registered names, trademarks, etc. in this publication does not imply, even in the absence of a specific statement, that such names are exempt from the relevant protective laws and regulations and therefore free for general use.

Product liability: The publisher cannot guarantee the accuracy of any information about dosage and application contained in this book. In every individual case the user must check such information by consulting the relevant literature.

*Cover Design*: WMXDesign GmbH, Heidelberg, Germany

Printed on acid-free paper

9 8 7 6 5 4 3 2 1

springer.com

# Preface

The group B coxsackieviruses have a long and colorful history, dating to the early days of virology as we now know it. In the late 1940s, ultracentrifugation and electron microscopy were new, high-tech tools and suckling mice were supplanting monkeys as the virus isolation vessel of choice. Viruses were, often as not, still referred to as "filterable agents." The rampage of paralytic poliomyelitis epidemics in the previous 20 or so years had spurred national investment in infectious disease research, resulting in an unprecedented period of virus discovery, eclipsed only a few years later once cell culture became the preferred method to isolate and identify mammalian viruses. The coxsackieviruses were isolated from feces of patients with paralytic poliomyelitis and nonparalytic poliomyelitis (aseptic meningitis), causing disease in suckling mice, but not in adult mice or monkeys. They were considered to be related to the polioviruses on the basis of their physical properties, such as virion size, acid and ether resistance, and temperature stability in 50% glycerol, and were classified into groups A and B by the nature of the disease induced in mice: flaccid paralysis by group A viruses and spastic paralysis by those of group B.

Our knowledge of the group B coxsackieviruses has progressed dramatically in the past 60 years. Some of the most recent advances include the identification of the coxsackievirus–adenovirus receptor, the dissection of genetic elements linked to virulence/attenuation, examination of the impact of recombination in virus evolution and diversity, and analysis of the role of viral proteins in regulating host-cell macromolecule synthesis and trafficking. The first edition of this work, published in 1997, described the molecular biology of coxsackie B viruses, as well as clinical, epidemiological, and immunological aspects of group B coxsackievirus disease. Much has been accomplished in the past 10 years, including determination of the crystal structure of a virus–receptor complex, significant advances in understanding the molecular details of virus–host interaction within the cell, and deeper insights into the systemic effects of virus infection and the host response. This second edition summarizes the current state of knowledge in group B coxsackievirus genomics and replication, receptor structure and function, host-cell interactions, the host immune response and immunopathology, viral virulence and pathogenesis, and the role of this important group of viruses in acute and chronic disease in humans.

2007                                                Steven Oberste, Kristen Drescher,
                                                                    and Steven Tracy

# Coxsackie B Viruses: An Introduction

B. W. J. Mahy

## History

The first isolates of what are now termed coxsackieviruses were made from the feces of two boys suffering from paralytic poliomyelitis who lived in the village of Coxsackie, New York (Dalldorf and Sickles 1948). These two isolates were not neutralized by antisera against polioviruses and provided the first evidence of the existence of a large number of human enteric viruses, many of which caused no apparent disease and so were called "enteric cytopathic human orphan" or echoviruses. However, the coxsackieviruses had in common the property of being highly pathogenic on injection into newborn mice and hamsters, and within 1 year following their discovery it was found that some isolates of coxsackievirus induced more severe pathological changes, such as generalized skeletal muscle destruction, than others (Gifford and Dalldorf 1951). On this basis, the viruses were divided into two groups termed A and B, with group B viruses causing the more severe symptoms. As further coxsackieviruses were isolated, they were assigned to one of the two groups and given sequential numbers (Dalldorf 1955).

In 1958, coxsackieviruses of group B were found to have caused epidemic myocarditis in newborn infants in South Africa (Gear and Measroch 1958), and it was realized that these viruses had a worldwide distribution. Coxsackie B viruses were found to be responsible for cases of pleurodynia in South Africa (Patz et al. 1953), including a laboratory accident in which a worker infected with coxsackie B2 virus developed pleurodynia and aseptic meningitis (Curnen 1950). Pleurodynia is also called epidemic myalgia, devil's grippe, or Bornholm's disease, named after an epidemic on the Danish island of Bornholm, and an outbreak of this disease in Oxford, England was shown to be due to coxsackie B3 virus (Warin et al. 1953).

Despite the subsequent discovery of more than 100 human and simian enteroviruses, 23 of which are classified as serotypes of coxsackievirus A, only six serotypes

---

B. W. J. Mahy
Coordinating Center for Infectious Diseases, Centers for Disease
Control and Prevention, Atlanta, GA, USA
bxm1@cdc.gov

of coxsackievirus B have been recognized. One of these serotypes, B5, is now believed to be virtually identical to the porcine enterovirus, *swine vesicular disease virus*, based on genome sequence analysis, and is now classified as a subspecies of Coxsackievirus B5, which is itself regarded as a strain of *Human enterovirus B*. Coxsackievirus B5 was first identified in 1952 in the feces of a patient with mild paralytic disease, and vesicular disease was first recognized in pigs in Italy in 1966 and subsequently shown to be caused by an enterovirus and named swine vesicular disease (Nardelli et al. 1968). The importance of swine vesicular disease virus (SVDV) is that, although it causes only low morbidity and mortality in pigs, it is highly contagious and induces lesions that are clinically indistinguishable from those seen in pigs infected with the economically important foot-and-mouth disease virus. Attempts have been made to induce clinical lesions in pigs with human coxsackie B5 virus, with little success, and the molecular basis for the differences in pathogenesis between the viruses is not presently understood (Seechurn et al. 1990; Zhang et al. 1993).

## Attachment and Growth in Cells

Human coxsackievirus B grows well in monkey kidney cells and in human cell lines such as A-549, HeLa, Hep-2, or RD cells. However, not all samples from a patient with coxsackievirus B infection will produce a cytopathic effect in cell culture, and virus isolation may prove difficult (Chonmaitree et al. 1982).

A receptor for attachment and subsequent entry of coxsackie B virus has been identified as a 46-kDa protein that also serves as the receptor for adenoviruses 2 and 5 (Bergelson et al. 1997) and is called the coxsackievirus and adenovirus receptor (CAR). However, there may be alternative receptors (Bergelson 2003), and in particular, although it was shown that although all six serotypes of coxsackievirus B will bind to CAR (Martino et al. 2000), coxsackieviruses B1, B3, and B5 can use decay accelerating factor (DAF) as a receptor for attachment (Shafren et al. 1995). DAF is the 70-kDa complement regulatory protein also known as CD55. Differences have been noted between clinical isolates of coxsackievirus B and various cell-adapted viruses, and the relative dependence of the coxsackievirus B on DAF or CAR may be altered, depending on the cell line used for propagation (Goodfellow et al. 2005).

Knowledge of these receptors is now being used to explore the use of recombinant soluble DAF and CAR molecules as inhibitors of coxsackievirus B-induced myocarditis and pancreatitis, using mouse models (Yanagawa et al. 2004).

## Structure and Replication

Preparations of coxsackievirus B1 derived from an infectious cDNA clone have been crystallized (Li et al. 1992) and the structure of the B3 virus determined to 3.5 Å resolution (Muckelbauer et al. 1995; Muckelbauer and Rossman 1997). As with other enteroviruses, each of the 12 pentamers on the icosahedron is surrounded by a canyon which is the binding site on the picornavirus capsid for specific cellular receptor molecules (Rossman and Palmenberg 1988).

Like other enteroviruses, coxsackievirus B contains a single-stranded RNA genome of positive sense with a genome-linked protein (VPg) linked to the 5'-end and a polyA tail at the 3'-end. The genome of coxsackievirus B1 is 7,389 nucleotides in length, and the open reading frame extends from nucleotide 742 and ends at nucleotide 7,287 (Iizuka et al. 1987). Following attachment and entry into the cell, the virion is uncoated and the released genome RNA acts directly as a messenger RNA in which the 5'-noncoding region serves as an internal initiation site for translation (Pelletier and Sonnenberg 1988). Studies using an infectious clone of coxsackievirus B1 suggest that a conserved 21 nucleotide region from positions 546 to 566 is important for translation initiation (Iizuka et al. 1991).

Once translation begins, host-cell protein synthesis is rapidly shut off, and the viral genome is translated into a large polyprotein, which is subsequently cleaved by virus-coded proteases into functional proteins. These include the proteins destined for incorporation into new virus capsids, as well as the RNA-dependent RNA polymerase and other enzymes involved in genome replication.

## Disease Associations

The coxsackieviruses of group B were originally singled out as a group because of their capacity to induce severe disease symptoms in newborn mice and hamsters, so it is not surprising that their association with a variety of diseases in humans has been established in several instances and is suspected in others. Those diseases which are generally accepted to result from coxsackievirus B infection include aseptic meningitis and acute myocarditis and pericarditis. It has been estimated that nearly 30% of all recently diagnosed cases of myocarditis are caused by infection with coxsackie B virus (Horwitz et al. 2006).

There is also good evidence that neonatal infection with coxsackievirus B may result in disseminated infection with meningitis, myocarditis, and occasionally fatal systemic infection. In such cases, the presence of viral RNA may be detected in the myocardium by sequence analysis of enteroviral polymerase chain reaction (PCR) products (Archard et al. 1998).

It is believed that a late complication of healed coxsackie B viral myocarditis is the development of dilated cardiomyopathy, and the presence of viral RNA has been demonstrated in biopsy tissue from the cardiac lesions (Archard et al. 1998; Fujioka and Kitaura 2001).

An association of coxsackievirus B infection with metabolic myopathy, reflecting impaired muscle energy metabolism, has been reported, and has been used to suggest a role for the virus as a cause of chronic fatigue syndrome (CFS). Quadriceps muscle biopsy samples taken from 48 CFS patients were examined by PCR using enterovirus-specific primers (Lane et al. 2003). The results were equivocal as only 20.8% of the patient samples were positive compared to none of the controls.

Perhaps the most significant disease association is that of coxsackievirus B with persistent infection of human pancreatic islet cells that mimics the loss of beta-cell function seen during the clinical course of autoimmune diabetes (Yin et al. 2002). There is epidemiological evidence both for and against this association. A 10-year study in Jefferson County, Alabama, showed an increased incidence of insulin-dependent diabetes mellitus (IDDM) in persons under 20 years of age following an epidemic of coxsackievirus B5 infection, which began in 1983 (Wagenknecht et al. 1991). A serological study conducted in Pittsburgh on children 18 years of age or younger showed a clear relation between IDDM and enterovirus IgM positivity (Helfand et al. 1995). Similarly, a study in Finland supported a link between IDDM and enterovirus infections in young children (Sadeharju et al. 2003). Further evidence for the association has come from studies in Germany (Moya-Suri et al. 2005), Scotland (Clements et al. 1995), and Belgium (Brilot and Geenen 2005). On the other hand, investigation of the pancreatic tissue from two fatal cases of IDDM failed to reveal any evidence for coxsackievirus sequences using PCR or Southern blot hybridization, (Buesa-Gomez et al. 1994). Recently a systematic review of published evidence for and against a relationship between coxsackievirus B serology and IDDM concluded that overall the results were inconsistent (Green et al. 2004). It has also been shown that coxsackie B virus may increase the severity of alcoholic chronic pancreatitis (DiMagno and DiMagno 2005).

Despite the uncertainty, an attractive hypothesis exists as to the possible molecular mechanism underlying IDDM. This disease is characterized by autoimmune destruction of insulin-producing pancreatic beta cells in the islets of Langerhans. It has clearly been shown in mice that a T-helper-based autoimmune response arises spontaneously against the enzyme glutamic acid decarboxylase concurrently with the onset of IDDM. Coxsackievirus B infection of the mice causes an increase in the expression of this autoantigen (Hou et al. 1993). Remarkably, a sequence of six identical amino acids (PEVKEK) is shared between glutamic acid decarboxylase and the 2C protein of coxsackievirus B, making molecular mimicry an attractive hypothesis for the induction of autoimmune diabetes following virus infection (Vreugdenhil et al. 1998; Kuhreja and Maclaren 2000; Chou et al. 2004). Whether or not coxsackievirus B acts as a trigger for the induction of IDDM, there are clearly other contributory factors involved such as the genetics of the host (See and Tilles 1998; Frisk and Tuvemo 2004; Hindersson et al. 2005) and differences in the strain of virus (Al-Hello et al. 2005). It seems likely that these many factors may underlie the contradictory reports from epidemiological studies.

## Therapy

It has been known for some time that interferon is an effective inhibitor of coxsackievirus B replication, so the effect of an interferon inducer, Ampligen (poly(I)-poly(C12U), was tested for its effect on induction of myocarditis in mice. It is believed that coxsackievirus B initially replicates in the pancreas and quickly spreads to the heart, inducing chronic autoimmunity. Ampligen given to infected mice at 20 mg/kg/day reduced the severity of virus-induced myocarditis by 98%, and was more effective than interferon itself or pegylated interferon (Padalko et al. 2004). Another approach to treatment of mice involved creating transgenic mice that constitutively express transforming growth factor-β (TGF-β). The expression of TGF-beta within pancreatic β cells prevented the mice from developing autoimmune myocarditis after infection with coxsackievirus B. In contrast, transgenic expression of interleukin-4 did not inhibit virus-induced heart disease (Horwitz et al. 2006).

So far as IDDM is concerned, the use of an antipicornaviral drug, pleconaril, has been tested directly in isolated human pancreatic islet cells. Two coxsackievirus B4 strains were used to infect the islet cells. The viruses replicated well in the islet cells, and in the presence of pleconaril one virus strain was inhibited to undetectable levels, whereas the other strain appeared to be resistant following an initial drop in titer.

This experiment emphasizes the importance of testing more than one virus strain when conducting tests for potential therapies for coxsackieviruses.

## Conclusions

The coxsackieviruses B were discovered at a time when tissue culture was in its infancy and classification and ordering of the viruses depended on serological techniques that were occasionally difficult to interpret. Now that genome sequence analysis has given much greater clarity to our understanding of these viruses, their relationships have become clear. We are also on the threshold of an understanding of the molecular basis of some of the diseases they cause, though much research remains to be done in this area. There is therefore a much firmer ground on which to base future therapeutic measures and eventually to limit the burden of these diseases in the future.

## References

Al-Hello H, Davydova B, Smura T, Kaialainen S, Ylipaasto P, Saario E, Hovi T, Rieder E, Roivainen M (2005) Phenotypic and genetic changes in coxsackievirus B5 following repeated passage in mouse pancreas in vivo. J Med Virol 75:566–574

Archard LC, Khan MA, Soteriou BA, Zhang H, Why HJ, Robinson NM, Richardson PJ (1998) Characterization of Coxsackie B virus RNA in myocardium from patients with dilated cardiomyopathy by nucleotide sequencing of reverse transcription-nested polymerase chain reaction products. Hum Pathol 29:578–584

Bergelson JM, Cunningham JA, Droguett G, Kurt-Jones EA, Krithivas A, Hong JS, Horwitz MS, Crowell RL, Finberg RW (1997) Isolation of a common receptor for Coxsackie B viruses and adenoviruses 2 and 5. Science 275:1320–1323

Bergelson JM (2003) Virus interactions with mucosal surfaces: alternative receptors, alternative pathways. Curr Opin Microbiol 6:386–391

Brilot F, Geenen V (2005) Role of viral infections in the pathogenesis of type 1 diabetes. Rev Med Liege 60:297–302

Buesa-Gomez J, de la Torre JC, Dyrberg T, Landin-Olsson M, Mauseth RS, Lernmark A, Oldstone MB (1994) Failure to detect genomic viral sequences in pancreatic tissues from two children with acute-onset diabetes mellitus. J Med Virol 42:193–197

Chonmaitre T, Menegus MA, Powell KR (1982) The clinical relevance of CSF viral culture. A two year experience with aseptic meningitis in Rochester, New York. JAMA 247:1843–184

Chou CC, Lin KH, Ke GM, Tung YC, Chao MC, Cheng JY, Chen BH (2004) Comparison of nucleotide sequence of p2C region in diabetogenic and non-diabetogenic coxsackie virus B5 isolates. Kaohsiung J Med Sci 20:525–532

Clements GB, Galbraith DN, Taylor KW (1995) Coxsackie B virus infection and onset of childhood diabetes. Lancet 346:221–223

Curnen EC (1950) Human disease associated with the coxsackie viruses. Bull N Y Acad Med 26:335–342

Dalldorf G (1955) The coxsackie viruses. Ann Rev Microbiol 9:277–296

Dalldorf G, Sickles GM (1948) An unidentified, filterable agent isolated from the feces of children with paralysis. Science 108:61–62

DiMagno MJ, DiMagno EP (2005) Chronic pancreatitis. Curr Opin Gastroenterol 21:544–554

Frisk G, Tuvemo T (2004) Enterovirus infections with beta cell tropic strains are frequent in siblings of children diagnosed with type 1 diabetes and in association with elevated levels of GAD65 antibodies. J Med Virol 73:450–459

Fujioka S, Kitaura Y (2001) Coxsackie B virus infection in idiopathic dilated cardiomyopathy: clinical and pharmacological implications. Biodrugs 15:791–799

Gear J, Measroch V (1958) Cases of meningo-encephalitis due to the coxsackie A-like ECHO 9 virus. S Afr Med J 32:1062–1066

Gifford R, Dalldorf G (1951) The morbid anatomy of experimental coxsackie virus infection. Am J Path 27:1047–1064

Goodfellow IG, Evans DJ, Blom AM, Kerrigan D, Miners JS, Morgan BP, Spiller OB (2005) Inhibition of coxsackie B virus infection by soluble forms of its receptors: binding affinities, altered particle formation, and competition with cellular receptors. J Virol 79:12016–12024

Green J, Casabonne D, Newton R (2004) Coxsackie B virus serology and type 1 diabetes mellitus: a systematic review of published case–control studies. Diabet Med 21:507–514

Helfand RF, Gary HE, Freeman CY, Anderson LJ, Pallansch MA (1995) Serologic evidence of an association between enteroviruses and the onset of type 1 diabetes mellitus. Pittsburgh Diabetes Research Group. J Infect Dis 172:1206–1211

Hindersson M, Eishebani A, Orn A, Tuvemo T, Frisk G (2005) Simultaneous type 1 diabetes onset in mother and son coincident with an enteroviral infection. J Clin Virol 33:158–167

Horwitz MS, Knudson M, Ilic A, Fine C, Sarvetnick N (2006) Transforming growth factor-beta inhibits coxsackievirus-mediated autoimmune myocarditis. Viral Immunol 19:722–733

Hou J, Sheikh S, Martin DL, Chatterjee NK (1993) Coxsackie virus B4 alters pancreatic glutamate decarboxylase expression in mice soon after infection. J Autoimmun 6:529–542

Iizuka N, Kuge S, Nomoto A (1987) Complete nucleotide sequence of the genome of coxsackievirus B1. Virology 156:64–73

Iizuka N, Yonekawa H, Nomoto A (1991) Nucleotide sequences important for translation initiation of enterovirus RNA. J Virol 65:4867–4873

Kukreja A, Maclaren NK (2000) Current cases in which epitope mimicry is considered as a component cause of autoimmune disease: immune-mediated (type 1) diabetes. Cell Mol Life Sci 57:534–541

Lane RJ, Soteriou BA, Zhang H, Archard LC (2003) Enterovirus related metabolic myopathy: a postviral fatigue syndrome. J Neurol Neurosurg Psychiatry 74:1382–1386

Li T, Zhang AQ, Iizuka N, Nomoto A, Arnold E (1992) Crystallization and preliminary X-ray diffraction studies of coxsackievirus B1. J Mol Biol 223:1171–1175

Martino TA, Petric M, Weingartl H, Bergelson JM, Opavsky MA, Richardson CD, Modlin JF, Finberg RW, Kain KC, Willis N, Gauntt CJ, Liu PP (2000) The coxsackie-adenovirus receptor is used by reference strains and clinical isolates representing all six serotypes of coxsackievirus group B and by swine vesicular disease virus. Virology 271:99–108

Moya-Suri V, Schlosser M, Zimmermann K, Rjasanowski I, Gurtler L, Mentel R (2005) Enterovirus RNA sequences in sera of schoolchildren in the general population and their association with type 1 diabetes-associated autoantibodies. J Med Microbiol 54:879–883

Muckelbauer JK, Kremer M, Minor I, Tong L, Zlotnick A, Johnson JE, Rossman MG (1995) Structure determination of coxsackievirus B3 to 3.5 A. Acta Crystallogr D Biol Crystallogr 51:871–887

Muckelbauer JK, Rossman MG (1997) The structure of coxsackievirus B3. Curr Top Microbiol Immunol 223:191–208

Nardelli L, Lodetti E, Gualandi GL, Burrows R, Goodridge D, Brown F, Cartwright B (1968) A foot-and-mouth syndrome in pigs caused by an enterovirus. Nature 219:1275–1276

Padalko E, Nuyens D, De Palma A, Verbekern E, Aerts JL, DeClercq E, Carmeliet P, Neyts J (2004) The interferon inducer Ampligen (polyI-polyC12U) markedly protects mice against coxsackie B3 virus-induced myocarditis. Antimicrob Agents Chemother 48:267–274

Patz IM, Measroch V, Gear J (1953) Bornholm disease, pleurodynia or epidemic myalgia: an outbreak in the Transvaal associated with coxsackie virus infection. S Afr Med J 27:397–402

Pelletier J, Sonenberg N (1988) Internal initiation of translation of eukaryotic mRNA directed by a sequence derived from poliovirus RNA. Nature 334:320–325

Rossman MG, Palmenberg AC (1988) Conservation of the putative receptor attachment site in picornaviruses. Virology 164:373–382

Sadeharju K, Hamalainen AM, Knip M, Lonnrot M, Koskela P, Virtanen SM, Ilonen J, Akerblom HK, Hyoty H; Finnish TRIGR Study Group (2003) Enterovirus infections as a risk factor for type 1 diabetes: virus analyses in a dietary intervention trial. Clin Exp Immunol 132:271–277

See DM, Tilles JG (1998) The pathogenesis of viral-induced diabetes. Clin Diagn Virol 9:85–88

Seechurn P, Knowles NJ, McCauley JW (1990) The complete nucleotide sequence of a pathogenic swine vesicular disease virus. Virus Res 16:255–274

Shafren DR, Bates RC, Agrez MV, Herd RL, Burns GF, Barry RD (1995) Coxsackieviruses B1, B3, and B5 use decay accelerating factor as a receptor for cell attachment. J Virol 69:3873–3877

Vreugdenhil GR, Geluk A, Ottenhoff TH, Melchers WJ, Roep BO, Galama JM (1998) Molecular mimicry in diabetes mellitus: the homologous domain in coxsackie B virus protein 2C and islet autoantigen GAD 65 is highly conserved in the coxsackie B-like enteroviruses and binds to the diabetes associated HLA-DR3 molecule. Diabetologia 41:40–46

Warin JF, Davies JBM, Sanders FK, Vizoso AD (1953) Oxford epidemic of Bornholm disease. BMJ 1:1345–1351

Wagenknect LE, Roseman JM, Herman WH (1991) Increased incidence of insulin-dependent diabetes mellitus following an epidemic of coxsackievirus B5. Am J Epidemiol 133:1024–1031

Yanagawa B, Spiller OB, Proctor DG, Choy J, Luo H, Zhang HM, Suarez A, Yang D, McManus BM (2004) Soluble recombinant coxsackievirus and adenovirus receptor abrogates coxsackievirus b3-mediated pancreatitis and myocarditis in mice. J infect Dis 189:1431–1439

Yin H, Berg AK, Westman J, Hellerstrom C, Frisk G (2002) Complete nucleotide sequence of a coxsackievirus B4 strain capable of establishing persistent infection in human pancreatic islet cells: effects on insulin release, proinsulin synthesis, and cell morphology. J Med Virol 68:544–557

Zhang G, Wilsden G, Knowles NJ, McCauley JW (1993) Complete nucleotide sequence of a coxsackie B5 virus and its relationship to swine vesicular disease virus. J Gen Virol 74:845–853

# Contents

**Section I    CVB Genetics**

**Coxsackieviruses and Quasispecies Theory:**
**Evolution of Enteroviruses** .......................................... 3
E. Domingo, V. Martin, C. Perales, and C. Escarmis

**Comparative Genomics of the Coxsackie B Viruses**
**and Related Enteroviruses** .......................................... 33
M. S. Oberste

**Group B Coxsackievirus Virulence** ............................. 49
S. Tracy and C. Gauntt

**Section II    CVB Entry and Replication**

**The Coxsackievirus and Adenovirus Receptor**....................... 67
P. Freimuth, L. Philipson, and S. D. Carson

**Coxsackievirus B RNA Replication: Lessons from Poliovirus** .......... 89
P. Sean and B. L. Semler

**CVB Translation: Lessons from the Polioviruses**..................... 123
J. M. Bonderoff and R. E. Lloyd

**Preferential Coxsackievirus Replication in Proliferating/**
**Activated Cells: Implications for Virus Tropism, Persistence,**
**and Pathogenesis** ................................................. 149
R. Feuer and J. L. Whitton

**Section III    Host-Virus Interaction**

**The Impact of CVB3 Infection on Host Cell Biology**.................. 177
D. Marchant, X. Si, H. Luo, B. McManus, and D. Yang

**Host Immune Responses to Coxsackievirus B3** ..................... 199
S. Huber

**Pediatric Group B Coxsackievirus Infections** ..................... 223
J. R. Romero

**CVB-Induced Pancreatitis and Alterations in Gene Expression** ........ 241
A. I. Ramsingh

**The CVB and Etiology of Type 1 Diabetes** ........................ 259
K. M. Drescher and S. M. Tracy

**Persistent Coxsackievirus Infection: Enterovirus Persistence
in Chronic Myocarditis and Dilated Cardiomyopathy** ................ 275
N. M. Chapman and K.-S. Kim

**Autoimmunity in Coxsackievirus Infection** ........................ 293
N. R. Rose

**CVB Infection and Mechanisms of Viral Cardiomyopathy** ............ 315
K. U. Knowlton

**Index** .......................................................... 337

# Contributors

J.M. Bonderoff
Department of Molecular Virology and Microbiology, Baylor College
of Medicine, One Baylor Plaza, Houston, TX 77030, USA

S.D. Carson
Department of Pathology and Microbiology, University of Nebraska
Medical Center, 986495 Nebraska Medical, Omaha, NE 68198-6495,
USA, scarson@unmc.edu

N.M. Chapman
University of Nebraska Medical Center, Omaha, NE 68198-6495, USA,
nchapman@unmc.edu

E. Domingo
Centro de Biología Molecular "Severo Ochoa" (CSIC-UAM),
Universidad Autónoma de Madrid, Cantoblanco, 28049 Madrid, Spain,
edomingo@cbm.uam.es

K.M. Drescher
Department of Medical Microbiology and Immunology, Creighton University
School of Medicine, Omaha, NE 68178, USA, kdresche@creighton.edu

C. Escarmis
Centro de Biología Molecular "Severo Ochoa" (CSIC-UAM), Universidad
Autónoma de Madrid, Cantoblanco, 28049 Madrid, Spain

R. Feuer
Department of Biology, San Diego State University,
5500 Campanille Drive, San Diego, CA 92182-4614, USA

P. Freimuth
Biology Department, Brookhaven National Laboratory, Upton, NY 11973, USA

C. Gauntt
Rancho de los Perdidos, Burnet, TX 78611, USA

S. Huber
University of Vermont, Burlington, VT 05405, USA, Sally.Huber@uvm.edu

K.-S. Kim
University of Nebraska Medical Center, Omaha, NE 68198-6495, USA

K.U. Knowlton
Department of Medicine and Institute of Molecular Medicine, University of California at San Diego, 9500 Gilman Drive, La Jolla, CA 92093-0613K, USA, kknowlton@ucsd.edu

R.E. Lloyd
Department of Molecular Virology and Microbiology, Baylor College of Medicine, One Baylor Plaza, Houston, TX 77030, USA, rlloyd@bcm.tmc.edu

H. Luo
The James Hogg iCAPTURE Centre, University of British Columbia, Providence Health Care, St. Pauls Hospital, Vancouver, BC, Canada

B.W.J. Mahy
Coordinating Center for Infectious Diseases, Centers for Disease Control and Prevention, Atlanta, GA, USA, bxml@cdc.gov

D. Marchant
The James Hogg iCAPTURE Centre, University of British Columbia, Providence Health Care, St. Pauls Hospital, Vancouver, BC, Canada

V. Martin
Centro de Biología Molecular "Severo Ochoa" (CSIC-UAM), Universidad Autónoma de Madrid, Cantoblanco, 28049 Madrid, Spain

B. McManus
The James Hogg iCAPTURE Centre, University of British Columbia, Providence Health Care, St. Pauls Hospital, Vancouver, BC, Canada, BMcManus@mrl.ubc.ca

M.S. Oberste
Centers for Disease Control and Prevention, 1600 Clifton Road NE, Mailstop G-17, Atlanta, GA 30333, USA, soberste@cdc.gov

C. Perales
Centro de Biología Molecular "Severo Ochoa" (CSIC-UAM), Universidad Autónoma de Madrid, Cantoblanco, 28049 Madrid, Spain

L. Philipson
Cell and Molecular Biology Department, Karolinska Institutet, Box 285, von Eulers väg 3, 17177 Stockholm, Sweden

P. Sean
Department of Microbiology and Molecular Genetics, School of Medicine, University of California, Irvine, CA 92697, USA

## Contributors

A.I. Ramsingh
Wadsworth Center, New York State Department of Health, 120 New Scotland Avenue, Albany, NY 12208, USA, ramsingh@wadsworth.org

J.R. Romero
University of Nebraska Medical Center, 986495 Nebraska Medical, Omaha, NE 68198-6495, USA, jrromero@unmc.edu

N.R. Rose
Johns Hopkins Center for Autoimmune Disease Research,
Johns Hopkins University, 615 N. Wolfe St., E5009, Baltimore, MD 21205, USA, nrrose@jhmi.edu

B.L. Semler
Department of Microbiology and Molecular Genetics, School of Medicine, University of California, Irvine, CA 92697, USA, blsemler@uci.edu

X. Si
The James Hogg iCAPTURE Centre, University of British Columbia, Providence Health Care, St. Pauls Hospital, Vancouver, BC, Canada

S. Tracy
Department of Pathology and Microbiology, University of Nebraska Medical Center, 986494 and 986495 Nebraska Medical Center, Omaha, NE 68198, USA, stracy@unmc.edu

J.L. Whitton
Molecular and Integrative Neurosciences Department, SP30-2110, The Scripps Research Institute, 10550 N Torrey Pines Road, La Jolla, CA 92037, USA, lwhitton@scripps.edu

D. Yang
The James Hogg iCAPTURE Centre, University of British Columbia, Providence Health Care, St. Pauls Hospital, Vancouver, BC, Canada

# Section I
# CVB Genetics

# Coxsackieviruses and Quasispecies Theory: Evolution of Enteroviruses

E. Domingo(✉), V. Martin, C. Perales, and C. Escarmis

1  Introduction to Quasispecies: Mutation Rates and Mutant Spectra ................................. 4
2  Why Quasispecies? ........................................................................................................ 7
3  Molecular Basis of Mutation and Recombination ......................................................... 8
   3.1  Role of Mutation and Recombination in RNA Virus Evolution ............................ 11
4  The Diversity of Enteroviruses: Species, Quasispecies, and Variation in
   Disease Manifestations .................................................................................................. 12
5  Mutation and Recombination as Mechanisms of Enterovirus Evolution ...................... 15
6  Evolution of Virulence in Enteroviruses ....................................................................... 18
7  Enterovirus Evolution in the Course of Persistent Infections ....................................... 20
8  Mutant Spectrum Complexity as a Factor in Viral Pathogenesis .................................. 20
9  Implications of Quasispecies Dynamics for Viral Disease Prevention and Treatment ...... 22
10 Overview, Connections, Challenges, and Trends .......................................................... 23
References ............................................................................................................................ 25

**Abstract** Enterovirus populations display quasispecies dynamics, characterized by high rates of mutation and recombination, followed by competition, selection, and random drift acting on heterogeneous mutant spectra. Direct experimental evidence indicates that high mutation rates and complex mutant spectra can serve for the adaptation of enteroviruses to complex environments. Studies with the RNA-dependent RNA polymerase of picornaviruses suggest that multiple enzyme sites may influence the template-copying fidelity (incorporation of incorrect vs correct nucleotide) during RNA replication. Mutation and recombination are an unavoidable consequence of the molecular mechanisms inherent to the process of viral genome replication and underlie the diversification of enterovirus genomes as they multiply in human and animal hosts. The diversity of disease manifestations associated with closely related enteroviruses is probably attributable to profound biological effects of some mutations that, because of their limited number, do not

E. Domingo
Centro de Biología Molecular "Severo Ochoa"(CSIC-UAM), Universidad Autónoma de Madrid, Cantoblanco, 28049 Madrid, Spain
edomingo@cbm.uam.es

necessarily affect the phylogenetic position of the virus. The combination of highly dynamic mutant spectra with unpredictable alterations of biological behavior by minimal genetic change defies classical classification schemes. The result is the need to update the grouping of enteroviruses quite frequently into genetic and serological types and subtypes. The tolerance of enterovirus genomes to remain replication-competent despite multiple mutation and recombination events encourages the engineering of live-attenuated vaccines. Also, the application of quasispecies theory to an understanding of the limits of viral genomes to accept mutations, together with an increasingly deeper understanding of the mechanisms of mutagenesis by nucleoside analogs, has paved the way for the application of lethal mutagenesis as a new antiviral strategy.

# 1 Introduction to Quasispecies: Mutation Rates and Mutant Spectra

Enteroviruses, as other RNA viruses, share a potential for adaptation and rapid evolution associated with two key features of their replication: high mutation rates and quasispecies dynamics. One of the first measurements of mutability of an RNA virus was carried out by Eggers and Tamm (Eggers and Tamm 1965) who calculated a rate of $1 \times 10^{-4}$ for the transition of coxsackievirus A9 from dependence to independence of 2-($\alpha$-hydroxybenzyl)-benzimidazole. The value is in line with measurements of mutation rates (the rate of occurrence of mutations during genome replication) and mutation frequencies (the frequency of mutations in a genome population) carried out with several RNA viruses (Drake and Holland 1999). A recent compilation of 19 values for riboviruses, retroviruses, and hepadnaviruses obtained in 15 different laboratories gives an average mutation frequency of $(2.6 \pm 6.6) \times 10^{-4}$ substitutions per nucleotide (Domingo 2007). The genome size of viruses that replicate via RNA is in the range of 3-33 kb. The combined mutation rate and genome size values imply that an average of 0.1-3 mutations per genome are expected to occur every time an RNA template is copied into a complementary RNA or DNA copy. Such template-copying events may take place hundreds of times in *each* infected cell from an infected organism!

Many newly arising mutations are detrimental to a virus, and more so when the virus is well adapted to the environment. Genomes with deleterious mutations are reduced in frequency (or even eliminated) through negative selection. Nevertheless, RNA viral populations that have been examined with some detail at the population level do not include genomes each with the same nucleotide sequence. Rather, viral populations replicating in infected hosts or in cell culture exist at any given time as very complex distributions of closely related (but many of them nonidentical) genomes. The unveiling of a complex population structure was not an easy task since current methods of nucleotide sequencing applied to the analysis of a viral population yield an average or consensus nucleotide sequence of the multitudes of genomes that are present in the sample. It was necessary to sequence the progeny of individual genomes (through molecular or biological cloning) from a population to uncover the presence of genomes whose sequence differs from the consensus. It is expected that with the adaptation of pyrosequencing techniques to the analysis of virus populations, or with the development of procedures to sequence individual genomes, a more accurate quantification of the population complexity of viruses might be achieved.

The complex mutant distributions subjected to a process of genetic variation, competition, and selection are termed viral quasispecies (Fig. 1). Quasispecies was developed as a general theory of molecular evolution by M. Eigen and P. Schuster three decades ago (Eigen and Schuster 1979). The theory addressed self-organization and adaptability of primitive RNA (or RNA-like) replicons that probably constituted an ancient RNA world at the onset of life on earth (Schuster and Stadler 1999). The initial theory was deterministic, a property suitable to place the problem of error-prone replication in general and mathematically solvable terms. In this formulation, mutant genomes were organized as steady-state, equilibrium distributions of infinite size, and mutant behavior was predictable (Eigen and Schuster 1979). It is obvious that in considering either primitive replicons or more advanced virus-like genetic elements, continued population equilibrium in a natural setting is unlikely. To accommodate reality to quasispecies theory, real quasispecies distributions could have been viewed as undergoing successions of brief equilibrium steps. However, this abstraction became unnecessary since theoretical studies on quasispecies have proceeded at a good pace, and extensions to finite genome populations replicating in changing environments have been formulated (Eigen 1987, 2000; Wilke et al. 2001a; Saakian and Hu 2006, among other studies). Quasispecies is one of several mathematical formulations of Darwinian evolutionary dynamics, suitable as a theoretical framework for systems displaying elevated mutation rates (Page and Nowak 2002).

The quasispecies concept had an independent origin in experimental observations made with bacteriophage Qβ. First, a mutation rate of $10^{-4}$ substitutions per nucleotide was calculated for the reversion of a single nucleotide replacement introduced in Qβ RNA by site-directed mutagenesis (Batschelet et al. 1976; Domingo et al. 1976). Second, passage of individual biological clones of the phage resulted in the rapid generation of mutant distributions. Third, when individual mutants were made to compete against the average population in growth-competition experiments, the average population always showed a selective advantage over the individual mutants (Domingo et al. 1978). The authors proposed that "A Qβ phage population is in a dynamic equilibrium, with viable mutants arising at a high rate on the one hand, and being strongly selected against on the other. The genome of Qβ phage cannot be described as a defined unique structure, but as a weighted average of a large number of different individual sequences" (Domingo et al. 1978). These features constitute a hallmark of quasispecies dynamics, later suggested to apply to foot-and-mouth disease virus (FMDV) and to vesicular stomatitis virus (VSV) in vivo (Domingo et al. 1980; Holland et al. 1982), and in general to animal and plant RNA viruses (reviews in Domingo et al. 2001; Domingo 2006).

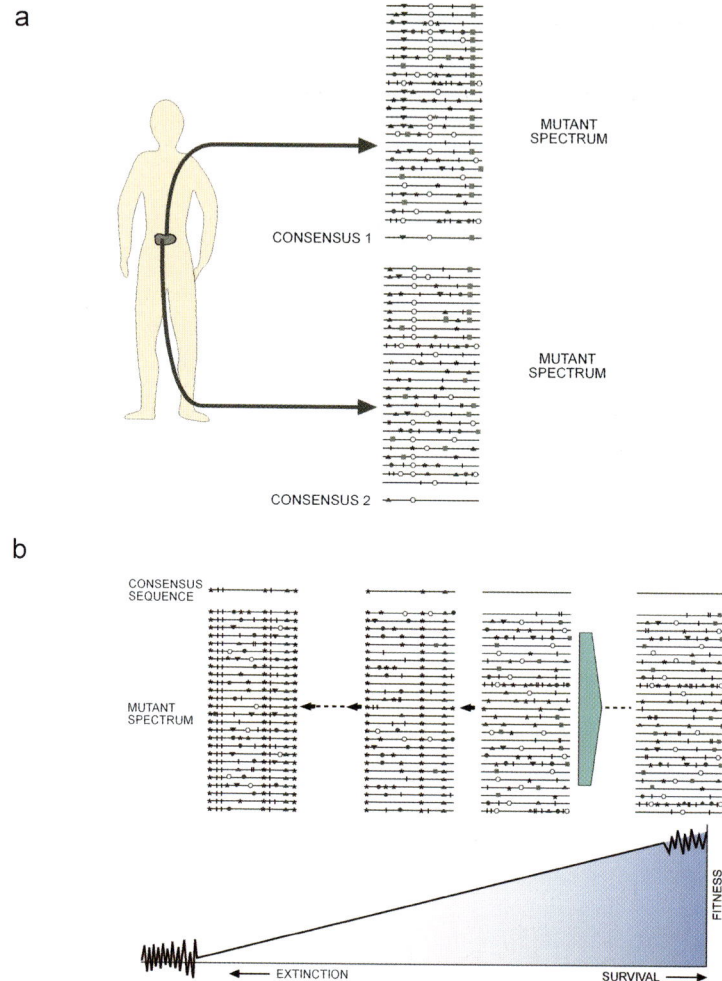

**Fig. 1** A simplified representation of viral quasispecies replicating in an infected host and of fitness variation as a result of the passage regime of the virus. **a** An infected individual includes multiple viral quasispecies (mutant distributions) at different sites of the organism, and also within the same organ; multiple viral sequences have been identified even in an infected cell. Here genomes are schematically represented by horizontal lines, and different types of mutations by symbols on the lines. Note that multiple consensus sequences of replicating units can be present, and that mutant spectra are highly dynamic. **b** Fitness variation as a result of the virus population size at each virus passage. Large population passages (*large arrow*) generally result in fitness gain, although stochastic fluctuations in fitness can be observed when further fitness increases are limited by the population size of the virus. In contrast, repeated bottleneck events (*small arrows*) result in accumulation of deleterious mutations and fitness decrease, although stochastic fluctuations in fitness can be observed when the virus reaches very low fitness. During replication in a well-defined and homogeneous biological and physical environment, competitive replication, without population size restrictions, promotes virus survival, while repeated bottleneck events approximate the viral population toward extinction. The schemes are based in many studies that have been reviewed in several chapters of (Domingo et al. 2001; Domingo 2006). (Modified from Domingo 2006, with permission)

## 2  Why Quasispecies?

Virologists use quasispecies to describe dynamic distributions of nonidentical but closely related mutant and recombinant viral genomes subjected to a continuous process of genetic variation, competition and selection, and which act as a unit of selection (Domingo et al. 2001). Observations that justify the terms of this definition are elaborated in the next sections. Relevant to enterovirus evolution is the consideration of recombination as a mechanism of mutant spectrum variation, supported by extensions of the original quasispecies theory (Boerlijst et al. 1996). The main contributions of quasispecies to virology stem from the recognition that replicating RNA genome populations are mutant spectra (also termed mutant clouds) rather than defined genomic sequences. The wild type is no longer represented by a defined genomic nucleotide sequence but by a distribution of mutant genomes. It is an experimental fact that mutation during RNA genome replication is a continuous rather than a sporadic event. This new view of viruses (and of any biological system characterized by error-prone replication) is compatible with alternative evolutionary models of mutation-selection balance from classical population genetics (Wilke 2005). One of the reasons why quasispecies (and not other theoretical formulations) has permeated experimental virology is its explanatory and experiment-provoking power. In particular, a number of studies have documented virus behavior dependent either on the complexity of the mutant spectrum or on interactions among components of a mutant spectrum. Such mutant spectrum-dependent behavior, with its multiple implications, would not have a straightforward interpretation based solely on consensus sequences. The objective of a theory is not to provide a detailed account of specific cases but to interpret coherently the experimental data and, in the case of quasispecies, to predict correctly features of virus evolution (Biebricher and Domingo 2007).

Several stochastic events intervene in virus evolution: the occurrence of mutations subjected to the quantum mechanical indetermination of base-pair interactions; the detachment of a viral polymerase molecule from a template molecule and binding to another template to yield a recombinant genome, which from current evidence are not regulated processes; the sampling of subsets of genomes in transmission events, either between different hosts or within a host (in the invasion of a new host compartment, sometimes even in the penetration into a single cell) variations in the biological environment of the host organisms, etc. (examples and reviews in Agol et al. 2001; Li and Roossinck 2004; Escarmís et al. 2006; Pfeiffer and Kirkegaard 2006; Domingo 2007).

Despite virus replication being frequently far from a population equilibrium, deterministic features of virus behavior have been observed in some experiments. In competitions between wild-type VSV and a marked, neutral mutant, a highly predictable nonlinear behavior was observed. At nearly constant periods of time (number of passages) after the onset of the competition, critical points were reached at which the wild type became dominant over the mutant (Quer et al. 1996). Interestingly, a number of environmental perturbations (presence of added or endogenously generated defective-interfering particles, enhanced mutagenesis by

5-fluorouracil [FU], or increase in temperature) prompted the dominance of the wild type over the mutant virus (Quer et al. 2001). Since the mutant VSV differed from the wild type in a few mutations, the results suggested that the mutant was more vulnerable to fitness decreases upon occurrence of additional mutations during replication. Either beneficial mutations are less likely, or detrimental mutations more likely, in the VSV mutant than in the wild type. In terms of Wrightian fitness landscapes (Wright 1982) the wild-type virus lay on a flat fitness surface while the mutant lay on a fitness peak (Domingo et al. 2001; Wilke et al. 2001b).

Another instance of deterministic behavior was a synchronous loss of memory genomes in parallel lineages of FMDV (Ruiz-Jarabo et al. 2003). Memory genomes are minority components of viral quasispecies that reflect those genomes that were dominant at a previous evolutionary phase of the same virus lineage. The presence of memory genomes is a direct consequence of quasispecies dynamics, and it may confer an advantage to a virus to respond to a selective constraint experienced during a previous phase of its evolution (Ruiz-Jarabo et al. 2000, 2002; Briones et al. 2003). It was suggested that the observed deterministic loss of memory was facilitated by an averaging effect of different mutations on viral fitness, and limited tolerance of RNA viruses to accept mutations, reflected in the same replacements seen in parallel evolutionary lineages (Ruiz-Jarabo et al. 2003).

Deterministic behavior in FMDV and VSV was probably favored by a large population size of the replicating virus in the relatively constant environment provided by an established cell line. This situation probably approximates a real viral quasispecies to the theoretical replicons in the initial quasispecies theory. Viral population size and environmental heterogeneity are important to interpret the behavior of viral quasispecies.

Quasispecies dynamics is currently approached experimentally with the tools of genetics, biochemistry, and structural biology, and by theoreticians with the tools of theoretical biophysics and computer science. This provides a promising substrate for transdisciplinary science that has already bridged seemingly disparate fields of activity such as replicon dynamics with management of patients afflicted with chronic viral disease (Sect. 9).

## 3 Molecular Basis of Mutation and Recombination

Virus evolution means a change in the genetic composition of the population over time, irrespective of the time frame involved. Viruses evolve within infected hosts, and their genetic composition can change in days or hours and evolve over long time periods to generate new biotypes or clades. Genetic change is a prerequisite for evolution. Enteroviruses use two main mechanisms of variation: mutation and recombination.

Mutation has as an immediate consequence the generation of quasispecies distributions (Sect. 1 and Fig. 1). The biochemical basis of high mutation rates in enteroviruses is the absence in their virus-coded RNA-dependent RNA polymerase

(RdRp) of a domain corresponding to a 3′ to 5′ exonuclease. This exonucleolytic activity is present in several cellular and viral DNA-dependent DNA polymerases (DpDp), and acts as a proofreading-repair activity to excise misincorporated nucleotides at the 3′ end of the growing nucleic acid chain. Misincorporations are frequent during nucleotide polymerization because rare tautomeric imino and enol forms of the standard bases can produce non-Watson-Crick pairs. The steric misalignment of bases (wobbling) can also lead to some types of point mutations. Base stacking can influence the pairing behavior of each template base, rendering very difficult predictions on the relative rate of incorporation of the correct nucleotide vs an incorrect one at any position of a viral genome (Menéndez-Arias 2002; Arnold et al. 2005; Castro et al. 2005; Friedberg et al. 2006).

It has been estimated that the error rate during template copying is in the range of $10^{-5}$-$10^{-6}$ mutations per nucleotide copied, with the contribution of base selection and proofreading repair; the error rate decreases to about $10^{-7}$ because of proteins present in the replication complex, and to about $10^{-10}$ with the additional participation of postreplicative mismatch correction mechanisms (Friedberg et al. 2006). Most significant for RNA virus variation, postreplicative mismatch correction pathways act on double-stranded DNA but not (or very inefficiently) on double-stranded RNA or DNA-RNA hybrids.

To maintain the genetic information of any replicating system (cell, virus, subviral entity) the template-copying accuracy must be higher the more complex the information content of the replicons is. Here complexity means the genome length, provided no redundant information is encoded. This important feature has a mathematical formulation derived from quasispecies theory and is expressed as an error threshold relationship, which sets the copying fidelity values in relation to viral fitness needed to prevent loss of genetic information (Eigen 2002; Biebricher and Eigen 2005). In the case of viruses, the loss of genetic information by enhanced mutagenesis results in virus extinction and it is currently explored as a new antiviral strategy termed lethal mutagenesis (see Sect. 9) (Eigen and Biebricher 1988; Eigen 2002; Biebricher and Eigen 2005, 2006). Thus, mutation must be reasonably controlled if it has to drive virus adaptation. Functional modules to ensure sufficient copying accuracy must have evolved as the complexity of the living systems increased.

In addition to mutations inherent to viral replication, nonreplicative mutation mechanisms can also alter viral genomes. These mechanisms include the activity of cellular deaminases, which have a physiological role in the cell to edit cellular DNA or RNA and can be recruited as part of the innate host response against viral infection. Cellular deaminases include some of the APOBEC cytidine deaminases, which mediate G → A and C → U hypermutation in retroviruses and hepatitis B virus, and the ADAR adenosine deaminases which mediate A → G and U → C hypermutation of riboviruses (reviews in Schaub and Keller 2002; Valente and Nishikura 2005; Chiu and Greene 2006). These activities can be regarded as a natural means to induce error catastrophe in viruses (Sect. 9). Remarkably, some specific sites in cellular genomes exploit high error rates for important physiological processes such as the generation of diversity in immunoglobulin genes, mediated by the highly conserved functions of some members of the APOBEC/AID protein family (Harris et al. 2002; Conticello et al. 2005). Nonreplicative mutations can also result from chemical damage to viral genomes: deamination, depurination, depyrimidination, reactions with oxygen radicals, effects of ionizing radiation, and photochemical reactions, among others. Although no quantitative evaluation of all these influences on viral mutation has been carried out, it is generally assumed that mutations associated with error-prone replication are the main source of mutational diversification of viral genomes.

Several forms of recombination have been characterized in RNA viruses: homologous, nonhomologous, replicative, and nonreplicative (Nagy and Simon 1997; Gmyl et al. 2003; Chetverin et al. 2005; Agol 2006). In homologous RNA recombination, there is extensive nucleotide sequence identity between the two parental genomes around the crossover site. In contrast, nonhomologous recombination is not associated with substantial nucleotide sequence identity. Replicative recombination requires RNA genome replication while nonreplicative recombination was observed upon cotransfection of cells with viral RNA fragments, which alone lack replicative activity (Gmyl et al. 2003; Gallei et al. 2004). The molecular mechanisms that promote these different types of recombination are still poorly understood, and a unifying picture is lacking. In many cases, however, the RNA replication machinery seems to be involved. A common form of homologous, replicative recombination is template switching, also termed copy choice. It involves detachment of the polymerase, with a growing product RNA, from a template molecule, and continuation of the copying activity at the equivalent position of another template molecule. In this type of RNA recombination, it is very likely that limited processivity (the capacity to remain on the same template molecule) together with nucleotide sequence context can have an important effect on the recombination rate.

Recombination can be either intermolecular, involving two different template molecules (as in the example of template switching described in the previous paragraph) or intramolecular, leading to genomes with internal deletions or duplications. The most extensively studied forms of the latter group are defective interfering (DI) RNAs, which can be produced at high frequency upon passage of some RNA viruses at high multiplicity of infection (MOI) (Roux et al. 1991). DI RNAs require a helper, standard virus for replication. High multiplicity passage of a biological clone of FMDV in cell culture resulted in dominance of two genetic variants of the parental FMDV, each harboring a different internal deletion, which were infectious and caused cell killing by complementation (García-Arriaza et al. 2004, 2006). Since each of the two defective genomes separately are incompetent in cell killing, the result has been viewed as a primitive evolutionary transition toward RNA genome segmentation. RNA genomes with internal deletions have been characterized in vivo, as in the case of noncytopathic coxsackievirus B3 variants with deletions at the 5'-untranslated region isolated from hearts of mice inoculated with the virus (Kim et al. 2005) (see the chapter by K.-S. Kim and N.M. Chapman, this volume). Recombination can be detected by the discordant position of different genes (or genomic regions) in phylogenetic trees relating the viruses under study (incongruent phylogenies). The occurrence of recombination must be supported by statistical evaluations (Martin et al. 2005) to distinguish it from convergent evolution by mutation of genomic regions of viruses in different evolutionary lineages.

## 3.1 Role of Mutation and Recombination in RNA Virus Evolution

Mutation universally and recombination in some viruses (including the picornaviruses) are active participants in the intrahost and interhost evolutionary events undergone by viruses. Mutation serves to adapt viruses to different environments, as extensively documented with the isolation of different types of escape mutants (to antibodies, to cytotoxic T lymphocytes, or to antiviral inhibitors) that permit virus survival through replication of subpopulations of mutant genomes. Also, mutants with altered tropism, host range, and virulence may mediate persistence of viruses at the population level (many examples have been reviewed in Domingo et al. 2001, 2006, 2007).

It has been argued that, because of the detrimental character of most mutations, high mutation rates are the result of a trade-off, and that they are the cost to be paid for rapid genome replication. At present, this remains an unproven hypothesis, and the available evidence points to a decisive contribution of high mutation rates to virus survival and pathogenesis. A key study has been conducted with a poliovirus mutant that displays approximately threefold higher copying fidelity than the wild-type virus (Pfeiffer and Kirkegaard 2005; Vignuzzi et al. 2006). The mutant encoded a RdRp with amino acid substitution G64S and was selected because it conferred poliovirus an increased resistance to the mutagenic nucleoside analog ribavirin. The high-fidelity mutant generated a narrower mutant spectrum than the wild type and, remarkably, it showed impaired capacity to produce neuropathology in mice. Significantly, when the complexity of the mutant spectrum was increased by chemical mutagenesis, the virus regained the capacity to cause neuropathology. These experiments have documented the requirement of a broad mutant spectrum - or complexity, a feature which is directly linked to the mutation rate - for adaptation to a complex environment (Pfeiffer and Kirkegaard 2005; Vignuzzi et al. 2006).

The molecular basis of nucleotide substrate discrimination during RNA and DNA synthesis is still poorly understood, but from many studies with viral and cellular polymerases, it is likely to be multifactorial (Friedberg et al. 2006; Ferrer-Orta et al. 2007). Velocity of RNA or DNA synthesis may be one such factors - as suggested by early work with the poliovirus RdRp (Ward et al. 1988) - but additional experimental work is needed to quantitate its influence relative to other factors. To theorize about the universal requirement of rapid replication for viruses, and high mutation rates being a consequence of it, is not justified by current virology. Indeed, viruses can be characterized by widely different time courses to complete their replication cycles, and have evolved capacities to produce acute, chronic, latent, slow, persistent, symptomatic, or inapparent infections. And sometimes the same virus - with its inherent replication machinery - can produce several types of infection. It is critical to build theoretical arguments on well-established facts. In the case we are addressing, it is important to dispel the notion that rapid replication evolved as a universal necessity for viruses, and that high mutation rates are an epiphenomenon of it.

Recombination in the evolution of RNA viruses has been viewed in two different ways. It can serve either to generate new genomic combinations, often with uncertain survival capacity (an exploratory evolutionary force), or to rescue fit genomes from less fit parents (a conservative evolutionary force). There is evidence for both types of effect of recombination. For example, the alphavirus Western equine encephalitis virus probably originated by recombination of a Sindbis-like virus with an Eastern equine encephalitis-like virus. In this and other reported cases (reviewed in Domingo 2007), it is likely that only a few out of many recombination events succeeded in generating a virus capable of evolutionary continuity. In contrast, some recombinant HIV-1 viruses rescued genomes harboring multiple drug-resistance mutations, thus permitting virus survival in the face of inhibitory activities.

Both mutation and recombination are blind processes dictated by molecular instructions contained in the replication machinery and its nucleotide substrates. In the case of mutation, one instruction is a direct consequence of the thermodynamical uncertainties that preside tautomeric shifts and base pairing in template substrate recognition (Sect. 3). Another instruction stems from the need of polymerases to accommodate the four standard nucleotides in their active site, incompatible with a perfect, tight, error-free recognition of each type of nucleotide individually. In the case of recombination, several polymerases have evolved a requirement for template switching during replication of the genomes that encode them (i.e., coronaviruses and retroviruses). Therefore, template switching is inherent to the normal activities that are essential to complete the life cycles of these viruses. Intrinsic instructions for mutation and recombination are likely to be unavoidable features that have permitted the rapid evolution of many life forms on earth. For viruses, the result of such molecular instructions (for mutation universally and for intermolecular recombination in many, but not all, cases) is the generation of large repertoires of variants that are continuously subjected to competition, selection, and random drift (the Darwinian principles of evolutionary dynamics). Most variants are extinguished, but this penalty at the individual level has as a counterpart increased survival probability at the population level. Enteroviruses are not an exception, and both mutation and recombination have contributed to shaping the enterovirus populations we isolate and analyze.

## 4 The Diversity of Enteroviruses: Species, Quasispecies, and Variation in Disease Manifestations

On the basis of their pathogenesis in humans and experimental animals, the enteroviruses were originally classified in four groups: poliovirus, coxsackie A viruses (CAV), coxsackie B viruses (CBV), and echoviruses (reviewed in Hyypiä et al. 1997). It was soon realized that there were considerable overlaps between biological properties of viruses of different groups, and differences among viruses assigned to the same group (reviewed in Oberste et al. 1999). Further subdivisions became necessary. In the current classification (Fauquet et al. 2005), enteroviruses are one of nine genera of the family Picornaviridae, and are subdivided in nine established and 17 tentative species. The human enteroviruses are classified into five species: poliovirus and human enteroviruses A-D. Members of the same species share greater than 70% amino acid sequence identity within P1 and nonstructural proteins 2C and 3D (Fig. 2). As a reflection of their extensive genetic and phenotypic diversity, the nine established species include at present a total of 78 strains, associated with a broad range of diseases affecting different animal host species.

The clinical and pathological manifestations of enteroviruses are remarkably diverse (Minor 1998): poliomyelitis; aseptic meningitis; paralytic disease; encephalitis; postfatigue syndrome; epidemic pleurodynia (Bornholm disease); congenital and neonatal infection; cardiac disease; herpangina; hand, foot, and mouth disease (a form of childhood exanthem), acute epidemic hemorrhagic conjunctivitis, respiratory disease, otitis media, diabetes, and more rarely gastroenteritis and hepatitis.

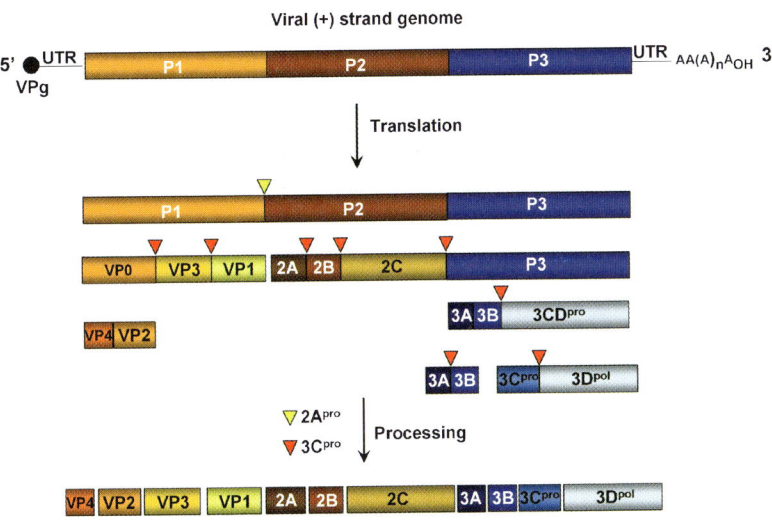

**Fig. 2** Scheme of the enterovirus genome and translation products. *Top* The genome is divided in two untranslated regions (*UTR*) at the 5' and 3' end regions, and three blocks encoding structural proteins (*P1*) and nonstructural proteins (*P2, P3*). VPg is the protein (*3B*) covalently linked to the 5' end of the RNA, and the $AA(A)_n A_{OH}$ is the 3'-terminal poly (A) tract. Below the genome the translation products and their processing are depicted in a simplified and schematic way. The polyprotein encompassing P1, P2, P3 is cotranslationally cleaved to yield several processing intermediates and the mature proteins. The cleavage sites that are the target of the virus-coded proteases $2A^{pro}$ and $3C^{pro}$ are indicated by inverted triangles. Processing of VPo to yield VP4 and VP2 is autocatalytic. $3D^{pol}$ is the enteroviral RNA-dependent RNA polymerase, responsible of the error-prone replication of these viruses. The genome structure and processing reactions are based on different chapters of Semler and Wimmer (2002)

The same virus (same according to standard classifications) can be associated with widely different diseases. Notable examples are the coxsackie B viruses, which have been associated with paralytic disease, encephalitis, febrile illness, hepatic or generalized neonatal infections, cardiac disease, mucocutaneous infections, and diabetes (Mills et al. 1989; Tracy et al. 2006; see also chapters by J.R. Romero, A. Matsumori, K. Knowlton, A. Ramsingh and K. Dreschner et al., this volume). Certainly, we are witnessing a growing interest in the clinical implications of enterovirus infections.

Human enterovirus 71 (EV71) outbreaks associated with mild (hand, foot, and mouth disease) and severe (poliomyelitis-like) disease have increased over the last decade, often with intervening quiescent periods of a few years in a given geographical area. Phylogenetic analyses have distinguished three major lineages of EV71: A, B, and C; B and C have been divided in several subgenogroups (Brown et al. 1999; Cardosa et al. 2003; Hosoya et al. 2006; Sanders et al. 2006). On the basis of the sequence of the VP1-coding region, 48 isolates from a single city (Sydney, Australia), obtained over a 19-year period, were divided into four subgenogroups (B2, B4, C1, C2), and several sublineages within C1 were distinguished (Sanders et al. 2006).

It seems inevitable that the current classification of enteroviruses will have to be modified in the future, as has happened in the past. Among other reasons, the available evidence is that each of the 78 strains corresponds to a virus population that circulated as a mutant swarm or viral quasispecies, with all the implications discussed in Sects. 1 and 2. Difficulties will be found whenever attempting to classify (in the form of species or other) entities which are essentially dynamic and whose key biological properties may depend on minor genetic variation (or combinations of minor changes) in a quite unpredictable fashion. Indeed, overwhelming evidence indicates that fundamental biological properties (such as virulence, host cell tropism, and host range) may depend on one or a few point mutations, or other modest lesions. Moreover, the same minor lesion can have a different biological impact depending on sequence context in the viral genome; constellations of compensatory mutations may be allowed in some sequence contexts (due to absent or limited negative selection) but not in others (due to strong negative selection). Even more, recent studies with poliovirus (Sect. 3.1) - in agreement with more indirect observations made with other viruses (i.e., the effect of mutant spectrum complexity either in disease outcome or in the response to treatment in infections with hepatitis C virus; reviewed in (Domingo 2006) - suggest that virulence may be determined in some cases by the amplitude of the mutant spectrum, and not by the consensus sequence (Pfeiffer and Kirkegaard 2005; Vignuzzi et al. 2006). Thus, the term "strain" hides a level of biological complexity that will render periodic classification adjustments inevitable, and associations between genomic sequences and biological behavior elusive. The phylogenetic position of a viral genome reflects accumulated sets of signatures influenced by ancient evolutionary events. Host range, tissue tropism, and antigenic profile of a virus are but a small part of all the features that determine a phylogenetic position. Thus, enteroviruses, simply by the fact of sharing a phylogenetic position, need not be associated with the same disease manifestations.

Difficulties of classification have been encountered with other viruses. Within the picornaviruses, the aphthovirus foot-and-mouth disease virus (FMDV) has been classically divided into seven serotypes and each of them into several serological subtypes. Two decades ago, when the number of subtypes reached more than 65, subtyping was discontinued because it became apparent that each of many new isolates could qualify as a new and different subtype. Comparison of reactivities of FMDV isolates from the same serotype with monoclonal antibodies (mAbs) indicated that the fine antigenic profile of each isolate was virtually unique (Mateu et al. 1988). Furthermore, in support of arguments given in the previous paragraph, single amino acid replacements at or near a major antigenic site of FMDV resulted in alterations of host-cell tropism (Baranowski et al. 2003). A precedent involving enteroviruses was the observation of the dynamics of gain and loss of reactive epitopes among clinical isolates of coxsackievirus B4 (Prabhakar et al. 1982, 1985).

There is an increasing number of data banks for virus nucleotide and protein sequences (reviewed in Domingo et al. 2006, 2007). It has been suggested that a second generation of data banks, including sequences of mutant spectra, could be contemplated, to try to incorporate to the everyday tasks of virologists at least part of the additional information converged by mutant spectra and their complexity (Domingo et al. 2006).

Computational approaches to define species-associated amino acid signatures of pathogenic viruses are likely to encounter difficulties arising from the multiple amino acid combinations compatible with host specificity (Chen et al. 2006). Again, each isolate for which amino acid signatures are sought is in reality a mutant cloud with its inherent statistical and biological indeterminations.

## 5 Mutation and Recombination as Mechanisms of Enterovirus Evolution

Mutation and recombination actively contribute to enterovirus genetic variation and diversification in nature. Since the early recognition of recombination between vaccine and wild-type poliovirus strains (Cammack et al. 1988; Furione et al. 1993), and recombination in other human enteroviruses (Santti et al. 1999), evidence has accumulated that enteroviruses evolve relentlessly through mutation and recombination as they circulate in human and animal hosts. Mutant and recombinant poliovirus strains have been isolated from healthy vaccinees, from vaccine-associated poliomyelitis cases, and from immunodeficient patients who excrete poliovirus for extended time periods (Cherkasova et al. 2003; Lukashev 2005; Agol 2006; Kimman and Boot 2006; Martin 2006; Shulman et al. 2006). Most frequently, the recombinants detected suggest intraspecies recombination events involving the nonstructural protein-coding region and regulatory regions of the genome (Santti et al. 1999; Lukashev et al. 2005; Chan and Abubakar 2006). Structural and functional restrictions at the RNA and protein levels are likely to limit the viability of enterovirus recombinants. Such restrictions embrace the need to preserve the geometry of the capsid, maintenance of the proteolytic activities and of their cleavage sites on the polyprotein, interactions between distant RNA regions to fold into functional structures, and so on. These restrictions may explain the lower frequency of enterovirus recombinants with crossover points within the P1 than within the P2, P3 genomic regions, although some cases of intraserotypic recombination in the capsid region have been reported (i.e., among coxsackie B viruses; Oberste et al. 2004). The preferential recombination in the P2, P3 genomic regions may be favored by a higher frequency of homologous recombination by a copy choice mechanism in the genomic regions where a high nucleotide sequence identity between the two parental viruses exists (King 1988).

Restrictions of different intensity may also contribute to differences in recombination frequency of closely related enterovirus groups. A survey of human enteroviruses A and B within a limited time period and in the same geographical region indicated much more frequent recombination events for species B than for species A (Simmonds and Welch 2006). The study used phylogenetic compatibility matrices on sequence alignments to define the sites of favored recombination; the latter were located throughout the nonstructural protein-coding regions examined. In this and other surveys based on analyses of field isolates, sampling limitations must also be considered. Some features of the evolution of picornavirus serotypes by recombination and mutation are shared with other nonenveloped positive-strand RNA viruses (Simmonds 2006). Shared properties were high frequency of recombination, evidence for positive selection, and differential codon usage in the capsid-coding region. These evolutionary mechanisms appear to be different from those underlying recent sequence diversification within picornavirus serotypes in which neutral or negative selection were dominant (Simmonds 2006). Multiple recombination events, to yield mosaic genomes in the P2 and P3 regions are frequent among enteroviruses of the same species (Oberste et al. 1999) (Fig. 3).

Phylogenetic evidence suggests the occurrence of intertypic recombinants between EV-71 and several HEV-A viruses, including CV-A16 (Chan and Abubakar 2006). Such recombination events may play a role in the emergence of EV-71 subgenotypes with different morbific potential. Some HEV-C strains are probably recombinants that have incorporated part of poliovirus genomes (Brown et al. 2003). Recent poliomyelitis outbreaks have been associated with recombinants between poliovirus and other human enteroviruses (Kew et al. 2002; Arita et al. 2005)

Several observations on recombination have been made with animal enteroviruses. A comparison of representatives of bovine enterovirus A and B (that belong to the same species in the current classification) suggests the occurrence of intraserotypic and interserotypic recombination, and the need to reclassify bovine enteroviruses A and B as two different species, each composed of multiple geno-/serotypes (Zell et al. 2006). Current evidence suggests that recombination in enteroviruses may be limited mainly by the requirement that two different parental genomes infect the same cell, and that the resulting recombinant(s) display sufficient replicative fitness to outcompete (or coexist with) the parental genomes in subsequent rounds of infection.

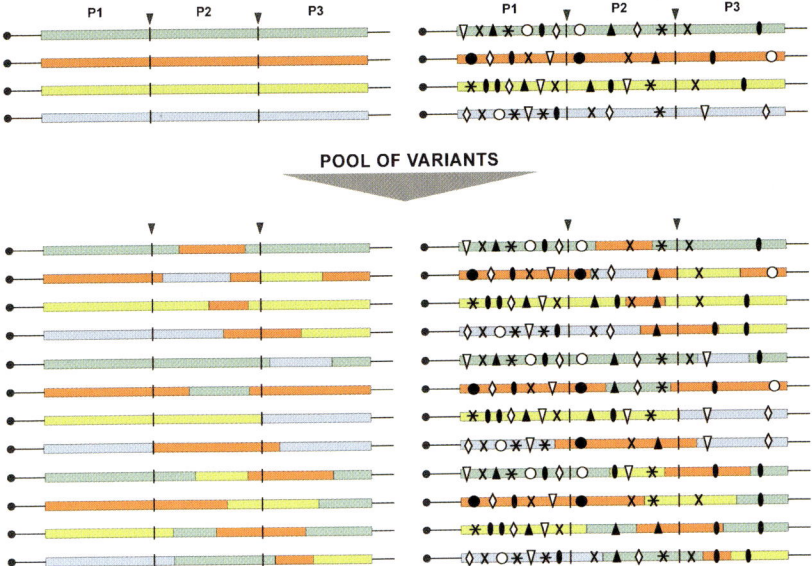

**Fig. 3** Schematic view of the complexity that can be attained by a population of enterovirus through recombination and mutation. *Top* four parental genomes (color coded, *left*) which themselves are mutant distributions (mutations depicted as symbols on the genomes, *right*). *Bottom* recombination acting on the pool of variants may produce multiple recombinants (*left*), which themselves will become dynamic mutant distributions (*right*)

The operation of mutation together with recombination in enteroviruses (Fig. 3) is illustrated by the dynamics of epitopic change among clinical isolates of coxsackievirus B4 (Prabhakar et al. 1982, 1985) and the occurrence of frequent recombination in this same group of coxsackieviruses (Oberste et al. 2004). Likewise, recombination in poliovirus vaccine strains occurs together with antigenic alterations due to amino acid substitutions that may have as their major effect the elimination of fitness-decreasing mutations present in the vaccine strains (Yakovenko et al. 2006). The recognition of high-frequency recombination has led to the view that enteroviruses exist as swarms of capsid genes, nonstructural protein genes, and 5′ UTRs, which evolve independently and recombine frequently to produce new variant forms as a substrate for natural selection. In this view, the virus replication machinery has evolved toward maximum adaptability and to avoid error catastrophe (Sect. 9) (review in Lukashev 2005). This dynamic view of evolution is in line with the concepts derived from quasispecies dynamics, and it expresses fundamental features of enterovirus evolution within infected hosts and at the epidemiological level (Gavrilin et al. 2000; Agol 2006; Domingo 2007). Yet, evolution of a genomic region is unlikely to be independent of evolution of the others, and recombination acts not on defined sequences but on mutant swarms (Domingo 2007). The right side of Fig. 3 is a modest representation of the real complexities inherent to recombination acting on dynamic quasispecies.

Mutant spectra (made of mutant and recombinant genomes) represent the first stage in the process of viral diversification in the infected host. When the immune response is vigorous and the virus is cleared after a limited time, the intrahost diversification can be of limited or null epidemiological significance. However, when viral replication is active and immune responses poor, diversification of quasispecies can be remarkable. A poliomyelitis patient with immunodeficiency generated at least five distinct sublineages derived from Sabin type 1 oral poliovirus vaccine, probably in a maximum of 567 days (Yang et al. 2005). In the process, the *ts* phenotype and key determinants of attenuation of the Sabin 1 virus were lost, and the diversified viruses showed increased virulence for susceptible, transgenic mice. All isolates were antigenic variants of Sabin 1, with multiple replacements at or near antigenic sites, and some genomes had a mosaic structure indicative of multiple intratypic recombination events. When a quasispecies is left to evolve unchecked, the potential to generate altered progeny can be astonishing.

Remarkably, the availability of genome domains for exchanges in recombination events can contribute to the emergence of new viral pathogens, a phenomenon that updates (in a form compressed in time) ancestral processes, according to the theory of the modular origin of viruses pioneered by Botstein with bacteriophages (Botstein 1980) and extended to animal and plant viruses (Botstein 1981; Zimmern 1988). Given the solid evidence for both the modular origin of viruses and the active exchange of genomic regions (modules) via recombination in many viruses (see also Urbanowicz et al. 2005 for emphasis on recombination as a feature of RNA genetics, it appears that evolution has exploited similar molecular tinkering mechanisms in widely different time frames. From a historical perspective, we note modular evolution as having originated the major groups of viruses (perhaps also organisms) we see today (Zimmern 1988). At present, we perceive ubiquitous recombination as a mechanism to generate short-term diversity and to confront immediate environmental demands. We have not studied sufficient generations of viruses (virology is only 115 years old, but simple replicons may be about $4 \times 10^9$ years old) to sort out mechanisms that dictate extinction vs survival, the latter affecting probably a minority of privileged forms.

## 6 Evolution of Virulence in Enteroviruses

Most viruses, and specifically picornaviruses, have multiple virulence determinants (review in Tracy et al. 2006). Virulence is often defined as the capacity to produce damage (disease or death) in the infected host. The existence of multiple virulence determinants is expected from the several viral gene products that can perturb basic cell functions. One such viral products is the viral capsid which must recognize cell surface molecules for entry into the cell. Receptor usage is yet another reflection of enterovirus diversity since several members of the immunoglobulin superfamily (coxsackievirus-adenovirus receptor [CAR], intracellular adhesion molecule 1, poliovirus receptor), integrins, decay accelerating factor (DAF), and heparan sulfate are among the cellular proteins used by enteroviruses as receptors; variant viruses generated either by mutation or recombination can use alternative receptors (Evans and Almond 1998; Baranowski et al. 2003; Bergelson 2003).

Modest genetic change occurring very rapidly in the course of disease outbreaks may enhance enterovirus virulence. In the case of poliovirus, mutations at the 5' UTR and in the VP1-coding region may greatly alter neurovirulence (Gromeier et al. 1999). Poliovirus attenuation has been achieved by introducing deletions at regulatory regions of the genome (Iizuka et al. 1989; Agol et al. 1996) or by domain shuffling with other picornaviruses (Gromeier et al. 1996). EV71 and coxsackievirus A16 (CA16) are often associated with hand, foot, and mouth disease. Yet, despite the close genetic relatedness of the two viruses, EV71, but not CA16, has been related to neurological disease. Molecular epidemiological surveys identified a single substitution in VP1 of EV-71 associated with neurovirulence that could be mediated by an alteration of receptor recognition (McMinn 2002). The effect of minimal genetic change explains that EV71 strains circulating during epidemics may vary greatly in neurotropism. Virulence of swine vesicular disease virus (SVDV) - a virus that probably emerged as a result of an infection of swine by human coxsackievirus B5 - was associated with single amino acid replacements in VP1 and in protease 2A. Two amino acid substitutions in VP1 and VP4 determined the association of mouse-adapted strains of coxsackievirus B4 with pancreatitis. In contrast, cardiovirulence of coxsackievirus B3 was mapped within a 5'-nontranslated region (5' NTR) that contains a predicted stem-loop structure (reviewed in Tracy et al. 2006).

A relevant question is the connection between viral fitness and virulence. A model study with FMDV in cell culture has unveiled the molecular basis for a lack of correlation between viral fitness and cell killing capacity (Herrera et al. 2007). The current picture of enterovirus evolution and disease potential suggests that virulence might be understood as a phenotypic trait - defined as causing disease or death - that will have a continuum of intensities as a consequence of being associated with multiple viral functions, the latter being sometimes dependent also on complex pools of mutant and recombinant genomes (Fig. 3). Either because of selection for other traits (independent of virulence) or because of random sampling events, some circulating variants might not cause any apparent disease; other variants, probably the most common, will produce normal disease states, while others will be associated with severe or lethal infection (McMinn 2002). In considering the spectra of disease symptoms associated with a single virus group, it is worth restating the thoughts of John J. Holland and colleagues more than a decade ago: "There is an unspoken assumption among many physicians and scientists that a particular RNA virus will generally cause a particular disease. This assumption may be true in a very broad practical sense, but it is important to understand that it can never be true in a formal scientific sense. Because a particular RNA virus simply does not exist, a particular RNA virus disease does not exist either. The science of infectious diseases is still in its infancy because we still understand so little of the fine details of host-pathogen interactions" (Holland et al. 1992).

## 7 Enterovirus Evolution in the Course of Persistent Infections

Several enteroviruses cause pathology in association with persistence in specific cells and tissues of infected organisms. Persistence both in vivo and in model systems in cell culture is the result of a complex interplay between viral and host influences. Early studies in cell culture documented evidence of coevolution of persistently infected cells and their resident virus. In a coxsackie A9-HeLa carrier cell system, cells with increased resistance to the virus, and viruses with increased virulence for HeLa cells were selected (Takemoto and Habel 1959). HeLa-coxsackie B3 carrier cell lines displayed a restriction for superinfection by coxsackie B5 virus (Crowell and Syverton 1961). In a persistent infection established with cloned WISH cells and plaque-purified echovirus 6, defective viruses that were unable to attach to the parental WISH cells were selected (Gibson and Righthand 1985).

A number of diseases and syndromes have been associated with enterovirus persistence: cardiopathies, diabetes, idiopathic inflammatory myopathy, chronic fatigue syndrome, Sjögren's syndrome, motor neuron disorders, and postpolio syndrome, with some debate (reviewed in Kandolf et al. 1987; Dalakas 1995; Carson et al. 1999; Tam and Messner 1999; Berger et al. 2000; Giraud et al. 2001; Ravits 2005); see chapter by K.-S. Kim and N.M. Chapman, this volume]. No consistent genetic changes have been observed associated with enterovirus persistence. Coxsackievirus $B1_T$ RNA from the muscle of mice afflicted with chronic inflammatory myopathy varied minimally and not consistently. However, contrary to the excess of positive-strand RNA observed in standard picornaviral infections, the amounts of positive-strand and negative-strand RNA in the muscle were similar, and the presence of double-stranded RNA may contribute to long-term persistence of viral RNA (Tam and Messner 1999). The relative evolutionary stasis of a persistent enterovirus is in sharp contrast with many other cases of deep evolutionary change in cell culture and in vivo (Colbere-Garapin et al. 1998, 2002; Tam and Messner 1999; Beaulieux et al. 2005). As suggested more than two decades ago by J.J. Holland and colleagues, variant forms of viruses not easily recognizable by standard diagnostic procedures could underlie several forms of persistent infections associated with chronic disease (Holland et al. 1982). Currently, the use of degenerate primers sets for RT-PCR amplification or decreasing hybridization temperatures during RNA amplification by RT-PCR has unveiled minority genomes with biased distributions of mutations (Suspène et al. 2005). It would be interesting to use such new genome screening procedures to try to identify atypical forms of enteroviruses that could be associated with diseases that, for the moment, are of unknown etiology.

## 8 Mutant Spectrum Complexity as a Factor in Viral Pathogenesis

The studies on viral virulence have led to an interesting distinction that may be relevant to enterovirus pathogenesis. A virus can evolve toward higher or lower virulence due to genetic determinants imprinted in individual genomes, independent of the presence of a mutant spectrum and its composition. This is the view more generally held as the genetic basis of virulence, and several examples were discussed in Sect. 6. However, more recent observations point to quasispecies complexity as an additional virulence determinant, independently of the genomic nucleotide sequence that dominates the population. This quasispecies-dependent behavior has its origins in the complementing or suppressive interactions that may occur among components of the mutant spectrum (reviewed in different chapters of Domingo et al. 2001, 2006).

Suppressive effects of mutant spectra were predicted by quasispecies theory (Eigen and Biebricher 1988), and the first case documented experimentally was reported by de la Torre and Holland, who described the suppression of a high-fitness vesicular stomatitis virus (VSV) by a mutant spectrum of VSV displaying a lower fitness (de la Torre and Holland 1990). In an evaluation of poliovirus vaccines, Chumakov and colleagues documented the presence of virulent variants in live attenuated vaccines, and that, unless the virulent variants were present above a critical threshold level, they did not produce neurological disease in the spinal cord infectivity test in monkeys (Chumakov et al. 1991). This and other cases of suppression of specific variants by related mutant spectra seen both in cell culture and in vivo (reviewed in Domingo 2006) illustrate that an atypical phenotype of a virus may be either suppressed or expressed in the course of an infection, largely depending on its proportion in the mutant spectrum as well as on the nature (composition) of the mutant spectrum.

The studies with a high copying fidelity mutant of poliovirus (Pfeiffer and Kirkegaard 2005; Vignuzzi et al. 2006) summarized in Sect. 3.1 have provided direct evidence of the influence of the mutant spectrum in the outcome of an infection in vivo in two respects. One is the requirement of a complex mutant spectrum to produce neuropathology, and, second, the demonstration that a variant that by itself could not reach the brain could do so when accompanied by a helper poliovirus population (see a comment on the implications of these findings in Biebricher and Domingo 2007).

From an evolutionary perspective, according to current models of parasite virulence (Lenski and May 1994; Poulin and Combes 1999; Brown et al. 2006), it may be to the advantage of a virus not to express in an immediate and irreversible fashion all its pathogenic potential. To ensure the presence of sufficient host individuals for long-term survival of the virus, pathogenesis must be modulated. What the current data suggest is that such a modulation can be attained by transitions between virulent and avirulent forms located quite close to each other in sequence space, and also by modifying the amplitude of mutant spectra around a dominant genome. The latter in turn can behave either as virulent or as attenuated by virtue of its surrounding mutant spectrum. Again, complexity of the mutant spectrum may depend on a single amino acid substitution in the viral polymerase (Pfeiffer and Kirkegaard 2005; Vignuzzi et al. 2006), implying a minimal movement in sequence space of the parental genome that greatly affects the occupation of sequence space by its progeny.

Although still largely unexplored for enteroviruses in their acute or persistent infections of human hosts, the complementing or suppressive effects among components of mutant spectra certainly deserve some attention as possible factors in pathogenesis.

## 9 Implications of Quasispecies Dynamics for Viral Disease Prevention and Treatment

The genetic flexibility and continuous variation of enteroviruses renders them candidates to be agents of emerging infectious disease (Palacios and Oberste 2005). Documented emergences are swine vesicular disease (from human coxsackievirus B5), acute hemorrhagic conjunctivitis (associated with enterovirus 70), or disease of the central nervous system (associated with enterovirus 71). The variation in clinical manifestations (Sects. 4 and 6) and the potential for disease emergence renders essential to design strategies to limit enterovirus circulation. For this purpose, there is a need to develop new antienterovirus compounds (Barnard 2006) so that combination treatments can be implemented. Simultaneous administration of multiple inhibitors directed to different viral targets is essential to prevent or delay the selection of inhibitor-resistant viral mutants. This is one of the main lessons learned from antiretroviral therapy, in the huge efforts to try to control AIDS (Cohen 2006; Yin et al. 2006) (as an overview of the problem of drug resistance in relation to quasispecies, see Domingo 2003).

Quasispecies dynamics has opened the way to a new antiviral strategy termed lethal mutagenesis (Loeb et al. 1999) (reviews in Anderson et al. 2004; Domingo 2005). It is based on the concept that for any replication system, there is a copying error threshold that, when violated, no genetic information can be maintained (Eigen 2002; Biebricher and Eigen 2005, 2006). The conceptual basis of lethal mutagenesis was already contained in the initial theoretical development of quasispecies by M. Eigen and P. Schuster more than 30 years ago (Eigen 1971; Eigen and Schuster 1979) and tested experimentally for the first time by J. Holland and colleagues (Holland et al. 1990). Studies with several virus-host systems in cell culture and in vivo have documented virus extinction by mutagenic agents, including nucleoside analogs (Anderson et al. 2004; Domingo 2005). The transition undergone by the mutant spectrum in its way toward extinction is characterized by a decrease in specific infectivity, an increase in the complexity of the mutant spectrum, and an invariant consensus sequence of the population. Interfering mutants play an important role in the transition of viruses to error catastrophe (González-López et al. 2004; Grande-Pérez et al. 2005; Perales et al. 2007). Mutagenesis, and not only inhibition of viral replication, is required to achieve extinction when violation of the error threshold is involved (Pariente et al. 2003; Domingo 2007).

Lethal mutagenesis is currently under intense investigation, including an ongoing clinical trial with AIDS patients treated with a deoxynucleoside analog (Harris et al. 2005) scheduled to initiate phase 2 trials in 2007. Lethal mutagenesis is amply supported by theoretical studies when the latter are based on realistic assumptions. Remarkably, the nucleoside analog ribavirin has been shown to act as mutagen for poliovirus (Crotty et al. 2000, 2001; Graci and Cameron 2006) and several other RNA viruses. Ribavirin tri-phosphate can act as a substrate analog, and the monophosphate is incorporated into viral RNA, resulting in increases in mutation frequency and decreases in viral infectivity, in some cases accompanied by viral extinction. Thus, mutagenesis is now well established as one of several antiviral mechanisms of ribavirin (Parker 2005). Therefore, a link has been established between anti-cancer chemotherapy by nucleoside analogs and lethal mutagenesis for viruses (Domingo 2007), which may provide a new generation of mutagenic nucleoside analogs that could be incorporated into antiviral combination treatments, including treatment of enterovirus infection.

A requirement of combination treatment applies also to use of small interfering RNAs as antiviral agents, to avoid selection of escape mutants (Saleh et al. 2004; Saulnier et al. 2006).

Quasispecies dynamics also necessitates that vaccines to prevent infection or disease by highly variable viruses present multiple B cell and T cell epitopes to the host immune system, to stimulate the immune system in a similar way as the authentic pathogen. This demand is best fulfilled by live-attenuated virus vaccines (Seder and Mascola 2003). Except in some cases in which their use is inadequate for safety reasons (i.e., in immunodeficient patients or for viruses that can integrate their genetic material in cellular DNA), live vaccines are the choice to evoke an effective prevention of either infection or disease. The remarkable tolerance of enteroviruses to accept multiple recombination and mutation events, while maintaining a good replicative capacity, makes them suitable candidates to derive vaccine strains whose attenuation level is stabilized by multiple genetic lesions (Dan and Chantler 2005; Macadam et al. 2006).

## 10 Overview, Connections, Challenges, and Trends

The evolution of enteroviruses is presided by universal Darwinian principles that apply also to other viruses and replicating entities in general. No conflict exists among different formulations of Darwinian evolutionary dynamics or with classical mutation-selection balance formulations of population genetics (Page and Nowak 2002; Wilke 2005). The behavior of enteroviruses can be considered extreme as a Darwinian system because the combined high rates of mutation and recombination produce an explosive adaptive capacity. This is reflected in an ever increasing identification of variant enteroviruses with new disease manifestations. This trend is likely to accelerate in coming years because of the improved methods to amplify and sequence genomes from biological specimens, including noncultivable enteroviruses. The astonishing biological consequences of enterovirus diversity can only be minimally appreciated, as we should bear in mind that *each* infected individual will contain multiple mutant clouds of the type depicted in Fig. 1 but with hundreds of thousands of different sequences. Furthermore, we should consider that each cloud will contain phenotypic variants dependent on a few mutations, and that our survey of enteroviruses is most frequently restricted to humans and a few animal species. It should not come as a surprise if we find that fewer and fewer defined symptoms are pathognomonic of an enterovirus-associated disease. Also, modifications in the current classification will be periodically needed unless the classification criteria are changed to assemble large numbers of viruses (perhaps similarly to the current orders) that embrace a variety of genetic and phenotypic features. Otherwise, the question will be frequently posed of when a mutant and recombinant virus becomes different from their ancestors. Will it be different because of some unusual genetic and phenotypic traits? Will its being accepted as a new genotype or serotype require a sustained presence (circulation) in one or several host species?

While we have an increasing understanding of enterovirus diversity - thanks primarily to application of phylogenetic methods to viral genomes isolated in surveys of clinical interest, also in connection with the program for the global eradication of poliovirus - other features of virus evolution are still largely unexplored and they may play a role in general adaptation and pathogenesis. These features are derived from quasispecies dynamics, as discussed in part in Sects. 2, 8, and 9, and pose a number of challenges that can be outlined as follows:

1. Elucidation of the molecular basis of template-copying fidelity of the viral RdRps, and quantification of possible differences in fidelity among picornaviral RdRps. We have learned from studies with several cellular and viral polymerases that copying fidelity can be modified by structural alterations in the enzymes, and that multiple polymerase sites (perhaps multiple domains) of the picornaviral RdRp are involved in nucleotide substrate recognition and discrimination (Arnold et al. 2005; Castro et al. 2005; Sierra et al. 2007).
2. Related to (1), the development of antienteroviral inhibitors, virus-specific mutagenic agents, and compounds capable of lowering the copying fidelity of viral polymerases, to be used for research on lethal mutagenesis.
3. To determine the possible participation of mutant spectrum complexity in pathogenesis. Mutant spectrum complexity can, in turn, be influenced by intervening changes in population size (possible occurrence of bottlenecks and their intensity) during virus replication in vivo.
4. To identify the presence of memory genomes, in particular in the course of persistent infections.
5. The definition of the constraints to variation acting at the level of viral RNA or viral proteins that limit virus diversification. Some of these constraints are inherent to viral RNA and protein functions, while others are imposed by multiple interactions of viral RNA and proteins with host nucleic acids, proteins, and lipids.
6. To discern the selective constraints acting on virus replication and that can promote rapid evolution of specific genomic sites: immune response, physiological alteration of the host, presence of antiviral agents, and others. It is hoped that research in these six aspects can distill new knowledge to find improved strategies to control enteroviral disease.

In current virology, progress of experimental research is closely linked to progress in theoretical biology, particularly the application of bioinformatics in its broadest sense. As emphasized elsewhere (Domingo 2007), we regard as key the continuous reference to experimental results to include realistic parameters in theoretical studies intended to gain an understanding of virus behavior. Theoretical investigations include models of viral population dynamics, in silico simulation of the behavior of virtual replicons with interactions among components of mutant spectra, transitions of the type occurring upon virus extinction through error catastrophe, modeling of quasispecies memory, and others. Examples of theoretical studies with unrealistic or biased parameters have been published that contradict experimental facts. Virology will not learn anything from such models, which may even distort solid knowledge. In contrast, experiment-driven, exploratory, conceptual, or summary models (the latter incorporating only a few key parameters), when based on reliable experimental determinations of mutation rates, population sizes, rounds of replication, fitness values of components of mutant spectra (which are never identical for a real mutant cloud) can be instrumental in the understanding of virus evolution. Fortunately, the literature also has many examples of such progress-driving theoretical treatments applied to virology.

Quasispecies theory, as it helps to explain enterovirus evolution, the main focus of this chapter, provides an adequate framework to understand the population dynamics of any replicative entity that produces error copies of its parental form on a regular basis. The scope of application of quasispecies theory goes from the origin of life to applied virology, and it embraces natural and artificial cellular and subcellular systems, from informationally minimal replicons to cancer cells. The broad applicability of quasispecies theory says much for its explanatory and experiment-provoking powers.

We close with the suggestion that quasispecies is only recently gaining ground over more traditional approaches to understand virus evolution and pathogenesis, in part because there was no clear awareness of the real complexity of viral populations. Currently, several new applications have come from an understanding of quasispecies dynamics (such as the necessity of combination therapy and multiepitopic vaccines, as summarized in Sect. 9) and other applications are under way (i.e., lethal mutagenesis). We deem extremely challenging the pursuit of an understanding of viruses as populations through the tools of quasispecies and of the sciences of complexity in general.

**Acknowledgements** This review has benefited from discussions with colleagues and students at Centro de Biología Molecular "Severo Ochoa" (CSIC-UAM) and Centro de Astrobiología (CSIC-INTA), as reflected in several literature references. We are indebted to all of them. Work was supported by grant BFU-2005-00863 from MEC, FIPSE 36377/03 and 36558/06 and Fundación Ramón Areces. CP is the recipient of a I3P contract from CSIC, financed by Fondo Social Europeo.

# References

Agol VI (2006) Molecular mechanisms of poliovirus variation and evolution. Curr Top Microbiol Immunol 299:211-259

Agol VI, Pilipenko EV, Slobodskaya OR (1996) Modification of translational control elements as a new approach to design of attenuated picornavirus strains. J Biotechnol 44:119-128

Agol VI, Belov GA, Cherkasova EA, Gavrilin GV, Kolesnikova MS, Romanova LI, Tolskaya EA (2001) Some problems of molecular biology of poliovirus infection relevant to pathogenesis, viral spread and evolution. Dev Biol (Basel) 105:43-50

Anderson JP, Daifuku R, Loeb LA (2004) Viral error catastrophe by mutagenic nucleosides. Annu Rev Microbiol 58:183-205

Arita M, Zhu SL, Yoshida H, Yoneyama T, Miyamura T, Shimizu H (2005) A Sabin 3-derived poliovirus recombinant contained a sequence homologous with indigenous human enterovirus species C in the viral polymerase coding region. J Virol 79:12650-2657

Arnold JJ, Vignuzzi M, Stone JK, Andino R, Cameron CE (2005) Remote site control of an active site fidelity checkpoint in a viral RNA-dependent RNA polymerase. J Biol Chem 280:25706-25716

Baranowski E, Ruiz-Jarabo CM, Pariente N, Verdaguer N, Domingo E (2003) Evolution of cell recognition by viruses: a source of biological novelty with medical implications. Adv Virus Res 62:19-111

Barnard DL (2006) Current status of anti-picornavirus therapies. Curr Pharm Des 12:1379-1390

Batschelet E, Domingo E, Weissmann C (1976) The proportion of revertant and mutant phage in a growing population, as a function of mutation and growth rate. Gene 1:27-32

Beaulieux F, Zreik Y, Deleage C, Sauvinet V, Legay V, Giraudon P, Kean KM, Lina B (2005) Cumulative mutations in the genome of Echovirus 6 during establishment of a chronic infection in precursors of glial cells. Virus Genes 30:103-112

Bergelson JM (2003) Virus interactions with mucosal surfaces: alternative receptors, alternative pathways. Curr Opin Microbiol 6:386-391

Berger MM, Kopp N, Vital C, Redl B, Aymard M, Lina B (2000) Detection and cellular localization of enterovirus RNA sequences in spinal cord of patients with ALS. Neurology 54:20-25

Biebricher CK, Domingo E (2007) The advantage of the high genetic diversity in RNA viruses. Future Virol 2:35-38

Biebricher CK, Eigen M (2005) The error threshold. Virus Res 107:117-127

Biebricher CK, Eigen M (2006) What is a quasispecies? Curr Top Microbiol Immunol 299:1-31

Boerlijst MC, Boenhoefer S, Nowak MA (1996) Viral quasispecies and recombination. Proc R Soc Lond B 263:1577-1584

Botstein D (1980) A theory of modular evolution for bacteriophages. Ann N Y Acad Sci 354:484-491

Botstein D (1981) A modular theory of virus evolution. In: Fields BN, Jaenisch R, Fox CF (eds) Animal virus genetics. Academic, New York pp 363-384

Briones C, Domingo E, Molina-París C (2003) Memory in retroviral quasispecies: experimental evidence and theoretical model for human immunodeficiency virus. J Mol Biol 331:213-229

Brown B, Oberste MS, Maher K, Pallansch MA (2003) Complete genomic sequencing shows that polioviruses and members of human enterovirus species C are closely related in the noncapsid coding region. J Virol 77:8973-8984

Brown BA, Oberste MS, Alexander JP Jr, Kennett ML, Pallansch MA (1999) Molecular epidemiology and evolution of enterovirus 71 strains isolated from 1970 to 1998. J Virol 73:9969-9975

Brown NF, Wickham ME, Coombes BK, Finlay BB (2006) Crossing the line: selection and evolution of virulence traits. PLoS Pathog 2:e42

Cammack N, Phillips A, Dunn G, Patel V, Minor PD (1988) Intertypic genomic rearrangements of poliovirus strains in vaccinees. Virology 167:507-514

Cardosa MJ, Perera D, Brown BA, Cheon D, Chan HM, Chan KP, Cho H, McMinn P (2003) Molecular epidemiology of human enterovirus 71 strains and recent outbreaks in the Asia-Pacific region: comparative analysis of the VP1 and VP4 genes. Emerg Infect Dis 9:461-468

Carson SD, Hobbs JT, Tracy SM, Chapman NM (1999) Expression of the coxsackievirus and adenovirus receptor in cultured human umbilical vein endothelial cells: regulation in response to cell density. J Virol 73:7077-7079

Castro C, Arnold JJ, Cameron CE (2005) Incorporation fidelity of the viral RNA-dependent RNA polymerase: a kinetic, thermodynamic and structural perspective. Virus Res 107:141-149

Cohen CJ (2006) Successful HIV treatment: lessons learned. J Manag Care Pharm 12:6-11
Colbere-Garapin F, Duncan G, Pavio N, Pelletier I, Petit I (1998) An approach to understanding the mechanisms of poliovirus persistence in infected cells of neural or non-neural origin. Clin Diagn Virol 9:107-113
Colbere-Garapin F, Peletier I, Ouzilou L (2002) Persistent infections by picornaviruses. In: Semler BL, Wimmer E (eds) Molecular biology of picornaviruses. ASM Press, Washington DC, pp 437-448
Conticello SG, Thomas CJ, Petersen-Mahrt SK, Neuberger MS (2005) Evolution of the AID/APOBEC family of polynucleotide (deoxy)cytidine deaminases. Mol Biol Evol 22:367-377
Crotty S, Maag D, Arnold JJ, Zhong W, Lau JYN, Hong Z, Andino R, Cameron CE (2000) The broad-spectrum antiviral ribonucleotide, ribavirin, is an RNA virus mutagen. Nat Med 6:1375-1379
Crotty S, Cameron CE, Andino R (2001) RNA virus error catastrophe: direct molecular test by using ribavirin. Proc Natl Acad Sci U S A 98:6895-6900
Crowell RL, Syverton JT (1961) The mammalian cell-virus relationship. VI. Sustained infection of HeLa cells by Coxsackie B3 virus and effect on superinfection. J Exp Med 113:419-435
Chan YF, Abubakar S (2006) Phylogenetic evidence for inter-typic recombination in the emergence of human enterovirus 71 subgenotypes. BMC Microbiol 6:74
Chen YM, Lan YC, Lai SF, Yang JY, Tsai SF, Kuo SH (2006) HIV-1 CRF07-BC infections, injecting drug users, Taiwan. Emerg Infect Dis 12:703-705
Cherkasova E, Laassri M, Chizhikov V, Korotkova E, Dragunsky E, Agol VI, Chumakov K (2003) Microarray analysis of evolution of RNA viruses: evidence of circulation of virulent highly divergent vaccine-derived polioviruses. Proc Natl Acad Sci U S A 100:9398-9403
Chetverin AB, Kopein DS, Chetverina HV, Demidenko AA, Ugarov VI (2005) Viral RNA-directed RNA polymerases use diverse mechanisms to promote recombination between RNA molecules. J Biol Chem 280:8748-8755
Chiu YL, Greene WC (2006) Multifaceted antiviral actions of APOBEC3 cytidine deaminases. Trends Immunol 27:291-297
Chumakov KM, Powers LB, Noonan KE, Roninson IB, Levenbook IS (1991) Correlation between amount of virus with altered nucleotide sequence and the monkey test for acceptability of oral poliovirus vaccine. Proc Natl Acad Sci U S A 88:199-203
Dalakas MC (1995). Enteroviruses and human neuromuscular diseases. In: Robart HA (ed) Human enterovirus infections. ASM Press, Washington, DC, pp 387-398
Dan M, Chantler JK (2005) A genetically engineered attenuated coxsackievirus B3 strain protects mice against lethal infection. J Virol 79:9285-9295
De la Torre JC, Holland JJ (1990) RNA virus quasispecies populations can suppress vastly superior mutant progeny. J Virol 64:6278-6281
Domingo E (2003) Quasispecies and the development of new antiviral strategies. Progress Drug Res 60:133-158
Domingo E (ed) (2005) Virus entry into error catastrophe as a new antiviral strategy. Virus Res 107:115-228
Domingo E (ed) (2006) Quasispecies: concepts and implications for virology. Curr Top Microbiol Immunol Vol. 299
Domingo E (2007) Virus evolution. In: Knipe DM Howley PM (eds) Field's virology, 5th edn. Lappincott Williams & Wilkins, Philadelphia, pp 389-421
Domingo E, Flavell RA, Weissmann C (1976) In vitro site-directed mutagenesis: generation and properties of an infectious extracistronic mutant of bacteriophage Qβ. Gene 1:3-25
Domingo E, Sabo D, Taniguchi T, Weissmann C (1978) Nucleotide sequence heterogeneity of an RNA phage population. Cell 13:735-744
Domingo E, Dávila M, Ortín J (1980) Nucleotide sequence heterogeneity of the RNA from a natural population of foot-and-mouth-disease virus. Gene 11:333-346
Domingo E, Biebricher C, Eigen M, Holland JJ (2001) Quasispecies and RNA virus evolution: principles and consequences. Landes Bioscience, Austin

Domingo E, Brun A, Núñez JI, Cristina J, Briones C, Escarmís C (2006) Genomics of Viruses. In: Hacker J, Dobrindt U (eds) Pathogenomics: genome analysis of pathogenic microbes. Wiley-VCH Verlag, Weinheim, pp 369-388

Drake JW, Holland JJ (1999) Mutation rates among RNA viruses. Proc Natl Acad Sci U S A 96:13910-13913

Eggers HJ, Tamm I (1965) Coxsackie A9 virus: mutation from drug dependence to drug independence. Science 148:97-98

Eigen M (1971) Self-organization of matter and the evolution of biological macromolecules. Naturwissenschaften 58:465-523

Eigen M (1987) New concepts for dealing with the evolution of nucleic acids. Cold Spring Harb Symp Quant Biol 52:307-320

Eigen M (2000) Natural selection: a phase transition? Biophys Chem 85:101-123

Eigen M (2002) Error catastrophe and antiviral strategy. Proc Natl Acad Sci U S A 99:13374-13376

Eigen M, Schuster P (1979) The hypercycle. A principle of natural self-organization. Springer, Berlin New York Heidelberg

Eigen M, Biebricher CK (1988) Sequence space and quasispecies distribution. In: Domingo E, Ahlquist P, Holland JJ (eds) RNA genetics, Vol. 3. CRC Press, Boca Raton, FL, pp 211-245

Escarmís C, Lázaro E, Manrubia SC (2006) Population bottlenecks in quasispecies dynamics. Curr Top Microbiol Immunol 299:141-170

Evans DJ, Almond JW (1998) Cell receptors for picornaviruses as determinants of cell tropism and pathogenesis. Trends Microbiol 6:198-202

Fauquet CM, Mayo MA, Maniloff J, Desselberger U, Ball LA (eds) (2005) Virus taxonomy. Eighth Report of the International Committee on Taxonomy of Viruses. Elsevier, San Diego

Ferrer-Orta C, Arias A, Perez-Luque R, Escarmis C, Domingo E, Verdaguer N (2007) Sequential structures provide insights into the fidelity of RNA replication. Proc Natl Acad Sci U S A 104:9463-9468

Friedberg EC, Walker GC, Siede W, Wood RD, Schultz RA, Ellenberger T (2006). DNA repair and mutagenesis. American Society for Microbiology, Washington, DC

Furione M, Guillot S, Otelea D, Balanant J, Candrea A, Crainic R (1993) Polioviruses with natural recombinant genomes isolated from vaccine-associated paralytic poliomyelitis. Virology 196:199-208

Gallei A, Pankraz A, Thiel HJ, Becher P (2004) RNA recombination in vivo in the absence of viral replication. J Virol 78:6271-6281

García-Arriaza J, Manrubia SC, Toja M, Domingo E, Escarmís C (2004) Evolutionary transition toward defective RNAs that are infectious by complementation. J Virol 78:11678-11685

García-Arriaza J, Ojosnegros S, Dávila M, Domingo E, Escarmis C (2006) Dynamics of mutation and recombination in a replicating population of complementing, defective viral genomes. J Mol Biol 360:558-572

Gavrilin GV, Cherkasova EA, Lipskaya GY, Kew OM, Agol VI (2000) Evolution of circulating wild poliovirus and of vaccine-derived poliovirus in an immunodeficient patient: a unifying model. J Virol 74:7381-7390

Gibson JP, Righthand VF (1985) Persistence of echovirus 6 in cloned human cells. J Virol 54:219-223

Giraud P, Beaulieux F, Ono S, Shimizu N, Chazot G, Lina B (2001) Detection of enteroviral sequences from frozen spinal cord samples of Japanese ALS patients. Neurology 56:1777-1778

Gmyl AP, Korshenko SA, Belousov EV, Khitrina EV, Agol VI (2003) Nonreplicative homologous RNA recombination: promiscuous joining of RNA pieces? RNA 9:1221-1231

González-López C, Arias A, Pariente N, Gómez-Mariano G, Domingo E (2004) Preextinction viral RNA can interfere with infectivity. J Virol 78:3319-3324

Graci JD, Cameron CE (2006) Mechanisms of action of ribavirin against distinct viruses. Rev Med Virol 16:37-48

Grande-Pérez A, Lazaro E, Lowenstein P, Domingo E, Manrubia SC (2005) Suppression of viral infectivity through lethal defection. Proc Natl Acad Sci U S A 102:4448-4452

Gromeier M, Alexander L, Wimmer E (1996) Internal ribosomal entry site substitution eliminates neurovirulence in intergeneric poliovirus recombinants. Proc Natl Acad Sci U S A 93:2370-2375

Gromeier M, Wimmer E, Gorbalenya AE (1999) Genetics, pathogenesis and evolution of picornaviruses. In: Domingo E, Webster R, GHolland JJ (eds) Origin and evolution of viruses. Academic, San Diego, pp 287-343

Harris RS, Petersen-Mahrt SK, Neuberger MS (2002) RNA editing enzyme APOBEC1 and some of its homologs can act as DNA mutators. Mol Cell 10:1247-1253

Harris KS, Brabant W, Styrchak S, Gall A, Daifuku R (2005) KP-1212/1461, a nucleoside designed for the treatment of HIV by viral mutagenesis. Antiviral Res 67:1-9

Herrera M, Garcia-Arriaza J, Pariente N, Escarmis C, Domingo E (2007) Molecular basis for a lack of correlation between viral fitness and cell killing capacity. PLoS Pathog 3:e53

Holland JJ, Spindler K, Horodyski F, Grabau E, Nichol S, VandePol S (1982) Rapid evolution of RNA genomes. Science 215:1577-1585

Holland JJ, Domingo E, de la Torre JC, Steinhauer DA (1990) Mutation frequencies at defined single codon sites in vesicular stomatitis virus and poliovirus can be increased only slightly by chemical mutagenesis. J Virol 64:3960-3962

Holland JJ, de La Torre JC, Steinhauer DA (1992) RNA virus populations as quasispecies. Curr Top Microbiol Immunol 176:1-20

Hosoya M, Kawasaki Y, Sato M, Honzumi K, Kato A, Hiroshima T, Ishiko H, Suzuki H (2006) Genetic diversity of enterovirus 71 associated with hand, foot and mouth disease epidemics in Japan from 1983 to 2003. Pediatr Infect Dis J 25:691-694

Hyypiä T, Hovi T, Knowles NJ, Stanway G (1997) Classification of enteroviruses based on molecular and biological properties. J Gen Virol 78:1-11

Iizuka N, Kohara M, Hagino-Yamagishi K, Abe S, Komatsu T, Tago K, Arita M, Nomoto A (1989) Construction of less neurovirulent polioviruses by introducing deletions into the 5' noncoding sequence of the genome. J Virol 63:5354-5363

Kandolf R, Ameis D, Kirschner P, Canu A, Hofschneider PH (1987) In situ detection of enteroviral genomes in myocardial cells by nucleic acid hybridization: an approach to the diagnosis of viral heart disease. Proc Natl Acad Sci U S A 84:6272-6276

Kew O, Morris-Glasgow V, Landaverde M, Burns C, Shaw J, Garib Z, Andre J, Blackman E, Freeman CJ, Jorba J, Sutter R, Tambini G, Venczel L, Pedreira C, Laender F, Shimizu H, Yoneyama T, Miyamura T, van Der Avoort H, Oberste MS, Kilpatrick D, Cochi S, Pallansch M, de Quadros C (2002) Outbreak of poliomyelitis in Hispaniola associated with circulating type 1 vaccine-derived poliovirus. Science 296:356-359

Kim KS, Tracy S, Tapprich W, Bailey J, Lee CK, Kim K, Barry WH, Chapman NM (2005) 5'-Terminal deletions occur in coxsackievirus B3 during replication in murine hearts and cardiac myocyte cultures and correlate with encapsidation of negative-strand viral RNA. J Virol 79:7024-7041

Kimman TG, Boot H (2006) The polio eradication effort has been a great success - let's finish it and replace it with something even better. Lancet Infect Dis 6:675-678

King AMQ (1988) Genetic recombination in positive strand RNA viruses. In: Domingo E, Holland JJ, Ahlquist P (eds) RNA genetics, Vol. II. CRC Press, Boca Raton, FL

Lenski RE, May RM (1994) The evolution of virulence in parasites and pathogens: reconciliation between two competing hypotheses. J Theor Biol 169:253-265

Li H, Roossinck MJ (2004) Genetic bottlenecks reduce population variation in an experimental RNA virus population. J Virol 78:10582-10587

Loeb LA, Essigmann JM, Kazazi F, Zhang J, Rose KD, Mullins JI (1999) Lethal mutagenesis of HIV with mutagenic nucleoside analogs. Proc Natl Acad Sci U S A 96:1492-1497

Lukashev AN (2005) Role of recombination in evolution of enteroviruses. Rev Med Virol 15:157-167

Lukashev AN, Lashkevich VA, Ivanova OE, Koroleva GA, Hinkkanen AE, Ilonen J (2005) Recombination in circulating human enterovirus B: independent evolution of structural and non-structural genome regions. J Gen Virol 86:3281-3290

Macadam AJ, Ferguson G, Stone DM, Meredith J, Knowlson S, Auda G, Almond JW, Minor PD (2006) Rational design of genetically stable, live-attenuated poliovirus vaccines of all three serotypes: relevance to poliomyelitis eradication. J Virol 80:8653-8663

Martin DP, Williamson C, Posada D (2005) RDP2: recombination detection and analysis from sequence alignments. Bioinformatics 21:260-262

Martin J (2006) Vaccine-derived poliovirus from long term excretors and the end game of polio eradication. Biologicals 34:117-122

Mateu MG, Da Silva JL, Rocha E, De Brum DL, Alonso A, Enjuanes L, Domingo E, Barahona H (1988) Extensive antigenic heterogeneity of foot-and-mouth disease virus of serotype C. Virology 167:113-124

McMinn PC (2002) An overview of the evolution of enterovirus 71 and its clinical and public health significance. FEMS Microbiol Rev 26:91-107

Menéndez-Arias L (2002) Molecular basis of fidelity of DNA synthesis and nucleotide specificity of retroviral reverse transcriptases. Prog in Nucl Acid Res Mol Biol 71:91-147

Mills DR, Priano C, DiMauro P, Binderow BD (1989) Q beta replicase: mapping the functional domains of an RNA-dependent RNA polymerase. J Mol Biol 205:751-764

Minor PD (1998) Picornaviruses: Topley and Wilson's microbiology and microbial infections. In: Mahy BWJ, Collier AC (eds) Virology I. Arnold, London, pp 485-510

Nagy PD, Simon AE (1997) New insights into the mechanisms of RNA recombination. Virology 235:1-9

Oberste MS, Maher K, Kilpatrick DR, Pallansch MA (1999) Molecular evolution of the human enteroviruses: correlation of serotype with VP1 sequence and application to picornavirus classification. J Virol 73:1941-1948

Oberste MS, Maher K, Pallansch MA (2004) Evidence for frequent recombination within species human enterovirus B based on complete genomic sequences of all thirty-seven serotypes. J Virol 78:855-867

Page KM, Nowak MA (2002) Unifying evolutionary dynamics. J Theor Biol 219:93-98

Palacios G, Oberste MS (2005) Enteroviruses as agents of emerging infectious diseases. J Neurovirol 11:424-433

Pariente N, Airaksinen A, Domingo E (2003) Mutagenesis versus inhibition in the efficiency of extinction of foot-and-mouth disease virus. J Virol 77:7131-7138

Parker WB (2005) Metabolism and antiviral activity of ribavirin. Virus Res 107:165-171

Perales C, Mateo R, Mateu MG, Domingo E (2007) Insights into RNA virus mutant spectrum and lethal mutagenesis events: replicative interference and complementation by multiple point mutants. J Mol Biol 369:985-1000

Pfeiffer JK, Kirkegaard K (2005) Increased fidelity reduces poliovirus fitness under selective pressure in mice. PLoS Pathogens 1:102-110

Pfeiffer JK, Kirkegaard K (2006) Bottleneck-mediated quasispecies restriction during spread of an RNA virus from inoculation site to brain. Proc Natl Acad Sci U S A 103:5520-5525

Poulin R, Combes C (1999) The concept of virulence: interpretations and implications. Parasitol Today 15:474-475

Prabhakar BS, Haspel VM, McClintock PR, Notkins AL (1982) High frequency of antigenic variants among naturally occurring human coxsackie B4 virus isolates identified by monoclonal antibodies. Nature 300:374-376

Prabhakar BS, Menegus MA, Notkins AL (1985) Detection of conserved and nonconserved epitopes on Coxsackievirus B4: frequency of antigenic change. Virology 146:302-306

Quer J, Huerta R, Novella IS, Tsimring L, Domingo E, Holland JJ (1996) Reproducible nonlinear population dynamics and critical points during replicative competitions of RNA virus quasispecies. J Mol Biol 264:465-471

Quer J, Hershey CL, Domingo E, Holland JJ, Novella IS (2001) Contingent neutrality in competing viral populations. J Virol 75:7315-7320

Ravits J (2005) Sporadic amyotrophic lateral sclerosis: a hypothesis of persistent (non-lytic) enteroviral infection. Amyotroph Lateral Scler Other Motor Neuron Disord 6:77-87

Roux L, Simon AE, Holland JJ (1991) Effects of defective interfering viruses on virus replication and pathogenesis in vitro and in vivo. Adv Virus Res 40:181-211

Ruiz-Jarabo CM, Arias A, Baranowski E, Escarmís C, Domingo E (2000) Memory in viral quasispecies. J Virol 74:3543-3547

Ruiz-Jarabo CM, Arias A, Molina-París C, Briones C, Baranowski E, Escarmís C, Domingo E (2002) Duration and fitness dependence of quasispecies memory. J Mol Biol 315:285-296

Ruiz-Jarabo CM, Miller E, Gómez-Mariano G, Domingo E (2003) Synchronous loss of quasispecies memory in parallel viral lineages: a deterministic feature of viral quasispecies. J Mol Biol 333:553-563

Saakian DB, Hu CK (2006) Exact solution of the Eigen model with general fitness functions and degradation rates. Proc Natl Acad Sci U S A 103:4935-4939

Saleh MC, Van Rij RP, Andino R (2004) RNA silencing in viral infections: insights from poliovirus. Virus Res 102:11-17

Sanders SA, Herrero LJ, McPhie K, Chow SS, Craig ME, Dwyer DE, Rawlinson W, McMinn PC (2006) Molecular epidemiology of enterovirus 71 over two decades in an Australian urban community. Arch Virol 151:1003-1013

Santti J, Hyypia T, Kinnunen L, Salminen M (1999) Evidence of recombination among enteroviruses. J Virol 73:8741-8749

Saulnier A, Pelletier I, Labadie K, Colbere-Garapin F (2006) Complete cure of persistent virus infections by antiviral siRNAs. Mol Ther 13:142-150

Schaub M, Keller W (2002) RNA editing by adenosine deaminases generates RNA and protein diversity. Biochimie 84:791-803

Schuster P, Stadler PF (1999) Nature and evolution of early replicons. In: Domingo E, Webster RG, Holland JJ (eds) Origin and evolution of viruses. Academic Press, San Diego, pp 1-24

Seder RA, Mascola JR (2003) Basic immunology of vaccine development. In: (Bloom BR, Lambert P-H (eds) The vaccine book. Academic Press, San Diego, pp 51-72

Semler BL, Wimmer E (eds) (2002) Molecular biology of picornaviruses. ASM Press, Washington DC

Shulman LM, Manor Y, Sofer D, Swartz T, Mendelson E (2006) Oral poliovaccine: will it help eradicate polio or cause the next epidemic? Isr Med Assoc J 8:312-315

Sierra M, Airaksinen A, González-López C, Agudo R, Arias A, Domingo E (2007) Foot-and-mouth disease virus mutant with decreased sensitivity to ribavirin: implications for error catastrophe. J Virol 81:2012-2024

Simmonds P (2006) Recombination and selection in the evolution of picornaviruses and other mammalian positive-stranded RNA viruses. J Virol 80:11124-11140

Simmonds P, Welch J (2006) Frequency and dynamics of recombination within different species of human enteroviruses. J Virol 80:483-493

Suspène R, Henry M, Guillot S, Wain-Hobson S, Vartanian JP (2005) Recovery of APOBEC3-edited human immunodeficiency virus G→A hypermutants by differential DNA denaturation PCR. J Gen Virol 86:125-129

Takemoto KK, Habel K (1959) Virus-cell relationship in a carrier culture of HeLa cells and Coxsackie A9 virus. Virology 7:28-44

Tam PE, Messner RP (1999) Molecular mechanisms of coxsackievirus persistence in chronic inflammatory myopathy: viral RNA persists through formation of a double-stranded complex without associated genomic mutations or evolution. J Virol 73:10113-10121

Tracy S, Chapman NM, Drescher KM, Kono K, Tapprich W (2006) Evolution of virulence in picornaviruses. Curr Top Microbiol Immunol 299:193-209

Urbanowicz A, Alejska M, Formanowicz P, Blazewicz J, Figlerowicz M, Bujarski JJ (2005) Homologous crossovers among molecules of brome mosaic bromovirus RNA1 or RNA2 segments in vivo. J Virol 79:5732-5742

Valente L, Nishikura K (2005) ADAR gene family and A-to-I RNA editing: diverse roles in post-transcriptional gene regulation. Prog Nucleic Acid Res Mol Biol 79:299-338

Vignuzzi M, Stone JK, Arnold JJ, Cameron CE, Andino R (2006) Quasispecies diversity determines pathogenesis through cooperative interactions in a viral population. Nature 439:344-348

Ward CD, Stokes MA, Flanegan JB (1988) Direct measurement of the poliovirus RNA polymerase error frequency in vitro. J Virol 62:558-562

Wilke CO (2005) Quasispecies theory in the context of population genetics. BMC Evol Biol 5:44

Wilke CO, Ronnewinkel C, Martinetz T (2001a) Dynamic fitness landscapes in molecular evolution. Phys Rep 349:395-446

Wilke CO, Wang JL, Ofria C, Lenski RE, Adami C (2001b) Evolution of digital organisms at high mutation rates leads to survival of the flattest. Nature 412:331-333

Wright S (1982) Character change, speciation, and the higher taxa. Evolution 36:427-443

Yakovenko ML, Cherkasova EA, Rezapkin GV, Ivanova OE, Ivanov AP, Eremeeva TP, Baykova OY, Chumakov KM, Agol VI (2006) Antigenic evolution of vaccine-derived polioviruses: changes in individual epitopes and relative stability of the overall immunological properties. J Virol 80:2641-2653

Yang CF, Chen HY, Jorba J, Sun HC, Yang SJ, Lee HC, Huang YC, Lin TY, Chen PJ, Shimizu H, Nishimura Y, Utama A, Pallansch M, Miyamura T, Kew O, Yang JY (2005) Intratypic recombination among lineages of type 1 vaccine-derived poliovirus emerging during chronic infection of an immunodeficient patient. J Virol 79:12623-12634

Yin PD, Das D, Mitsuya H (2006) Overcoming HIV drug resistance through rational drug design based on molecular, biochemical, and structural profiles of HIV resistance. Cell Mol Life Sci 63:1706-1724

Zell R, Krumbholz A, Dauber M, Hoey E, Wutzler P (2006) Molecular-based reclassification of the bovine enteroviruses. J Gen Virol 87:375-385

Zimmern D (1988) Evolution of RNA viruses. In: Domingo E, Holland JJ, Ahlquist P (eds) RNA Genetics, Vol. 2, CRC Press, FL, pp 211-240

# Comparative Genomics of the Coxsackie B Viruses and Related Enteroviruses

M. S. Oberste

| | | |
|---|---|---|
| 1 | Overview | 34 |
| 2 | Picornavirus Genomics | 34 |
| | 2.1   5'-NTR Diversity | 36 |
| | 2.2   3'-NTR Diversity | 36 |
| | 2.3   Polyprotein | 37 |
| | 2.4   Capsid Sequence Diversity | 37 |
| | 2.5   Nonstructural Region | 39 |
| | 2.6   *Cis*-Acting Replication Element | 41 |
| 3 | Conclusions | 42 |
| 4 | Future Directions | 42 |
| References | | 43 |

**Abstract** Genomic analysis of the group B coxsackieviruses (CVB) has improved our understanding of CVB evolution, epidemiology, and pathogenesis. Comparison of capsid sequence alignments and virion structures allows correlation of capsid diversity with surface features, such as loops, the receptor canyon, and antigenic sites. Pairwise sequence comparisons and phylogenetic analyses can be used to rapidly identify and classify enteroviruses. Enteroviruses are monophyletic by type only within the capsid region. The CVBs as a group are monophyletic in the capsid region, probably due to their shared use of the coxsackievirus-adenovirus receptor (other members of HEV-B use different receptors). Outside the capsid region, enteroviruses are monophyletic only by species (not by type), reflecting a high frequency of intertypic recombination within a species. Further genomic studies, accompanied by well-characterized clinical outcome/disease data, will facilitate fine-scale mapping of genetic determinants that contribute to virulence.

---

M. S. Oberste
Centers for Disease Control and Prevention, 1600 Clifton Road NE,
Mailstop G-17, Atlanta, GA 30333, USA
soberste@cdc.gov

## 1 Overview

The coxsackie B viruses (CVBs) were identified as a unique enterovirus group on the basis of the characteristic disease they caused in suckling mice inoculated intracerebrally (Pallansch and Roos 2006). Sequence analyses have shown that echoviruses and many newer enteroviruses are closely related to the CVBs; as a result, these viruses are classified together in the species *Human enterovirus B* (HEV-B; genus *Enterovirus*, family *Picornaviridae*) (Stanway et al. 2005). Complete genome sequences are available for at least one representative of all 54 recognized types within HEV-B, except EV78, and multiple genomes are available for some types (total $n=96$) (Stanway et al. 2005) (http://www.picornaviridae.com). Comparison of capsid sequence alignments and virion structures allows correlation of capsid diversity with surface features, such as loops, the receptor canyon, and antigenic sites. Pairwise sequence comparisons and phylogenetic analyses can be used to rapidly identify and classify enteroviruses. Such analyses reveal that (1) enteroviruses are monophyletic by type only within the capsid region (Oberste et al. 1999); (2) the CVBs as a group are monophyletic in the capsid region (Hyypiä et al. 1997; Oberste et al. 1999; Pöyry et al. 1996), probably due to their shared use of the coxsackievirus-adenovirus receptor (other members of HEV-B use different receptors); and (3) outside the capsid region, enteroviruses are monophyletic only by species (not by type), reflecting a high frequency of intertypic recombination within a species (Andersson et al. 2002; Brown et al. 2003; Hyypiä et al. 1997; Lukashev et al. 2003, 2004, 2005; Oberste et al. 2004a, 2004b, 2004c, 2004d; Pöyry et al. 1996; Santti et al. 1999).

## 2 Picornavirus Genomics

While the genetic basis of complex phenotypes, such as transmissibility, host range, and receptor usage may not be clearly understood, all intrinsic properties of a picornavirus must ultimately derive from the viral genome. The Genomics Age for eukaryotic virology began in 1981, with the publication of the complete genome sequence of poliovirus type 1 (Mahoney strain) (Kitamura et al. 1981; Racaniello and Baltimore 1981). This accomplishment permitted the direct mapping of genetically and functionally defined viral proteins and facilitated the development of reverse genetic systems to help probe the molecular details of poliovirus replication, translation, and protein function (Racaniello and Baltimore 1981; Sarnow 1989; Semler et al. 1984; van der Werf et al. 1986). Similar approaches were quickly applied to studies of other virus families (Knipe et al. 2006), and other picornavirus genome sequences also followed soon afterward, representing all genera of *Picornaviridae*, and sometimes helping to define new genera (Cohen et al. 1987; Doherty et al. 1999; Forss et al. 1984; Hyypiä et al. 1992; Krumbholz et al. 2002; Oberste et al. 2003; Palmenberg et al. 1984; Stanway et al. 1984, 2005; Wutz et al. 1996; Yamashita et al. 1998). The quantity and quality of picornavirus sequence data, and the ease with which it can be generated, have increased substantially with the introduction of PCR and improvements in sequencing technology over the last 25 years (Fig. 1) (http://www.picornaviridae.com).

This chapter will discuss lessons learned from studies on nucleotide and amino acid sequence conservation and divergence among the CVBs, and related enteroviruses, focusing on members of the species HEV-B, with reference to other enterovirus species to illustrate specific points when necessary. It must be borne in mind that the available enterovirus sequences are generally derived from prototype reference strains or a small number of more recent clinical isolates. Each of these isolates represents only a temporal and geographic snapshot in enterovirus evolution and may or may not be representative of their particular serotype or of enteroviruses as a whole. Generalizable patterns may be discerned only from the careful analysis of a large number of sequences obtained from viruses with a wide temporal and geographic distribution.

HEV-B is composed of 56 serotypes - approximately half of all known enterovirus serotypes - and includes coxsackievirus A9 (CVA9), the coxsackie B viruses (six types: CVB1-6), the echoviruses (E; 28 types: E1-7, 9, 11, 13-21, 24-27, 29-33), and most of the newer, numbered enteroviruses (EV; 21 types: EV69, 73-75, 77-88, 93, 97, 98, 100-101) (Stanway et al. 2005). Swine vesicular disease virus (SVDV) is also a member of HEV-B. SVDV infects and causes disease in pigs, but it is serotypically identical to CVB5 (Brown et al. 1973; Knowles and McCauley 1997; Zhang et al. 1999). The CVB genomes vary in length from 7,389 nucleotides (CVB1, strain Japan) to 7,403 nucleotides (CVB2, strain Ohio-1 and CVB5, strain 2000/CSF/KOR), with the typical picornavirus genome organization of a single, long, open reading frame flanked by 5'- and 3'-nontranslated regions (NTRs) that function in viral replication and translation (Racaniello 2001). The range of genome lengths within HEV-B is 7,389 (CVB1, strain Japan) to 7,453 (E9-strain DM) (Oberste et al. 2004a).

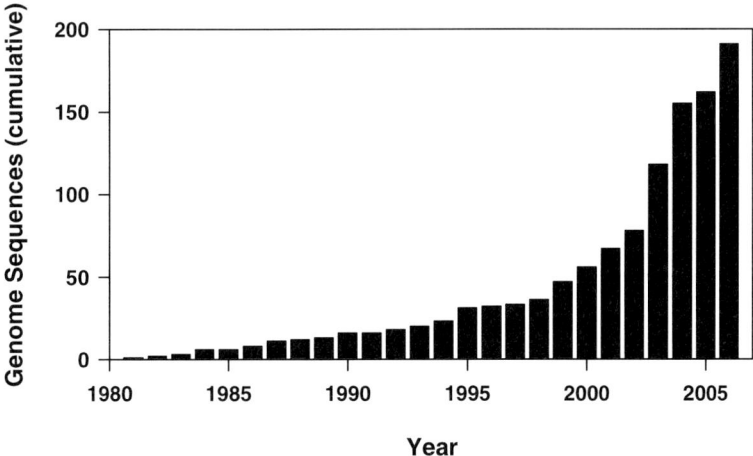

**Fig. 1** The cumulative number of picornavirus complete genome sequences, by year

## 2.1 5′-NTR Diversity

The HEV-B 5′-NTR sequences are 738-750 nucleotides long and differ from one another by 5%-23% (Oberste et al. 2004a). Nearly 50% percent of the 5′-NTR residues are invariant among all of the viruses and almost 30% of the variable sites are concentrated in the hypervariable region, the 80-110 residues immediately upstream of the initiation codon (Oberste et al. 2004a). In the 5′-NTR, the viruses in HEV-A and HEV-B are all closely related to one another and intermix without regard to species, forming enterovirus 5′-NTR group II, whereas HEV-C (including the polioviruses) and HEV-D form group I (Brown et al. 2003; Hyypiä et al. 1997; Oberste et al. 2004a; Santti et al. 1999). A number of mutations that attenuate the cardiovirulent phenotype of certain CVB3 strains or the neurovirulence of polioviruses have been mapped to the 5′-NTR (Dunn et al. 2000, 2003; Evans et al. 1985; Kawamura et al. 1989; Macadam et al. 1991) (see also the chapter by K. Knowlton). Structural elements that are important for the function of the internal ribosome entry site are well conserved among all enteroviruses (see also the chapters by Sean and Semler and Marchant et al.). While RNA secondary structures are the primary functional units of the 5′-NTR, there also exist short segments of extraordinarily high primary sequence identity (Oberste and Pallansch 2005). These short segments have been exploited by numerous investigators as targets for molecular diagnostic assays, such as nucleic acid hybridization and RT-PCR (Oberste and Pallansch 2005; Rotbart and Romero 1995).

## 2.2 3′-NTR Diversity

The enterovirus 3′-NTR, the site of initiation of negative-strand RNA synthesis, is required for efficient genome replication (Brown et al. 2004; Mirmomeni et al. 1997; Rohll et al. 1995). The 3′-NTRs of the HEV-B viruses are similar in length, 102-109 nucleotides, and are 70%-99% identical to one another but only 42%-62% identical to those of representatives of other human enterovirus species (Oberste et al. 2006). While 3′-NTR sequences vary widely among the various enterovirus species (Brown et al. 2003; Oberste et al. 2004a, 2004b, 2004c), the existence of highly conserved secondary structures suggests that these structures, rather than the primary sequences, are the functional unit involved in replication (Mirmomeni et al. 1997; Pilipenko et al. 1992; Pilipenko et al. 1996). The predicted structures consist of three stem-loops termed X, Y, and Z in HEV-A and HEV-B, two stem-loops (X and Y) in HEV-C and HEV-D, and one stem-loop in the human rhinoviruses (Mirmomeni et al. 1997). Stem-loops X and Y form a tertiary structure through a so-called kissing interaction of their loop residues (Melchers et al. 1997; Mirmomeni et al. 1997; Pilipenko et al. 1992). The Z domain is apparently dispensable for replication in culture but may play a role in viral pathogenesis in vivo (Merkle et al. 2002).

## 2.3 Polyprotein

Picornavirus proteins are expressed from a single open reading frame, resulting in a polyprotein of approximately 2200 amino acids that is processed by viral proteases to yield the mature viral proteins (Racaniello 2001). The polyprotein is functionally divided into three regions: P1, P2, and P3 (Rueckert and Wimmer 1984). P1 encodes the virion structural proteins (capsid), while protein processing, replication, and host-cell interaction functions are encoded in P2 and P3 (see Sect. 6).

## 2.4 Capsid Sequence Diversity

The mature virion proteins, 1A-1D, are also known as VP4, VP2, VP3, and VP1, respectively (Rueckert and Wimmer 1984). The icosahedral capsid is composed of 60 copies of each of these proteins, five copies of VP1 at each fivefold axis of symmetry, and three copies each of VP2 and VP3 at each threefold axis, with VP4 internal to the capsid shell. The canyon surrounding the fivefold axis is the principal site of receptor interaction, with the dominant neutralizing epitopes arrayed on surface projections around the edges of the canyon. Most of the residues in these conformational epitopes are contributed by VP1 and VP2, but some are also contributed by VP3 (Huber et al. 1993; Mateu 1995; Reimann et al. 1991; Usherwood and Nash 1995). The capsid region is the most variable part of the polyprotein, both within and between species, whereas the non-capsid region sequences are much more highly conserved (Fig. 2). Despite the overall divergence, there are short conserved motifs throughout the capsid, often in structurally important regions. The HEV-B capsid protein (P1) sequences vary in length from 848 to 868 amino acids (Oberste et al. 2004a). Capsid sequences of a given serotype are collinear, but there are often insertions or deletions when comparing sequences of strains of different types. VP1, VP2, and VP3 vary in length, between types, and between species, but VP4 is collinear for all enteroviruses and rhinoviruses. The length differences tend to accumulate in regions of known diversity among the enterovirus capsid proteins - most of these variable regions are loops that are exposed on the surface of the virion, rather than being in the beta-barrel structural elements. The largest regions of high diversity are the VP2 puff region, the region surrounding the VP3 knob, both prominent surface projections, and the N- and C-terminal regions of VP1 (Fig. 2). The amino terminus of VP1 is buried in the native virion but changes conformation on virion uncoating, exposing an epitope that is highly conserved among all enteroviruses (Hovi and Roivainen 1993; Samuelson et al. 1994).

Within HEV-B, the complete P1 sequences are at least 68% identical to one another (Oberste et al. 2004a), and viruses of the same type are generally at least

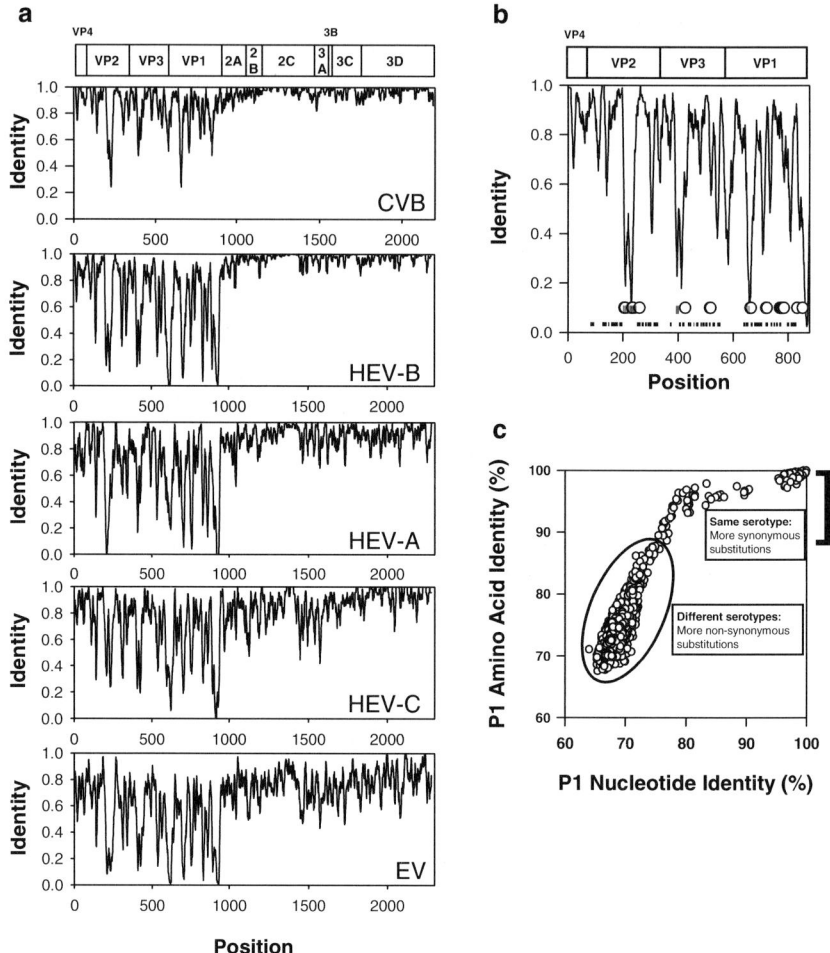

Fig. 2 Human enterovirus polyprotein amino acid sequence variation. **a** Overall polyprotein diversity among human enteroviruses (all EV, HEV-A-C, CVB); HEV-D is not shown because there are only two complete genome sequences available. Amino acid sequence identity in a sliding window of ten residues is plotted as a continuous curve. The individual plots depict diversity among (i) CVB + SVDV, (ii) all HEV-B, (iii) all HEV-A, (iv) all HEV-C, and (v) all enteroviruses. **b** HEV-B capsid diversity. Amino acid sequence identity in a sliding window of ten residues is plotted as a continuous curve. *Open circles* indicate regions that are in the receptor canyon of CVB3 (Muckelbauer et al. 1995). *Short vertical lines* indicate residues that form the α-helix and β-barrel structures of CVB3. **c** HEV-B P1 diversity. Amino acid identity is plotted vs nucleotide sequence identity for each pair of HEV-B sequences. The square bracket indicates the range of diversity among viruses of the same type

90% identical in complete capsid sequence (Oberste et al. 2001, 2005). The greatest sequence variation occurs in VP1, which varies by up to 43% among members of HEV-B (Oberste et al. 1999, 2004a). P1 sequences are monophyletic, both within serotype and within species (Fig. 3a) (Brown et al. 2003; Oberste et al. 1999, 2004a, 2004b, 2004c). The HEV-B viruses differ from one another by up to 28% in VP2, 35% in VP3, and up to 30% in the VP4 sequences. The individual capsid proteins, VP1, VP2, and VP3, are also monophyletic by serotype and species, suggesting that recombination is rare within the capsid.

## 2.5 Nonstructural Region

Proteins derived from the P2 and P3 regions are involved in genome replication and protein processing, and some of these proteins are also involved in other important functions during viral replication, such as disruption of cellular processes. Most of these functions were determined using poliovirus, CVB3, or human rhinoviruses, but it is presumed that the proteins function similarly in most or all of the enteroviruses. Many of the proteins are discussed in greater detail elsewhere in this volume, but they will be briefly introduced here.

Protein 2A is a cysteine protease that cleaves in *cis* to liberate the P1 protein from the genome polyprotein (Ryan and Flint 1997; Toyoda et al. 1986) and is also involved in shutoff of host-cell protein synthesis (Kräusslich et al. 1987); however, the precise mechanism of host-cell shutoff has not been fully resolved (Belsham and Jackson 2000). The 2B protein plays a role in RNA replication by participating in the formation of membranous replication vesicles (Aldabe and Carrasco 1995) and intracellular transmembrane pores (van Kuppeveld et al. 2002); these membrane alterations may also contribute to release of mature virions (van Kuppeveld et al. 1995, 1997a,1997b). Vesicle formation has also been attributed to 2BC and 2C (Aldabe and Carrasco 1995; Cho et al. 1994). 2C has RNA-binding, NTPase, and cysteine-rich sequence motifs that are highly conserved (Gorbalenya et al. 1988, 1989; Gorbalenya and Koonin 1989); the RNA-binding motif facilitates binding of 2C or 2BC to the 3' end of negative-strand RNA (Banerjee and Dasgupta 2001; Banerjee et al. 1997, 2001; Klein et al. 1999; Mirzayan and Wimmer 1992, 1994) and the cysteine-rich motif is involved in binding zinc (Pfister et al. 2000), but the role of the NTPase activity remains unknown.

Proteins derived from the P3 region provide the major enzymatic activities of the viral replication complex, contributing the primer protein, 3B (VPg), probably in the form of 3AB which is known to associate with intracellular membranes (Datta and Dasgupta 1994; Semler et al. 1982; Towner et al. 1996), as well as the RNA-dependent RNA polymerase (3D and/or 3CD) (Flanegan and Baltimore 1977), VPg uridylylation activity (3D) (Paul et al. 1998, 2000), and determinants involved in RNA binding and interaction with cellular proteins that are recruited into the viral replication complex (3C and or 3CD) (Andino et al. 1990a, 1990b; Blair et al. 1998; Herrold and Andino 2001; Parsley et al. 1999). The 3C and 3CD proteins also provide the chymotrypsin-like serine protease activity that is responsible for the majority of viral protein processing (Ryan and Flint 1997).

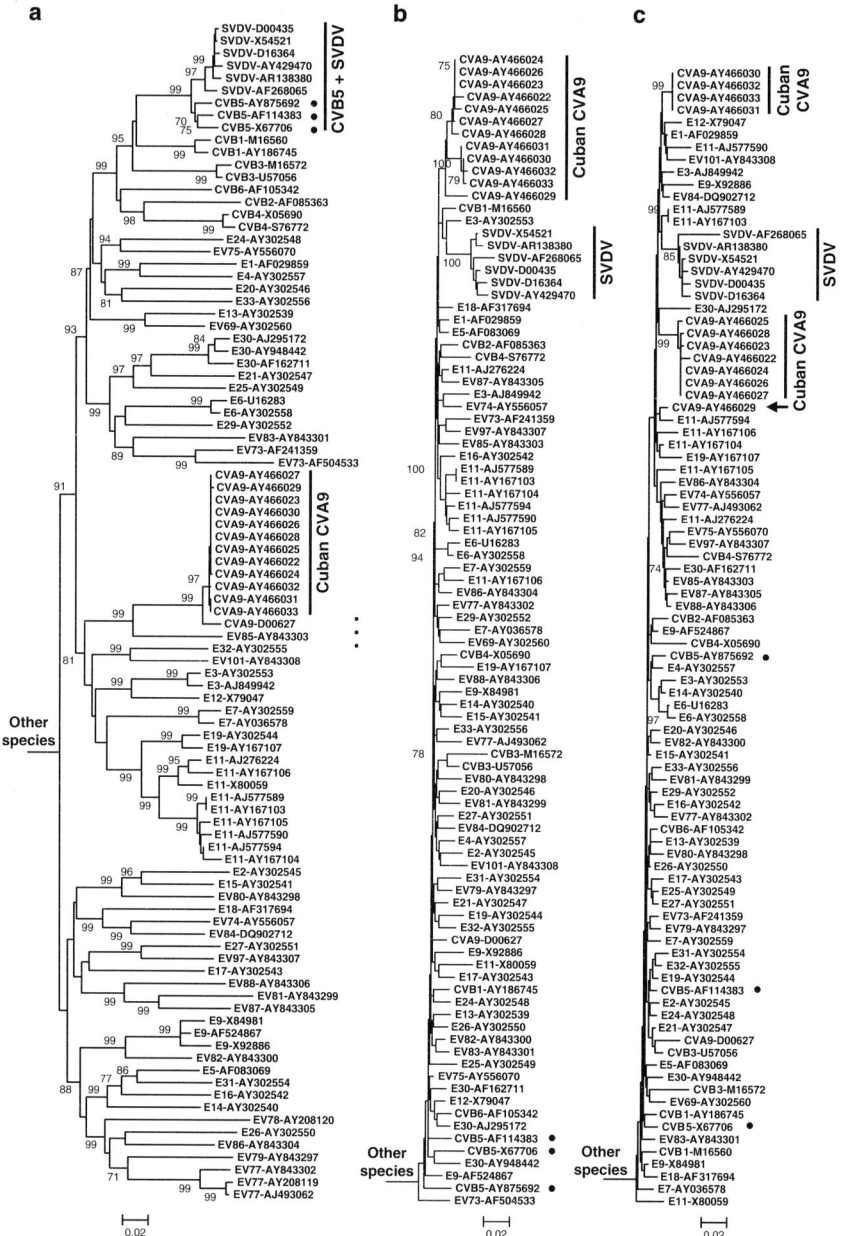

**Fig. 3** Phylogenetic relationships based on aligned HEV-B amino acid sequences. Trees were constructed separately for P1 (**a**), P2 (**b**), and P3 (**c**), using the neighbor-joining algorithm implemented in MEGA, version 3.1 (Kumar et al. 2001), with the JTT substitution model (Jones et al. 1992); they are plotted to the same scale for each region. The scale bar indicates the number of amino acid substitutions per site

Because of the extensive RNA recombination that occurs outside the capsid-coding region, the CVB nonstructural proteins cannot be considered separately from those of the other viruses in HEV-B; that is, all members of HEV-B draw their P2 and P3 regions from a common genetic pool that is constantly exchanged by RNA recombination within a given capsid lineage (Andersson et al. 2002; Lindberg et al. 2003; Lukashev et al. 2003, 2004, 2005; Oberste et al. 2004a, 2004d; Santti et al. 1999). The non-capsid proteins are fully collinear among all of the HEV-B viruses (P2 = 578 aa; P3 = 756 aa). The P2 and P3 regions are highly conserved within HEV-B, more so than among members of other human enterovirus species (Fig. 2) (Brown et al. 2003; Oberste et al. 2004a, 2004b, 2004c). The 2A proteins are the most variable in P2-P3, differing by up to 19% within HEV-B. The 3B protein also varies by up to 18% (two amino acid differences, of 22 total), but the other mature nonstructural proteins (2B, 2C, 3A, 3C, and 3D) vary by no more than 14% (3A), and there are numerous examples of identical amino acid sequences for some nonstructural proteins among viruses of heterologous serotypes (Oberste et al. 2004a). The deduced 2C and 3D protein sequences are the most highly conserved, with no more than 6% variation in either protein.

Diversity of the nonstructural proteins is probably constrained by enzyme structure/function, as enzymes tend to be more sensitive to mutation than are structural proteins. In P2 and P3, the interspecies phylogenetic diversity (e.g., HEV-A vs HEV-B) is similar to P1, but the intraspecies diversity is much lower (Fig. 3). Unlike the P1 region, P2 and P3 sequences are not monophyletic by type and the CVBs are not monophyletic as a group. In P2 and P3, sequence monophyly can be taken as evidence of epidemiologic linkage, provided surveillance is sufficiently sensitive; that is, two viruses that share nearly identical sequences in this region must have diverged very recently from a common ancestor, as recombination has not yet occurred. For example, the SVDVs are monophyletic as a group in P2, but delinked from CVB5s (Fig. 3b), and SVDVs remain monophyletic in P3 (Fig. 3c), suggesting that they emerged as a swine pathogen following a single introduction (Zhang et al. 1999). Similarly, a group of Cuban CVA9 isolates are monophyletic in P1 and P2 but not in P3 (Fig. 3b,c), indicating that they are beginning to diverge from one another by recombination with other HEV-B strains that are cocirculating in Cuba (Fig. 3c).

## 2.6 Cis-*Acting Replication Element*

A distinct RNA structural element, the *cis*-acting replication element (*cre*), has been shown to be required for enterovirus replication (Goodfellow et al. 2000; Rieder et al. 2000). The poliovirus *cre* is a four-part stacked stem and conserved loop located in the region encoding 2C. The AAACA motif in the loop, which is required for *cre* function (Goodfellow et al. 2000; van Ooij et al. 2006), is completely conserved among all of the enterovirus 2C sequences (Brown et al. 2003; Oberste et al. 2004a, 2004b, 2004c). The expanded and generalized version of this motif, $RN_3AARN_6R$, which models stem 1 of Goodfellow et al. (2000) as part of the loop (Yang et al. 2002), is also conserved. The structure of the predicted stem region is also well conserved among enteroviruses, with complete sequence conservation of a five-base-pair stem immediately adjacent to the 14-residue loop (Brown et al. 2003).

## 3 Conclusions

In all enterovirus species, nucleotide sequence evolution is largely the result of recombination and synonymous substitutions, resulting in relative conservation of the encoded polypeptide sequences, except in the capsid region where diversity is almost exclusively driven by nucleotide substitutions, with amino acid sequences relatively conserved within a type, but highly variable between viruses of different types. If recombination occurs at all within the capsid, the evidence is quickly obscured by rapid accumulation of nucleotide substitutions. With the exception of viruses of known epidemiologic linkage, all serotypes with multiple complete sequences show evidence of recombination; therefore, all enterovirus strains can be considered recombinants relative to nonlinked strains.

While non-capsid sequences may influence pathogenicity or tropism (e.g., by affecting replication or translation), the principal identity of an enterovirus (its antigenic structure, receptor binding, etc.) is controlled by the capsid. In general, an enterovirus might be viewed as a capsid sequence in search of non-capsid sequences of the highest fitness to provide a selective replicative advantage. The 5'-NTR and P2-P3-3'-NTR sequences of a given isolate represent only a snapshot of that particular isolate or of a closely related lineage, within a narrow temporal and geographic window. This view of the role of recombination in enterovirus evolution would predict that the specific genomic combinations and sequences in the P2-P3 regions of the prototype strains from 50 years ago are not likely to be present in currently circulating strains of the same serotype. Conversely, sequences related to those of a given prototype strain may be found in different serotypes within the same species among currently circulating enteroviruses. The observed genomic sequences agree well with these predictions. The designation of a serotype prototype strain is purely arbitrary, but it provides a context for the analysis of other clinical isolates of that serotype. That is, the prototype strains are simply a snapshot in time, arbitrarily chosen as a reference.

## 4 Future Directions

Despite recent progress in enterovirus genomics, there are many areas in which additional genomic studies can enhance our understanding of enterovirus basic biology and disease association. Genomic sequences from a large collection of isolates of a given type (or related types, e.g., the CVBs) with well-characterized clinical outcome/disease will facilitate fine-scale mapping of genetic determinants that contribute to virulence. The combination of more capsid sequences and additional three-dimensional virion structures will permit comparative mapping of receptor interaction sites and broaden our understanding of virus-host interaction at the cell surface. Large-scale comparative genomics of wild strains, as well as directed cell-based and cell-free in vitro studies, will help develop a better understanding of the factors that facilitate and constrain enterovirus recombination. Finally, better methods to rapidly generate complete genome sequences, especially directly from original clinical material, will make all of these studies easier, cheaper, and more practical.

# References

Aldabe R, Carrasco L (1995) Induction of membrane proliferation by poliovirus proteins 2B and 2BC. Biochem Biophys Res Comm 206:64-76
Andersson P, Edman K, Lindberg AM (2002) Molecular analysis of the echovirus 18 prototype. Evidence of interserotypic recombination with echovirus 9. Virus Res 85:71-83
Andino R, Rieckhof GE, Baltimore D (1990a) A functional ribonucleoprotein complex forms around the 5' end of poliovirus RNA. Cell 63:369-380
Andino R, Rieckhof GE, Trono D, Baltimore D (1990b) Substitutions in the protease ($3C^{pro}$) gene of poliovirus can suppress a mutation in the 5' noncoding region. J Virol 64:607-612
Banerjee R, Dasgupta A (2001) Interaction of picornavirus 2C polypeptide with the viral negative-strand RNA. J Gen Virol 82:2621-2627
Banerjee R, Echeverri A, Dasgupta A (1997) Poliovirus-encoded 2C polypeptide specifically binds to the 3'-terminal sequences of viral negative-strand RNA. J Virol 71:9570-9578
Banerjee R, Tsai W, Kim W, Dasgupta A (2001) Interaction of poliovirus-encoded 2C/2BC polypeptides with the 3' terminus negative-strand cloverleaf requires an intact stem-loop b. Virol 280:41-51
Belsham GJ, Jackson RJ (2000) Translation initiation on picornavirus RNA. In: Sonenberg N, Hershey JWB, Mathews MB (eds) Translational control of gene expression. Cold Spring Harbor Laboratory Press, Cold Spring Harbor, NY, pp 869-900
Blair WS, Parsley TB, Bogerd HP, Towner JS, Semler BL, Cullen BR (1998) Utilization of a mammalian cell-based RNA binding assay to characterize the RNA binding properties of picornavirus 3C proteinases. RNA 4:215-225
Brown BA, Maher K, Oberste MS, Pallansch MA (2003) Complete genomic sequencing shows that polioviruses and members of human enterovirus species C are closely related in the non-capsid coding region. J Virol 77:8973-8984
Brown DM, Kauder SE, Cornell CT, Jang GM, Racaniello VR, Semler BL (2004) Cell-dependent role for the poliovirus 3' noncoding region in positive-strand RNA synthesis. J Virol 78:1344-1351
Brown F, Talbot P, Burrows R (1973) Antigenic differences between isolates of swine vesicular disease virus and their relationship to coxsackie B5 virus. Nature 245:315-316
Cho MW, Teterina N, Egger D, Bienz K, Ehrenfeld E (1994) Membrane rearrangement and vesicle induction by recombinant poliovirus 2B and 2BC in human cells. Virology 202:129-145
Cohen JI, Ticehurst JR, Purcell RH, Buckler-White A, Baroudy BM (1987) Complete nucleotide sequence of wild-type hepatitis A virus: comparison with different strains of hepatitis A virus and other picornaviruses. J Virol 61:50-59

Datta U, Dasgupta A (1994) Expression and subcellular localization of poliovirus VPg-precursor protein 3AB in eukaryotic cells: evidence for glycosylation in vitro. J Virol 68: 4468-4477

Doherty M, Todd D, McFerran N, Hoey EM (1999) Sequence analysis of a porcine enterovirus serotype 1 isolate: relationships with other picornaviruses. J Gen Virol 80:1929-1941

Dunn JJ, Bradrick SS, Chapman NM, Tracy SM, Romero JR (2003) The stem loop II within the 5' nontranslated region of clinical coxsackievirus B3 genomes determines cardiovirulence phenotype in a murine model. J Inf Dis 187:1552-1561

Dunn JJ, Chapman NM, Tracy S, Romero JR (2000) Genomic determinants of cardiovirulence in coxsackievirus B3 clinical isolates: localization to the 5' nontranslated region. J Virol 74:4787-4794

Evans DMA, Dunn G, Minor PD, Schild GC, Cann AJ, Almond JW (1985) Increased neurovirulence associated with a single nucleotide change in a non-coding region of the Sabin type 3 polio vaccine genome. Nature 314:548-550

Flanegan JB, Baltimore D (1977) Poliovirus-specific primer-dependent RNA polymerase able to copy poly(A). Proc Natl Acad Sci U S A 74:3677-3680

Forss S, Strebel K, Schaller H (1984) Nucleotide sequence and genome organization of foot-and-mouth disease virus. Nucl Acids Res 12:6587-6601

Goodfellow I, Chaudhry Y, Richardson A, Meredith J, Almond JW, Barclay W, Evans DJ (2000) Identification of a *cis*-acting replication element within the poliovirus coding region. J Virol 74:4590-4600

Gorbalenya AE, Blinov VM, Donchenko AP, Koonin EV (1989) An NTP-binding motif is the most conserved sequence in a highly diverged monophyletic group of proteins involved in positive-strand viral RNA replication. J Mol Evol 28:256-268

Gorbalenya AE, Koonin EV (1989) Viral proteins containing the purine NTP-binding sequence pattern. FEBS Lett 243:103-114

Gorbalenya AE, Koonin EV, Donchenko AP, Blinov VM (1988) A conserved NTP-motif in putative helicases. Nature 333:22

Herrold J, Andino R (2001) Poliovirus RNA replication requires genome circularization through a protein-protein bridge. Mol Cell 7:581-591

Hovi T, Roivainen M (1993) Peptide antisera targeted to a conserved sequence in poliovirus capsid protein VP1 cross-react widely with members of the genus *Enterovirus*. J Clin Microbiol 31:1083-1087

Huber S, Polgar J, Moraska A, Cunningham M, Schwimmbeck P, Schulteiss P (1993) T lymphocyte responses in CVB3-induced murine myocarditis. Scan J Inf Dis 88:67-78

Hyypiä T, Horsnell C, Maaronen M, Khan M, Kalkkinen N, Auvinen P, Kinnunen L, Stanway G (1992) A distinct picornavirus group identified by sequence analysis. Proc Natl Acad Sci U S A 89:8847-8851

Hyypiä T, Hovi T, Knowles NJ, Stanway G (1997) Classification of enteroviruses based on molecular and biological properties. J Gen Virol 78:1-11

Jones DT, Taylor WR, Thornton JM (1992) The rapid generation of mutation data matrices from protein sequences. Comp Appl Biosci 8:275-282

Kawamura N, Kohara M, Abe S, Komatsu T, Tago K, Arita M, Nomoto A (1989) Determinants in the 5' noncoding region of poliovirus Sabin 1 RNA that influence the attenuation phenotype. J Virol 63:1302-1309

Kitamura N, Semler BL, Rothberg PG, Larsen GR, Adler CJ, Dorner AJ, Emini EA, Hanecak R, Lee JJ, van der Werf S, Anderson CW, Wimmer E (1981) Primary structure, gene organization and polypeptide expression of poliovirus RNA. Nature 291:547-553

Klein M, Eggers HJ, Nelsen-Salz B (1999) Echovirus 9 strain Barty non-structural protein 2C has NTPase activity. Virus Res 65:155-160

Knipe DM, Howley PM, Griffin DE, Lamb RA, Martin MA, Roizman B, Straus SE, editors (2006) Fields virology, 5th edn. Lippincott Williams & Wilkins, Philadelphia

Knowles NJ, McCauley JW (1997) Coxsackievirus B5 and the relationship to swine vesicular disease virus. Curr Top Microbiol Immunol 223:153-167

Kräusslich H-G, Nicklin MJ, Toyoda H, Etchison D, Wimmer E (1987) Poliovirus protease 2A induces cleavage of eucaryotic initiation factor 4F polypeptide p220. J Virol 61:2711-2718

Krumbholz A, Dauber M, Henke A, Birch-Hirschfeld E, Knowles NJ, Stelzner A, Zell R (2002) Sequencing of porcine enterovirus groups II and III reveals unique features of both virus groups. J Virol 76:5813-5821

Kumar S, Tamura K, Jakobsen IB, Nei M (2001) MEGA2: molecular evolutionary genetics analysis software. Bioinformatics 17:1244-1245

Lindberg AM, Andersson P, Savolainen C, Mulders MN, Hovi T (2003) Evolution of the genome of *Human enterovirus B*: incongruence between phylogenies of the VP1 and 3CD regions indicates frequent recombination within the species. J Gen Virol 84:1223-1235

Lukashev AN, Lashkevich VA, Ivanova OE, Koroleva GA, Hinkkanen AE, Ilonen J (2003) Recombination in circulating enteroviruses. J Virol 77:10423-10431

Lukashev AN, Lashkevich VA, Koroleva GA, Ilonen J, Hinkkanen AE (2004) Recombination in uveitis-causing enterovirus strains. J Gen Virol 85:463-470

Lukashev AN, Lashkevich VA, Ivanova OE, Koroleva GA, Hinkkanen AE, Ilonen J (2005) Recombination in circulating *Human enterovirus B*: independent evolution of structural and non-structural genome regions. J Gen Virol 86:3281-3290

Macadam AJ, Pollard SR, Ferguson G, Dunn G, Skuce R, Almond JW, Minor PD (1991) The 5' non-coding region of the type 2 poliovirus vaccine strain contains determinants of attenuation and temperature sensitivity. Virology 181:451-458

Mateu MG (1995) Antibody recognition of picornaviruses and escape from neutralization. Virus Res 38:1-24

Melchers WJ, Hoenderop JG, Bruins Slot HJ, Pleij CW, Pilipenko EV, Agol VI, Galama JM (1997) Kissing of the two predominant hairpin loops in the coxsackie B virus 3' untranslated region is the essential structural feature of the origin of replication required for negative-strand RNA synthesis. J Virol 71:686-696

Merkle I, van Ooij MJM, van Kuppeveld FJM, Glaudemans DHRF, Galama JMD, Henke A, Zell R, Melchers WJG (2002) Biological significance of a human enterovirus B-specific RNA element in the 3' nontranslated region. J Virol 76:9900-9909

Mirmomeni MH, Hughes PJ, Stanway G (1997) An RNA tertiary structure in the 3' untranslated region of enteroviruses is necessary for efficient replication. J Virol 71:2363-2370

Mirzayan C, Wimmer E (1992) Genetic analysis of an NTP-binding motif in poliovirus polypeptide 2C. Virology 189:547-555

Mirzayan C, Wimmer E (1994) Biochemical studies on poliovirus polypeptide 2C: evidence for ATPase activity. Virology 199:176-187

Muckelbauer JK, Kremer M, Minor I, Diana G, Dutko FJ, Groarke J, Pevear DC, Rossman MG (1995) The structure of coxsackievirus B3 at 3.5Å resolution. Structure 3:653-667

Oberste MS, Maher K, Kilpatrick DR, Pallansch MA (1999) Molecular evolution of the human enteroviruses: correlation of serotype with VP1 sequence and application to picornavirus classification. J Virol 73:1941-1948

Oberste MS, Maher K, Michele SM, Uddin M, Belliot G, Pallansch MA (2005) Enteroviruses 76, 89, 90, and 91 represent a novel group within the species *Human enterovirus A*. J Gen Virol 86:445-451

Oberste MS, Maher K, Pallansch M (2004a) Evidence for frequent recombination within *Human enterovirus B* based on complete genomic sequences of all 37 serotypes. J Virol 78:855-867

Oberste MS, Maher K, Pallansch MA (2003) Genetic evidence that SV2 and six other simian picornaviruses represent a new genus in Picornaviridae. Virology 314:283-293

Oberste MS, Maher K, Schnurr D, Peters H, Sessions W, Kirk C, Chatterjee N, Fuller S, Hanauer JM, Pallansch MA (2004b) Enterovirus 68 is associated with respiratory illness and shares features with both the enteroviruses and the rhinoviruses. J Gen Virol 85:2577-2584

Oberste MS, Maher K, Williams AJ, Dybdahl-Sissoko N, Brown BA, Gookin MT, Peñaranda S, Mishrik NG, Uddin M, Pallansch MA (2006) Species-specific RT-PCR amplification of human enteroviruses: a tool for rapid species identification of uncharacterized enteroviruses. J Gen Virol 87:119-128

Oberste MS, Pallansch MA (2005) Enterovirus molecular detection and typing. Rev Med Microbiol 16:163-171

Oberste MS, Peñaranda S, Maher K, Pallansch MA (2004c) Complete genome sequences of members of the species Human enterovirus A. J Gen Virol 85:1597-1607

Oberste MS, Peñaranda S, Pallansch MA (2004d) RNA recombination plays a major role in genomic change during circulation of coxsackie B viruses. J Virol 78:2948-2955

Oberste MS, Schnurr D, Maher K, al-Busaidy S, Pallansch MA (2001) Molecular identification of new picornaviruses and characterization of a proposed enterovirus 73 serotype. J Gen Virol 82:409-416

Pallansch MA, Roos R (2006) Enteroviruses: polioviruses, coxsackieviruses, echoviruses, and newer enteroviruses. In: Knipe DM, Howley PM, Griffin DE, Lamb RA, Martin MA, Roizman B, Straus SE (eds) Fields virology, 5th edn. Lippincott Williams & Wilkins, Philadelphia, pp 725-776

Palmenberg AC, Kirby EM, Janda MR, Drake NI, Potratz KF, Collett MC (1984) The nucleotide and deduced amino acid sequences of the encephalomyocarditis viral polyprotein coding region. Nucl Acids Res 12:2969-2985

Parsley TB, Cornell CT, Semler BL (1999) Modulation of the RNA binding and protein processing activities of poliovirus polypeptide 3CD by the viral RNA polymerase domain. J Biol Chem 274:12867-12876

Paul AV, Rieder E, Kim DW, van Boom JH, Wimmer E (2000) Identification of an RNA hairpin in the poliovirus RNA that serves as the primary template in the in vitro uridylylation of VPg. J Virol 74:10359-10370

Paul AV, van Boom JH, Filippov D, Wimmer E (1998) Protein-primed RNA synthesis by purified poliovirus RNA polymerase. Nature 393:280-284

Pfister T, Jones KW, Wimmer E (2000) A cysteine-rich motif in poliovirus protein $2C^{ATPase}$ is involved in RNA replication and binds zinc in vitro. J Virol 74:334-343

Pilipenko EV, Maslova SV, Sinyakov AN, Agol VI (1992) Towards identification of cis-acting elements involved in the replication of enterovirus and rhinovirus RNAs: a proposal for the existence of tRNA-like terminal structures. Nucl Acids Res 20:1739-1745

Pilipenko EV, Poperechny KV, Maslova SV, Melchers WJ, Slot HJ, Agol VI (1996) Cis-element, oriR, involved in the initiation of (-) strand poliovirus RNA: a quasi-globular multi-domain RNA structure maintained by tertiary ('kissing') interactions. EMBO J 15:5428-5436

Pöyry T, Kinnunen L, Hyypia T, Brown B, Horsnell C, Hovi T, Stanway G (1996) Genetic and phylogenetic clustering of enteroviruses. J Gen Virol 77:1699-1717

Racaniello VR (2001) Picornaviridae: the viruses and their replication. In: Knipe DM, Howley PM, Griffin DE, Lamb RA, Martin MA, Roizman B, Straus SE (eds) Fields virology, 5th edn. Lippincott Williams & Wilkins Philadelphia, pp 685-722

Racaniello VR, Baltimore D (1981) Cloned poliovirus complementary DNA is infectious in mammalian cells. Science 214:916-919

Reimann B-Y, Zell R, Kandolf R (1991) Mapping of a neutralizing antigenic site of coxsackievirus B4 by construction of an antigen chimera. J Virol 65:3475-3480

Rieder E, Paul AV, Kim DW, van Boom JH, Wimmer E (2000) Genetic and biochemical studies of poliovirus *cis*-acting replication element *cre* in relation to VPg uridyldylation. J Virol 74:10371-10380

Rohll JB, Moon DH, Evans DJ, Almond JW (1995) The 3' untranslated region of picornaviruses RNA: features required for efficient genome replication. J Virol 69:7835-7844

Rotbart HA, Romero JR (1995) Laboratory diagnosis of enteroviral infections. In: Rotbart HA (ed) Human enterovirus infections. ASM Press, Washington DC, pp 401-418

Rueckert RR, Wimmer E (1984) Systematic nomenclature of picornavirus proteins. J Virol 50:957-959

Ryan MD, Flint M (1997) Virus-encoded proteinases of the picornavirus super-group. J Gen Virol 78:699-723

Samuelson A, Forsgren M, Johansson B, Wahren B, Sällberg M (1994) Molecular basis for serological cross-reactivity between enteroviruses. Clin Diag Lab Immunol 1:336-341

Santti J, Hyypia T, Kinnunen L, Salminen M (1999) Evidence of recombination among enteroviruses. J Virol 73:8741-8749

Sarnow P (1989) Role of 3'-end sequences in infectivity of poliovirus transcripts made in vitro. J Virol 63:467-470

Semler BL, Anderson CW, Hanecak R, Dorner LF, Wimmer E (1982) A membrane-associated precursor to poliovirus VPg identified by immunoprecipitation with antibodies directed against a synthetic hexapeptide. Cell 28:405-412

Semler BL, Dorner AJ, Wimmer E (1984) Production of infectious poliovirus from cloned cDNA is dramatically increased by SV40 transcription and replication signals. Nucl Acids Res 12:5123-5141

Stanway G, Brown F, Christian P, Hovi T, Hyypiä T, King AMQ, Knowles NJ, Lemon SM, Minor PD, Pallansch MA, Palmenberg AC, Skern T (2005) Picornaviridae. In: Fauquet CM, Mayo MA, Maniloff J, Desselberger U, Ball LA (eds) Virus taxonomy: eighth report of the International Committee on the Taxonomy of Viruses. Elsevier, Amsterdam, pp 757-778

Stanway G, Hughes PJ, Mountford RC, Minor PD, Almond JW (1984) The complete nucleotide sequence of a common cold virus: human rhinovirus 14. Nucl Acids Res 12:7859-7875

Towner JS, Ho TV, Semler BL (1996) Determinants of membrane association for poliovirus protein 3AB. J Biol Chem 271:26810-26818

Toyoda H, Nicklin MJ, Murray MG, Anderson CW, Dunn JJ, Studier FW, Wimmer E (1986) A second virus-encoded proteinase involved in proteolytic processing of poliovirus polyprotein. Cell 45:761-770

Usherwood EJ, Nash AA (1995) Lymphocyte recognition of picornaviruses. J Gen Virol 76: 499-508

Van der Werf S, Bradley J, Wimmer E, Studier FW, Dunn JJ (1986) Synthesis of infectious poliovirus RNA by purified T7 RNA poliovirus. Proc Nat Acad Sci U S A 83:2330-2334

Van Kuppeveld FJ, Galama JM, Zoll J, Melchers WJ (1995) Genetic analysis of a hydrophobic domain of coxsackie B3 virus protein 2B: a moderate degree of hydrophobicity is required for a *cis*-acting function in viral RNA synthesis. J Virol 69:7782-7790

Van Kuppeveld FJ, Hoenderop JG, Smeets RL, Willems PH, Dijkman HB, Galama JM, Melchers WJ (1997a) Coxsackievirus protein 2B modifies endoplasmic reticulum membrane and plasma membrane permeability and facilitates virus release. EMBO J 16:3519-3532

Van Kuppeveld FJ, Melchers WJ, Kirkegaard K, Doedens JR (1997b) Structure-function analysis of coxsackie B3 virus protein 2B. Virology 227:111-118

Van Kuppeveld FJ, Melchers WJ, Willems PHGM, Gadella TWJ (2002) Homomultimerization of the coxsackievirus 2B protein in living cells visualized by fluorescence resonance energy transfer microscopy. J Virol 76:9446-9456

Van Ooij MJM, Vogt DA, Paul A, Castro C, Kuijpers J, Van Kuppeveld FJM, Cameron CE, Wimmer E, Andino R, Melchers WJG (2006) Structural and functional characterization of the coxsackievirus B3 CRE92C): role of DRE(2C) in negative- and positive-strand RNA synthesis. J Gen Virol 87:103-113

Wutz G, Auer H, Nowotny N, Skern T, Keuchler E (1996) Equine Rhinovirus serotypes 1 and 2: relationship to each other and to aphthoviruses and cardioviruses. J Gen Virol 77:1719-1730

Yamashita T, Sakae K, Tsuzuki H, Suzuki Y, Ishikawa N, Takeda N, Miyamura T, Yamazaki S (1998) Complete nucleotide sequence and genetic organization of Aichi virus, a distinct member of the Picornaviridae associated with acute gastroenteritis in humans. J Virol 72:8408-8412

Yang Y, Rijnbrand R, McKnight KL, Wimmer E, Paul AV, Martin A, Lemon SM (2002) Sequence requirements for viral RNA replication and VPg uridylylation directed by the internal *cis*-acting replication element (*cre*) of human rhinovirus type 14. J Virol 76:7485-7494

Zhang G, Haydon DT, Knowles NJ, McCauley JW (1999) Molecular evolution of swine vesicular disease virus. J Gen Virol 80:639-651

# Group B Coxsackievirus Virulence

S. Tracy(✉) and C. Gauntt

1 Introduction .................................................................................................. 49
2 Virulence in the CVB ................................................................................... 50
3 Virulence and Attenuating Mutations ........................................................... 52
4 How Might Virulent CVB Phenotypes Occur? ............................................ 53
    4.1 CVB and Pancreatitis ............................................................................ 55
    4.2 CVB and Neonatal Encephalomyocarditis Syndrome ........................... 56
    4.3 CVB and Type 1 (Insulin-Dependent) Diabetes ................................... 57
5 Understanding Virulence in the CVB ........................................................... 58
6 Summary ....................................................................................................... 59
References .......................................................................................................... 60

**Abstract** That which is understood of virulence phenotypes in the picornaviruses derives in large part from studies of artificially attenuating phenotypes rather than through examination of naturally occurring virus strains. The CVB replicate well in a variety of different murine and human cell cultures, making them excellent viruses with which to engage the problem of how the host environment interacts with specific viral genetics to promote varying efficiencies of viral replication. It is not known how highly virulent CVB strains may arise but evidence suggests such strains are not the norm.

## 1 Introduction

The group B coxsackieviruses (six serotypes, hereafter CVB1-6) have been recognized for more than 50 years (see review in Dalldorf 1955) as causative agents of diverse human maladies; many are considered to be clinically mild but others, such

---

S. Tracy
Department of Pathology and Microbiology, University of Nebraska,
986495 Nebraska Medical Center, Omaha, NE 68198, USA
stracy@unmc.edu

as aseptic meningitis, myocarditis, and pancreatitis, are significant issues. There is also a putative etiologic relationship between the CVB and human type 1 diabetes (T1D), which is now increasingly supported, like myocarditis and pancreatitis, by experimental evidence from relevant murine models. Along with the polioviruses (PV) and human rhinoviruses (HRV), the CVBs are among the best studied and understood of the various human enteroviruses (HEVs). Unlike the PVs and HRVs, however, CVBs readily infect numerous diverse cell cultures of human and mouse origin as well as in mice, thereby making them excellent probes for investigations into the viral genetics that impact expression of specific CVB virulence phenotypes.

## 2 Virulence in the CVB

A virulent CVB phenotype is defined here as the ability of a CVB strain to induce pathogenic changes in a normal host, that is, a disease, whereas an avirulent strain is defined as one that is unable to accomplish this in the same normal host. A body of literature discusses attenuated enterovirus strains and the mechanisms underlying the attenuation (e.g., Racaniello 2006; Racaniello 1988; Tu et al. 1995; Knowlton et al. 1996; Evans et al. 1985; Tracy et al. 2006a; Kim et al. 2006); this topic is not discussed at any length here. In the present discussion, attenuation is defined more broadly but ultimately means that the virus strain is less able or unable to induce the disease that the parental strain can cause under the same conditions.

A brief review of the literature reveals a long-standing interest in understanding factors-usually although not always, those of the host-that modulate viral disease. This is, of course, quite natural: we focus on causation of disease for the simple reason that we fear disease and wish to avoid it. And the logical sequelae of this work have been numerous studies designed to discover how to attenuate virus strains and then to understand the mechanism of attenuation in order to perfect vaccine strains. Of course, academic research on this topic has also been substantially supported by the economics of understanding and suppressing disease at the funding level from diverse commercial, private, and governmental agencies, which in large part explains why there are specific viral diseases on which much research is carried out while others remain essentially orphans. Most remarkably, there has been much less interest in understanding the basic biology responsible for what naturally makes a virus strain virulent. This is certainly true for the human picornaviruses.

The chance of any CVB infection, and thus any CVB strain, causing a serious disease is quite small. This is self-evident, or we would all be stricken with various CVB-induced disease states. This holds for other HEV as well. Poliomyelitis, caused by the PV, arguably one of the more feared and epidemic diseases of the twentieth century, occurred in fewer than 1% of the infected population (Khetsuriani et al. 2003; Minor 2003), indicating that despite its fearsome reputation, the great majority of PV infections held few long-term serious consequences. This is in part due to the stochastic nature of virus infections, as well as how the host has learned to deal with them. Before an enterovirus like CVB can replicate and become a possible cause of disease, consider just a few of the things that it must do to induce a serious disease, for example myocarditis, in a human. The virus must be shed to the environment by a previous host; this means the infection had to have been productive. It can be argued that shock-and-awe infections by a virus would be serious detriment to successful, long-term existence of the virus in a population; with few exceptions, this is indeed the case. [It is interesting to speculate how host revulsion caused by serious disease in a fellow being may affect how viruses evolve disease patterns. For example, recent work indicates that diseased lobsters are shunned by healthy lobsters (Behringer et al. 2006), suggesting that host responses to disease symptoms may be deeply and evolutionarily seated]. The majority of CVB infections, most commonly spread by a fecal-oral route of transmission, are indeed mild or unnoticed (Modlin 1990). The virus must also engage a new host. This has its own perils; due to outbreeding, host organisms are seldom identical (Wilke et al. 2006). And it can further be argued that in an increasingly hygienic human society, a virus that relies largely on a fecal-oral route of transmission such as CVB and other HEVs, will have a lower chance of engaging a new host (Viskari et al. 2000). What mothers have always said and what hospitals increasingly are more rigorously enforcing regarding washing hands regularly and properly, applies here. Of course, if the host is immune to the specific CVB serotype, infection will be limited and ablated rapidly without significant virus replication. Assuming the CVB strain is not one of the rarer, highly virulent strains (Tracy et al. 2000), the initial infecting virus dose will also be more of an issue as enterovirus replication efficiency is linked to virulence (Svitkin et al. 1985; Kanno et al. 2006). Dose of infection is a primary bottleneck, impacting the percentage of the initial viral population that may be passed to the next host (Domingo et al. 2006). It is easy to understand that the stochastic nature of viral infection tends to work largely in favor of the host, and therefore CVB disease can be considered a minority issue in the majority of infections.

That said, because a CVB infection is initiated by the virus, the viral genome is therefore a key player in the whole process: understanding this part of the equation is crucial to understanding the issue of disease causation. Successful replication of an enterovirus in the host cell demands the death of the host cell [to the best of our knowledge, although persistent infections with lowered cytopathic effects are known and may be more common than suspected (e.g., Kim et al. 2005)]. This cell death permits release of progeny virions and this requires that one must revisit the definition of virulence: by extension of the definition to the cellular host, for which death is highly probable outcome, one cannot really describe *any* CVB strain as avirulent. Extrapolated to the host in general, then, what is avirulence? In light of experimental results, avirulence must be considered a term relative to the conditions of the experimental infection. Here is an example. The CVB3 strain GA is a clinical isolate which under normal experimental conditions, replicates in mouse pancreas and heart tissues without inducing detectable pathogenic effects (Tracy et al. 2000) and as a consequence has been considered to be avirulent (Lee et al. 2005). However, CVB3/GA, like any other CVB strain of any serotype, induces lysis of HeLa cell cultures as well as many other cell cultures: there is nothing wrong with its fundamental ability to kill host cells. But only when CVB3/GA is inoculated at more than $10^6$ infectious units per mouse, does disease (pancreatitis) result (Kanno et al. 2006). Pancreatitis is caused by any CVB strain in mice, and usually quite readily at 10- to 1,000-fold fewer infectious virus particles per inoculative dose than CVB3/GA. CVB3/GA replicates more slowly and to lower titers than other virulent strains in the mouse host and in cell cultures, necessitating the injection of significantly more virus into a mouse before the virus can induce sufficient pancreatitis and/or cause diabetes before the rise in the antiviral adaptive immune response (Kanno et al. 2006). Thus, in an immunocompetent host, an infection by an HEV strain that produces a population that naturally replicates more slowly than what occurs from other strains, requires an initial infectious dose to be higher than other viruses in order to achieve a pathogenic virus load in the host as rapidly. This is, we would argue, a key issue in CVB virulence from the standpoint of the CVB strain itself.

## 3 Virulence and Attenuating Mutations

Despite the many lessons learned from experimental approaches, defining mechanisms of artificially induced attenuation has not to date defined the genetics used by nature to differentiate between CVB strains of high and low virulence phenotypes. Returning briefly to the hard life of an enterovirus in nature, it is clear that a virus strain which is attenuated for replication efficiency will not in the very great majority of cases compete successfully against strains that do. Enteroviruses, like many RNA viruses, spend little time on this issue, however, living as they do on the edge of genetic extinction; the high mutation rate of the enteroviral RNA-dependent RNA polymerase (Drake and Holland 1999) provides a potential advantage to any viral population, as does the viral swarm generated by this mechanism in which many different and potentially fit (depending upon the environmental conditions) viral genomes exist (Domingo et al. 2006). Here is an example. Following vaccination, the artificially induced U residue, the key attenuating mutation in Sabin PV3 vaccine strain at nucleotide 472 in the 5′ nontranslated region (NTR) of the genome, rapidly reverts to the wild-type C (Evans et al. 1985; Cann et al. 1984). This attenuating mutation-from the viral perspective-is undesirable and discarded rather promptly during replication in the human vaccinee, permitting the revertant virus strain's progeny competitive access once again to the real world. This rapid adaptation stems from the error-prone viral RNA polymerase; without it, enteroviruses lose fitness (Pfeiffer and Kirkegaard 2005). This is not limited to enteroviruses, of course; similar results have been observed in the compact and efficient bacteriophage Qβ genome in which the adherence to the master sequence is so tight that neutral mutations revert to wild-type mutations (Biebricher and Eigen 2006).

The CVBs have been studied in some depth in terms of attenuating mutations. It is established that nearly any change one makes to the genome results in an attenuated replication phenotype and often also engenders changes in antigenic specificity (reviewed in (Tracy et al. 2006a; Kim et al. 2006)). The CVB have been used to express foreign peptides and proteins (e.g., Chapman et al. 2000; Hofling et al. 2000) and such chimeric virus strains replicate to lower titers than the parental strains, which is effectively a disease-attenuating mechanism. However, the HEVs do not augment (that is, add information to) their genomes during replication in the host as sometimes found in other virus families, meaning such an attenuation route is naturally moot. A single amino acid change (asparagine to an aspartic acid residue) in the puff region of the VP2 capsid protein of one strain of CVB3 attenuates the myocarditic phenotype of this strain (Knowlton et al. 1996). It is interesting that the VP2 puff rims the repeating depression known as the canyon (Mucklebaur and Rossmann 1997), where mutations might deleteriously interact with receptor binding into the canyon; indeed a series of 39 mutations were made in this region of CVB3, the great majority of which were lethal (B.A. Coller, S. Tracy, unpublished data; Coller 1993). An examination of numerous CVB polyprotein sequences fails to demonstrate that naturally occurring virulence phenotypes (as measured in mice) correlate with VP2 puff mutations. A single nucleotide mutation in the 5' nontranslated region (5' NTR) of the CVB3 strain termed CVB3/0, is necessary and sufficient to attenuate a myocarditic phenotype (Tu et al. 1995). Like the PV3 attenuating mutation, this too is a transition but from U (wild-type) to C (attenuated) at nucleotide 234 of the genome. However, as 234C does not appear in any other known HEV genome (Chapman et al. 1997), the CVB3/0 strain must be considered to be a rare capture of a mutant strain. So, do attenuated strains like these, which can be created in the laboratory, ever exist in nature and if so, why are they not selected? The ever-renewing quasispecies populations in each new host means that there is an abundant supply of different mutations in these populations. Thus, diverse attenuating and lethal mutations have a high likelihood of occurring in these populations but must also be noncompetitive, real outsiders in the mutant cloud (Biebricher and Eigen 2006; Domingo et al. 1996). As RNA virus mutation rates are high and as many coding and noncoding changes decrease the resultant progeny viruses' ability to compete, only those most closely related to the master sequence (the majority population) will normally be captured by molecular cloning efforts. It is important to keep in mind, therefore, that such captures do not represent the potential diversity of these viral genomes.

## 4  How Might Virulent CVB Phenotypes Occur?

The CVB are commonly isolates every year (e.g., see Centers for Disease Control 2000, 2006) but do truly virulent CVB strains circulate commonly? While it would appear not, a systematic study has yet to be carried out. What can we infer? The majority of clinical CVB isolates in one study (Willian et al. 2000), when isolated and tested in susceptible mouse strains, did not induce myocarditis when inoculated at a commonly used dose of $10^4$-$10^5$ infectious units per mouse. We have periodically tested various CVB strains of different serotypes, isolated over several decades, in mice at this titer and seldom found myocarditis (on the order of <10%; S. Tracy, unpublished data). Induction of pancreatitis is nearly always observed in most, if not all, murine strains, although the extent of this disease varies as a function of the virus strain being examined. It has been discussed above that the majority of PV infections during the epidemic years did not result in paralytic disease and while this may well have been due to a variety of reasons, it is also clear that not all PV strains were as neurovirulent as others (Sabin 1955). Further, the majority of HEV infectious are not clinically notable. Therefore, the strong inference is that truly virulent CVB populations (those that cause notable disease, worthy of clinical attention) are indeed not the rule. This is to say, the dominant phenotype in these circulations is seldom virulent.

Why would this be so? Previously (Tracy et al. 2006a) we have suggested that cardiovirulent CVB strains may arise commonly, for example, in infected neonates, in whom the adaptive immune response is only beginning to develop. In this environment, virus strains that gain access to the heart may adapt to replicating well in the heart, causing myocarditis and often in very young humans, death. Fecal shedding could occur as the child has little or no development of adaptive immunity to check the virus, which would (in an adult) otherwise make such an adaptation in effect a dead end for the specific virus population. One might also imagine other cases in which this could occur, for example in immunosuppressed individuals. One might alternately argue that CVB strains that replicate well are simply unable to be suppressed in time before gaining access to heart tissue and inducing clinical havoc. As either approach can be modeled in mice, they should also be considered potentially valid for humans. But lacking such a special environment in which to replicate relatively unhindered by a rapidly activating adaptive immune response, the most common CVB strains (which do not induce myocarditis) appear not to have the time to adapt to the cardiac environment and cause disease in the host before being eliminated. Evolution of a more virulent strain of virus could occur in an individual following an infection, although how this is related to the viral genome has not been clearly enunciated to date. As the distance in sequence space between distinct quasispecies is significant, it is unlikely that a stable population would suddenly move from relative banality to significant virulence within the few days granted a new infection by the adaptive immune response prior to sufficient immune build-up. However, in biology, much is possible and this possibility cannot be entirely excluded (Biebricher and Eigen 2006).

The earlier discussion of how artificially attenuated HEV strains rapidly revert to wild-type virulence is also instructive. We recently observed that CVB can evolve into a dramatically different genome during replication in heart or certain cell cultures by deleting the terminal 5' nucleotides of the genome, an event that drastically impacts replication efficiency but permits long-term persistence in both mouse and human heart tissue as well as in cell culture (Kim et al. 2005; N. Chapman, K.-S. Kim, S. Tracy, unpublished data). [Interestingly, while enteroviruses-or at least the CVB-do not add to their genomes, they can naturally delete portions of the RNA; see also, e.g., McClure et al. (1980); Cole et al. (1971)].

The rate of appearance of these partially deleted genomes has not been measured, but they become readily assayable within days of inoculation of mice and completely dominant within 1-2 weeks. Thus, we may at present consider the initiating event to be occurring at some low steady state all the time or perhaps only once the virus establishes a productive replication in the heart (and perhaps elsewhere). The latter explanation is favored, for efforts to duplicate generation of such mutations in diverse cell cultures has been possible only in some, not all cultures. This strongly implicates the cellular host environment as having a significant impact upon this development. Although generation of markedly attenuated CVB genomes does not fit the definition of evolving a *virulent* strain, it is-similar to the development of the myocarditic phenotypes-certainly a movement away from the most common CVB phenotype. It is interesting to speculate that a CVB genome with a naturally occurring 5' terminal deletion such as these might connive to revisit the competitive enterovirus world through a secondary infectious and recombinational event (Oberste et al. 2004).

The stochastic nature of any CVB infection and the diverse events that impact it, are key issues in how virulent (significantly disease-causing) viral strains are selected. Although a relatively benign CVB3 strain (CVB3/GA) has been characterized (Lee et al. 2005), which is unable even to induce pancreatitis at standardly used inoculum titers in susceptible mice, such a strain is a rarity: we at least have not found another like it. At the other end of the spectrum, there are CVB3 strains of high virulence, capable of causing both widespread pancreatitis and severe (in some mice, fatal) myocarditis (e.g., Klump et al. 1990; Tracy et al. 1992; Lee et al. 1997). Although these are more common, they are significantly in the minority. Between these extremes exist the great majority of CVB3 strains (the same must be said for the other CVB serotypes as well), which can and do replicate and cause pancreatitis at varying levels depending on the strain, age, and sex of the mouse and the dose of virus inoculum. If these represent the dominant phenotype in nearly every circulating population, then we hypothesize that a move into a dominant virulent quasispecies in most individuals is likely to be a final event. Such virulent strains can be captured and cloned, but represent mere snapshots-relatively rare snapshots, actually-of what the virus can become in a specialized (and uncommon) host environment. Populations of highly virulent CVB strains are not common because they are selected only under specific conditions; barring a dramatic alteration in their environment resulting in egress to the environment, such populations are terminally evolved or for all intents and purposes, they are strains destined to hit a dead-end in the host. A somewhat similar fate is thought to occur in secondary hosts of alphavirus infections (Weaver 2006).

## 4.1 *CVB and Pancreatitis*

There is a significant body of literature that is consistent with a common etiologic role for the CVB in the induction of human pancreatitis (Ursing 1973; Ozsvar et al.

1992; Lal et al. 1988; Kennedy et al. 1986; Gooby-Toedt et al. 1996; Arnesjo et al. 1976). Pancreatitis is a common result in mice following CVB inoculation (Tracy et al. 2000; Ramsingh 1997; Huber and Ramsingh 2004). Our understanding of the relationship between the CVB and murine pancreatitis induction may, however, be skewed by results obtained using the intraperitoneal route of infection (Bopegamage et al. 2005), although even with oral dosage of mice, the pancreas is infected. As the pancreas is a common target in both species and promotes high titer CVB replication [at least in mice; Tracy et al. (2000)], pancreatitis may be the most common, CVB-induced pathologic state in humans ranging from minor to clinically serious disease states. As the pancreas is inordinately difficult to biopsy, it cannot be stated how common minor pancreatic inflammation may actually be as a consequence of CVB exposure. At present, the strong inference is that the pancreas is a primary target of CVB replication. This can at times become quite serious (e.g., (Ursing 1973; Kennedy et al. 1986)) and even normal or minor CVB-induced pancreatitis episodes can be exacerbated through, for example, alcohol abuse (Clemens and Jerrells 2004). The genetics of CVB-induced murine pancreatitis have been extensively studied by Ramsingh (see the chapter by A. Ramsingh, this volume) and reviewed elsewhere (Tracy et al. 2006a).

## 4.2 CVB and Neonatal Encephalomyocarditis Syndrome

Group B coxsackievirus disease can be devastating in newborns in the first 1-2 weeks of life (Kaplan et al. 1983; Lu et al. 2005; reviewed in Modlin 1990), with fatality rates nearing 50% and with death commonly associated with heart failure. For example, in one sample of pediatric heart tissue obtained at death, the infectious titer of the CVB2 strain was $10^8$ TCID50/g of tissue (S. Tracy, unpublished data), indicating a massive infection. Virus is widely disseminated in such cases, including the central nervous system. Without protective neutralizing antibody from mother's milk and with an adaptive immune system still in its own developmental infancy, a newborn child is at serious risk. In this environment, the infecting CVB strain can replicate in a relatively unrestrained fashion, much as in a cell culture. From the quasispecies arising in the child, a specific new master sequence could rapidly be selected, and as such could be termed a myocarditic strain.

Aseptic meningitis is also a common outcome of CVB infection in young children, although it is more commonly linked to echovirus than CVB infections. While not modeled in mice, recent work has shown that expression of a human receptor protein for echovirus type 1 (EV1) in mice, results in EV1 replication both in the brain and in the heart (Hughes et al. 2003). In this study, $10^9$ infectious EV1 particles were needed to induce disease, leading the authors to speculate that the EV1 strain employed was not virulent. Nonetheless, this model presents new possibilities to further study how highly virulent virus strains may be selectable or evolve that are directly related to this systemic, fulminant syndrome in children.

Enteroviral myocarditis is a disease of adults as well as children, but if we presume that myocarditic strains are not commonly circulating (compared to other strains that do not cause myocarditis), how do normal immunocompetent adults acquire this disease? Myocarditic CVB strains may circulate briefly, following shedding by an individual with a poor or developing immune system in which the virus population expanded (Tracy et al. 2006a; Kaplan et al. 1983) with the spread of such strains constrained by herd type-specific immunity; such strains appear and can be isolated because they escape the host's adaptive immune response into the environment where they can acquire new hosts. Myocarditic CVB strains could also arise in immunocompetent host, but in this case even with damage to the heart (thereby defining a cardiovirulent strain), the new virus population would have severely limited or no access to new hosts due to the host's own protective adaptive immune response which would prevent fecal shedding.

## 4.3 CVB and Type 1 (Insulin-Dependent) Diabetes

The cause of type 1 diabetes (T1D) remains in large part unknown. Although individual human genetics plays some role, the majority of cases are thought to have an environmental trigger that may or may not require supporting host genetics (Rotter et al. 1990; Atkinson and Eisenbarth 2001; Barnett et al. 1981; Atkinson and Maclaren 1994; Akerblom et al. 2002). A common indictment of the CVB is as an initiator of T1D, for which supporting data come from linkages between sudden onset cases and previous or concurrent viral disease, isolation of CVB from newly diagnosed individuals, and serological correlations (Smith et al. 1998; Maria et al. 2005; Hyoty et al. 1998; Helfand et al. 1995; Williams et al. 2006; Hyoty and Taylor 2002). Recent experimental studies in the nonobese diabetic (NOD) mouse have suggested that CVB causation of T1D may be linked tightly to development of an autoimmune attack on pancreatic islets of Langerhans (insulitis) (see also the chapters by K. Drescher and S. Tracy, this volume). Young NOD mice show no insulitis, although this becomes evident by 6-8 weeks of age, a development that will on its own proceed to cause T1D in most mice by 20-25 weeks of age. Inoculation of young NOD mice with CVB does not result in productive infection of pancreatic islets (Tracy et al. 2002) and most strikingly, protects mice from developing T1D as they age. Failure to replicate productively in islets is not caused by a lack of receptor (coxsackievirus-adenovirus receptor, or CAR) expression in islets. However, infection of older, prediabetic NOD mice with CVB can rapidly initiate T1D well in advance of naturally occurring T1D in mock-infected control mice of the same age (Drescher et al. 2004). In these mice, CVB protein and RNA can be detected in islets prior to T1D onset. Thus, the development of insulitis presents a new islet environment to the CVB that is not available in young mice without insulitis. Depending upon the replication efficiency of the infecting CVB strain (Kanno et al. 2006) at any given inoculum, T1D may be the outcome.

We see in the foregoing three examples of whether or not CVB can induce disease, that the initiation event is largely dependent upon issues outside the realm of virus replication. The odds are indeed stacked against CVB-induced serious disease, except when specific replication-enabling environmental conditions are present during the infection. Assuming the usual infectious dose to be small, it is the quasispecies that arises in the newly infected host that provides the genetic basis that allows the virus to take advantage of the new environment. This new virus population may assume dominance either through the presence of an already adapted virus strain in the inoculum or through rapid acquisition of the necessary genetic changes needed to replicate optimally in the new host. At a minimum and as a rule for the CVBs, this consists of replication in the gut and pancreas with few or no clinically detectable symptoms, followed by shedding to the environment, most commonly in feces. In rare cases (e.g., a newborn child, an individual with lesions in the adaptive immune system, or an individual with developing autoimmune insulitis), the host environment presents new evolutionary vistas to the infecting CVB.

## 5 Understanding Virulence in the CVB

What we know of how the CVB genetically determine virulent phenotypes has been recently reviewed (Kim et al. 2006; Tracy et al. 2006b). This is to be distinguished from what has been researched regarding artificially induced attenuating mechanisms in picornaviruses (e.g., Tu et al. 1995; Evans et al. 1985; Duke et al. 1990). The best model to date has been the work by Dunn and colleagues (2000, 2003) in which clinical CVB3 isolates, as opposed to strains selected in cell culture or mouse passage, were used in an effort to map the site(s) that determined a myocarditic phenotype. This work narrowed the field of candidates to a single short sequence in the CVB3 5' NTR defined as domain II, a stem-loop structure. Further work characterizing the poorly virulent CVB3/GA genome (Lee et al. 2005) demonstrated findings consistent with the domain II structure being involved in determination of this phenotype: computer modeling and chemical probing of the RNA (Lee et al. 2005; Bailey and Tapprich 2007) reveals this sequence to fold differently than the folds observed from myocarditic genomes. Preliminary work has shown that replacement of the myocarditic CVB3/28 (Tracy et al. 2002) domain II with that from CVB3/GA ablated the myocarditic phenotype while leaving intact the ability to readily induce pancreatitis (K. Kono, N.M. Chapman, S. Tracy, unpublished data). The complementary construct, in which the CVB3/GA domain II was removed and replaced with that from the myocarditic strain CVB3/28, did not induce the extent of myocarditis observed with the CVB3/28 strain alone (T. Kanno, S. Tracy, unpublished data). This last piece of data indicates for the first time that domain II may not in fact be the sole arbiter of a CVB3 myocarditic phenotype. The GA domain II is sufficient to ablate myocardicity in the CVB3/28-based chimeric strain, while the domain II CVB3/28 sequence in the CVB3/GA genome is insufficient to make this chimeric strain as myocarditic as the parental CVB3/28. Nonetheless, this chimeric strain does induce lesions (although many fewer than the parental myocarditic strain) in heart muscle, a phenotype not observed with CVB3/GA or the CVB3/28-SLIIGA chimeric strain, even following inoculation of $10^9$ TCID50 per mouse. While these results cumulatively indicate that domain II plays a significant role in determining a myocarditic CVB3 phenotype, it appears that other factors may complement or mitigate the role of domain II, and at present these are not known. Numerous other studies have indicated that changes in viral proteins may also play a role in determining a virulence phenotype (reviewed in Tracy et al. 2006a). However, in these cases, these inferences are drawn from studying the impact of attenuating mutations or in one case, mouse-adapted virulence. It is of course clear that specific changes in protein structure could well impact a virulence phenotype. Are faster replication rates of specific viral strains due only to changes in the 5' NTR, or could they also be due to, for example, discrete differences in the 3Dpol? There are at present too few data points to argue for any single genetic mechanism underlying the expression of a CVB virulent phenotype.

Understanding and definitively mapping the genetics underlying virulence phenotypes in the CVB genome may also be frustrated by diverse paths in sequence space that may be available to these genomes on their way to becoming virulent. Biebricher and Eigen (Biebricher and Eigen 2006) have enunciated an intriguing argument that in the vastness of sequence space, there are quite possibly numerous solutions to the same mechanistic end, citing the similarity in capsid structure and genome organization of the bacteriophage in the two groups of the Levivirus family in lieu of identity at the level of genomic RNA sequences. Thus, two discrete sequence families can give rise to highly complex protein assemblies that are distinctly similar. There are other examples, such as the closely similar capsid structures and protein sequences of the cardioviruses, encephalomyocarditis virus, and Mengo virus, which are largely derived from RNA genomes exhibiting maximum use of codon wobble. Clearly, similar viral protein structures as well as individual proteins themselves can be achieved by different coding sequences.

## 6 Summary

The concept of CVB virulence (or for that matter, any HEV) comes down to primary definition. Clearly, great variation exists in terms of replication rate between different CVB strains. This is usually most clearly observed when studying replication in primary cell cultures or in murine organs. Cell cultures such as HeLa appear to be equally obliging to nearly any CVB strain. As every CVB strain studied lyses any number of primary or established human and murine cell cultures following inoculation, single cell virulence can be considered a way of life for these viruses. Failure to infect cells rests initially upon expression of the appropriate receptor but even with receptor expression, CVB induce disease in remarkably few organs.

[An even better argument can be made for the polioviruses in this regard (Racaniello 2006; Mendelsohn et al. 1989)]. This argues strongly for control of expression of the viral phenotype by the host cell/tissue. For example, mouse pancreatic exocrine or acinar tissue is easily destroyed by CVB replication but leaves endocrine tissue (islets of Langerhans) intact (Tracy et al. 2000; Ramsingh 1997). A hindrance to the study of how specific CVB strains that differ in pathogenic outcomes in mice remains the availability of clearly defined, relevant cell cultures.

**Acknowledgements** This work was supported in part by research grants from the American Diabetes Association and the Juvenile Diabetes Research Foundation International.

# References

Akerblom HK, Vaarala O, Hyoty H, Ilonen J, Knip M (2002) Environmental factors in the etiology of type 1 diabetes. Am J Med Genet 115:18-29
Arnesjo B, Eden T, Ihse I, Nordenfelt E, Ursing B (1976) Enterovirus infections in acute pancreatitis-a possible etiological connection. Scand J Gastroenterol 11:645-649
Atkinson MA, Eisenbarth GS (2001) Type 1 diabetes: new perspectives on disease pathogenesis and treatment. Lancet 358:221-229
Atkinson MA, Maclaren N (1994) The pathogenesis of insulin-dependent diabetes mellitus. New Engl J Med 331:1428-1436
Bailey JM, Tapprich WE (2007) Structure of the 5' nontranslated region of the coxsackievirus B3 genome: chemical modification and comparative sequence analysis. J Virol 81:650-668
Barnett AH, Eff C, Leslie R, Pyke D (1981) Diabetes in identical twins. A study of 200 pairs. Diabetologia 20:87-93
Behringer DC, Butler MJ, Shields JD (2006) Ecology: avoidance of disease by social lobsters. Nature 441:421
Biebricher CK, Eigen M (2006) What is a quasispecies? Curr Top Microbiol Immunol 299:1-32
Bopegamage S, Kovacova J, Vargova A et al (2005) Coxsackie B virus infection of mice: inoculation by the oral route protects the pancreas from damage, but not from infection. J Gen Virol 86:3271-3280
Cann AJ, Stanway G, Hughes PJ et al (1984) Reversion to neurovirulence of the live-attenuated Sabin type 3 oral poliovirus vaccine. Nucleic Acids Res 12:7787-7792
Centers for Disease Control (2000) Non-polio enterovirus surveillance-United States 1993-1996. MMWR Morb Mortal Wkly Rep 46:748-750
Centers for Disease Control (2006) Enterovirus surveillance-United States, 2002-2004. MMWR Morb Mortal Wkly Rep 55:153-156
Chapman N, Romero J, Pallansch M, Tracy S (1997) Sites other than nucleotide 234 determine cardiovirulence in natural isolates of coxsackievirus B3. J Med Virol 52:258-261
Chapman N, Kim KS, Tracy S et al (2000) Coxsackievirus expression of the murine secretory protein IL-4 induces increased synthesis of IgG1 in mice. J Virol 74:7952-7962
Clemens DL, Jerrells TR (2004) Ethanol consumption potentiates viral pancreatitis and may inhibit pancreas regeneration: preliminary findings. Alcohol 33:183-189
Cole CN, Smoler D, Wimmer E, Baltimore D (1971) Defective interfering particles of poliovirus. I. Isolation and physical properties. J Virol 7:478-485
Coller BA (1993) Biologic and mutagenesis studies on non-poliovirus enteroviruses. Dissertation, University of Nebraska Medical Center
Dalldorf G (1955) The coxsackie viruses. Annu Rev Microbiol 9:277-296

Domingo E, Escarmis C, Sevilla N et al (1996) Basic concepts in RNA virus evolution. FASEB J 10:859-864

Domingo E, Martin V, Perales C, Grande-Perez A, Garcia-Arriaza J, Arias A (2006) Viruses as quasispecies: biological implications. Curr Top Microbiol Immunol 299:51-82

Drake JW, Holland JJ (1999) Mutation rates among RNA viruses. Proc Natl Acad Sci U S A 6:13910-13913

Drescher KM, Kono K, Bopegamage S, Carson SD, Tracy S (2004) Coxsackievirus B3 infection and type 1 diabetes development in NOD mice: insulitis determines susceptibility of pancreatic islets to virus infection. Virology 329:381-394

Duke G, Osorio J, Palmenberg A (1990) Attenuation of Mengo virus through genetic engineering of the 5' noncoding poly(C) tract. Nature 343:474-476

Dunn G, Bradrick S, Chapman N, Tracy S, Romero J (2003) The stem loop II within the 5' non-translated region of clinical coxsackievirus B3 genomes determines cardiovirulence phenotype in a murine model. J Infect Dis 15:1552-1561

Dunn JJ, Chapman NM, Tracy S, Romero JR (2000) Natural genetics of cardiovirulence in coxsackievirus B3 clinical isolates: Localization to the 5' non-translated region. J Virol 74:4787-4794

Evans D, Dunn G, Minor PD et al (1985) Increased neurovirulence associated with a single nucleotide change in a noncoding region of the Sabin type 3 poliovaccine genome. Nature 314:548-550

Gooby-Toedt D, Byrd J, Omori D (1996) Coxsackievirus-associated pancreatitis mimicking metastatic carcinoma. South Med J 89:441-443

Helfand R, Gary HJ, Freeman C, Anderson LJ, Pallansch M (1995) Serologic evidence of an association between enteroviruses and the onset of type 1 diabetes mellitus. Pittsburgh Diabetes Research Group. J Infect Dis 172:1206-1211

Hofling K, Tracy S, Chapman N, Leser SL (2000) Expression of the antigenic adenovirus type 2 hexon protein L1 loop region in a group B coxsackievirus. J Virol 74:4570-4578

Huber S, Ramsingh AI (2004) Coxsackievirus-induced pancreatitis. Viral Immunol 17:358-369

Hughes SA, Thaker HM, Racaniello VR (2003) Transgenic mouse model for echovirus myocarditis and paralysis. Proc Natl Acad Sci U S A 100:15906-15911

Hyoty H, Taylor KW (2002) The role of viruses in human diabetes. Diabetologia 45:1353-1361

Hyoty H, Hiltunen M, Lonnrot M (1998) Enterovirus infections and insulin dependent diabetes mellitus-evidence for causality. Clin Diagnost Virol 9:77-84

Kanno T, Kono K, Drescher KM, Chapman NM, Tracy S (2006) Group B coxsackievirus diabetogenic phenotype correlates with viral replication efficiency. J Virol 80:5637-5643

Kaplan MH, Klein SW, McPhee J, Harper RG (1983) Group B coxsackievirus infections in infants younger than three months of age: a serious childhood illness. Rev Infect Dis 5:1019-1032

Kennedy JD, Talbot IC, Tanner MS (1986) Severe pancreatitis and fatty liver progressing to cirrhosis associated with coxsackie B4 infection in a three year old with alpha-1-antitrypsin deficiency. Acta Paediatr Scand 75:336-339

Khetsuriani N, Prevots D, Quick L et al (2003) Persistence of vaccine-derived polioviruses among immunodeficient persons with vaccine-associated paralytic poliomyelitis. J Infect Dis 188:1845-1852

Kim K, Kanno T, Chapman NM, Tracy S (2006) Genetic determinants of virulence in the group B coxsackieviruses. Future Virol 1:597-604

Kim KS, Tracy S, Tapprich W et al (2005) 5'-Terminal deletions occur in coxsackievirus B3 during replication in murine hearts and cardiac myocyte cultures and correlate with encapsidation of negative-strand viral RNA. J Virol 79:7024-7041

Klump W, Bergmann I, Muller B, Ameis D, Kandolf R (1990) Complete nucleotide sequence of infectious Coxsackievirus B3 cDNA: two initial 5' uridine residues are regained during plus-strand RNA synthesis. J Virol 64:1573-1583

Knowlton K, Jeon ES, Berkley N, Wessely R, Huber S (1996) A mutation in the puff region of VP2 attenuates the myocarditic phenotype of an infectious cDNA of the Woodruff variant of coxsackievirus B3. J Virol 70:7811-7818

Lal S, Fowler D, Losasso C, Berg G (1988) Coxsackie virus-induced acute pancreatitis in a long-term dialysis patient. Am J Kidney Dis 11:434-436

Lee C, Maull E, Chapman N, Tracy S, Wood G, Gauntt C (1997) Generation of an infectious cDNA of a highly cardiovirulent coxsackievirus B3(CVB3m) and comparison to other infectious CVB3 cDNAs. Virus Res 50:225-235

Lee CK, Kono K, Haas E et al (2005) Characterization of an infectious cDNA copy of the genome of a naturally-occurring, avirulent coxsackievirus B3 clinical isolate. J Gen Virol 86:197-210

Lu JC, Koay KW, Ramers CB, Milazzo AS (2005) Neonate with coxsackie B1 infection, cardiomyopathy and arrhythmias. J Natl Med Assoc 97:1028-1030

Maria H, Elshebani A, Anders O, Torsten T, Gun F (2005) Simultaneous type 1 diabetes onset in mother and son coincident with an enteroviral infection. J Clin Virol 33:158-167

McClure M, Holland JJ, Perrault J (1980) Generation of defective interfering particles in picornaviruses. Virology 100:408-418

Mendelsohn CL, Wimmer E, Racaniello VR (1989) Cellular receptor for poliovirus: molecular cloning, nucleotide sequence, and expression of a new member of the immunoglobulin superfamily. Cell 56:855-865

Minor PD (2003) Polio vaccines and the cessation of vaccination. Exp Rev Vaccines 2:99-104

Modlin JF (1990) Coxsackieviruses, echoviruses, and newer enteroviruses. In: Principles and practice of infectious diseases. Mandell GL, Douglas RG Jr, Bennett JE III (eds) Churchill Livingstone, New York, pp 1367-1383

Mucklebaur JK, Rossmann MG (1997) The structure of coxsackievirus B3. Curr Top Microbiol Immunol 223:191-208

Oberste MS, Penaranda S, Pallansch M (2004) RNA recombination plays a major role in genomic change during circulation of coxsackie B viruses. J Virol 78:2948-2955

Ozsvar Z, Deak J, Pap A (1992) Possible role of Coxsackie B virus infection in pancreatitis. Int J Pancreatol 11:105-108

Pfeiffer JK, Kirkegaard K (2005) Increased fidelity reduces poliovirus fitness and virulence under selective pressure in mice. PLoS Pathog 1:e11

Racaniello V (1988) Poliovirus neurovirulence. Adv Virus Res 34:217-246

Racaniello VR (2006) One hundred years of poliovirus pathogenesis. Virology 344:9-16

Ramsingh AI (1997) Coxsackievirus and pancreatitis. Front Biosci 2:53-62

Rotter J, Vadheim C, Rimoin D (1990) Genetics of diabetes mellitus. In: Rifkin H, Porte D (eds)Diabetes mellitus: theory and practice, 4th edn. Elsevier, Amsterdam, pp 378-413

Sabin AB (1955) Characteristics and genetic potentialities of experimentally produced and naturally occurring variants of poliomyelitis virus. Ann N Y Acad Sci 61:924-939

Smith C, Clements G, Riding M, Collins P, Bottazo G, Taylor K (1998) Simultaneous onset of type 1 diabetes mellitus in identical infant twins with enterovirus infection. Diabet Med 15:515-517

Svitkin Y, Maslova S, Agol V (1985) The genomes of attenuated and virulent poliovirus strains differ in their in vitro translation efficiencies. Virology 147:243-252

Tracy S, Chapman N, Tu Z (1992) Coxsackievirus B3 from an infectious cDNA copy of the genome is cardiovirulent in mice. Arch Virol 122:399-409

Tracy S, Hofling K, Pirruccello S, Lane PH, Reyna SM, Gauntt C (2000) Group B coxsackievirus myocarditis and pancreatitis in mice: connection between viral virulence phenotypes. J Med Virol 62:70-81

Tracy S, Drescher KM, Chapman NM et al (2002) Toward testing the hypothesis that group B coxsackieviruses (CVB) trigger insulin-dependent diabetes: inoculating nonobese diabetic mice with CVB markedly lowers diabetes incidence. J Virol 76:12097-12111

Tracy S, Chapman NM, Drescher KM, Kono K, Tapprich W (2006a) Evolution of virulence in picornaviruses. Curr Topics Microbiol Immunol 299:193-210

Tracy S, Chapman N, Drescher KM, Kono K, Tapprich W (2006b) Evolution of virulence in picornaviruses. Curr Top Microbiol Immunol 299:193-209

Tu Z, Chapman N, Hufnagel G et al (1995) The cardiovirulent phenotype of coxsackievirus B3 is determined at a single site in the genomic 5' non-translated region. J Virol 69:4607-4618

Ursing B (1973) Acute pancreatitis in coxsackie B infection. BMJ 13:524-525

Viskari HR, Koskela P, Lonnrot M et al (2000) Can enterovirus infections explain the increasing incidence of type 1 diabetes? Diabetes Care 23:414-416

Weaver SC (2006) Evolutionary influences in arboviral disease. Curr Top Microbiol Immunol 299:285-314

Wilke CO, Forester R, Novella IS (2006) Quasispecies in time-dependent environments. Curr Top Microbiol Immunol 299:33-50

Williams CH, Oikarinen S, Tauriainen S, Salminen K, Hyoty H, Stanway G (2006) Molecular analysis of an Echovirus 3 strain isolated from an individual concurrently with appearance of islet cell and IA-2 autoantibodies. J Clin Microbiol 44:441-448

Willian S, Chapman NM, Leser JS, Romero JR, Tracy S (2000) Mutations in a conserved enteroviral RNA sequence: correlation between predicted RNA structural alteration and diminished viability. Arch Virol 145:2061-2086

# Section II
# CVB Entry and Replication

# The Coxsackievirus and Adenovirus Receptor

P. Freimuth, L. Philipson, and S. D. Carson(✉)

| | |
|---|---|
| 1 Introduction | 68 |
| 2 The CAR Gene, mRNA, and Protein | 68 |
| 3 CAR Expression in Cells and Tissues | 76 |
| 4 CAR Accessibility, CAR Interaction with CVB, and CVB Cell Penetration | 78 |
| 5 Puzzles for the Next Edition | 81 |
| 6 Postscript | 82 |
| References | 83 |

**Abstract** The coxsackievirus and adenovirus receptor (CAR) has been studied extensively since its identification and isolation in 1997. The CAR is an immunoglobulin superfamily protein with two extracellular Ig-like domains, a single membrane-spanning sequence, and a significant cytoplasmic domain. It is structurally and functionally similar to the junctional adhesion molecules. The amino terminal domain, distal from the membrane, has been structurally characterized alone, bound to the adenovirus fiber knob, and, in full-length CAR, docked in the canyon structure of the coxsackievirus capsid. Although the past decade has produced a burst of new knowledge about CAR, significant questions concerning its function in normal physiology and coxsackievirus-related pathology remain unanswered.

S. D. Carson
Department of Pathology and Microbiology, University of Nebraska Medical Center,
986495 Nebraska Medical Center, Omaha, NE 68198-6495, USA
scarson@unmc.edu

## Abbreviations

| | |
|---|---|
| Ad | Adenovirus |
| CAR | Coxsackievirus and adenovirus receptor |
| CVB | Coxsackievirus group B |
| DAF | Decay Accelerating Factor |
| Ig | Immunoglobulin |
| PVR | Poliovirus receptor |

## 1 Introduction

When reviewed in the last edition of this volume (Kuhn 1997), the search for the receptor(s) used by the B group coxsackieviruses (CVB) had uncovered several CVB-binding proteins, but the suspected primary receptor remained elusive. Remarkably, the primary receptor for CVBs was identified and reported by three independent groups shortly after the 1997 chapter was finished. The receptor, shared by CVBs and adenoviruses, has been called CAR (coxsackievirus and adenovirus receptor), and its gene is designated CXADR. (Note that a search of PubMed returns several proteins called CAR, only one of which, with splice variants, is the protein encoded by CXADR.) Kuhn's thorough review is highly recommended reading for those interested in the historical details of the search for the CVB receptor.

Since 1997, the CAR structure has been partially determined, and molecular details of its interaction with the canyon structure in the CVB capsid, as well as the adenovirus fiber knob, have been described. The protein has been assigned a function in cell-cell adhesion, and is generally localized to intercellular junctions, primarily those of epithelial cells. Messenger RNA splice variants have been identified, and expression of two carboxyl-terminal variants of the CAR protein has been studied in cells and tissues. Studies of CAR developmental and cellular biology have been reported by multiple laboratories, and CAR-deficient mice have been engineered. The relationship between a CVB-binding protein, decay accelerating factor, and CAR-dependent infection has been more clearly defined. Due to interest in utilizing adenoviruses as vehicles for gene therapy, there is considerable literature dealing with CAR and its role as receptor for adenoviruses, from which one can infer relevance to infection by CVB. While all of this work is inherently interesting and important for understanding CAR biology, this chapter will focus primarily on those aspects of CAR most relevant to its pathological function as a receptor for CVB.

## 2 The CAR Gene, mRNA, and Protein

The CAR cDNA was cloned independently by two groups using different techniques (Bergelson et al. 1997a; Tomko et al. 1997). Concurrently, the CAR protein was isolated and partially sequenced by a third group (Carson et al. 1997), and nucleotide sequences corresponding to the amino acid sequence were identified in the GenBank EST database. All three groups had identified the same protein, which bound CVB, was recognized by an antibody previously shown to protect cells from CVB infection, and conferred permissivity to CVBs when expressed in cells normally resistant to infection. Experiments confirmed that the CAR protein was also used as a receptor by Ad, binding via the fiber knob proteins.

The human CAR gene was localized to chromosome *21q11.2* (Bowles et al. 1999). The NCBI database (www.ncbi.nlm.nih.gov) places it at *21q11.1* where Bowles et al. localized an apparent pseudogene. In the mouse, mCAR is on chromosome 16 (Chen et al. 2003). On Northern blots, the principal mRNAs correspond to 6- and 2.4-kb species in humans and 6- and 1.4-kb species in mice (Tomko et al. 1997). The longer form is predominant and probably represents a partially spliced product. Other mRNAs corresponding to alternative-splice variants have also been characterized (Bergelson et al. 1998; Thoelen et al. 2001; Dorner et al. 2004). From the sequence (GenBank acc.no. AF 200465 and 2422862-65; Andersson et al. 2000) the transcription unit is around 57 kb (Hattori et al. 2000). Five regions close to the promoter are similar in the human and mouse genomes (GenBank Acc.no. 242861), which may have implications for transcriptional regulation. Seven exons were reported for the human gene (Bowles et al. 1999), but at least eight exons have been identified in the mouse gene (Chen et al. 2003). The mouse transcript contains an additional splice site 27 kb downstream of exon 7; the human transcript shares exon sequences at this position, but lacks the acceptor splice site (Andersson et al. 2000). Several CAR pseudogenes have also been described (Bowles et al. 1999).

The CAR mRNA full-length open reading frame encodes 365 amino acids with a short leader sequence (19 residues), an ectodomain of about 216 residues, a transmembrane domain of about 23 amino acids, and dependent on alternative splicing, a cytoplasmic tail of either 107 or 94 residues. Thus, at least two CAR proteins differing in the carboxyl-terminus have been identified and are referred to as mCAR1 and mCAR2 for the mouse and hCAR1 and hCAR2 for the human. Figure 1 shows a schematic representation of the human CAR and its two principal isoforms. The predicted mature protein has a molecular mass of about 38 kDa (hCAR1), but due to two N-linked glycans (at Asn106 and Asn201), the Mr on SDS-gels is about 46 k. Two membrane-proximal cytoplasmic cysteines are subject to fatty acid acylation (van't Hof and Crystal 2002). The carboxyl-terminal amino acid sequence of the cytoplasmic tail, its length, and the number of tyrosines that could be involved in signal transduction depend on alternative splicing (Andersson et al. 2000). PDZ-binding motifs (-TTV-COOH, or -SIV-COOH) that can interact with intracellular PDZ-proteins are present at the extreme C-terminal end of both mCAR and hCAR isoforms (Fig. 1).

Sequence analysis predicted that the CAR protein is a member of the immunoglobulin (Ig) superfamily, with two extracellular Ig-type domains located at the N-terminal half of the protein, a single membrane-spanning helix, and a C-terminal cytoplasmic tail (Bergelson et al. 1997a; Tomko et al. 1997). This structural model was supported by and confirmed in subsequent biochemical and structural studies. The membrane-distal domain (referred to as D1, IG1, or V-like) at the N-terminus of the mature (processed) CAR protein is classified as an Ig variable-type domain,

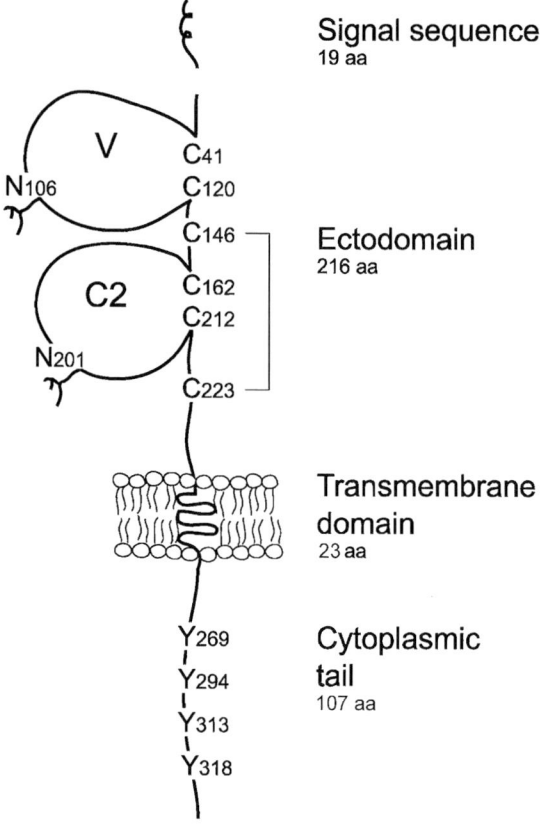

**Fig. 1** Schematic structure of the hCAR1 protein. The ectodomain consists of two Ig-loops (D1 and D2) of the variable type (*V*, formed between Cys41 andCys120) and constant type (*C2*, encompassing Cys 162-Cys212). The D2 domain contains an extra disulfide bridge formed between Cys146 and Cys223 not typical of a C2-like Ig domain. Numbering refers to the whole open reading frame. Alternative splicing results in different C-terminal sequences for hCAR1 and hCAR2. (Modified from Philipson and Pettersson 2004, with permission)

containing a single intradomain disulfide bond and a single site for N-linked carbohydrate modification. CAR D1 was expressed independently in *Escherichia coli*, and its structure solved to near atomic resolution by X-ray crystallography (van Raaij et al. 2000) and NMR spectroscopy (Fig. 2) (Jiang et al. 2004). The β-sandwich fold characteristic of Ig variable-type domains is clearly evident in these structures, with one anti-parallel β sheet composed of strands A, B, E, and D packing against the other, larger antiparallel β sheet composed of strands

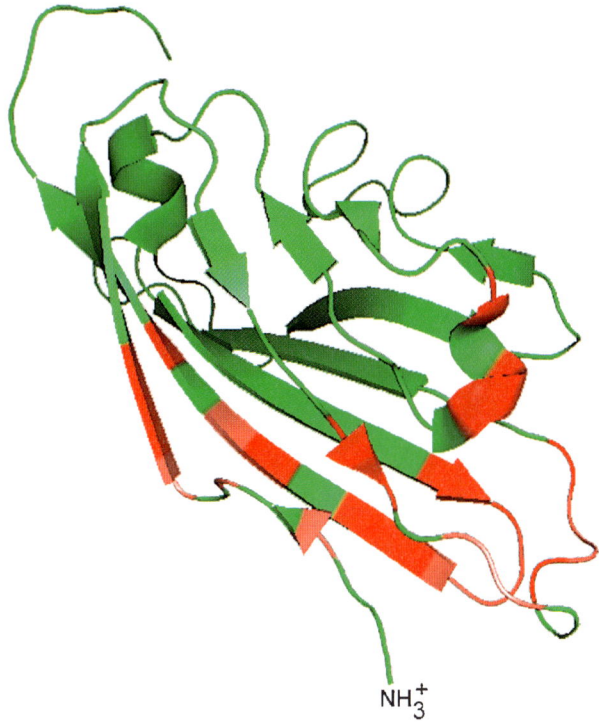

**Fig. 2** PDB-based structure of CAR D1 as determined by NMR (1RSF.pdb) (Jiang et al. 2004). Residues interacting with CVB (Fig. 3) are shown in *red* (large footprint) or *salmon* (smaller footprint) based on interactive footprint area in (He et al 2001). Other structural (PDB) files for online viewing of CAR-adenovirus fiber knob interactions (1KAC.pdb) and the CAR-CVB alpha-carbon traces (1JEW.pdb) can be accessed online at (www.expasy.org). The image was rendered with PyMol (www.pymol.org) and completed with Adobe PhotoShop

C, C′, C″, F and G. To date there has been no high-resolution structural analysis of the other CAR extracellular domain (referred to as D2, IG2, or C2-like), although a NMR structure appears imminent (Jiang and Caffrey 2005). Based on sequence analysis, CAR D2 is classified as an Ig constant (C2) type domain and was reasonably modeled as such in the cryoEM reconstruction of full-length CAR bound to CVB3 (He et al. 2001). CAR D2 has a single site for modification with N-linked carbohydrate and two intradomain disulfide bonds. Density corresponding to the membrane-spanning helix and cytoplasmic tail also was observed in the cryoEM reconstruction (Fig. 3) (He et al. 2001), but structural features of these regions were not well resolved and no data on the structure of these regions at higher resolution has been reported.

Structural analysis of CAR D1 supports the proposed function of CAR as a mediator of cell-adhesion (Honda et al. 2000) in the junctional complexes of epithelial cells in many tissues (Cohen et al. 2001b; Walters et al. 2002). Aggregation

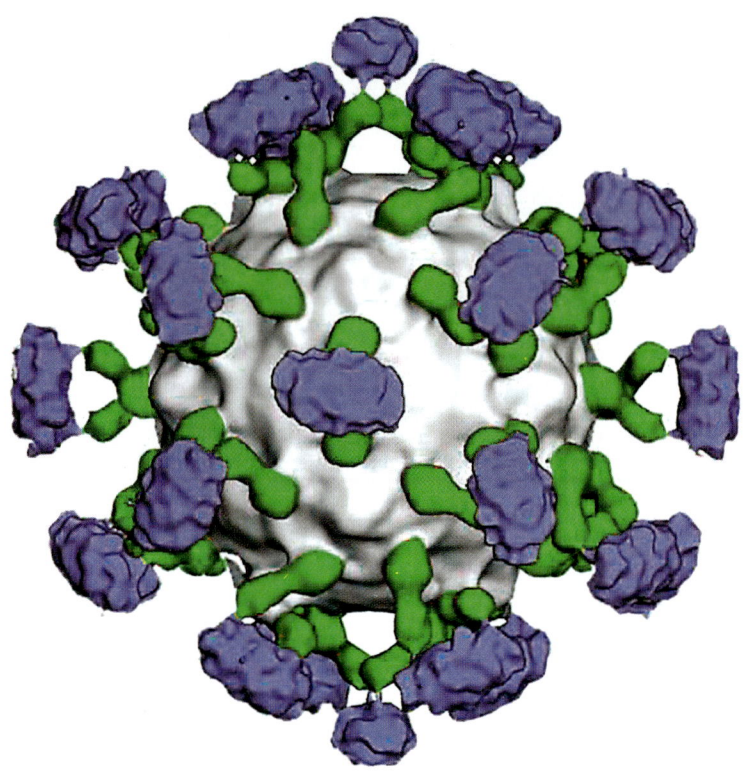

**Fig. 3** CryoEM structure of coxsackievirus (*gray*) decorated with CAR D1+D2 (*green*) dimerized via the transmembrane and cytoplasmic domains (*blue*). The CAR D1 (Ig V-like domain) is docked in the CVB canyon. (Modified from He et al. 2001, with permission)

of CAR-expressing cells can be inhibited by specific antibodies to CAR D1 or by isolated adenovirus fiber knob (Honda et al. 2000). At physiological pH, CAR D1 self-associates to form dimers with a dissociation constant of 16 micromolar, and the crystal structure of CAR D1 determined at pH 5.2 shows a dimeric form of CAR D1 in the asymmetric unit with the two CAR D1 molecules in lateral contact and in head-to-tail orientation (van Raaij et al. 2000). In contrast, NMR spectroscopy indicated that CAR D1 is monomeric at pH 3 (Jiang et al. 2004), supporting the conclusion that ion pairing plays an important role in stabilization of the CAR D1 homodimer interface. The negatively charged side chains of interfacial residues D54 and E56 would become protonated at pH 3, disrupting the ionic interaction with the positively charged side chain of interfacial residue K123 (Jiang et al. 2004). D54, E56, and K123 are located within a cluster of hydrophobic residues that becomes buried in the homodimer interface of CAR D1. This self-interactive surface is located on the face of CAR D1 defined by β strands G, F, C, C′, and C″, and is opposite from the face of CAR D1 that is modified by carbohydrate linked to N106 on β strand E.

The crystal structure of CAR D1 bound to the fiber knob (head) domain of adenovirus-12 indicated that the GFCC'C" face of CAR D1, which forms the homodimer interface, is also buried in the interface with knob (Bewley et al. 1999). In contrast, cryoEM reconstructions of CVB3 in complex with full-length CAR protein (Fig. 3) and with the CAR D1+D2 fragment indicated that the CAR D1 distal BC and FG loops and the lateral surface defined by β strands A and G interact with CVB3 (Figs. 2 and 3) (He et al. 2001). This contact region is consistent with the canyon model of picornavirus interaction with Ig-type receptors proposed earlier by Rossmann and colleagues (Rossmann et al. 1985), which postulates that canyons, depressions in the capsid surface surrounding the five-fold axes, accommodate the distal tips of the receptors. Canyons are a conserved structural feature of picornaviruses and have been shown to correspond to the sites where poliovirus receptor (PVR) binds to poliovirus and ICAM-1 binds to rhinovirus (reviewed in Rossmann et al. 2002). Virus structural proteins VP1, VP2, and VP3 form the CVB3 canyon walls, as in other studied picornaviruses. Contact residues are contributed by all three subunits, although VP1 dominates the interaction with CAR D1 (He et al. 2001). The surfaces of CAR D1 that bind to CVB3, fiber knob, and CAR D1 itself partially overlap in a region that includes the charged residues D54, E56, and K123 mentioned above, possibly accounting for the earlier observation that adenoviruses and coxsackieviruses compete for the identical binding sites on the cell plasma membrane (Lonberg-Holm et al. 1976). Direct involvement of these three residues in the presumably essential functions of CAR in epithelial cell adhesion and in neuronal organization during development (Honda et al. 2000) likely accounts for their strict conservation (in bovine, mouse, rat, and zebrafish CAR D1) and for the ability of CAR from rodents and other animals to bind human adenoviruses and coxsackieviruses (Tomko et al. 1997; Bergelson et al. 1998). Use of a highly conserved molecule such as CAR as a cellular receptor may have provided an important selective advantage during the evolution of these viruses.

In cryoEM analysis of CVB3-CAR complexes, it was noted that the CAR D1 and CAR D1+D2 extracellular fragments bound to CVB3 with reduced affinity or stability relative to the binding of full-length CAR protein (He et al. 2001). This was observed both in the EM, where only subsaturating amounts of the CAR D1 and D1+D2 fragment were detected bound to the CVB3 capsid, and in plaque-reduction assays where full-length CAR was more effective than either of the extracellular fragments in blocking infection of tissue culture cells. Absence of carbohydrate modification is unlikely to account for the reduced affinity of CAR D1 (produced in *E. coli*) for CVB3, since the glycosylated extracellular fragment (CAR D1+D2) also exhibited reduced affinity for CVB3 compared to the full-length CAR molecule. In the cryoEM reconstruction of full-length CAR-CVB3 complexes, density corresponding to the membrane-spanning and cytoplasmic tail regions of CAR was detected and appeared to be shared by adjacent CAR molecules related by icosahedral two-fold axes (Fig. 3) (He et al. 2001). This shared density suggests a model where the membrane-spanning helices and/or the cytoplasmic tails of adjacent CAR molecules physically associate to form CAR dimers in the cell plasma membrane. The higher avidity resulting from bivalent interaction of CAR dimers with two binding sites on the CVB3 capsid could account for the ability of the full-length protein (extracted from the plasma membrane with non-ionic detergents) to saturate binding sites on CVB3 and to more effectively inhibit CVB3 infectivity. If the shared density indeed reflects association of the membrane-spanning or cytoplasmic tails of full-length CAR, this association did not alter the orientation of the extracellular domain relative to the CVB3 capsid, since the extracellular region of full-length CAR was essentially superimposable with the CAR D1+D2 fragment bound to the CVB3 capsid (He et al. 2001). It is important to note that the proposed CAR dimers resulting from association of the membrane-spanning helices or cytoplasmic tails are distinct from the CAR dimers described earlier that result from association of the D1 domains (van Raaij et al. 2000). In the latter case, the D1 domains from CAR molecules on adjacent cells interact laterally in head-to-tail orientation to mediate cell adhesion, while in the former case, the dimers form within a single membrane system and are independent of the D1-D1 interaction.

The extracellular domain of CAR thus participates in three well-characterized intermolecular associations (CVB, Ad, and D1-D1). Additional interactions with immunoglobulins (of unknown significance; Carson and Chapman 2001), JAML (reported to participate in neutrophil migration; Zen et al. 2005), and JAM-C (observed at cell-cell junctions in testis; Mirza et al. 2006) have been reported, but neither the molecular details of these interactions nor their significance for CVB infection have yet been reported.

In addition to CAR1 and CAR2 transcripts, three minor forms of alternatively spliced CAR mRNA have also been identified (Thoelen et al. 2001) and shown to produce secreted soluble proteins when expressed in HeLa cells from plasmids (Dorner et al. 2004). All of the predicted proteins lack the membrane-spanning domain. The two longer forms (called CAR4/7 and CAR 3/7) retain an intact D1 domain and are able to inhibit infection by CVB3 (Dorner et al. 2004). CAR antigens with apparent Mr near 40 k, 37.5 k, and 31 k have been reported in CVB-infected HeLa cells, vesicles shed from cultured cells, and in malignant pleural effusions, respectively (Carson 2000; Bernal et al. 2002; Carson 2004). Based on the reported Mr ranges, the Mr 40-k and 37.5-k antigens may not be different. The Mr 31-k CAR was suggested to result from secretion (i.e., an alternative splice product) or shedding (i.e., proteolytic release of the extracellular domains D1+D2), while the 37.5-k form was shown to be associated with membrane vesicles and the likely result of proteolytic cleavage within the cytoplasmic domain. As with the soluble CAR with intact D1 expressed in HeLa, these soluble forms of CAR may interfere with CVB (or Ad) infection.

The cytoplasmic tail of CAR seems to be unnecessary for either CVB or adenovirus infection (Leon et al. 1998; Wang and Bergelson 1999), but it may be important for function(s) in cell signaling and growth. The CAR cytoplasmic domain apparently has a role in tumor cell growth, but the nature of the effect varies among reports using different cell types (Okegawa et al. 2001; Bruning et al. 2005). Binding of Ad fiber knob to CAR on respiratory cells stimulates production of inflammatory products (Tamanini et al. 2006). Since inflammation is a key component of CVB-associated pathology, CAR-mediated signaling may prove to be very important in CVB-associated diseases. In addition to signaling, the cytoplasmic domain also influences CAR distribution within the cell (Cohen et al. 2001a; van't Hof and Crystal 2002), and the carboxyl-terminal PDZ-binding motif(s) association with PDZ-domain proteins has been established.

CAR has been shown to colocalize with tight junction proteins (e.g., occludin; Fig. 4; Raschperger et al. 2006), and the tight junction protein ZO-1, a PDZ-domain protein, can be immunoprecipitated together with CAR (Cohen et al. 2001b). The direct complex formation has not been established perhaps due to weak interaction between the two molecules. A yeast-two-hybrid system revealed interaction with the LNX proteins and the interacting regions were identified (Sollerbrant et al. 2003; Mirza et al. 2005). The second PDZ domain in the LNX proteins interacted with the C-terminal in both splice forms of CAR, but also with a shared internal sequence in the cytoplasmic tail. The LNX proteins have not yet been identified as tight junction proteins but may be involved in proteolysis in the Notch pathway. More recently, a direct interaction between CAR and the tight junction MUPP-1 protein was reported and the interacting PDZ region identified (Coyne et al. 2004). CAR is also able to interact with several other proteins containing PDZ domains (e.g., MAGI-1b, PICK 1 and PSD 95; Excoffon et al. 2004).

The overall homology between human, mouse, rat, dog, and pig CAR is about 90% (Tomko et al. 1997; Bergelson et al. 1998; Fechner et al. 1999). The D1 domain is more conserved (91%-94%) than the D2 domain (83%-89%), whereas the cytoplasmic tail is about 95% identical among these species (Fechner et al. 1999). The transmembrane domain is less conserved, being 77% identical between human and mouse (Tomko et al. 1997). Even the zebrafish CAR has similar domains with overall 52% amino acid identity to hCAR, and it can function as a receptor for CVB and adenoviruses (Petrella et al. 2002).

With the discovery of CTX (a *Xenopus laevis* cortical thymocyte protein; Chretien et al. 1998) and A33 (found on intestinal epithelial cells; Heath et al. 1997), proteins similar to CAR, a new subfamily of Ig-like molecules was identified

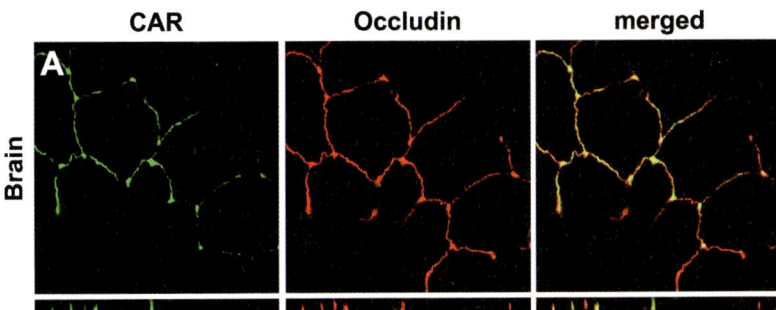

**Fig. 4** CAR and occludin colocalize in tight junctions of choroid plexus epithelium. (From Raschperger et al. 2006, with permission)

and named the CTX-subfamily of the immunoglobulin superfamily of proteins (Chretien et al. 1998; Du Pasquier et al. 1999). The proteins of the CTX subfamily consist of two Ig-loops of the V and C2 type, with an extra disulfide link in the C2-type domain, a transmembrane region, and a cytoplasmic tail of variable length. While the list of CTX subfamily members continues to grow, notable members include CLMP (Raschperger et al. 2004), ESAM (Hirata et al. 2001), and at least four junctional adhesion molecules (JAMs) (Johnstone et al. 2000; Mandell and Parkos 2005). Of these proteins, only CAR and JAM-A, a receptor for reovirus (Barton et al. 2001), have so far been identified as virus receptors.

## 3 CAR Expression in Cells and Tissues

The CVB susceptibility of cells in vitro is clearly related to the measurable presence of CAR (e.g., Shafren et al. 1997), so it has been reasonably presumed that the tissue tropism of CVB in vivo is related to differential expression of the receptor among cell types. The exceptions, however, suggest that the situation is more complicated in vivo. Some tissues with readily measurable CAR are not associated with significant CVB pathology (e.g., liver; Wessely et al. 2001), and cytoplasmic host proteins may inhibit the ability of CVB to replicate in some cells (Cheung et al. 2005). In contrast, CVBs have been documented in cells of some organs that have not been reported to express CAR at readily detectable levels (e.g., Anderson et al. 1996; Mena et al. 1999). CAR is a regulated protein that is expressed during fetal development (particularly in brain and muscle; Nalbantoglu et al. 1999; Honda et al. 2000; Ito et al. 2000), with expression in some tissues persisting until after birth and diminishing during the neonatal period. Persistent expression of the CVB receptor may contribute to the increased susceptibility of neonates to CVB infection (Dalldorf and Sickles 1948). Homozygous CAR-deficient mice die at embryonic day 12-14 without forming a functional heart muscle (Asher et al. 2005b; Dorner et al. 2005). Mice with cardiomyocyte-specific CAR deletion near embryonic day 9.5 died with defective hearts, whereas mice with CAR deleted from cardiomyocytes near embryonic day 11.5 survived to adulthood (Chen et al. 2006).

A few studies of CAR expression and cellular distribution have used in situ hybridization, but most have used antibodies against CAR for immunohistochemistry. There are potential issues with both methods. Since the major mRNA detected by Northern blot is much larger than the translated form, in situ RNA analysis may not necessarily detect the translated mRNA. The monoclonal antibody Rmcb (Hsu et al. 1988) binds the native first Ig motif (D1) of human CAR and inhibits CVB infection, but detects CAR reproducibly on blots only from nondenaturing gels. It appears to work well for immunohistochemistry using frozen tissues or lightly fixed cultured cells. Other monoclonal antibodies (e.g., MoAb E1.2D3; Carson et al. 1999) appear to recognize CAR on Western blots, but poorly, if at all, in immunohistochemical applications. Several polyclonal antibodies against the extracellular domains or various regions in the cytoplasmic tail have been used to study CAR (e.g., Honda et al. 2000; Cohen et al. 2001b; Rauen et al. 2002; Shaw et al. 2004; Mirza et al. 2005), as have commercially available antibodies (e.g., Drescher et al. 2004). Considering the conserved nature of the extensive immunoglobulin protein superfamily, and the CTX subfamily, proof of antibody specificity must be rigorously established. For example, the 21 carboxyl-terminal amino acids of hCAR1 are 76% identical with the carboxyl-terminal sequence of ESAM, and one of our monoclonal antibodies against the extracellular domain of CAR binds multiple other proteins on Western blots of HeLa cell lysates (S.D. Carson, unpublished data). Nevertheless, these tools have been successfully used to study CAR expression in cells and tissues.

Though present in low amounts in the adult heart, CAR is increased in hearts during inflammation or healing, and in hearts with dilated cardiomyopathy (Ito et al. 2000; Noutsias et al. 2001; Fechner et al. 2003). CAR has been immunochemically visualized in the intercalated discs of cardiomyocytes (Shaw et al. 2004). These results are consistent with the association between CVB infection and viral cardiomyopathy (Baboonian et al. 1997). The developing brain expresses increased CAR compared to the adult brain of mice (Honda et al. 2000), and CAR is abundantly expressed in the mouse embryo with a preference for the nervous system, including all neuroepithelial cells, in ganglions, and peripheral nerves (Tomko et al. 2000). CAR mRNA was expressed in mouse brain, with the highest levels near birth and diminishing thereafter, though the timing of peak expression varied between regions of the brain (Honda et al. 2000). CAR can also be detected in epithelial cells of embryonic liver, lung, heart, eye, digestive system, pancreas, kidney, and the submandibular glands (Raschperger et al. 2006). A rapid downregulation of CAR occurs at birth, after which CAR is only sparsely detected in most of these tissues. The expression in epithelial cells in kidney, intestine, and liver was not downregulated after birth, but the staining in pancreas and the submandibular glands had partially disappeared (Raschperger et al. 2006). In the nervous system of the adult mouse, expression appears restricted to cells close to the ependymal region lining the ventricular system, which is consistent with adenovirus infection from the lateral ventricles (Tomko et al. 2000). CAR was also detected in trachea and bronchi but absent in alveoli in the adult animals.

Although CAR was not detected in endothelial or mesenchymal cells in the adult mouse (Raschperger et al. 2006), CAR has been detected in cultured human umbilical vein endothelial cells, where expression was related to cell density and could be downregulated by treatment with cytokines (Carson et al. 1999; Vincent et al. 2004). CAR was detected in CD31+ cells in damaged areas of the heart, but not in the endothelium of vessels in undamaged tissue (Fechner et al. 2003), suggesting that endothelial CAR expression in vivo may be restricted to regions of vessel growth or tissue repair. Apparent conflicts among studies of CAR expression in pancreas also remain to be resolved. CAR has been reported to be present in only the ductal epithelium (Raschperger et al. 2006), in pancreatic acinar tissue (Mena et al. 2000), which is susceptible to CVB infection, in islets (Meyers et al. 2004), and in both acinar cells and islets (Drescher et al. 2004; Kanno et al. 2006), both of which were also shown to be infected by

CVB. Such strikingly different results from good laboratories indicate that much remains to be learned.

In cultures of polarized epithelial cells, CAR is preferentially expressed at the basolateral side of the tight junctions, where it may be sequestered and inaccessible to virus, especially from the apical side (Walters et al. 1999; Cohen et al. 2001b). Although CAR may function as a homophilic cell adhesion molecule, it should be noted that CAR also has been detected on membrane regions other than tight junctions, for example on the luminal (apical) surface of the prostate epithelium (Rauen et al. 2002; Bao et al. 2005). CAR also may be expressed at high concentrations on the basal surface of hepatocytes, since adenoviral knob domain injected into the mouse bloodstream is rapidly taken up by the liver in a CAR-dependent manner (Zinn et al. 1998; Awasthi et al. 2004). These alternate cellular locations for CAR, its apparent capacity to form heterophilic associations (e.g., with JAML), and its potential role in cell signaling indicate that CAR can serve functions in addition to homophilic cell adhesion.

## 4 CAR Accessibility, CAR Interaction with CVB, and CVB Cell Penetration

Exploitation of CAR as a receptor by adenoviruses and coxsackie B viruses is difficult to reconcile with a strict localization of CAR in epithelial tight junctions, since virus particles would probably have limited access to this membrane subdomain (Pickles et al. 1998; Walters et al. 1999). An analysis of the early steps in adenovirus infection of polarized epithelial cells revealed, however, that adenovirus released from the basal-lateral aspects of infected cells can access the CAR and disrupt tight junctions (Walters et al. 2002). Astonishingly, even the isolated fiber can open the junctions and access the receptor (Rentsendorj et al. 2005). Clearly the CAR in junctions can be rendered accessible to Ad by Ad, but similar dynamics with CVB have not been reported. Such a finding would be surprising since the fiber knob CAR-binding sites on Ad are presented away from the capsid (perhaps capable of probing clefts between cells), while the CAR binding site on CVB lies in the canyon on the capsid surface. The fact that CAR can be expressed on the apical surface of epithelial cells in at least some tissues (e.g., prostate; Rauen et al. 2002; Bao et al. 2005) suggests the possibility that cells with this phenotype also may exist in other tissues and that such cells might be relatively more susceptible to infection by virus particles invading from the apical surface. Moreover, some tight junction proteins must be released from intercellular contacts by stimuli that increase cell layer permeability; e.g., JAM-A has been shown to move out of junctions onto the apical cell surface in response to inflammatory cytokines (Ozaki H 1999; Martinez-Estrada et al. 2005). So, whether due to variability of cellular expression and apical distribution or to induced redistribution away from cell junctions, it is probable that CAR can be available outside of cell-cell contacts and accessible to CVB. Other mechanisms for (some) CVB to gain access to CAR have been described and are discussed below.

The interaction of CVB with CAR leads to the formation of viral A-particles, an event common to all enteroviruses and characterized by a partial exposure of the RNA and release of the internal VP4 from the virus (Crowell and Philipson 1971; Huang et al. 2000). The A-particle, or eclipsed virus, has a lower buoyant density, can be readily eluted from cells (if internalization is blocked), and has lost its capacity to bind to and infect cells. This initial step in viral penetration of the cell was originally identified on CVB eluted shortly after interaction with cells in suspension cultures (Crowell et al. 1971). In fact, CVB exposed to the soluble extracellular portion of CAR in solution can form viral A-particles, with higher efficiency for CAR dimers than for CAR monomers (Goodfellow et al. 2005).

It is interesting to consider the success of cryoEM imaging of CAR-dimers bound to CVB (He et al. 2001) in light of the expected CAR-induced triggering of conversion to viral A-particles. Since the cryoEM data are convincing, and the receptor-induced viral eclipse is well established, there are apparently a few details of the CAR-CVB interaction yet to be elucidated. The study of CAR splice variants (Dorner et al. 2004) led to the conclusion that CAR D1 (CAR 3/7) is sufficient for binding CVB and inhibition of infection, but unfortunately did not test the formation of A-particles.

The observation that soluble CAR can function as a virus trap leading to inactive A-particles has been suggested as an approach for CVB therapy in which the extracellular domain of CAR might be used to eliminate virus at early times of infection. The elegant approach with selective transgenic expression of CAR on erythrocytes may also help to attenuate the infection (Asher et al. 2005a). However, the effects of high concentrations of circulating CAR on tight junction integrity, erythrocyte clumping or adhesion, neutrophil function, or physiology in general remain to be addressed.

The processes of CVB infection subsequent to CAR binding and A-particle formation are subjects of recent studies and are already controversial. Chung et al. (2005) examined the infection of HeLa cells by a single CVB3 strain. They found CVB3 particles colocalized with clathrin (i.e., with coated vesicles) and showed that internalization was dependent on dynamin and Hsc70. CAR was lost from the cell surface and not recycled, and CVB3 particles eventually colocalized with early endosome markers. In contrast, CVB4 internalization in cultured pancreas cells was found to be dependent on lipid rafts, independent of clathrin-coated pits, and targeted to the Golgi (Triantafilou and Triantafilou 2004). Coyne and Bergelson (2006) used a CVB variant (CVB3-RD) that binds DAF (decay accelerating factor, CD55) as well as CAR to study infection of polarized epithelial cells. They found that CVB internalization was dependent on caveolin-1, and independent of the clathrin pathway of endocytosis. CAR was not internalized and remained at cell-cell junctions. CVB conversion to A particles occurred at the tight junction (with CAR), and further CVB uncoating occurred after internalization. Ultimately, the CVB colocalized with endoplasmic reticulum markers. While it is easy to rationalize that these results differ because the studies used different cell lines and different

CVBs, it may be more productive to consider these different tissues, different virus strains, and different pathways to infection and the potential implications regarding known but unexplained issues of viral tropism and pathogenicity.

DAF is one of several macromolecules, including nucleolin and heparan sulfate proteoglycans (de Verdugo et al. 1995; Zautner et al. 2003), that have been shown to bind CVB and suggested to serve as potential receptors alternative to CAR. DAF has been most thoroughly studied, and lessons learned with DAF may well apply to other cell surface molecules that can bind CVB.

Surveying the literature suggests that CVB (at least CVB1, 3 and 5; Shafren et al. 1995) might be subdivided into three generalized DAF-binding phenotypes: CVB that do not bind DAF (which apparently includes most or all CVB2, 4, and 6, as well as strains of CVB1, 3, and 5), CVB that bind DAF but require CAR for infection, and CVB that bind DAF and apparently infect cells in the absence of CAR. The first and second phenotypes have been observed in clinical isolates of CVBs (e.g., Bergelson et al. 1997b) and can be influenced by the host cell type used during propagation in vitro (Reagan et al. 1984). The third phenotype has been derived from CVB serially passaged through CAR-deficient cells, and may, or may not, still utilize CAR as a receptor when it is available (Spiller et al. 2002; Goodfellow et al. 2005). Precisely how DAF serves as a receptor capable of supporting infection in the absence of CAR remains to be clarified.

The second phenotype, CVB that bind DAF but require CAR for infection, has become much more interesting as the relationship between binding to DAF and binding to CAR has been revealed. In vitro binding assays, using isolated components, showed that a CVB3 strain that survived covalent attachment to the binding surface bound CAR with about 1,000 times greater affinity than it bound DAF (Goodfellow et al. 2005). Binding avidity increased with the valency (dimers) of either. When polarized epithelial cells that express CAR in cell-cell junctions were found to be somewhat resistant to infection, but less resistant to CVB that bind DAF as well as CAR, the relevance of DAF-binding CVB strains was established (Cohen et al. 2001b; Shieh and Bergelson 2002). These results suggested that the CAR in cell junctions was sequestered and inaccessible to virus, but that cell surface DAF provided an initial attachment site that improved CVB access to the CAR. As reported by Coyne and Bergelson (2006), CVB3-RD binding to DAF initiates a series of signals and responses within Caco-2 cells. Their experiments revealed that CVB-induced DAF clustering resulted in activation of both Abl and Fyn. The Abl pathway was associated with activation of Rac (as well as Rho and Cdc42), actin restructuring, and translocation of the bound CVB to cell-cell junctions where it associated with CAR. Within the same timeframe, and preceding virus arrival, tight junction disruption was documented by measurements of monolayer permeability. Fyn activation by DAF clustering contributed to phosphorylation of caveolin-1, which was important for virus internalization. Events following CVB delivery to CAR were discussed above in this section.

The cumulative knowledge from work on the DAF-binding, CAR-dependent CVB strains provides good rationale for the existence of these strains and exposes a new level of complexity in the cell biology of the host-pathogen interaction.

It will be interesting to learn whether other CVB-binding cell surface molecules initiate similar responses. These findings also begin to expose unexplained aspects of infection by CVB strains that do not bind DAF and those that do not bind CAR.

## 5 Puzzles for the Next Edition

The advances made since the last edition are reflected in the bibliography. Most of the work discussed here was published after Kuhn's chapter appeared in 1997 (Kuhn 1997). The progress has been remarkable, and while new events in the process have been discovered, current knowledge does not satisfactorily explain the end-to-end mechanisms of CVB interactions with cells and how they result in infection.

From studies of CAR binding to CVB, the apparent dissociation constants have been measured (Goodfellow et al. 2005) and amino acid residues mediating the association have been revealed (He et al. 2001), yet these very reports illustrate how little we know about receptor-mediated mechanisms that result in CVB uncoating, VP4 insertion into the cell membrane, and release of the genome. He et al. decorated CVB with CAR dimers, but the interaction appears not to have triggered virus eclipse. In contrast, two of the strains used by Goodfellow et al. were apparently destabilized by attachment to a planar substrate. These observations might suggest that the CVB capsid is stable to symmetrical saturation by CAR, but unstable to asymmetrical forces applied by partial occupancy with CAR (metaphorically, a molecular bathysphere). What is the optimal number of CAR per virion to trigger eclipse?

Though some cellular sequelae of CVB binding to DAF have been elucidated (Coyne and Bergelson 2006), how DAF binding to CVB affects capsid stability or structure remains an open question. Goodfellow et al. (2005) showed that DAF binding to CVB-RD can result in a particle with buoyant density normally observed for A-particles, but in which VP4 is retained. Moreover, elution of A-particles of DAF-dependent CVB from CAR-negative cells, but only partial eclipse after DAF dimers were bound to virus (Goodfellow et al. 2005), indicates that the change in CVB buoyant density and loss of VP4 can be separable steps and implicates a second active moiety on the cells that is missing from the DAF-dimer preparation. Existing data remain inadequate to explain even these initial events in the CVB-host interaction.

Since empty capsids can be isolated after virus internalization, it is reasonable to conclude that the viral genome is released without complete capsid disassembly. Just as steps in the transition from CVB to A-particle are missing from our knowledge, we remain largely ignorant concerning the when, where, and how of genome release from the internalized A-particle. These steps must logically be associated with the trafficking of internalized particles, for which data have recently been reported. Unfortunately, as discussed above, the data appear to be conflicting as to whether internalization involves clathrin coated pits (Chung et al. 2005), lipid rafts, consistent with caveolae (Triantafilou and Triantafilou 2004), or caveolae (Coyne and Bergelson 2006). Perhaps all three are correct, in which case each combination of CVB strain, cell type, and tissue may represent a unique system that requires independent characterization. This becomes an interesting consideration in view of efforts to understand the variations in CVB tropism and pathogenicity. For internalization, whether the virus uses a vesicle normally targeted to a favorable replication site or redirects a vesicle to that site is not known (data thus far indicate that a vesicular transport compartment is a reasonable argument). The mechanism by which the genome escapes the vesicle to access the replication machinery and the fate of the capsid once free of the genome have yet to be described.

Leaving CVB infection of CAR-deficient tissues (e.g., via DAF) for much later consideration (i.e., not in this edition), the final puzzle for this chapter is how CVB gains access to a receptor (CAR) that is apparently sequestered in intercellular junctions. In particular, CVB has to breach the intestinal epithelium before it can cause pancreatitis, myocarditis, encephalitis, or pathology in any other tissue. Disruption of tight junctions by DAF-binding strains of CVB (Coyne and Bergelson 2006) provides one potential mechanism for virus to gain access to CAR, but can only be applied to DAF-binding strains. These data, however, also present a puzzle: antibodies to DAF resulted in Abl activation, similar to CVB binding, and presumably tight junction disruption, similar to CVB binding. If antibodies to DAF activate this DAF-dependent pathway, why do the antibodies to DAF not increase infection by CVB that bind only CAR?

Perhaps CVB that do not bind DAF gain access to CAR in areas of local physical damage, where CAR could be exposed, an event easily envisioned in the intestine (Kesisoglou et al. 2006). Or perhaps low levels of CAR are available on partially differentiated epithelial cells that must be present during epithelial renewal. Disruption of tight junctions by inflammatory cytokines, with dispersion of junctional adhesion molecules onto the apical surface (Ozaki H 1999), provides another potential mechanism for exposure of CAR to CVB. Existing data require that CVB, other than strains selected for growth on CAR-deficient cells, must bind CAR to infect cells and the host. That humans and other animals are routinely infected by CVBs is empirical evidence that CAR, whether in cell junctions or presented on the cell surface, must be available to the virus.

## 6 Postscript

Just as with Kuhn's 1997 effort, significant results have appeared between submission of this chapter and its publication. As anticipated in The CAR Gene, mRNA, and Protein, the structure of CAR D2 has been solved by NMR and is available as 2NPL.pdb at www.ncbi.nlm.nih.gov. Unexpectedly, Rossmann's group has published the cryoEM structure of CVB3 complexed with DAF (JVI Accepts, published online ahead of print on 5 September 2007; J Virol.DOI:10.1128/JVI.00931-07).

# References

Anderson DR, Wilson JE, Carthy CM, Yang D, Kandolf R, McManus BM (1996) Direct interactions of coxsackievirus B3 with immune cells in the splenic compartment of mice susceptible or resistant to myocarditis. J Virol 70:4632-4645

Andersson B, Tomko RP, Edwards K et al (2000) Putative regulatory domains in the human and mouse CVADR genes. Gene Func Dis 2:11-15

Asher DR, Cerny AM, Finberg RW (2005a) The erythrocyte viral trap: transgenic expression of viral receptor on erythrocytes attenuates coxsackievirus B infection. Proc Natl Acad Sci U S A 102:12897-12902

Asher DR, Cerny AM, Weiler SR et al (2005b) Coxsackievirus and adenovirus receptor is essential for cardiomyocyte development. Genesis 42:77-85

Awasthi V, Meinken G, Springer K, Srivastava SC, Freimuth P (2004) Biodistribution of radioiodinated adenovirus fiber protein knob domain after intravenous injection in mice. J Virol 78:6431-6438

Baboonian C, Davies MJ, Booth JC, McKenna WJ (1997) Coxsackie B viruses and human heart disease. In: Tracy S, Chapman NM, Mahy BWJ (eds) The Coxsackie B Viruses. Springer, Berlin New York Heidelberg, pp 31-52

Bao Y, Peng W, Verbitsky A et al (2005) Human coxsackie adenovirus receptor (CAR) expression in transgenic mouse prostate tumors enhances adenoviral delivery of genes. Prostate 64:401-407

Barton ES, Forrest JC, Connolly JL et al (2001) Junction adhesion molecule is a receptor for reovirus. Cell 104:441-451

Bergelson JM, Cunningham JA, Droguett G et al (1997a) Isolation of a common receptor for coxsackie B viruses and adenoviruses 2 and 5. Science 275:1320-1323

Bergelson JM, Krithivas A, Celi L et al (1998) The murine CAR homolog is a receptor for coxsackie B viruses and adenoviruses. J Virol 72:415-419

Bergelson JM, Modlin JF, Wieland-Alter W, Cunningham JA, Crowell RL, Finberg RW (1997b) Clinical coxsackievirus B isolates differ from laboratory strains in their interaction with two cell surface receptors. J Infect Dis 175:697-700

Bernal RM, Sharma S, Gardner BK et al (2002) Soluble coxsackievirus adenovirus receptor is a putative inhibitor of adenoviral gene transfer in the tumor milieu. Clin Cancer Res 8:1915-1923

Bewley MC, Springer K, Zhang Y-B, Freimuth P, Flanagan JM (1999) Structural analysis of the mechanism of adenovirus binding to its human cellular receptor, CAR. Science 286:1579-1583

Bowles KR, Gibson J, Wu J, Shaffer LG, Towbin JA, Bowles NE (1999) Genomic organization and chromosomal localization of the human coxsackievirus B adenovirus receptor gene. Hum Genet 105:354-359

Bruning A, Stickeler E, Diederich D et al (2005) Coxsackie and adenovirus receptor promotes adenocarcinoma cell survival and is expressionally activated after transition from preneoplastic precursor lesions to invasive adenocarcinomas. Clin Cancer Res 11:4316-4320

Carson SD (2000) Limited proteolysis of the coxsackievirus and adenovirus receptor (CAR) on HeLa cells exposed to trypsin. FEBS Lett 484:149-152

Carson SD (2004) Coxsackievirus and adenovirus receptor (CAR) is modified and shed in membrane vesicles. Biochemistry 43:8136-8142

Carson SD, Chapman NM (2001) Coxsackievirus and adenovirus receptor (CAR) binds immunoglobulins. Biochemistry 40:14324-14329

Carson SD, Chapman NM, Tracy SM (1997) Purification of the putative coxsackievirus B receptor from HeLa cells. Biochem Biophys Res Commun 233:325-328

Carson SD, Hobbs JT, Tracy SM, Chapman NM (1999) Expression of the coxsackievirus and adenovirus receptor in cultured human umbilical vein endothelial cells: regulation in response to cell density. J Virol 73:7077-7079

Chen JW, Ghosh R, Finberg RW, Bergelson JM (2003) Structure and chromosomal localization of the murine coxsackievirus and adenovirus receptor gene. DNA Cell Biol 22:253-259

Chen JW, Zhou B, Yu Q-C et al (2006) Cardiomyocyte-specific deletion of the coxsackievirus and adenovirus receptor results in hyperplasia of the embryonic left ventricle and abnormalities of sinuatrial valves. Circ Res 98:923-930

Cheung PK, Yuan J, Zhang HM et al (2005) Specific interactions of mouse organ proteins with the 5′ untranslated region of coxsackievirus B3: potential determinants of viral tissue tropism. J Med Virol 77:414-424

Chretien I, Marcuz A, Courtet M et al (1998) CTX, a Xenopus thymocyte receptor, defines a molecular family conserved throughout vertebrates. Eur J Immunol 28:4094-4104

Chung SK, Kim JY, Kim IB, Park SI, Paek KH, Nam JH (2005) Internalization and trafficking mechanisms of coxsackievirus B3 in HeLa cells. Virology 333:31-40

Cohen CJ, Gaetz J, Ohman T, Bergelson JM (2001a) Multiple regions within the coxsackievirus and adenovirus receptor cytoplasmic domain are required for basolateral sorting. J Biol Chem 276:25392-25398.

Cohen CJ, Shieh JT, Pickles RJ, Okegawa T, Hsieh J-T, Bergelson JM (2001b) The coxsackievirus and adenovirus receptor is a transmembrane component of the tight junction. Proc Natl Acad Sci U S A 98:15191-15196

Coyne CB, Bergelson JM (2006) Virus-induced Abl and Fyn kinase signals permit coxsackievirus entry through epithelial tight junctions. Cell 124:119-131

Coyne CB, Voelker T, Pichla SL, Bergelson JM (2004) The coxsackievirus and adenovirus receptor interacts with the multi-PDZ domain protein-1 (MUPP-1) within the tight junction. J Biol Chem 279:48079-48084

Crowell RL, Landau BJ, Philipson L (1971) The early interaction of coxsackievirus B3 with HeLa cells. Proc Soc Exp Biol Med 137:1082-1088

Crowell RL, Philipson L (1971) Specific alterations of coxsackievirus B3 eluted from HeLa cells. J Virol 8:509-515

Dalldorf G, Sickles G (1948) An unidentified, filterable agent isolated from the feces of children with paralysis. Science 108:61-62

De Verdugo UR, Selinka H-C, Huber M et al (1995) Characterization of a 100 kilodalton binding protein for the six serotypes of coxsackie B viruses. J Virol 69:6751-6757

Dorner A, Xiong D, Couch K, Yajima T, Knowlton KU (2004) Alternatively spliced soluble coxsackie-adenovirus receptors inhibit coxsackievirus infection. J Biol Chem 279:18497-18503

Dorner AA, Wegmann F, Butz S et al (2005) Coxsackievirus-adenovirus receptor (CAR) is essential for early embryonic cardiac development. J Cell Sci 118:3509-3521

Drescher KM, Kono K, Bopegamage S, Carson SD, Tracy S (2004) Coxsackievirus B3 infection and type 1 diabetes development in NOD mice: insulitis determines susceptibility of pancreatic islets to virus infection. Virology 329:381-394

Du Pasquier L, Courtet M, Chretien I (1999) Duplication and MHC linkage of the CTX family of genes in Xenopus and in mammals. Eur J Immunol 29:1729-1739

Excoffon KJDA, Hruska-Hageman A, Klotz M, Traver GL, Zabner J (2004) A role for the PDZ-binding domain of the coxsackie B virus and adenovirus receptor (CAR) in cell adhesion and growth. J Cell Sci 117:4401-4409

Fechner H, Haack A, Wang H et al (1999) Expression of coxsackie adenovirus receptor and alphav-integrin does not correlate with adenovector targeting in vivo indicating anatomical vector barriers. Gene Ther 6:1520-1535

Fechner H, Noutsias M, Tschoepe C et al (2003) Induction of coxsackievirus-adenovirus-receptor expression during myocardial tissue formation and remodeling: identification of a cell-to-cell contact-dependent regulatory mechanism. Circulation 107:876-882

Goodfellow IG, Evans DJ, Blom AM et al (2005) Inhibition of coxsackie B virus infection by soluble forms of its receptors: binding affinities, altered particle formation, and competition with cellular receptors. J Virol 79:12016-12024

Hattori M, Fujiyama A, Taylor TD et al (2000) The DNA sequence of human chromosome 21. Nature 405:311-319

He Y, Chipman PR, Howitt J et al (2001) Interaction of coxsackievirus B3 with the full length coxsackievirus-adenovirus receptor. Nat Struct Biol 8:874-878

Heath JK, White SJ, Johnstone CN et al (1997) The human A33 antigen is a transmembrane glycoprotein and a novel member of the immunoglobulin superfamily. Proc Natl Acad Sci U S A 94:469-474

Hirata K, Ishida T, Penta K et al (2001) Cloning of an immunoglobulin family adhesion molecule selectively expressed by endothelial cells. J Biol Chem 276:16223-16231

Honda T, Saitoh H, Masuko M et al (2000) The coxsackievirus-adenovirus receptor protein as a cell adhesion molecule in developing mouse brain. Mol Brain Res 77:19-28

Hsu KH, Lonberg-Holm K, Alstein B, Crowell RL (1988) A monoclonal antibody specific for the cellular receptor for the group B coxsackieviruses. J Virol 62:1647-1652

Huang Y, Hogle JM, Chow M (2000) Is the 135S poliovirus particle an intermediate during cell entry? J Virol 74:8757-8761

Ito M, Kodama M, Masuko M et al (2000) Expression of coxsackievirus and adenovirus receptor in hearts of rats with experimental autoimmune myocarditis. Circ Res 86:275-280

Jiang S, Caffrey M (2005) NMR assignment and secondary structure of the coxsackievirus and adenovirus receptor domain 2. Protein Pept Lett 12:537-539

Jiang S, Jacobs A, Laue TM, Caffrey M (2004) Solution structure of the coxsackievirus and adenovirus receptor domain 1. Biochemistry 43:1847-1853

Johnstone CN, Tebbutt NC, Abud HE et al (2000) Characterization of mouse A33 antigen, a definitive marker for basolateral surfaces of intestinal epithelial cells. Am J Physiol Gastrointest Liver Physiol 279:G500-G510

Kanno T, Kim K, Kono K, Drescher KM, Chapman NM, Tracy S (2006) Group B coxsackievirus diabetogenic phenotype correlates with replication efficiency. J Virol 80:5637-5643

Kesisoglou F, Schmiedlin-Ren P, Fleisher D, Roessler B, Zimmermann EM (2006) Restituting intestinal epithelial cells exhibit increased transducibility by adenoviral vectors. J Gene Med 8:1379-1392

Kuhn RJ (1997) Identification and biology of cellular receptors for the coxsackie B viruses group. In: Tracy S, Chapman NM, Mahy BWJ (eds) The Coxsackie B Viruses. Springer, Berlin New York Heidelberg, pp 209-226

Leon RP, Hedflund T, Meech SJ et al (1998) Adenoviral-mediated gene transfer in lymphocytes. Proc Natl Acad Sci U S A 95:13159-13164

Lonberg-Holm K, Crowell RL, Philipson L (1976) Unrelated animal viruses share receptors. Nature 259:679-681

Mandell KJ, Parkos CA (2005) The JAM family of proteins. Adv Drug Deliv Rev 57:857-867

Martinez-Estrada OM, Manzi L, Tonetti P, Dejana E, Bazzoni G (2005) Opposite effects of tumor necrosis factor and soluble fibronectin on junctional adhesion molecule-A in endothelial cells. Am J Physiol Lung Cell Mol Physiol 288:L1081-L1088

Mena I, Fischer C, Gebhard JR, Perry CM, Harkins S, Whitton JL (2000) Coxsackievirus infection of the pancreas: evaluation of receptor expression, pathogenesis, and immunopathology. Virology 271:276-288

Mena I, Perry CM, Harkins S, Rodriguez F, Gebhard J, Whitton JL (1999) The role of B lymphocytes in coxsackievirus B3 infection. Am J Pathol 155:1205-1215

Meyers SE, Brewer L, Shaw DP et al (2004) Prevalent human coxsackie B-5 virus infects porcine islet cells primarily using the coxsackie-adenovirus receptor. Xenotransplantation 11:536-546

Mirza M, Hreinsson J, Strand M-L et al (2006) Coxsackievirus and adenovirus receptor (CAR) is expressed in male germ cells and forms a complex with the differentiation factor JAM-C in mouse testis. Exp Cell Res 312:817-830

Mirza M, Raschperger E, Philipson L, Pettersson RF, Sollerbrant K (2005) The cell surface protein coxsackie- and adenovirus receptor (CAR) directly associates with the ligand-of-numb protein X2 (LNX2). Exp Cell Res 309:110-120

Nalbantoglu J, Pari G, Karpati G, Holland PC (1999) Expression of the primary coxsackie and adenovirus receptor is downregulated during skeletal muscle maturation and limits the efficacy of adenovirus-mediated gene delivery to muscle cells. Hum Gene Ther 10:1009-1019

Noutsias M, Fechner H, de Jonge H et al (2001) Human coxsackie-adenovirus receptor is colocalized with integrins alpha(v)beta(3) and alpha(v)beta(5) on the cardiomyocyte sarcolemma and upregulated in dilated cardiomyopathy: implications for cardiotropic viral infections. Circulation 104:275-280

Okegawa T, Pong R-C, Li Y, Bergelson JM, Sagalowsky AI, Hsieh J-T (2001) The mechanism of the growth-inhibitory effect of coxsackie and adenovirus receptor (CAR) in human bladder cancer: a functional analysis of CAR protein structure. Cancer Res 61:6592-6600

Ozaki HIK, Horiuchi H, Arai H, Kawamoto T, Okawa K, Iwamatsu A, Kita T (1999) Cutting edge: combined treatment of TNF-alpha and IFN-gamma causes redistribution of junctional adhesion molecule in human endothelial cells. J Immunol 163:553-557

Petrella J, Cohen CJ, Gaetz J, Bergelson JM (2002) A zebrafish coxsackievirus and adenovirus receptor homologue interacts with coxsackie B virus and adenovirus. J Virol 76:10503-10506

Philipson L, Pettersson RF (2004) The coxsackie-adenovirus receptor - a new receptor in the immunoglobulin family involved in cell adhesion. Curr Top Microbiol Immunol 273:87-111

Pickles RJ, McCarty D, Matsui H, Hart PJ, Randell SH, Boucher RC (1998) Limited entry of adenovirus vectors into well-differentiated airway epithelium is responsible for inefficient gene transfer. J Virol 72:6014-6023

Raschperger E, Engstrom U, Pettersson RF, Fuxe J (2004) CLMP, a novel member of the CTX family and a new component of epithelial tight junctions. J Biol Chem. 279:796-804

Raschperger E, Thyberg J, Pettersson S, Philipson L, Fuxe J, Pettersson RF (2006) The coxsackie and adenovirus receptor (CAR) is an in vivo marker for epithelial tight junctions, with a potential role in regulating permeability and tissue homeostasis. Exp Cell Res 312:1566-1580

Rauen KA, Sudilovsky D, Le JL et al (2002) Expression of the coxsackie adenovirus receptor in normal prostate and in primary and metastatic prostate carcinoma: potential relevance to gene therapy. Cancer Res 62:3812-3818

Reagan KJ, Goldberg B, Crowell RL (1984) Altered receptor specificity of coxsackievirus B3 after growth in rhabdomyosarcoma cells. J Virol 49:635-640

Rentsendorj A, Agadjanian H, Chen X et al (2005) The Ad5 fiber mediates nonviral gene transfer in the absence of the whole virus, utilizing a novel cell entry pathway. Gene Ther 12:225-237

Rossmann MG, Arnold E, Erickson JW et al (1985) Structure of a human common cold virus and functional relationship to other picornaviruses. Nature 317:145-153

Rossmann MG, He Y, Kuhn RJ (2002) Picornavirus-receptor interactions. Trends Microbiol 10:324-331

Shafren DR, Bates RC, Agrez MV, Herd RL, Burns GF, Barry RD (1995) Coxsackieviruses B1, B3, and B5 use decay accelerating factor as a receptor for cell attachment. J Virol 69:3873-3877

Shafren DR, Williams DT, Barry RD (1997) A decay-accelerating factor-binding strain of coxsackievirus B3 requires the coxsackievirus-adenovirus receptor protein to mediate lytic infection of rhabdomyosarcoma cells. J Virol 71:9844-9848

Shaw CA, Holland PC, Sinnreich M et al (2004) Isoform-specific expression of the coxsackie and adenovirus receptor (CAR) in neuromuscular junction and cardiac intercalated discs. BMC Cell Biol 5:42

Shieh JTC, Bergelson JM (2002) Interaction with decay-accelerating factor facilitates coxsackievirus B infection of polarized epithelial cells. J Virol 76:9474-9480

Sollerbrant K, Raschperger E, Mirza M et al (2003) The coxsackievirus and adenovirus receptor (CAR) forms a complex with the PDZ domain-containing protein ligand-of-numb protein-X (LNX). J Biol Chem 278:7439-7444

Spiller OB, Goodfellow IG, Evans DJ, Hinchcliffe SJ, Morgan BP (2002) Coxsackie B viruses that use human DAF as a receptor infect pig cells via pig CAR and do not use pig DAF. J Gen Virol 83:45-52

Tamanini A, Nicolis E, Bonizzato A et al (2006) Interaction of adenovirus type 5 fiber with the coxsackievirus and adenovirus receptor activates inflammatory response in human respiratory cells. J Virol 80:11241-11254

Thoelen I, Magnusson C, Tagerud S, Polacek C, Lindberg M, Ranst MV (2001) Identification of alternative products encoded by the human coxsackie-adenovirus receptor gene. Biochem Biophys Res Commun 287:216-222

Tomko RP, Johansson CB, Totrov M, Abagyan R, Frisen J, Philipson L (2000) Expression of the adenovirus receptor and its interaction with the fiber knob. Exp Cell Res 255:47-55

Tomko RP, Xu R, Philipson L (1997) HCAR and MCAR: the human and mouse cellular receptors for subgroup C adenoviruses and group B coxsackieviruses. Proc Natl Acad Sci U S A 94:3352-3356

Triantafilou K, Triantafilou M (2004) Lipid-raft-dependent coxsackievirus B4 internalization and rapid targeting to the Golgi. Virology 326:6-19

Van't Hof W, Crystal RG (2002) Fatty acid modification of the coxsackievirus and adenovirus receptor. J Virol 76:6382-6386

Van Raaij MJ, Chouin E, van der Zandt H, Bergelson JM, Cusack S (2000) Dimeric structure of the coxsackievirus and adenovirus receptor D1 domain at 1.7 A resolution. Struct Fold Des 8:1147-1155

Vincent T, Pettersson RF, Crystal RG, Leopold RL (2004) Cytokine-mediated down-regulation of coxsackievirus-adenovirus receptor in endothelial cells. J Virol 78:8047-8058

Walters RW, Freimuth P, Moninger TO, Ganske I, Zabner J, Welsh MJ (2002) Adenovirus fiber disrupts CAR-mediated intercellular adhesion allowing virus escape. Cell 110:789-799

Walters RW, Grunst T, Bergelson JM, Finberg RW, Welsh MJ, Zabner J (1999) Basolateral localization of fiber receptors limits adenovirus infection from the apical surface of airway epithelia. J Biol Chem 274:10219-10226

Wang X, Bergelson J (1999) Coxsackievirus and adenovirus receptor cytoplasmic and transmembrane domains are not essential for coxsackievirus and adenovirus infection. J Virol 73:2559-2562

Wessely R, Klingel K, Knowlton KU, Kandolf R (2001) Cardioselective infection with coxsackievirus B3 requires intact type I interferon signaling: implications for mortality and early viral replication. Circulation 103:756-761

Zautner AE, Korner U, Henke A, Badorff C, Schmidtke M (2003) Heparan sulfates and coxsackievirus-adenovirus receptor: each one mediates coxsackievirus B3 PD infection. J Virol 77:10071-10077

Zen K, Liu Y, McCall IC et al (2005) Neutrophil migration across tight junctions is mediated by adhesive interactions between coxsackie and adenovirus receptor and a junctional adhesion molecule-like protein on neutrophils. Mol Biol Cell 16:2694-2703

Zinn KR, Douglas JT, Smyth CA et al (1998) Imaging and tissue biodistribution of 99mTc-labeled adenovirus knob (serotype 5). Gene Ther 5:798-808

# Coxsackievirus B RNA Replication: Lessons from Poliovirus

P. Sean and B. L. Semler(✉)

| | | |
|---|---|---|
| 1 | Introduction | 90 |
| 2 | Background | 90 |
| | 2.1 Cap-Independent Translation | 91 |
| 3 | Viral Proteins Involved in RNA Replication | 91 |
| | 3.1 Proteins of P2:2A, 2BC, 2B, and 2C | 92 |
| | 3.2 Proteins of P3:3AB, 3A, 3B, 3CD, 3C, and 3D$^{pol}$ | 94 |
| 4 | RNA Secondary Structures Involved in Enterovirus RNA Replication | 97 |
| | 4.1 Stem-Loop I and Negative-Strand 3′ Stem-Loop I | 98 |
| | 4.2 *cre* and VPg Uridylylation | 99 |
| | 4.3 3′ NCR and Poly(A) Tract | 101 |
| 5 | Cellular Proteins Involved in RNA Replication | 103 |
| | 5.1 PCBP | 103 |
| | 5.2 PABP | 104 |
| | 5.3 hnRNP C | 104 |
| 6 | Membranous Vesicles | 105 |
| 7 | Transition from Translation to RNA Replication | 106 |
| 8 | Enterovirus RNA Replication | 108 |
| | 8.1 Negative-Strand RNA Synthesis | 110 |
| | 8.2 Positive-Strand RNA Synthesis | 111 |
| 9 | Summary | 113 |
| | References | 113 |

**Abstract** The replication of coxsackievirus RNA occurs with rapid onset, starting approximately 2.5 h after infection. The mechanisms entailing the RNA replication of enteroviruses, like coxsackievirus and poliovirus, are highly conserved. These processes require two steps of RNA amplification: (i) complete synthesis of the negative-strand RNA using input RNA as the template and (ii) synthesis of the positive-strand RNA using the intermediate negative-strand RNA as the template. Successful enterovirus RNA replication requires all of the viral nonstructural proteins in their mature and precursor forms, as well as RNA secondary structures in the template. The encoded nonstructural proteins are responsible for RNA replication through

B. L. Semler
School of Medicine, Department of Microbiology and Molecular Genetics,
University of California, Irvine, CA 92697, USA
blsemler@uci.edu

multiple protein-protein interactions between viral and/or host proteins to mediate RNA synthesis, induce membranous vesicles, and deliver the replication complex to the template. The RNA secondary structures at the 5' and 3' termini of the template position the RNA replication complex at the initiation site(s) for both negative- and positive-strand RNA synthesis, thus providing binding sites for viral and host proteins that may functionally circularize the genome during RNA synthesis. Although considerable knowledge has been gained regarding the mechanism of enterovirus RNA synthesis, the complete steps in RNA replication have not been fully determined. The aim of this review is to summarize the current state of our knowledge and to present a model that encompasses the identified steps of enterovirus RNA replication.

# 1 Introduction

As one of the pathogenic agents of viral cardiomyopathies, coxsackievirus B3 (CVB3) is an essential virus to study (for review, see Kim et al. 2001). Much of what we know of coxsackievirus RNA replication is based on the vast information already known from a closely related virus, poliovirus, considered to be the prototypic picornavirus. Coxsackievirus and poliovirus make up the *Enterovirus* genus within the family Picornaviridae. Genomic RNAs from both viruses are approximately 7,500 nucleotides in length, and their genomic organizations are identical. The encoded nonstructural proteins of poliovirus and coxsackievirus carry out essentially the same functions in viral replication. A convincing example of the similarity of poliovirus and coxsackievirus gene expression was reported 20 years ago. In these studies, segments of the 5' noncoding region (5' NCR) of poliovirus were replaced with those of coxsackievirus, resulting in chimeric viruses displaying near wild-type poliovirus growth phenotypes in tissue culture (Semler et al. 1986; Johnson and Semler 1988). Although these studies were carried out before the discovery of important secondary structural elements required for translation and RNA replication in the 5' NCR, they demonstrated the ability of poliovirus proteins to utilize the RNA secondary structure of coxsackievirus to mediate viral functions.

The similar mechanisms of translation and RNA replication shared by poliovirus and coxsackievirus, as illustrated by the chimeric studies, indicate that poliovirus is an ideal model system to study coxsackievirus replication. Published biochemical and genetic studies have shown that the proteins of poliovirus and coxsackievirus participate in the same steps of RNA replication and induce the same alterations of cellular components. The current literature reveals a general consensus that the mechanism of RNA replication elucidated for poliovirus is applicable to coxsackievirus.

# 2 Background

Picornaviruses have a positive-sense RNA genome, which can be immediately translated upon infection. The genome of picornaviruses is small, averaging about 7,500 nucleotides in length. Within the RNA genome, about 900 nucleotides are noncoding and flank the single open reading frame. With such a limited coding capacity, picornaviruses utilize virus encoded proteins and host proteins within the infected cell to carry out important viral functions. Additionally, the RNA secondary structures that form in the genome act as contact points for viral and host proteins to mediate translation and RNA replication. To maximize the functions of the encoded proteins, most if not all of the precursor and mature proteins of enteroviruses are multifunctional. Successful picornavirus infection requires interaction of RNA genomic elements, virus encoded proteins, and cellular proteins to: (1) mediate cap-independent translation, (2) induce membranous vesicle formation, and (3) prime RNA replication.

## 2.1 Cap-Independent Translation

RNA replication of the picornavirus genome is intrinsically linked to translation since the majority of proteins/enzymes involved in RNA replication are virus-encoded. After uncoating of the capsid, the viral RNA is translated in the cytoplasm via a cap-independent mechanism (Pelletier and Sonenberg 1988; Jang et al. 1988). Canonical cap-dependent translation of cellular mRNAs involves recognition of the 5' cap by the eIF4F cap-binding complex and recruitment of ribosomes to the RNA (Merrick 1990); however, picornaviruses lack a 7-methyl G cap at the 5' end. Instead, there is a viral protein termed VPg that is covalently linked to the 5' terminus of the picornavirus genome (Lee et al. 1977; Flanegan et al. 1977). Additional evidence demonstrating a cap-independent mechanism for enterovirus RNA translation is that during an infection, the virus encoded proteinase 2A cleaves eIF4G, the scaffolding protein in the cap-binding complex (Etchison et al. 1983; Krausslich et al. 1987). Thus, the cleavage of eIF4G acts to inhibit eIF4F-mediated ribosome scanning in cap-dependent translation. Finally, the 5' NCRs of picornaviruses are composed of long sequences that contain extensive secondary structures and multiple start codons. All of these features are inhibitory for the canonical cap-binding and ribosome scanning mechanism that has been determined for cellular mRNAs (Hellen and Sarnow 2001). All of the combined evidence points toward an alternative mechanism for translation of the viral genome, and it has been determined that the secondary structures of the RNA in the 5' NCR act as an internal ribosome entry site (IRES) for cap-independent translation (Pelletier and Sonenberg 1988; Jang et al. 1988).

## 3 Viral Proteins Involved in RNA Replication

The enterovirus genome contains a single open reading frame that is translated via IRES-mediated initiation to generate a polyprotein of approximately 250 kDa. The polyprotein then undergoes proteolytic processing by its own genetically encoded proteinases, 2A and 3C in *cis* and in *trans* to generate precursors and mature proteins (for review, see Leong et al. 2002). As noted above, the genome organization

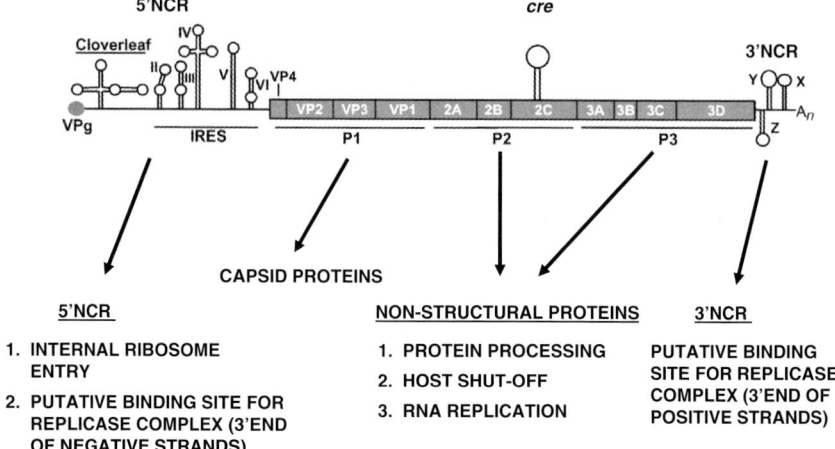

**Fig. 1** Genome organization of coxsackievirus B3. The genome of coxsackievirus is approximately 7,400 nucleotides (taken from van Ooij et al. 2006, with permission). RNA secondary structures that form within the noncoding regions of the genome and in the 2C coding region are important for cap-independent translation, RNA replication, and protein priming for RNA synthesis. After translation, the polyprotein undergoes multiple cleavage events by viral-encoded proteinases to generate functional precursor and mature proteins. Successful RNA replication utilizes the nonstructural proteins in their precursor and mature forms

of poliovirus and coxsackievirus is identical; unlike some other members of *Picornaviridae*, they lack a leader protein (L protein) at the amino terminus of the polyprotein (Fig. 1). The genome is divided into three regions: P1, P2, and P3. The proteins of P1 make up the capsid or structural proteins, whereas P2 and P3 comprise the nonstructural proteins. Only the RNA-dependent RNA polymerase, $3D^{pol}$, and the primer, VPg (3B), have been shown to have a direct role in enterovirus RNA replication; however, successful replication requires all of the nonstructural proteins of P2 and P3 in both the mature and precursor forms.

## 3.1 Proteins of P2: 2A, 2BC, 2B, and 2C

### 3.1.1 2A

Protein 2A is a cysteine proteinase that has a structure similar to that of chymotrypsin and is responsible for cleaving the polyprotein at the P1/P2 junction (Toyoda et al. 1986; Bazan and Fletterick 1988; Konig and Rosenwirth 1988). An intact 2A is needed for efficient poliovirus RNA replication, but its direct role has not been determined (Yu et al. 1995; Jurgens et al. 2006). Other substrates of 2A include host factors important for cellular translation. As noted above, 2A cleaves eIF4G, thus inhibiting cellular cap-dependent cellular translation (Gradi et al. 1998; Liebig et al. 2002). Picornaviruses are not affected by eIF4G cleavage because they utilize a cap-independent, IRES-mediated translation mechanism (Pelletier and Sonenberg 1988; Jang et al. 1988). Besides cleavage of eIF4G, 2A also cleaves poly(A) binding protein (PABP), a binding partner of eIF4G (Joachims et al. 1999; Kerekatte et al. 1999). PABP is an RNA-binding protein that interacts with the 3' poly(A) tract of cellular mRNAs and eIF4G simultaneously to functionally circularize the template to synergistically enhance translation (Tarun and Sachs 1996). Thus, 2A acts to downregulate overall cellular translation and sequester the translational machinery for enterovirus gene expression (Lloyd et al. 1987). In addition, 2A has been shown to cleave TATA-binding protein (TBP), a protein important in host cell RNA transcription; however, this cleavage does not shut down host transcription (Yalamanchili et al. 1997).

### 3.1.2 2BC, 2B, and 2C

Protein 2BC is the precursor to the mature 2B and 2C proteins. Experimental data showed that expression of poliovirus 2BC or 2B disrupts the Golgi complex and the secretory pathway, alters membrane permeabilization, and increases $Ca^{2+}$ levels in the transfected cell (Doedens and Kirkegaard 1995; Aldabe et al. 1997; Sandoval and Carrasco 1997). Expression of poliovirus 2BC or the amino-terminal portion of 2C was able to induce proliferation and rearrangement of membranous vesicles in tissue culture, like that seen during an enterovirus infection (Cho et al. 1994; Aldabe et al. 1996; Teterina et al. 1997). The poliovirus 2B protein has been shown to interact with membranous vesicles that are induced during poliovirus infection (Aldabe and Carrasco 1995). Likewise, expression of coxsackievirus 2B in mammalian cells induces similar cellular effects as seen with poliovirus 2B (van Kuppeveld et al. 1997a). The amino terminus of poliovirus 2B is predicted to form a cationic amphipathic α-helix, a motif indicative of ionophores (van Kuppeveld et al. 1997a). Mutations made to poliovirus 2B showed a defect in RNA synthesis. This defect was not complementable upon co-infection with wild-type poliovirus (Johnson and Sarnow 1991). Ectopic expression of coxsackievirus 2B in HeLa cells reduced the $Ca^{2+}$ levels in both the endoplasmic reticulum and Golgi complex, which causes an inhibition of caspase activation (Campanella et al. 2004; Cornell et al. 2006).

Poliovirus 2C has been shown to exhibit ATPase activity and interact with membranous vesicles induced from the rearrangement of the Golgi and endoplasmic reticulum (Cho et al. 1994; Mirzayan and Wimmer 1994; Echeverri and Dasgupta 1995). Although poliovirus 2C harbors a helicase motif, no helicase activity has been attributed to it (Gorbalenya et al. 1990; Pfister and Wimmer 1999). Poliovirus 2C contains two RNA-binding domains, which are zinc-binding Cys-rich motifs, located at the amino- and carboxy-termini (Rodriguez and Carrasco 1995; Pfister et al. 2000). At the amino-terminus of 2C is the membrane-binding determinant, which also overlaps a putative amphipathic helix (Paul et al. 1994; Echeverri and

Dasgupta 1995). Besides interacting with membranous vesicles, 2C was shown to interact with the negative-strand 3′ stem-loop I RNA (Banerjee et al. 1997). Addition of guanidine-HCl inhibits poliovirus negative-strand RNA synthesis, and this inhibition was mapped genetically to the 2C coding region (Pincus et al. 1986; Barton and Flanegan 1997; Pfister and Wimmer 1999). This suggests a role for 2C in negative strand RNA synthesis, possibly through its ability to hydrolyze ATP.

## 3.2 Proteins of P3:3AB, 3A, 3B, 3CD, 3C, and 3D$^{pol}$

### 3.2.1 3AB

The exact role of the protein 3AB has not been fully determined; however, biochemical data indicate that 3AB is a multifunctional protein. The reported functions of 3AB include interactions with cellular membranous vesicles, the virally encoded RNA-dependent RNA polymerase (RdRp) 3D$^{pol}$, and the proteinase/polymerase precursor 3CD (Molla et al. 1994; Towner et al. 1996; Hope et al. 1997). Protein 3AB also forms a ribonucleoprotein complex on poliovirus stem-loop I with 3CD (Xiang et al. 1995). Addition of 3AB to in vitro translation reactions of poliovirus RNA showed a stimulation of 3CD autoproteolysis (Molla et al. 1994). During in vitro 3D$^{pol}$ elongation reactions, addition of 3AB stimulated the activity of the polymerase (Lama et al. 1994; Plotch and Palant 1995; Richards and Ehrenfeld 1998). In 3D$^{pol}$ elongation reactions, addition of poliovirus 3AB allowed extension of the primer in the absence of prehybridization of primer and template (DeStefano and Titilope 2006). In this latter study, the authors also showed that poliovirus 3AB had helix destabilizing activity, indicative of poliovirus 3AB being a nucleic acid chaperone protein (DeStefano and Titilope 2006). The association of 3AB with membranous vesicles occurs through the hydrophobic domain in the 3A portion of the protein, and this interaction is thought to anchor the replication complex to cellular membranous vesicles. It is not known whether 3AB functions as a mature polypeptide or in the context of a precursor form (3BC, 3BCD, P3). Towner and colleagues demonstrated that *trans* addition of 3AB to in vitro RNA replication assays was unable to rescue a replication defect of poliovirus RNA containing an amino acid mutation, F69H; however, *trans* addition of the P3 precursor was able to rescue RNA replication in vitro (Towner et al. 1998). It is also hypothesized that 3AB, rather than the mature VPg (3B), is delivered to the replication complexes for VPg uridylylation (Liu et al. 2007). Recently, it was also shown that that 3AB can serve as a substrate for VPg uridylylation by 3D$^{pol}$ (Richards et al. 2006).

### 3.2.2 3A

Protein 3A is the mature N-terminal protein from the 3AB precursor; 3A is the least conserved protein among picornaviruses (Choe et al. 2005). It was predicted that

1977; Van Dyke and Flanegan 1980). Like all polymerases, 3D$^{pol}$ of poliovirus and coxsackievirus has a right-hand conformation with the associated thumb, finger, and palm domains (Hansen et al. 1997; Thompson and Peersen 2004). 3D$^{pol}$ also has unwinding activity, similar to a helicase, that may be important for enzyme processivity through structured RNA template domains or for strand displacement of duplexes (Cho et al. 1993). 3D$^{pol}$ was shown to bind to RNA cooperatively, with 10 nucleotides as the minimum length for binding (Beckman and Kirkegaard 1998). The crystal structure of 3D$^{pol}$ shows that it may form oligomers in a head to tail orientation (Hansen et al. 1997; Hobson et al. 2001). Amino acid sequences thought to be involved in oligomerization have been shown to be important for polymerase activity (Hobson et al. 2001). The oligomerization of purified 3D$^{pol}$ results in crystal-like lattices that may be associated with membranous vesicles in poliovirus-infected cells (Lyle et al. 2002). The amino acid sequence KKKRD on the finger domain of 3D$^{pol}$ can act as a nuclear localization signal (NLS) to allow for both 3D$^{pol}$ and 3CD to relocalize to the nucleus during a poliovirus infection (Sharma et al. 2004). The localization of 3CD into the nucleus allows for cleavage of cellular transcription factors (Sharma et al. 2004). In addition to chain elongation, 3D$^{pol}$ catalyzes the covalent linkage of UMP onto VPg (3B) using the *cre* as the template (Paul et al. 1998, 2000). Although the data point toward VPg as the substrate for uridylylation, it was shown that the 3AB binding site on 3D$^{pol}$ overlaps that of VPg(3B) suggesting that uridylylation of VPg might occur through the precursor 3AB (Hope et al. 1997; Xiang et al. 1998; Lyle et al. 2002a). The interaction of 3D$^{pol}$ and 3AB might act to recruit the RdRp to membranous vesicles, the sites for RNA replication.

## 4 RNA Secondary Structures Involved in Enterovirus RNA Replication

As mentioned above, approximately 10% of the enterovirus 7,500-nucleotide-long genomic sequence does not code for protein. With such a small genome, enteroviruses have developed a mechanism to utilize the nucleotide sequences flanking the open reading frame to mediate viral functions. Approximately 750 nucleotides make up the 5′ noncoding region (5′ NCR) and about 70 (for poliovirus) and 120 (for coxsackievirus) nucleotides make up the 3′ noncoding region (3′ NCR) (Kitamura et al. 1981; Racaniello and Baltimore 1981; Lindberg et al. 1987). Directly downstream of the 3′ NCR is the poly(A) tract which is essential for viral viability (Yogo and Wimmer 1972; Spector and Baltimore 1975). An internal stem-loop element, termed the *cis*-acting replication element or *cre*, was also discovered to form within the 2C coding region of the enterovirus genome (Goodfellow et al. 2000; van Ooij et al. 2006). Enteroviruses utilize the complex RNA secondary structures, including stem-loops and bulges that form on the nucleotide sequences, as contact points for both cellular and viral protein binding. The multiple protein-RNA interactions on the secondary structures mediate important viral activities including translation and RNA replication.

## 4.1 Stem-Loop I and Negative-Strand 3' Stem-Loop I

A highly conserved cruciform-like secondary structure, termed stem-loop I or cloverleaf, forms at the very 5' end of enterovirus genomic RNAs (Andino et al. 1990; Zell et al. 2002). Mutagenesis studies have shown that stem-loop I is important for viral RNA synthesis and VPg uridylylation (Andino et al. 1990; Parsley et al. 1997; Lyons et al. 2001). Poliovirus stem-loop I contains four cytosine residues required for interaction with the cellular poly(rC) binding protein (PCBP) (Parsley et al. 1997). The interaction of both PCBP and 3CD with stem-loop I results in an RNP structure that has been termed ternary complex (Fig. 2). This complex is involved in the initiation of RNA synthesis (Gamarnik and Andino 1997; Parsley et al. 1997; Barton et al. 2001). The interaction of 3CD and PCBP with stem-loop I is a conserved interaction that is also observed for coxsackievirus (Bell et al. 1999). Like poliovirus, coxsackievirus stem-loop I is essential for RNA synthesis. Stem-loop I deletion mutations made to coxsackievirus RNA allowed for translation of the polyprotein but not RNA synthesis (Hunziker et al. 2007). Interestingly, it was reported that an intact coxsackievirus stem-loop I may not be required for viral RNA viability in the infected cell. Tracy and colleagues showed that passage of coxsackievirus B3 in murine cardiomyocytes generated variants that were not cytolytic in HeLa cells; however, the viral RNA was persistent. Analysis of these coxsackievirus RNAs revealed 5' terminal deletion mutations, from nucleotides 7-49. The terminal deletions encompass the region of stem-loop I required for binding to PCBP. The authors hypothesized that the remaining sequences in stem-loop I were still capable of forming replication complexes with 3CD (or 3C) (Kim et al. 2005). In addition to PCBP, poliovirus 3AB can also bind to stem-loop I to form an alternative ribonucleoprotein complex with 3CD (Xiang et al. 1995).

Upon synthesis of the complementary negative-strand RNA, a secondary structure forms at the 3' end of the RNA (Andino et al. 1990). The negative-strand 3' stem-loop I is a mirror image of the 5' positive-strand stem-loop I and may act as

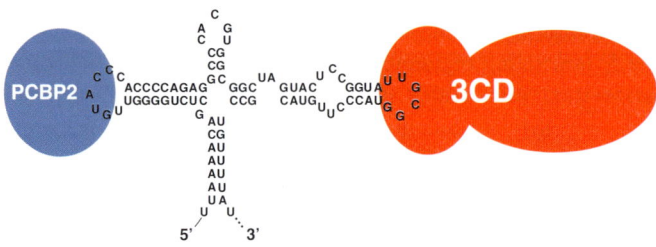

**Poliovirus stem-loop I**

**Fig. 2** Ternary complex with poliovirus stem-loop I. Ternary complex formation with the enterovirus 5' stem-loop I RNA structure and cellular PCBP and viral 3CD is an important step required for negative-strand RNA synthesis. Failure to form the ternary complex inhibits negative-strand RNA synthesis, possibly due to the lack of genome circularization, VPg uridylylation, or RNA stability

the site for initiation of positive-strand RNA synthesis. Through UV-crosslinking assays, it was shown that during a poliovirus infection of HeLa cells, two cellular proteins of 36- and 38-kDa molecular masses interacted with the negative-strand 3′ stem-loop I (Roehl and Semler 1995). The 36-kDa protein was later identified as hnRNP C (Brunner et al. 2005). The poliovirus 2C protein has also been shown to interact with the negative strand 3′ stem-loop I (Banerjee et al. 1997). This interaction might act to recruit negative-stranded RNA intermediates to membranous vesicles during positive-strand RNA synthesis.

## 4.2 cre *and VPg Uridylylation*

A stem-loop structure forms within the coding region of enterovirus RNAs corresponding to the 2C amino acid sequence, termed *cis*-acting replication element or *cre* (Fig. 3) (Goodfellow et al. 2000; Paul et al. 2000; van Ooij et al. 2006). The *cre* was first discovered in another picornavirus, human rhinovirus 14, but this *cre* is located in the coding region for VP1 (McKnight and Lemon 1998). Cre functions as a template for 3D$^{pol}$ catalyzed uridylylation of VPg or VPg-containing precursors.

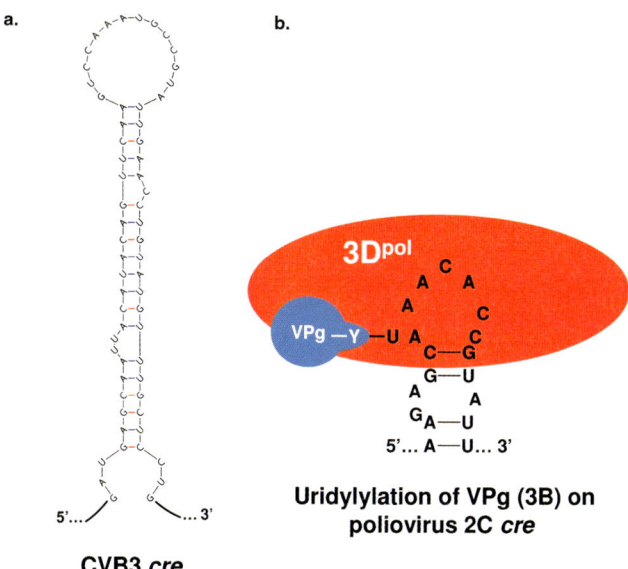

**Fig. 3** Enterovirus 2C *cis*-replication element and VPg-uridylylation. **a** The *mfold* predicted secondary structure of the coxsackievirus 2C *cis*-replication element, *cre* (van Ooij et al. 2006; Mathews et al. 1999; Zuker 2003). The sequences shown are nucleotides 4365-4425 in the CVB3 genome. **b** 3D$^{pol}$ catalyzes the formation of a phosphodiester bond between tyrosine (Y) of VPg (3B) and a uridine residue using the adenosine residue of the *cre* as a template (for additional details, see Paul 2002)

At the apex of the poliovirus *cre* stem is the consensus loop sequence, 5′-AAACA-3′, with the first two adenylate residues required for 3D$^{pol}$ to covalently link UMP nucleotides to VPg (Paul et al. 2003). Interestingly, disruption of the poliovirus *cre* sequences or structure showed a defect in positive- but not negative-strand RNA synthesis (Murray and Barton 2003; Goodfellow et al. 2003; Morasco et al. 2003). This would suggest that (1) for negative-strand synthesis VPg uridylylation was *cre* independent and (2) for positive-strand synthesis, the functional *cre* on the positive strand is involved in *trans*. However, recent data from van Ooij and colleagues suggested that the CVB3 *cre* has a role in both positive- and negative-strand RNA synthesis (van Ooij et al. 2006).

The uridylylation of VPg is the first step in RNA synthesis. It was shown that efficient uridylylation of VPg requires *cre* within a full-length RNA, UTP, 3CD, and 3D$^{pol}$ (Paul et al. 2000). Mutagenesis of the adenylate nucleotides within the consensus 5′-AAACA-3′ of the poliovirus *cre* indicates that only the first adenine is used as the template for uridylylation, perhaps via a slide-back mechanism (Paul et al. 2003). Uridylylation ceases following the covalent linkage of two UMP nucleotides on VPg, potentially due to a small active site on 3D$^{pol}$ and not due to the small apex of the *cre* loop (Paul et al. 2003). Using computer-predicted modeling of the interaction between VPg and 3D$^{pol}$ and in vitro uridylylation assays, Kirkegaard and colleagues demonstrated that VPg docked on the backside of the thumb of 3D$^{pol}$ and not on the active site (palm) for chain elongation as the site for VPg uridylylation (Tellez et al. 2006). This finding is in contrast to a study on a related picornavirus, foot-and-mouth disease virus (FMDV), in which Verdaguer and colleagues showed through co-crystallization of FMDV's 3D$^{pol}$ and VPg that the interaction was through the front or chain elongation-active site of the polymerase (Ferrer-Orta et al. 2004).

Lyons and colleagues showed that an intact poliovirus stem-loop I was required for VPg uridylylation (Lyons et al. 2001). Although stem-loop I has no direct role in uridylylation, it may act to arrange the required proteins in a step-wise fashion for uridylylation. The uridylylation of VPg is a *trans*-dominant reaction in vivo (Crowder and Kirkegaard 2005). Crowder and Kirkegaard showed that lethal mutations made to poliovirus *cre* or VPg inhibited wild-type poliovirus growth upon co-transfection in tissue culture, thus exhibiting a *trans*-dominant negative effect (Crowder and Kirkegaard 2005).

The finding that an intact *cre* may not be required for negative-strand RNA synthesis suggests the utilization of the 3′ poly(A) tract of poliovirus RNA for VPg uridylylation (Paul et al. 1998). Interestingly, it was recently shown that 3D$^{pol}$ can be uridylylated in vitro, as well as 3CD and 3AB (Richards et al. 2006). In vitro uridylylation of 3AB may suggest the utilization of a precursor form of VPg prior to initiation of viral RNA synthesis (Richards et al. 2006; Liu et al. 2007). Most recently, Wimmer and colleagues, using model lipid membrane vesicles, showed that when poliovirus 3AB is anchored in these membranes, VPg can be uridylylated by 3D$^{pol}$ only when proteolytically active 3CD is used as a co-factor (Fujita et al. 2007). Previously, it was shown that only membrane-bound 3AB was susceptible to 3C/3CD cleavage, suggesting that VPg can be uridylylated after cleavage of the membrane associated precursor by 3CD (Lama et al. 1994).

## 4.3  3′ NCR and Poly(A) Tract

At the 3′ terminus of enterovirus RNA, secondary structures form a tRNA-like or L-shaped conformation. Two secondary structures are predicted to form within the poliovirus 3′ NCR, termed stem-loop X and Y, while the 3′ NCR of coxsackievirus RNA was predicted to contain three stem-loops, X, Y, and Z (Fig. 4) (Pilipenko et al. 1992). Stem-loop Z in coxsackievirus RNA replication is dispensable for RNA replication because stem-loop Z deletion mutations displayed wild-type growth characteristics in tissue culture (Merkle et al. 2002). A tertiary interaction is predicted to form between stem-loop X and Y for both poliovirus and coxsackievirus 3′ NCRs (Pilipenko et al. 1996; Melchers et al. 1997; Mirmomeni et al. 1997). Nucleotide mutations made to disrupt this tertiary interaction of poliovirus stem-loop X and Y resulted in a virus displaying a delay in viral RNA synthesis (Pilipenko et al. 1996). For coxsackievirus, disruption of the predicted tertiary interaction of stem-loop X and Y was lethal to the virus (Melchers et al. 1997). Evidence for an alternative tertiary structure for the 3′ NCR of poliovirus RNA has been reported by Sarnow and colleagues (Jacobson et al. 1993). This structure is predicted to include part of the 3D coding region as well.

The requirement for the 3′ NCR in enterovirus RNA replication was tested in studies undertaken by Semler and colleagues. These investigators showed that a poliovirus mutant with a deletion of the entire 3′ NCR was still able to replicate its RNA in HeLa cells, although the level of RNA synthesis was slightly reduced and delayed when compared to wild type (Todd et al. 1997; Brown et al. 2003). Interestingly, the replication defect seen with the 3′ NCR deletion mutant poliovirus occurred primarily at the level of positive-strand RNA synthesis (Brown et al. 2003). This would suggest that the secondary structures of the 3′ NCR are not absolutely essential for RNA synthesis but may allow for efficient RNA synthesis through arrangement of the enterovirus RNA or the replication complex on membranous vesicles (Todd et al. 1997).

The poly(A) tract at the very 3′ end of the enterovirus genome is presumed to be the start site for negative-strand RNA synthesis. For poliovirus, the 3′ poly(A) tract is about 60-80 nucleotides in length. Unlike cellular mRNAs, the 3′ poly(A) tract is not added by poly(A) polymerase; rather, it is genetically encoded (Yogo and Wimmer 1972; Dorsch-Hasler et al. 1975). A minimum of $(A)_{12}$ on poliovirus transcripts allows for efficient infectivity, whereas transcripts harboring 3′ $(A)_{100}$ tracts had a specific infectivity equivalent to that of virion RNA (Sarnow 1989). Poliovirus RNA with less than $A_{(8)}$ failed to synthesize negative-strand RNA in vitro (Herold and Andino 2001). Upon synthesis of the nascent negative strand, the 5′ end contains complementary poly(U) sequences, but the number of uridylate residues is less (40-60 nt) than that of its poly(A) template (Yogo and Wimmer 1973; Spector and Baltimore 1975). The mechanism of extending the poly(A) tail on the positive-strand RNA beyond the complementary poly(U) on the negative strand is not known. Recently, van Ooij and colleagues proposed two hypotheses: a stuttering of $3D^{pol}$ on the poly(U) template to allow for poly(A) extension or polyadenylation by a terminal nucleotidyl transferase-like enzyme (van Ooij et al. 2006).

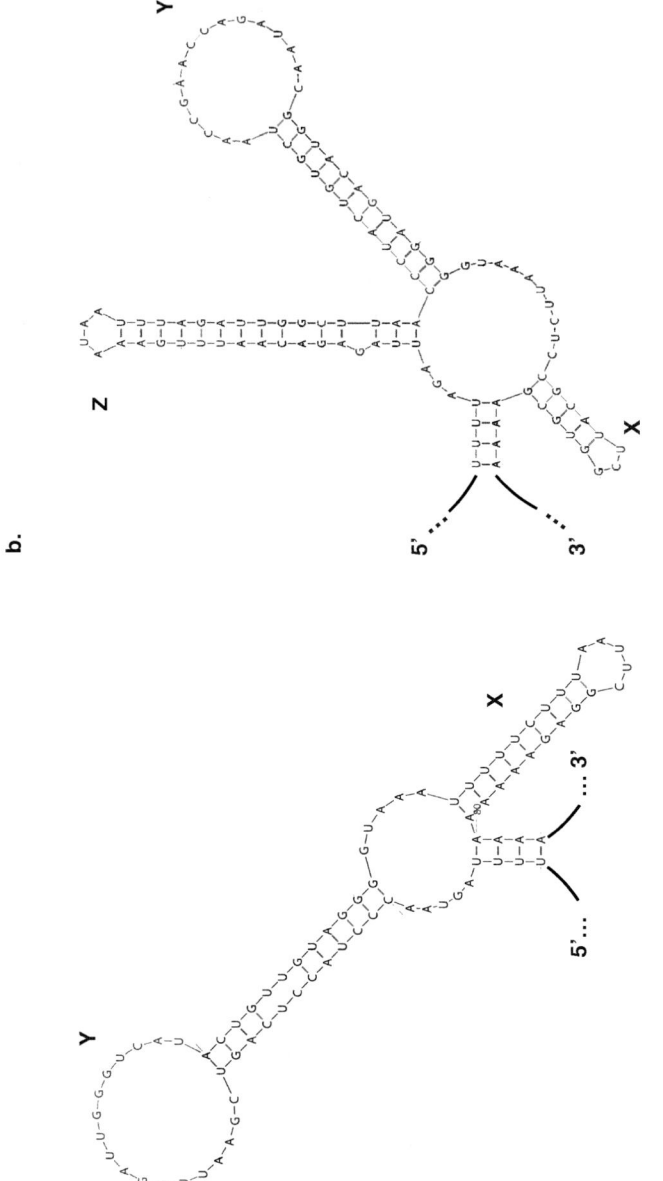

**Fig. 4** Secondary structures of the 3' NCR of poliovirus and coxsackievirus. The secondary structures of 3' NCR of enterovirus as predicted by *mfold* (Mathews et al. 1999; Zuker 2003). The nucleotide constraints were determined by chemical and enzymatic structure probing (Pilipenko et al. 1992). **a** The secondary structure of the poliovirus 3' NCR is predicted to form two stem-loops, termed X and Y. The structure shown represents nucleotide 7358-7431 plus $A_{(9)}$ residues for the poliovirus type 1 genomic RNA. **b** The 3' NCR of coxsackievirus is predicted to form three stem-loops, termed X, Y, and Z. The structure shown represents nucleotide 7293-7397 plus $A_{(4)}$ residues for coxsackievirus B3 genomic RNA

The importance of the 3' poly(A) tract in infectivity suggests a critical role in poliovirus replication. It was shown that the poly(A) tail can serve as a template for VPg uridylylation in poliovirus RNA replication; however, this reaction is not as efficient as using the *cre* as the template (Paul et al. 1998, 2000). The simultaneous interaction of the cellular PABP on the 3' poly(A) tract and 3CD and PCBP on stem-loop I has been suggested to facilitate circularization of the poliovirus genome (Herold and Andino 2001). Genome circularization has been suggested to be a required step in negative-strand RNA synthesis (Herold and Andino 2001; Barton et al. 2001). Flanegan and colleagues have shown that replication of poliovirus transcripts with poly(A) tails extending from 12 to 20 adenylate residues dramatically increases negative-strand RNA synthesis; however, binding of PABP to such transcripts did not increase. The authors suggested that the increase in negative-strand synthesis was independent of PABP binding and that a longer 3' poly(A) tract allows for efficient VPg uridylylation and initiation of negative-strand RNA synthesis (Silvestri et al. 2006).

## 5 Cellular Proteins Involved in RNA Replication

Due to their limited coding capacity, enteroviruses have evolved to exploit the host cell to mediate viral functions. Besides using the cellular translational machinery to express viral proteins, enteroviruses utilize cellular host factors to mediate viral RNA replication. Not surprisingly, most of the cellular proteins that have been identified to have a role in RNA replication are RNA-binding proteins. The interaction between cellular proteins and viral RNA has been shown to stabilize enterovirus RNAs and possibly recruit nonstructural viral proteins to sites of RNA replication.

### *5.1 PCBP*

To date, the only cellular protein identified to have a definitive role in enterovirus RNA replication is poly(rC) binding protein (PCBP) (Parsley et al. 1997; Gamarnik and Andino 1997a; Walter et al. 2002). PCPBs are RNA-binding proteins that preferentially bind to single-stranded stretches of cytidines (for review see Makeyev and Liebhaber 2002). In mammalian cells, there are four isoforms, PCBP 1-4, but only PCBP1 and PCBP2 have been experimentally shown to have roles in the life cycles of enteroviruses. PCBP binds to poly(rC) stretches in the 3' NCR of specific cellular mRNAs and stabilizes these messenger RNAs, in the case of α-globin, or modulates translation, in the case of lipoxygenase (Weiss and Liebhaber 1994; Ostareck et al. 1997). Similar to cellular mRNAs, it has been shown that the interaction of PCBPs with poliovirus stem-loop I contributes to the overall stability of the viral RNA in vitro (Murray et al. 2001). The interaction of PCBP with stem-loop I, along with the viral protein 3CD, forms the ternary complex, a required step in negative-strand RNA synthesis (Gamarnik and Andino 1997; Parsley et al. 1997;

Bell et al. 1999). In HeLa cytoplasmic extracts depleted of PCBPs, poliovirus RNA replication was inhibited; however, addition of recombinant PCBPs rescued RNA replication to mock-depleted levels (Walter et al. 2002).

## 5.2 PABP

PABP is a cellular RNA-binding protein that binds to the 3′ poly(A) tract of cellular mRNAs with a high affinity (Grange et al. 1987; Gorlach et al. 1994). PABP binds through two RNA-recognition motifs (RRM) to stabilize the cognate mRNAs. As mentioned previously, PABP has been shown to interact simultaneously with eIF4G, the scaffolding protein in the cap-binding complex, and the poly(A) tract of cellular mRNAs to functionally circularize the mRNA for synergistic enhancement of translation (Otero et al. 1999). Likewise, PABP has also been suggested to be a bridging protein that might promote the circularization of poliovirus RNA (Herold and Andino 2001). During an enterovirus infection, PABP is a target for 2A- and 3C-mediated cleavage (Kerekatte et al. 1999; Joachims et al. 1999). It would seem counterintuitive for enteroviruses to cleave a protein hypothesized to be important in negative-strand RNA synthesis; however, it was shown that the viral proteinases selectively cleave PABP enriched in ribosome fractions of the cell (Kuyumcu-Martinez et al. 2002). Thus, PABP in RNA replication could still be intact and functional.

## 5.3 hnRNP C

Another cellular RNA-binding protein, heterogeneous ribonuclear protein C (hnRNP C), has been identified as a possible component of poliovirus RNA replication complexes (Brunner et al. 2005). hnRNP C is a nuclear RNA-binding protein that is involved in mRNA biogenesis (Dreyfuss et al. 1993). However, during poliovirus infection of HeLa cells, Roehl and Semler showed that hnRNP C interacts with the negative-strand 3′ stem-loop I RNA of poliovirus (Roehl and Semler 1995; Brunner et al. 2005). Interestingly, in GST-pulldown assays, hnRNP C was shown to interact with multiple nonstructural proteins of poliovirus, including 3D$^{pol}$, 3CD, P2, and P3 (Brunner et al. 2005). While hnRNP C has been shown to have multiple interactions with poliovirus replication elements, its direct role in RNA synthesis has not been established. Recently, Brunner, Ertel, and Semler observed that in SKOV3 cells, an ovarian carcinoma cell line that has reduced levels of hnRNP C, infection with poliovirus displayed a delayed kinetics of replication (Holcik et al. 2003, Brunner et al., unpublished observations). Upon subsequent transient transfection with plasmids expressing hnRNP C, there was an increase in the virus yield. Ongoing studies will determine the extent to which hnRNP C plays a role in RNA synthesis in cells infected with either poliovirus or coxsackievirus.

## 6 Membranous Vesicles

Successful replication of enterovirus RNA requires an induction of membranous vesicles via nonstructural proteins, resulting in a rearrangement of cellular components into rosette-like vesicles (Fig. 5) (Bienz et al. 1992). It has been postulated that the membranous vesicles act as scaffolds for favorable RNA synthesis and protection of the nascent RNAs from nucleases (Fogg et al. 2003). The membranous vesicles may also allow for proper positioning of the replication proteins to carry out important reactions (Fogg et al. 2003). Membranous vesicles are also hypothesized to concentrate viral replication proteins to efficiently catalyze RNA synthesis (Tershak 1984). Addition of lipophilic agents disrupts membranous vesicles and inhibits poliovirus RNA replication in cell-free extracts (Molla et al. 1993). The exact mechanistic steps of vesicle formation are not clear, but cellular protein markers from the Golgi complex and endoplasmic reticulum were detected in vesicles from poliovirus-infected cells (Schlegel et al. 1996). Expression of the poliovirus nonstructural proteins from P2 and P3 regions induce vesicle formation that is independent of RNA replication (Teterina et al. 2001). The enterovirus 2B, 2C, and 2BC, and 3A proteins are primarily responsible for vesicle induction (Cho et al. 1994; Aldabe and Carrasco 1995; Suhy et al. 2000). Enterovirus 2B, 2C and 3A proteins have been shown to directly interact with the membrane structures, so potentially they may act to anchor the viral replication complex to the surface of

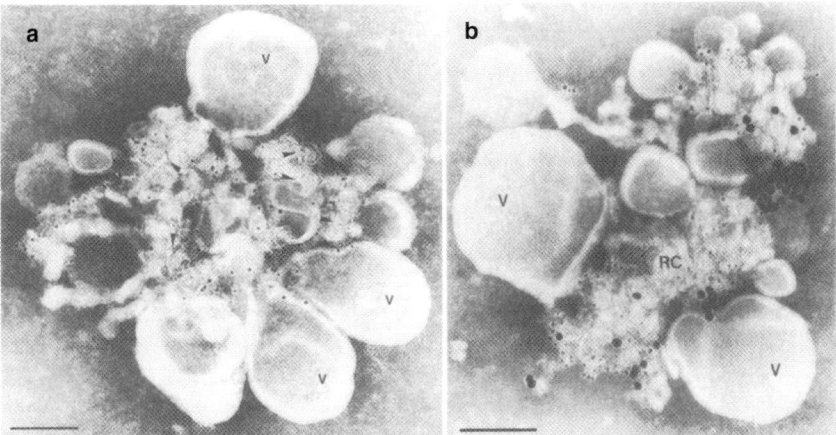

**Fig. 5** Induction of rosette-like membranous vesicles by enterovirus infection. An electron microscopic image of rosette-like membranous vesicles (*V*) that are induced during poliovirus infection at 4 h postinfection (taken from Bienz et al. 1992, with permission). The bar is a length marker for the EM image, representing 100 nm. Protein markers on the induced vesicles indicate components from the endoplasmic reticulum, the Golgi complex, and cellular autophagosomes. **a** The dark speckles on the vesicles are poliovirus 2C proteins labeled with gold. **b** At the center of the vesicles is the poliovirus replication complex (*RC*)

the membranes (Echeverri and Dasgupta 1995; Towner et al. 1996; Teterina et al. 1997; van Kuppeveld et al. 1997b).

Enteroviruses may utilize different steps in cellular secretion, coinciding with the stages of virus infection, to induce vesicle formation (for review, see Belov and Ehrenfeld 2007). Previously, components of cellular autophagosomes have been implicated in membranous vesicle formation (Suhy et al. 2000). Kirkegaard and colleagues showed that poliovirus infection of HeLa cells or co-expression of poliovirus 2BC and 3A induces co-localization of LC3 and LAMP1, markers for mature autophagosomes (Jackson et al. 2005). Poliovirus yield increased three- to fourfold when HeLa cells were treated with inducers of autophagy such as rapamycin and tamoxifen; so potentially, the membranous vesicles that originate from autophagosome may function in release of the virus (Jackson et al. 2005).

Bienz and colleagues demonstrated that early in poliovirus infection, vesicle formation is through the endoplasmic reticulum, COPII-mediated pathway and did not include the Golgi complex (Rust et al. 2001). The authors showed through immunofluorescence assays that during poliovirus infection, the COPII protein markers, Sec13 and Sec31, localized to the membranous vesicles that were induced, similar to those of the anterograde traffic vesicles (Rust et al. 2001). Belov and Ehrenfeld suggested that late in infection, 3A and 3CD were responsible for induction of membranous vesicles, mediated through ADP-ribosylation factor (ARF) (Belov et al. 2007). ARF is a cellular protein involved in Golgi-mediated retrograde transport, but during poliovirus infection it was shown to translocate to the vesicle membranes of RNA replication complexes (Belov et al. 2005). Expression of 3A and 3CD recruits ARF to membranes through their interaction with two guanine exchange factors (GEFs), GBF1 and BIG1/2, respectively. GEFs then catalyze the conversion of an inactive ARF-GDP to an active ARF-GTP that then associates with the membranes to induce vesicle formation. The possible role of ARF in vesicle formation has been corroborated by the addition of a fungal metabolite, brefeldin A (BFA), to inhibit poliovirus replication (Maynell et al. 1992; Cuconati et al. 1998). BFA inhibits the activity of ARF by stabilizing the ARF-GDP, thus preventing recycling of active ARF-GTP, and overexpression of GBF1 rescues BFA inhibition of poliovirus RNA replication (Morinaga et al. 1996; Belov et al. 2007).

## 7 Transition from Translation to RNA Replication

Prior to viral RNA replication, ribosomes translating the polyprotein must be cleared from the RNA genome. Potentially, the ribosome(s) on the translating RNA going downstream would interfere with the 3D$^{pol}$ containing replication complex coming upstream, thus stalling negative-strand RNA synthesis (Fig. 6). Using preinitiation replication complexes, Barton and colleagues have shown that freezing ribosomes on translating RNAs with cycloheximide inhibited RNA replication, while dislocating the ribosomes with puromycin allowed RNA synthesis to occur (Barton et al. 1999).

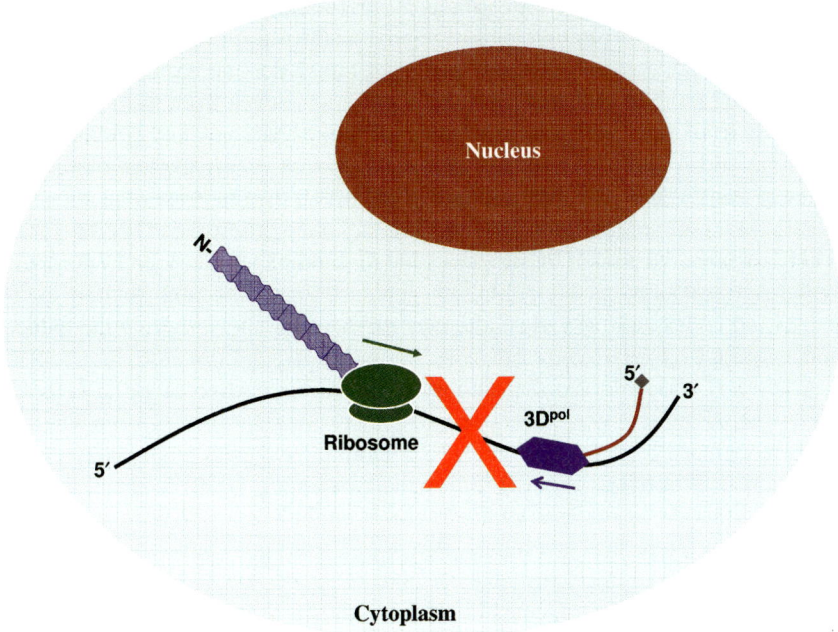

**Fig. 6** Enterovirus translation and RNA synthesis on the same RNA template. A paradox of positive-strand RNA viruses is the utilization of the RNA template for both translation and RNA replication. An identified mechanism switches the utilization of the RNA template so that the translation machinery does not collide with the RNA replication complex. The ribosome (*green*) is traversing from the 5' end of the positive-strand RNA (*black*) and synthesizing polypeptides (*light blue*). If the ribosome is not cleared from the RNA template, a collision may occur between the ribosome and the RNA replication complex (*dark blue*) moving upstream from the 3' end, thus stalling both processes, as denoted by the large *red X*. The newly synthesized RNA (*maroon*) with the primer, VPg (*grey*) is shown extending from the replication complex. 3D$^{pol}$ is the RNA-dependent RNA polymerase

In addition to clearing ribosomes, another potential prerequisite for viral RNA synthesis is the requirement for the positive-strand template to have been previously translated (Novak and Kirkegaard 1994). Novak and Kirkegaard showed that co-transfection of a mutant poliovirus RNA with an amber codon in the 2A coding region and a transcript that provides the nonstructural proteins in *cis* failed to replicate the mutant poliovirus RNA. They identified a *cis*-translation required (CTR) region, from the 2A to 3C coding region of poliovirus RNA, that needed to be translated first before the RNA could be used as a template in negative-strand RNA synthesis (Novak and Kirkegaard 1994). In another *trans*-complementation experiment, Barton and colleagues showed that prior translation of the template was not a prerequisite for RNA replication (Murray et al. 2004). The authors showed that a poliovirus transcript completely lacking the IRES was a sufficient template for negative- and positive-strand RNA synthesis in vitro upon *trans*-complementation with a transcript expressing the nonstructural proteins of poliovirus (Murray et al. 2004).

After successive rounds of IRES-mediated translation, an accumulation of the nonstructural proteins is thought to induce a gradual switch from the utilization of poliovirus RNA for translation to RNA replication (Gamarnik and Andino 1998). Gamarnik and Andino have suggested that this switch occurs through the binding of newly translated 3CD to stem-loop I RNA, which then increases the affinity of PCBP2 for the same stem-loop element. The increased affinity of PCBP2 for stem-loop I decreases the pool of PCBP2 available for binding to stem-loop IV for IRES-mediated translation of the poliovirus genome (Gamarnik and Andino 2000). Semler and colleagues offered an alternate mechanism, also involving the reduced binding of PCBP to poliovirus stem-loop IV, for the switch from translation to RNA replication. It was observed that during a poliovirus infection of HeLa cells, the level of PCBP protein decreases along with a concomitant increase in a fragment that is recognized by polyclonal anti-PCBP antibodies (Perera et al. 2007). Amino acid sequence analysis of PCBP2 revealed potential 3C/3CD cleavage sites, suggesting that the appearance of the fragment might be a 3C/3CD cleavage product of PCBP2. To further investigate, Perera and colleagues generated a truncated recombinant PCBP2 that was able to bind to stem-loop I and mediate RNA replication; however, the truncated PCBP2 was unable to bind to stem-loop IV and stimulate IRES-mediated poliovirus translation (Perera et al. 2007).

Following translation of the viral genome, enterovirus RNA is then translocated via an undetermined mechanism to membranous vesicles, which are the sites for both negative- and positive-strand RNA synthesis (Bolten et al. 1998). It has been shown that poliovirus 3AB interacts with both the membranous vesicles and viral proteins involved in RNA replication, so potentially 3AB functions to recruit the replication complex to the surface of the membranous vesicles (Semler et al. 1982; Towner et al. 1996; Hope et al. 1997). A potential role for 3AB in the recruitment of the replication complex might involve 3CD because 3CD has been shown to interact with 3AB and both termini of positive-strand poliovirus RNA (Harris et al. 1994; Xiang et al. 1998). Protein 2C has also been shown to interact with the membranous vesicles and negative-strand poliovirus RNA, so 2C might act as an RNA anchoring protein on the surface of the membranous vesicles (Echeverri and Dasgupta 1995; Banerjee et al. 1997).

## 8 Enterovirus RNA Replication

Enterovirus RNA replication can be thought of as a two step process: (1) the first step is the synthesis of the negative complementary strand from the parental positive genomic mRNA and (2) the second step is the synthesis of a nascent daughter RNA strand that is then used for either cap-independent translation or packaged into virions for subsequent infections (Fig. 7). In addition, this daughter strand could be used for additional rounds of negative-strand RNA synthesis. At present, the published

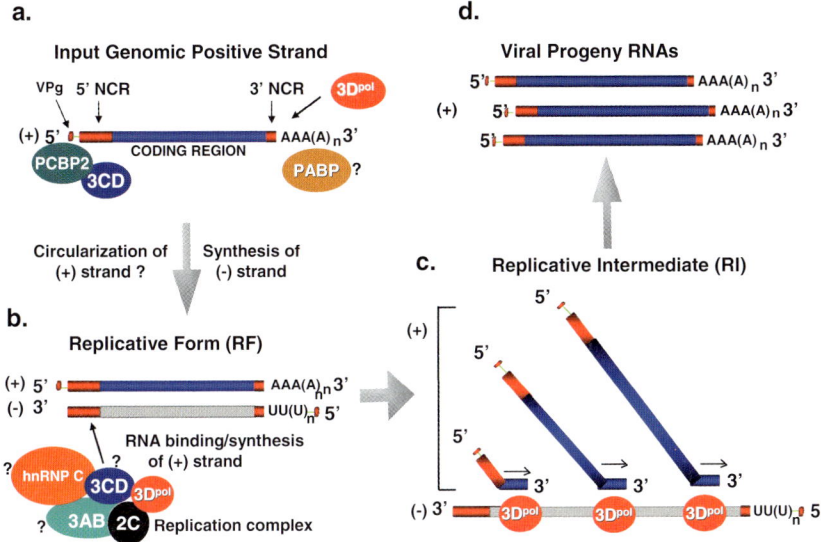

**Fig. 7** Negative- and positive-strand enterovirus RNA synthesis. The complete mechanistic steps of negative- and positive-strand RNA synthesis are still not understood. **a** Initiation of negative-strand RNA synthesis involves the interaction of cellular and viral proteins on the noncoding region of the template. The replication complex synthesizes the nascent negative-strand RNA beginning at the 3' poly(A) tract of the genomic RNA. **b** After completion of negative-strand RNA synthesis, the RNAs are present as heteroduplexes, called replicative forms. Initiation of positive-strand synthesis may involve binding of viral and/or cellular proteins to the negative-strand RNA template. **c** Asymmetric synthesis of positive-strand RNAs compared to negative-strand RNAs is a result of replicative intermediate structures, with up to six nascent positive-strand RNAs being simultaneously synthesized on one negative-strand template. **d** The newly synthesized nascent positive-strand enterovirus RNAs are packaged into virions

literature does not provide a complete mechanistic picture of the steps involved in negative- and positive-strand RNA synthesis. Significant limitations to elucidating these steps arise from the redundant functions of viral proteins and overlapping signals for RNA replication in the genome. Site-directed mutagenesis of viral proteins to study the process of positive-strand synthesis will likely affect a prior step in negative-strand RNA synthesis. Likewise, mutations made to viral RNA sequences to study the effects on negative-strand synthesis may also affect positive-strand synthesis. More recently, the utilization of an in vitro RNA replication assay using HeLa cytoplasmic extract has allowed us to tease out the mechanistic steps of enterovirus RNA replication (Molla et al. 1991). Using this in vitro approach, investigators have begun to unravel the stepwise interactions of cellular and viral proteins and RNA sequence elements; however, these ongoing studies do not provide a complete picture, and the data are clearly open to different interpretations. In the following sections, we describe some of the steps shown to be necessary for RNA replication. We will present a model that combines the data presently available.

## 8.1 Negative-Strand RNA Synthesis

The onset of enterovirus RNA synthesis occurs rapidly, approximately 1.5-2 h after infection (Baltimore et al. 1966). Two important events are required before negative-strand synthesis can occur. First, the translating ribosomes must be cleared from enterovirus RNA. If the translating ribosome traveling 5′ to 3′ are not cleared from the RNA, they might collide with the 3D$^{pol}$ traveling upstream, causing an abrupt termination of negative-strand synthesis. The second event that needs to occur is formation of the ternary complex. It has been shown that failure to form a ribonucleoprotein complex consisting of PCBP, 3CD, and poliovirus stem-loop I RNA inhibits negative-strand RNA synthesis (Parsley et al. 1997; Lyons et al. 2001; Teterina et al. 2001).

So it begs the question: if initiation of RNA replication occurs at the 3′ end of the RNA, why is the ternary complex at the 5′ end so critical? The significance of the ternary complex may be in its ability to functionally circularize the template and deliver the replication complex to the 3′ poly(A) tract for negative-strand RNA synthesis. Circularization of the poliovirus template may be mediated through 3CD and/or PCBP2 bound to stem-loop I with PABP on the 3′ poly(A) tract (Herold and Andino 2001; Barton et al. 2001). Deletion of stem-loop I RNA inhibits negative-strand synthesis in vitro, possibly due to VPg not being uridylylated, or due to the disruption of the long-range association between the ternary complex and the 3′ poly(A) tract (Barton et al. 2001, Lyons et al. 2001). Thus, a circularized template could act to deliver the replication complex that assembles on the 5′ end to the 3′ end for negative-strand RNA synthesis. Such a mechanism also provides specificity for the RNA replication process to ensure that only viral RNAs are being copied. If the replication complex only recognized poly(A) tracts, poly(A)-containing cellular mRNAs might be copied. Another advantage of a circularized template is the maintenance of terminal RNA sequences, which are important replication signals in both negative- and positive-strand RNA synthesis.

After ternary complex formation and circularization, the replication complex would then be delivered to the poly(A) tract. The polymerase precursor, 3CD may have a role in the recruitment or positioning of the replication complex for negative-strand RNA synthesis. The 3CD protein is the only viral protein shown to interact with both termini of the poliovirus genome and to have multiple interactions with nonstructural viral and cellular proteins (Harris et al. 1994; Molla et al. 1994; Xiang et al. 1998; Herold and Andino 2001; Brunner et al. 2005). At the 3′ poly(A) tract, 3D$^{pol}$ uses the adenylate residues as the template for VPg uridylylation (Paul et al. 1998). Following uridylylation, the 3D$^{pol}$/VPgpUpU containing replication complex elongates, adding complementary uridines encoded by the 3′ poly(A) tract of the positive strand, to the nascent negative-strand RNA (Yogo and Wimmer 1973; Flanegan and Baltimore 1977). Once 3D$^{pol}$ reaches the 5′ end of the poliovirus RNA template, negative-strand synthesis is complete; however, the replication complex does not appear to read through the circular RNA to extend the nascent strand past the 5′ end.

After negative-strand RNA synthesis, both the positive-strand template and the negative-strand nascent RNAs are predicted to be heteroduplexed. The heteroduplex RNAs are known as the replicative form (RF). In the replicative form, the negative strand is slightly shorter than the positive strand, due to internal priming of negative strand on the 3' poly(A) tract (Yogo and Wimmer 1973; Nomoto et al. 1977; Pettersson et al. 1978). The heteroduplexed RNAs are transiently separated into single strands during positive-strand RNA synthesis via an unknown mechanism. The newly synthesized negative strand (in the RF) is retained in the membranous vesicles and immediately used as the template for positive-strand RNA synthesis. Also, it is not known if the RNA replication complex would still be associated with the positive strand for further negative-strand RNA synthesis or retained on negative strand to commence positive-strand RNA synthesis.

## 8.2 Positive-Strand RNA Synthesis

Like negative-strand RNA synthesis, positive-strand RNA synthesis requires nonstructural proteins of P2 and P3 and the membranous vesicles that they induce. In addition to the nonstructural viral proteins, two RNA elements are required for positive-strand synthesis; the positive strand 5' stem-loop I and the *cre* that forms on the positive strand in the 2C coding region (Andino et al. 1990; Murray and Barton 2003; Goodfellow et al. 2003; Morasco et al. 2003). An intact 5' stem-loop I is required for positive-strand synthesis because upon negative-strand RNA synthesis, it is the template used to generate the complementing negative strand 3' stem-loop I, which is the site where the RNA replication complex begins positive-strand RNA synthesis (Andino et. 1990; Sharma et al. 2005). The requirement for an intact *cre* for positive-strand synthesis suggests that in positive-strand synthesis, uridylylation of VPg occurs on the *cre* and is then recruited in *trans* to the 3' end of the negative-strand template (Murray et al. 2003; Goodfellow et al. 2003; Morasco et al. 2003). The uridylylation of VPg or VPg-containing precursors on the *cre* is a very efficient reaction and will generate excess uridylylated VPg, a possible mechanism leading to the asymmetric levels of positive-strand poliovirus RNAs compared to negative strands seen in the infected cell (Giachetti and Semler 1991; Novak and Kirkegaard 1991; Paul et al. 2000).

The initiation of positive-strand RNA synthesis begins with the recruitment of the uridylylated-VPg containing replication complex to the negative-strand 3' stem-loop I. The recruitment could involve 2C because it was shown that 2C directly interacts with the negative strand 3' stem-loop I (Banerjee et al. 1997). Another possibility is that the interaction of hnRNP C with the negative-strand 3' stem-loop I will act to recruit 3CD, a protein already shown to stimulate *cre*-mediated VPg uridylylation (Paul et al. 2000; Brunner et al. 2005). Once recruited to the start site of positive-strand RNA synthesis, VPgpUpU anneals with two adenylate residues that are in the 3' stem-loop I of negative-strand poliovirus RNA (Sharma et al. 2005). Flanegan and colleagues showed that deletion of the 3' terminal adenosines greatly decreases positive-strand RNA synthesis (Sharma et al. 2005). Positive-strand RNA synthesis ceases after the replication complex reaches the 5′ terminus on the negative-strand template; however, it should be noted that synthesis of a single copy of nascent positive-strand RNA from a negative-strand template is an overly simplified model. During poliovirus positive-strand RNA synthesis, multiple positive-strand RNAs are simultaneously being synthesized from a negative-strand template.

The partially double-stranded RNA structure of positive- and negative-strand RNA is known as the replicative intermediate (RI) (Baltimore and Girard 1966). The exact mechanism establishing the replicative intermediate during poliovirus positive-strand synthesis has not been determined. We hypothesize that the negative-strand stem-loop I is able to recruit several VPgpUpU containing replication complexes, possibly through 2C or hnRNP C (Banerjee et al. 1997; Brunner et al. 2005). The availability of multiple VPgpUpU primers for the replicative intermediate might be facilitated by the efficient uridylylation of VPg on the *cre* (Paul et al. 2000).

One obstacle that the replicative intermediate must overcome is hybridization of the growing nascent positive-strand RNAs with the negative-strand template. Heteroduplexed RNAs might sterically inhibit the replication complex from translocating downstream on the template and synthesizing positive-strand RNAs. Strand separation may be achieved through extensive secondary structures that form on the nascent RNA, thus inhibiting single-stranded regions from annealing with the template. Another mechanism that may inhibit annealing of the nascent positive-strand with the template is utilization of a helicase. Helicase activity will separate the heteroduplexed RNAs and allow multiple replication complexes to interact with the negative-strand to synthesize new RNAs. A potential viral encoded helicase is the RdRp itself, since 3D$^{pol}$ was demonstrated to have unwinding activity (Cho et al. 1993). Ehrenfeld and colleagues showed in vitro that 3D$^{pol}$ was able to synthesize nascent RNA on a template that was stably prehybridized with an antisense RNA of 1,000 nucleotides (Cho et al. 1993). The multifunctional 3AB might also act as a helicase because it was shown to have helix-destabilizing activity (DeStefano and Titilope 2006). Although 2C appears to harbor a helicase motif, no helicase activity has been attributed to it (Pfister and Wimmer 1999). An alternative mechanism for formation of the replicative intermediate is through the protein lattice generated by oligomerization of 3D$^{pol}$ (Lyle et al. 2002b). The 3D$^{pol}$ lattice is supported by the membranous vesicles, so potentially one negative-strand template can weave through the stationary 3D$^{pol}$ lattice to simultaneously generate multiple positive-strand RNAs.

The mechanism establishing the replicative intermediate might explain the abundance of positive-strand poliovirus RNA detected over negative-strand RNA during an enterovirus infection. The ratio of positive-strand RNA compared to negative-strand RNA is approximately 40:1 in poliovirus-infected HeLa cells (Giachetti and Semler 1991; Novak and Kirkegaard 1991; Brown et al. 2004). The presence of an asymmetric abundance of positive-strand poliovirus RNA could be due to a mechanistic difference in protein utilization or nucleic acid requirement between the two processes. One possibility is the inefficiency of VPg uridylylation on the 3′ poly(A) tract during negative-strand synthesis compared to VPg uridylylation on the *cre* in positive-strand synthesis (Paul et al. 1998, 2000). Or perhaps, positive-strand synthesis is a more efficient reaction because negative-strand synthesis is rate limiting, requiring translation and processing of the nonstructural proteins and membranous vesicle formation before synthesis. In contrast, positive-strand RNA synthesis has the required components readily available. This phenomenon might also be attributed to enterovirus utilization of the positive- and negative-strand RNAs. The newly synthesized positive-strand RNAs are multifunctional; they can be used as messenger RNAs for translation, templates for negative-strand synthesis, or packaged into new virions, while the negative strand simply acts as a template for positive-strand RNA synthesis.

## 9 Summary

Enterovirus infection of cells is an efficient and productive event. By 2 h after infection, the enterovirus genome is undergoing RNA replication and by 8 h, cultured cells are lysed. The lessons learned form poliovirus RNA replication start from the observation that RNA synthesis is intrinsically linked to viral protein expression. Not only are the viral nonstructural proteins used in the direct replication of genomic RNA, but they also alter host functions such as host gene expression and membrane rearrangement. Poliovirus and coxsackievirus use multiple RNA-protein interactions to mediate important reactions in their life cycle, including cap-independent translation, possible circularization of the genome, protein priming for RNA replication, and synthesis of viral RNA. The conserved mechanism(s) of RNA replication shared by poliovirus and coxsackievirus have allowed investigators to test and confirm important steps in RNA replication between the two viruses. The redundant mechanism of RNA replication has also allowed coxsackievirus researchers to pursue other aspects of infection, such as apoptosis and cell signaling.

**Acknowledgements** We are grateful to Kenneth Ertel, Sarah Daijogo, and Janet Rozovics for critical comments and helpful suggestions on the manuscript. Research in the authors' laboratory is supported by Public Health Service grants AI 26765 and AI 22693 from the National Institutes of Health.

## References

Aldabe R, Carrasco L (1995) Induction of membrane proliferation by poliovirus proteins 2C and 2BC. Biochem Biophys Res Commun 206:64-76
Aldabe R, Barco A, Carrasco L (1996) Membrane permeabilization by poliovirus proteins 2B and 2BC. J Biol Chem 271:23134-23137
Aldabe R, Irurzun A, Carrasco L (1997) Poliovirus protein 2BC increases cytosolic free calcium concentrations. J Virol 71:6214-6217
Ambros V, Baltimore D (1978) Protein is linked to the 5' end of poliovirus RNA by a phosphodiester linkage to tyrosine. J Biol Chem 253:5263-5266

Andino R, Rieckhof GE, Baltimore D (1990) A functional ribonucleoprotein complex forms around the 5′ end of poliovirus RNA. Cell 63:369-380

Baltimore D, Girard M (1966) An intermediate in the synthesis of poliovirus RNA Proc Natl Acad Sci U S A 56:741-748

Baltimore D, Girard M, Darnell JE (1966) Aspects of synthesis of poliovirus RNA and formation of virus particles. Virology 29:179-189

Banerjee R, Echeverri A, Dasgupta A (1997) Poliovirus-encoded 2C polypeptide specifically binds to the 3′-terminal sequences of viral negative-strand RNA. J Virol 71:9570-9578

Barton DJ, Flanegan JB (1997) Synchronous replication of poliovirus RNA: initiation of negative-strand RNA requires the guanidine inhibited activity of protein 2C. J Virol 71:8482-8489

Barton DJ, Morasco BJ, Flanegan JB (1999) Translating ribosomes inhibit poliovirus negative-strand RNA synthesis. J Virol 73:10104-10112

Barton DJ, O'Donnell BJ, Flanegan JB (2001) 5′ cloverleaf in poliovirus RNA is a cis-acting replication element required for negative-strand synthesis. EMBO J 20:1439-1448

Bazan JF, Fletterick RJ (1988) Viral cysteine proteases are homologous to the trypsin-like family of serine proteases: structural and functional implications. Proc Natl Acad Sci U S A 85:7872-7876

Beckman MTL, Kirkegaard K (1998) Site size of cooperative Single-stranded RNA binding by poliovirus RNA-dependent RNA polymerase. J Biol Chem 273:6724-6730

Bell YC, Semler BL, Ehrenfeld E (1999) Requirements for RNA replication of a poliovirus replicon by coxsackievirus B3 RNA polymerase. J Virol 73:9413-9421

Belov GA, Fogg MH, Ehrenfeld E (2005) Poliovirus proteins induce membrane association of GTPase ADP-ribosylation factor. J Virol 79:7207-7216

Belov GA, Ehrenfeld E (2007) Involvement of cellular membrane traffic proteins in poliovirus replication. Cell Cycle 6:36-38

Belov GA, Altan-Bonnet N, Kovtunovych G, Jackson CL, Lippincott-Schwartz J, Ehrenfeld E (2007) Hijacking components of the cellular secretory pathway for replication of poliovirus RNA. J Virol 81:558-567

Bienz K, Egger D, Pfister T, Troxler M (1992) Structural and functional characterization of the poliovirus replication complex. J Virol 66:2740-2747

Blair WS, Parsley TB, Bogerd HP, Towner JS, Semler BL, Cullen BR (1998) Utilization of a mammalian cell-based RNA binding assay to characterize the RNA binding properties of picornavirus 3C proteinases. RNA 4:215-225

Bolten R, Egger D, Gosert R, Schaub G, Landmann L, Bienz K (1998) Intracellular localization of poliovirus plus- and minus-strand RNA visualized by strand-specific fluorescent in situ hybridization. J Virol 72:8578-8585

Brown DM, Kauder SE, Cornell CT, Jang GM, Racaniello VR, Semler BL (2004) Cell-dependent role for the poliovirus 3′ noncoding region in positive-strand RNA synthesis. J Virol 78:1344-1351

Brunner JE, Nguyen JHC, Roehl HH, Ho TV, Swiderek KM, Semler BL (2005) Functional interaction of heterogeneous nuclear ribonucleoprotein C with poliovirus RNA synthesis initiation complexes. J Virol 79:3254-3266

Campanella M, de Jong AS, Lanke KW, Melchers WJ, Willems PH, Pinton P, Rizzuto R, van Kuppeveld F (2004) The coxsackievirus 2B protein suppresses apoptotic host cell responses by manipulating intracellular Ca2+ homeostasis. J Biol Chem 279:18440-18450

Cho MW, Richards OC, Dmitrieva TM, Agol V, Ehrenfeld E (1993) RNA duplex unwinding activity of poliovirus RNA-dependent RNA polymerase 3Dpol. J Virol 67:3010-3018

Cho MW, Teterina N, Egger D, Bienz K, Ehrenfeld E (1994) Membrane rearrangement and vesicle induction by recombinant poliovirus 2C and 2BC in human cells. Virology 202:129-145

Choe SS, Dodd DA, Kirkegaard K (2005) Inhibition of cellular protein secretion by picornaviral 3A proteins. Virology 337:18-29

Clark ME, Lieberman PM, Berk AJ, Dasgupta A (1993) Direct cleavage of human TATA-binding protein by poliovirus protease 3C in vivo and in vitro. Mol Cell Biol 13:1232-1237

Cornell CT, Kiosses WB, Harkins S, Whitton JL (2006) Inhibition of protein trafficking by coxsackievirus B3: multiple viral proteins target a single organelle. J Virol 80:6637-6647

Crawford NM, Baltimore D (1983) Genome-linked protein VPg of poliovirus is present as free VPg and VPg-pUpU in poliovirus-infected cells. Proc Natl Acad Sci U S A 80:7452-7455

Crowder S, Kirkegaard K (2005) Trans-dominant inhibition of RNA viral replication can slow growth of drug-resistant viruses. Nat Genet 37:701-709

Cuconati A, Xiang W, Lahser F, Pfister T, Wimmer E (1998) A protein linkage map of the P2 nonstructural proteins of poliovirus. J Virol 72:1297-1307

DeStefano JJ, Titilope O (2006) Poliovirus protein 3AB displays nucleic acid chaperone and helix-destabilizing activities. J Virol 80:1662-1671

Dodd DA, Giddings TH, Kirkegaard K (2001) Poliovirus 3A protein limits interleukin-6 (IL-6), IL-8, and beta interferon secretion during viral infection. J Virol 758158-75865

Doedens JR, Giddings TH, Kirkegaard K (1997) Inhibition of endoplasmic reticulum-to-Golgi traffic by poliovirus protein 3A: genetic and ultrastructural analysis. J Virol 71:9054-9064

Doedens JR, Kirkegaard K (1995) Inhibition of cellular protein secretion by poliovirus proteins 2B and 3A. EMBO J 14:894-907

Dorsch-Hasler K, Yogo Y, Wimmer E (1975) Replication of picornaviruses. I. Evidence from in vitro RNA synthesis that poly(A) of the poliovirus genome is genetically coded. J Virol 16:1512-1517

Dreyfuss G, Matunis MJ, Pinol-Roma S, Burd CG (1993) hnRNP proteins and the biogenesis of mRNA. Annu Rev Biochem 62:289-321

Echeverri AC, Dasgupta A (1995) Amino terminal regions of poliovirus 2C protein mediate membrane binding. Virology 208:540-553

Etchison D, Milburn SC, Edery I, Sonenberg N, Hershey JW (1982) Inhibition of HeLa cell protein synthesis following poliovirus infection correlates with the proteolysis of a 220,000-dalton polypeptide associated with eucaryotic initiation factor 3 and a cap binding protein complex. J Biol Chem 257:14806-14810

Ferrer-Orta C, Arias A, Perez-Luque R, Escarmis C, Domingo E, Verdaguer N (2004) Structure of foot-and-mouth disease virus RNA-dependent RNA polymerase and its complex with a template-primer RNA. J Biol Chem 279:47212-47221

Flanegan JB, Baltimore D (1977) Poliovirus-specific primer-dependent RNA polymerase able to copy poly(A) Proc Natl Acad Sci U S A 74:3677-3680

Flanegan JB, Pettersson RF, Ambros V, Hewlett NJ, Baltimore D (1977) Covalent linkage of a protein to a defined nucleotide sequence at the 5′-terminus of virion and replicative intermediate RNAs of poliovirus. Proc Natl Acad Sci U S A 74:961-965

Flint SJ, Enquist LW, Racaniello VR, Skalka AM (2004) Principles of virology: molecular biology pathogenesis, control of animal viruses, 2$^{nd}$ edn. ASM Press, Washington DC

Fogg MH, Teterina NL, Ehrenfeld E (2003) Membrane requirements for uridylylation of the poliovirus VPg protein and viral RNA synthesis in vitro. J Virol 77:11408-11416

Fujita K, Krishnakumar SS, Franco D, Paul AV, London E, Wimmer E (2007) Membrane topography of the hydrophobic anchor sequence of poliovirus 3A and 3AB proteins and the functional effect of 3A/3AB membrane association upon RNA replication. Biochemistry 46:5185-5199

Gamarnik AV, Andino R (1997) Two functional complexes formed by KH domain containing proteins with the 5′ noncoding region of poliovirus RNA. RNA 3:882-892

Gamarnik AV, Andino R (1998) Switch from translation to RNA replication in a positive-stranded RNA virus. Genes Dev 12:2293-2304

Gamarnik AV, Andino R (2000) Interactions of viral protein 3CD and poly(rC) binding protein with the 5′ untranslated region of the poliovirus genome. J Virol 74:2219-2226

Giachetti C, Hwang SS, Semler BL (1992) cis-acting lesions targeted to the hydrophobic domain of a poliovirus membrane protein involved in RNA replication. J Virol 66:6045-6057

Giachetti C, Semler BL (1991) Role of a viral membrane polypeptide in strand-specific initiation of poliovirus RNA synthesis. J Virol 65:2647-2654

Goodfellow I, Chaudhry Y, Richardson A, Meredith J, Almond JW, Barclay W, Evans DJ (2000) Identification of a cis-acting replication element within the poliovirus coding region. J Virol 74:4590-4600

Goodfellow IG, Kerrigan D, Evans DJ (2003) Structure and function analysis of the poliovirus cis-acting replication element (CRE). RNA 9:124-137

Goodfellow IG, Polacek C, Andino R, Evans DJ (2003) The poliovirus 2C cis-acting replication element-mediated uridylylation of VPg is not required for synthesis of negative-sense genomes. J Gen Virol 84:2359-2363

Gorbalenya AE, Koonin EV, Wolf YI (1990) A new superfamily of putative NTP-binding domains encoded by genomes of small DNA and RNA viruses. FEBS Lett 262:145-148

Gorlach M, Burd CG, Dreyfuss G (1994) The mRNA poly(A)-binding protein: localization, abundance, RNA-binding specificity. Exp Cell Res 211:400-407

Gradi A, Svitkin YV, Imataka H, Sonenberg N (1998) Proteolysis of human eukaryotic translation initiation factor eIF4GII, but not eIF4GI, coincides with the shutoff of host protein synthesis after poliovirus infection. Proc Natl Acad Sci U S A 95:11089-11094

Grange T, de Sa CM, Oddos J, Pictet R (1987) Human mRNA polyadenylate binding protein: evolutionary conservation of a nucleic acid binding motif. Nucleic Acids Res 15:4771-4787

Hansen JL, Long AM, Schultz SC (1997) Structure of the RNA-dependent RNA polymerase of poliovirus. Structure 5:1109-1122

Harris KS, Reddigari SR, Nicklin MJ, Hammerle T, Wimmer E (1992) Purification and characterization of poliovirus polypeptide 3CD, a proteinase and a precursor for RNA polymerase. J Virol 66:7481-7489

Harris KS, Xiang W, Alexander L, Lane WS, Paul AV, Wimmer E (1994) Interaction of poliovirus polypeptide 3CDpro with the 5' and 3' termini of the poliovirus genome. Identification of viral and cellular cofactors needed for efficient binding. J Biol Chem 269:27004-27014

Hellen CU, Sarnow P (2001) Internal ribosome entry sites in eukaryotic mRNA molecules. Genes Dev 15:1593-1612

Herold J, Andino R (2001) Poliovirus RNA replication requires genome circularization through a protein-protein bridge. Mol Cell 7:581-591

Hobson SD, Rosenblum ES, Richards OC, Richmond K, Kirkegaard K, Schultz SC (2001) Oligomeric structures of poliovirus polymerase are important for function. EMBO J 20:1153-1163

Holcik M, Gordon BW, Korneluk RG (2003) The internal ribosome entry site-mediated translation of antiapoptotic protein XIAP is modulated by the heterogeneous nuclear ribonucleoproteins C1 and C2. Mol Cell Biol 23:280-288

Hope DA, Diamond SE, Kirkegaard K (1997) Genetic dissection of interaction between poliovirus 3D polymerase and viral protein 3AB. J Virol 71:9490-9498

Hunziker IP, Cornell CT, Whitton JL (2007) Deletions within the 5'UTR of coxsackievirus B3: consequences for virus translation and replication. Virology 360:120-128

Jackson WT, Giddings TH, Taylor MP, Mulinyawe S, Rabinovitch M, Kopito RR, Kirkegaard K (2005) Subversion of cellular autophagosomal machinery by RNA viruses. PLoS Biol 3:861-871

Jacobson SJ, Konings DA, Sarnow P (1993) Biochemical and genetic evidence for a pseudoknot structure at the 3' terminus of the poliovirus RNA genome and its role in viral RNA amplification. J Virol 67:2961-2971

Jang SK, Krausslich HG, Nicklin MJ, Duke GM, Palmenberg AC, Wimmer E (1988) A segment of the 5' nontranslated region of encephalomyocarditis virus RNA directs internal entry of ribosomes during in vitro translation. J Virol 62:2636-2643

Joachims M, Van Breugel PC, Lloyd RE (1999) Cleavage of poly(A)-binding protein by enterovirus proteases concurrent with inhibition of translation in vitro. J Virol 73:718-727

Johnson KL, Sarnow P (1991) Three poliovirus 2B mutants exhibit noncomplementable defects in viral RNA amplification and display dosage-dependent dominance over wild-type poliovirus. J Virol 65:4341-4349

Johnson VH, Semler BL (1988) Defined recombinants of poliovirus and coxsackievirus: sequence-specific deletions and functional substitutions in the 5′-noncoding regions of viral RNAs. 162:47-57

Jurgens CK, Barton DJ, Sharma N, Morasco BJ, Ogram SA, Flanegan JB (2006) 2A$^{pro}$ is a multifunctional protein that regulates the stability, translation and replication of poliovirus RNA. Virology 345:346-357

Kerekatte V, Keiper BD, Badorff C, Cai A, Knowlton KU, Rhoads RE (1999) Cleavage of Poly(A)-binding protein by coxsackievirus 2A protease in vitro and in vivo: another mechanism for host protein synthesis shutoff? J Virol 73:709-717

Kim KS, Hufnagel G, Chapman NM, Tracy S (2001) The group B coxsackieviruses and myocarditis. Rev Med Virol 11:355-368

Kim KS, Tracy S, Tapprich W, Bailey J, Lee CK, Kim K, Barry WH, Chapman NM (2005) 5′-Terminal deletions occur in coxsackievirus B3 during replication in murine hearts and cardiac myocyte cultures and correlate with encapsidation of negative-strand viral RNA. J Virol 79:7024-7041

Kitamura N, Semler BL, Rothberg PG, Larsen GR, Adler CJ, Dorner AJ, Emini EA, Hanecak R, Lee JJ, van der Werf S, Anderson CW, Wimmer E (1981) Primary structure, gene organization and polypeptide expression of poliovirus RNA. Nature 291:547-553

König H, Rosenwirth B (1988) Purification and partial characterization of poliovirus protease 2A by means of a functional assay. J Virol 62:1243-1250

Kräusslich HG, Nicklin MJ, Toyoda H, Etchison D, Wimmer W (1987) Poliovirus proteinase 2A induces cleavage of eucaryotic initiation factor 4F polypeptide p220. J Virol 61:2711-2718

Kuyumcu-Martinez NM, Joachims M, Lloyd RE (2002) Efficient cleavage of ribosome-associated Poly(A)-binding protein by enterovirus 3C protease. J Virol 76:2062-2074

Kuyumcu-Martinez NM, Van Eden ME, Younan P, Lloyd RE (2004) Cleavage of poly(A)-binding protein by poliovirus 3C protease inhibits host cell translation: a novel mechanism for host translation shutoff. Mol Cell Biol 2004:24:1779-1790

Lama J, Paul AV, Harris KS, Wimmer E (1994) Properties of purified recombinant poliovirus protein 3AB as substrate for viral proteinases and as co-factor for RNA polymerase 3Dpol. J Biol Chem 269:66-70

Lawson MA, Semler BL (1991) Poliovirus thiol proteinase 3C can utilize a serine nucleophile within the putative catalytic triad. Proc Natl Acad Sci U S A 88:9919-9923

Lee YF, Nomoto A, Detjen BM, Wimmer E (1977) A protein covalently linked to poliovirus genome RNA. Proc Natl Acad Sci U S A 74:59-63

Leong LE-C, Cornell CT, Semler BL (2002) Processing determinants and functions of cleavage products of picornavirus polyproteins. In: Semler BL and Wimmer E (eds) Molecular biology of picornaviruses. ASM Press Washington DC, pp 187-97

Liebig HD, Seipelt J, Vassilieva E, Gradi A, Kuechler E (2002) A thermosensitive mutant of HRV2:2A proteinase: evidence for direct cleavage of eIF4GI and eIF4GII. FEBS Lett 523:53-57

Lindberg AM, Stalhandske PO, Pettersson U (1987) Genome of coxsackievirus B3. Virology 156:50-63

Liu Y, Franco D, Paul AV, Wimmer E (2007) Tyrosine 3 of poliovirus terminal peptide VPg(3B) has an essential function in RNA replication in the context of its precursor protein 3AB. J Virol 81:5669-5684

Lloyd RE, Jense HG, Ehrenfeld E (1987)Restriction of translation of capped mRNA in vitro as a model for poliovirus-induced inhibition of host cell protein synthesis: relationship to p220 cleavage. J Virol 61:2480-2488

Lyle JM, Clewell A, Richmond K, Richards OC, Hope DA, Schultz SC, Kirkegaard K (2002a) Similar structural basis for membrane localization and protein priming by an RNA-dependent RNA polymerase. J Biol Chem 277:16324-16331

Lyle JM, Bullitt E, Bienz K, Kirkegaard K (2002b) Visualization and functional analysis of RNA-dependent RNA polymerase lattices. Science 296:2218-2222

Lyons T, Murray KE, Roberts AW, Barton DJ (2001) Poliovirus 5′-terminal cloverleaf RNA is required in cis for VPg uridylylation and the initiation of negative-strand RNA synthesis. J Virol 75:10696-10708

Makeyev AV, Liebhaber SA (2002) The poly(C)-binding proteins: a multiplicity of functions and a search for mechanisms. RNA 8:265-278

Marcotte LL, Wass AB, Gohara DW, Pathak HB, Arnold JJ, Filman DJ, Cameron CE, Hogle JM (2007) Crystal structure of poliovirus 3CD protein: virally encoded protease and precursor to the RNA-dependent RNA polymerase. J Virol 81:3583-3596

Mathews DH, Sabina J, Zuker M, Turner DH (1999) Expanded sequence dependence of thermodynamic parameters improves prediction of RNA secondary structure. J Mol Biol 288:911-940

Maynell LA, Kirkegaard K, Klymkowsky MW (1992) Inhibition of poliovirus RNA synthesis by brefeldin A. J Virol 66:1985-1994

McKnight KL, Lemon SM (1998) The rhinovirus type 14 genome contains an internally located RNA structure that is required for viral replication. RNA 4:1569-1584

Melchers WJ, Hoenderop JG, Bruins Slot HJ, Pleij CW, Pilipenko EV, Agol VI, Galama JM (1997) Kissing of the two predominant hairpin loops in the coxsackie B virus 3′ untranslated region is the essential structural feature of the origin of replication required for negative-strand RNA synthesis. J Virol 71:686-696

Merkle I, van Ooij MJ, van Kuppeveld FJ, Glaudemans DH, Galama JM, Henke A, Zell R, Melchers WJ (2002) Biological significance of a human enterovirus B-specific RNA element in the 3′ nontranslated region. J Virol 76:9900-9909

Merrick WC (1990) Overview: mechanism of translation initiation in eukaryotes. Enzyme 44:7-16

Mirmomeni MH, Hughes PJ, Stanway G (1997) An RNA tertiary structure in the 3′ untranslated region of enteroviruses is necessary for efficient replication. J Virol 71:2363-2370

Mirzayan C, Wimmer E (1994) Biochemical studies on poliovirus polypeptide 2C: evidence for ATPase activity. Virology 199:176-187

Molla A, Paul AV, Wimmer E (1991) Cell-free, de novo synthesis of poliovirus. Science 254:1647-1651

Molla A, Paul AV, Wimmer E (1993) Effects of temperature and lipophilic agents on poliovirus formation and RNA synthesis in a cell-free system. J Virol 67:5932-5938

Molla A, Harris KS, Paul AV, Shin SH, Mugavero J, Wimmer E (1994) Stimulation of poliovirus proteinase 3Cpro-related proteolysis by the genome-linked protein VPg and its precursor 3AB. J Biol Chem 269:27015-27020

Morasco BJ, Sharma N, Parilla J, Flanegan JB (2003) Poliovirus cre(2C)-dependent synthesis of VPgpUpU is required for positive- but not negative-strand RNA synthesis. J Virol 77:5136-5144

Morinaga N, Tsai SC, Moss J, Vaughan M (1996) Isolation of a brefeldin A-inhibited guanine nucleotide-exchange protein for ADP ribosylation factor (ARF) 1 and ARF3 that contains a Sec7-like domain. Proc Natl Acad Sci U S A 93:12856-12860

Mosimann SC, Cherney MM, Sia S, Plotch S, James MN (1997) Refined X-ray crystallographic structure of the poliovirus 3C gene product. J Mol Biol 273:1032-1047

Murray KE, Barton DJ (2003) Poliovirus CRE-dependent VPg uridylylation is required for positive-strand RNA synthesis but not for negative-strand RNA synthesis. J Virol 77:4739-4750

Murray KE, Roberts AW, Barton DJ (2001) Poly(rC) binding proteins mediate poliovirus mRNA stability. RNA 7:1126-1141

Murray KE, Steil BP, Roberts AW, Barton DJ (2004) Replication of poliovirus RNA with complete internal ribosome entry site deletions. J Virol 78:1393-1402

Nomoto A, Detjen B, Pozzatti R, Wimmer E (1977) The location of the polio genome protein in viral RNAs and its implication for RNA synthesis. Nature 268:208-213

Novak JE, Kirkegaard K (1991) Improved method for detecting poliovirus negative strands used to demonstrate specificity of positive-strand encapsidation and the ratio of positive to negative strands in infected cells. J Virol 65:3384-3387

Novak JE, Kirkegaard K (1994) Coupling between genome translation and replication in an RNA virus. Genes Dev 8:1726-1737

Ostareck DH, Ostareck-Lederer A, Wilm M, Thiele BJ, Mann M, Hentze MW (1997) mRNA silencing in erythroid differentiation: hnRNP K and hnRNP E1 regulate 15-lipoxygenase translation from the 3' end. Cell 89:597-606

Otero LJ, Ashe MP, Sachs AB (1999) The yeast poly(A)-binding protein Pab1p stimulates in vitro poly(A)-dependent and cap-dependent translation by distinct mechanisms. EMBO J 18:3153-3163

Parsley TB, Towner JS, Blyn LB, Ehrenfeld E, Semler BL (1997) Poly (rC) binding protein 2 forms a ternary complex with the 5'- terminal sequences of poliovirus RNA and the viral 3CD proteinase. RNA 3:1124-1134

Paul A (2002) Possible unifying mechanism of picornavirus genome replication. In: Semler BL Wimmer E (eds) Molecular biology of picornaviruses. ASM Press Washington DC, pp 227-246

Paul AV, Molla A, Wimmer E (1994) Studies of a putative amphipathic helix in the N-terminus of poliovirus protein 2C. Virology 199:188-199

Paul AV, Peters J, Mugavero J, Yin J, van Boom JH, Wimmer E (2003) Biochemical and genetic studies of the VPg uridylylation reaction catalyzed by the RNA polymerase of poliovirus. J Virol 77:891-904

Paul AV, Rieder E, Kim DW, van Boom JH, Wimmer E (2000) Identification of an RNA hairpin in poliovirus RNA that serves as the primary template in the in vitro uridylylation of VPg. J Virol 74:10359-10370

Paul AV, van Boom JH, Filippov D, Wimmer E (1998) Protein-primed RNA synthesis by purified poliovirus RNA polymerase. Nature 393:280-284

Pelletier J, Sonenberg N (1988) Internal initiation of translation of eukaryotic mRNA directed by a sequence derived from poliovirus RNA. Nature 334:320-325

Perera R, Daijogo S, Walter BL, Nguyen JH, Semler BL (2007) Cellular protein modification by poliovirus: the two faces of poly(rC)-binding protein. J Virol 81:8919-8932

Pettersson RF, Ambros V, Baltimore D (1978) Identification of a protein linked to nascent poliovirus RNA and to the polyuridylic acid of negative-strand RNA J Virol 27:357-365

Pfister T, Jones KW, Wimmer E (2000) A cysteine-rich motif in poliovirus protein 2C(ATPase) is involved in RNA replication and binds zinc in vitro. J Virol 74:334-343

Pfister T, Wimmer E (1999) Characterization of the nucleoside triphosphatase activity of poliovirus protein 2C reveals a mechanism by which guanidine inhibits poliovirus replication. J Biol Chem 274:6992-7001

Pilipenko EV, Maslova SV, Sinyakov AN, Agol VI (1992) Towards identification of cis-acting elements involved in the replication of enterovirus and rhinovirus RNAs: a proposal for the existence of tRNA-like terminal structures. Nucleic Acids Res 20:1739-1745

Pilipenko EV, Poperechny K, Maslova SV, Melchers WJG, Bruins Slot HJ, Agol VI (1996) Cis-element, oriR, involved in the initiation of (-) strand poliovirus RNA: a quasi-globular multi-domain RNA structure maintained by tertiary ('kissing') interactions. EMBO J 15:5428-5436

Pincus SE, Diamond DC, Emini EA, Wimmer E (1986) Guanidine-selected mutants of poliovirus: mapping of point mutations to polypeptide 2C. J Virol 57:638-646

Plotch SJ, Palant O (1995) Poliovirus protein 3AB forms a complex with and stimulates the activity of the viral RNA polymerase, 3Dpol. J Virol 69:7169-7179

Racaniello VR, Baltimore D (1981) Molecular cloning of poliovirus cDNA and determination of the complete nucleotide sequence of the viral genome. Proc Natl Acad Sci U S A 78:4887-4891

Richards OC, Ehrenfeld E (1998) Effects of poliovirus 3AB protein on 3D polymerase-catalyzed reaction. J Biol Chem 273:12832-12840

Richards OC, Spagnolo JF, Lyle JM, Vleck SE, Kuchta RD, Kirkegaard K (2006) Intramolecular and intermolecular uridylylation by poliovirus RNA-dependent RNA polymerase. J Virol 80:7405-7415

Rieder E, Paul AV, Kim DW, van Boom JH, Wimmer E (2000) Genetic and biochemical studies of poliovirus cis-acting replication element cre in relation to VPg uridylylation. J Virol 74:10371-10380

Rodriguez PL, Carrasco L (1995) Poliovirus protein 2C contains two regions involved in RNA binding activity. J Biol Chem 270:10105-10112

Roehl HH, Semler BL (1995) Poliovirus infection enhances the formation of two ribonucleoprotein complexes at the 3' end of viral negative-strand RNA. J Virol 69:2954-2961

Rothberg PG, Harris TJ, Nomoto A, Wimmer E (1978) O4-(5'-uridylyl)tyrosine is the bond between the genome-linked protein and the RNA of poliovirus. Proc Natl Acad Sci U S A 75:4868-4872

Rust RC, Landmann L, Gosert R, Tang BL, Hong W, Hauri HP, Egger D, Bienz K (2001) Cellular COPII proteins are involved in production of the vesicles that form the poliovirus replication complex. J Virol 75:9808-9818

Sandoval IV, Carrasco L (1997) Poliovirus infection and expression of the poliovirus protein 2B provoke the disassembly of the Golgi complex, the organelle target for the antipoliovirus drug Ro-090179. J Virol 71:4679-4693

Sarnow P (1989) Role of 3'-end sequences in infectivity of poliovirus transcripts made in vitro. J Virol 63:467-470

Semler BL, Anderson CW, Hanecak R, Dorner LF, Wimmer E (1982) A membrane-associated precursor to poliovirus VPg identified by immunoprecipitation with antibodies directed against a synthetic heptapeptide. Cell 28:405-412

Semler BL, Johnson VH, Tracy S (1986) A chimeric plasmid from cDNA clones of poliovirus and coxsackievirus produces a recombinant virus that is temperature-sensitive. Proc Natl Acad Sci U S A 83:1777-1781

Schlegel A, Giddings TH, Ladinsky MS, Kirkegaard K (1996) Cellular origin and ultrastructure of membranes induced during poliovirus infection. J Virol 70:6576-6588

Sharma R, Raychaudhuri S, Dasgupta A (2004) Nuclear entry of poliovirus protease-polymerase precursor 3CD: implications for host cell transcription shut-off. Virology 320:195-205

Sharma N, O'Donnell BJ, Flanegan JB (2005) 3'-Terminal sequence in poliovirus negative-strand templates is the primary cis-acting element required for VPgpUpU-primed positive-strand initiation. J Virol 79:3565-3577

Silvestri LS, Parilla JM, Morasco BJ, Ogram SA, Flanegan JB (2006) Relationship between poliovirus negative-strand RNA synthesis and the length of the 3' poly(A) tail. Virology 345:509-519

Spector DH, Baltimore D (1975) Polyadenylic acid on poliovirus RNA II poly(A) on intracellular RNAs. J Virol 15:1418-1431

Suhy DA, Giddings TH Jr, Kirkegaard K (2000) Remodeling the endoplasmic reticulum by poliovirus infection and by individual viral proteins: an autophagy-like origin for virus-induced vesicles. J Virol 74:8953-8965

Takegami T, Kuhn RJ, Anderson CW, Wimmer E (1983) Membrane-dependent uridylylation of the genome-linked protein VPg of poliovirus. Proc Natl Acad Sci U S A 80:7447-7451

Tarun SZ, Sachs AB (1996) Association of the yeast poly(A) tail binding protein with translation initiation factor eIF-4G. EMBO J 15:7168-7177

Tellez AB, Crowder S, Spagnolo JF, Thompson AA, Peersen OB, Brutlag DL, Kirkegaard K (2006) Nucleotide channel of RNA-dependent RNA polymerase used for intermolecular uridylylation of protein primer. J Mol Biol 357:665-675

Tershak DR (1984) Association of poliovirus proteins with the endoplasmic reticulum. J Virol 52:777-783

Teterina NL, Egger D, Bienz K, Brown DM, Semler BL, Ehrenfeld E (2001) Requirements for assembly of poliovirus replication complexes and negative-strand RNA synthesis. J Virol 75:3841-3850

Teterina NL, Gorbalenya AE, Egger D, Bienz K, Ehrenfeld E (1997) Poliovirus 2C protein determinants of membrane binding and rearrangements in mammalian cells. J Virol 71:8962-8972

Thompson AA, Peersen OB (2004) Structural basis for proteolysis-dependent activation of the poliovirus RNA-dependent RNA polymerase. EMBO J 23:3462-3471

Todd S, Towner JS, Brown DM, Semler BL (1997) Replication-competent picornaviruses with complete genomic RNA 3' noncoding region deletions. J Virol 71:8868-8874

Towner JS, Ho TV, Semler BL (1996) Determinants of membrane association for poliovirus protein 3AB. J Biol Chem 271:26810-26818

Towner JS, Mazanet MM, Semler BL (1998) Rescue of defective poliovirus RNA replication by 3AB-containing precursor polyproteins. J Virol 72:7191-7200

Toyoda H, Nicklin MJ, Murray MG, Anderson CW, Dunn JJ, Studier FW, Wimmer E (1986) A second virus-encoded proteinase involved in proteolytic processing of poliovirus polyprotein. Cell 45:761-770

Van Dyke TA, Flanegan JB (1980) Identification of poliovirus polypeptide P63 as a soluble RNA-dependent RNA polymerase. J Virol 35:732-740

van Kuppeveld FJ, Hoenderop JG, Smeets RL, Willems PH, Dijkman HB, Galama JM, Melchers WJ (1997a) Coxsackievirus protein 2B modifies endoplasmic reticulum membrane and plasma membrane permeability and facilitates virus release. EMBO J 16:3519-3532

van Kuppeveld FJ, Melchers WJ, Kirkegaard K, Doedens JR (1997b) Structure-function analysis of coxsackie B3 virus protein 2B. Virology 227:111-118

Van Ooij MJ, Vogt DA, Paul A, Castro C, Kuijpers J, van Kuppeveld FJ, Cameron CE, Wimmer E, Andino R, Melchers WJ (2006) Structural and functional characterization of the coxsackievirus B3 CRE(2C): role of CRE(2C) in negative- and positive-strand RNA synthesis. J Gen Virol 87:103-113

Van Ooij MJ, Polacek C, Glaudemans DH, Kuijpers J, van Kuppeveld FJ, Andino R, Agol VI, Melchers WJ (2006) Polyadenylation of genomic RNA and initiation of antigenomic RNA in a positive-strand RNA virus are controlled by the same cis-element. Nucleic Acids Res 34:2953-2965

Walter BL, Parsley TB, Ehrenfeld E, Semler BL (2002) Distinct poly(rC) binding protein KH domain determinants for poliovirus translation initiation and viral RNA replication. J Virol 76:12008-12022

Weiss IM, Liebhaber SA (1994) Erythroid cell-specific determinants of alpha-globin mRNA stability. Mol Cell Biol 14:8123-8132

Wessels E, Notebaart RA, Duijsings D, Lanke K, Vergeer B, Melchers WJ, van Kuppeveld FJ (2006) Structure-function analysis of the coxsackievirus protein 3A: identification of residues important for dimerization, viral RNA replication, and transport inhibition. J Biol Chem 281:28232-28243

Xiang W, Cuconati A, Hope D, Kirkegaard K, Wimmer E (1998) Complete protein linkage map of poliovirus P3 proteins: interaction of polymerase 3Dpol with VPg and with genetic variants of 3AB. J Virol 72:6732-6741

Xiang W, Harris K S, Alexander L, Wimmer E (1995) Interaction between the 5'-terminal cloverleaf and 3AB/3CDpro of poliovirus is essential for RNA replication. J Virol 69:3658-3667

Yalamanchili P, Banerjee R, Dasgupta A (1997) Poliovirus-encoded protease 2APro cleaves the TATA-binding protein but does not inhibit host cell RNA polymerase II transcription in vitro. J Virol 71:6881-6886

Yogo Y, Wimmer E (1972) Polyadenylic acid at the 3'-terminus of poliovirus RNA Proc Natl Acad Sci U S A 69:1877-1882

Yogo Y, Wimmer E (1973) Poly (A) and poly (U) in poliovirus double stranded RNA Nat New Biol 242:171-174

Ypma-Wong MF, Dewalt PG, Johnson VH, Lamb JG, Semler BL (1988) Protein 3CD is the major poliovirus proteinase responsible for cleavage of the P1 capsid precursor. Virology 166:265-270

Yu SF, Benton P, Bovee M, Sessions J, Lloyd RE (1995) Defective RNA replication by poliovirus mutants deficient in 2A protease cleavage activity. J Virol 69:247-252

Zell R, Sidigi K, Bucci E, Stelzner A, Gorlach M (2002) Determinants of the recognition of enteroviral cloverleaf RNA by coxsackievirus B3 proteinase 3C. RNA 8:188-201

Zuker M (2003) Mfold web server for nucleic acid folding and hybridization prediction. Nucleic Acids Res 31:3406-3415

# CVB Translation: Lessons from the Polioviruses

J. M. Bonderoff and R. E. Lloyd(✉)

| | | |
|---|---|---|
| 1 | Introduction | 124 |
| 2 | Poliovirus and Coxsackievirus IRES Structure and Function | 124 |
| 3 | Proteins Required for Enteroviral IRES-Mediated Translation | 127 |
| | 3.1 Canonical Translation Factors | 127 |
| | 3.2 Noncanonical Translation Factors | 130 |
| 4 | 5′-3′ Interactions in Viral Translation | 132 |
| 5 | Shutoff of Cellular Translation | 133 |
| 6 | Shutoff of Viral Translation | 136 |
| 7 | Cell-Specific Restriction of Translation | 138 |
| 8 | Concluding Remarks | 140 |
| References | | 140 |

**Abstract** Our understanding of coxsackie B virus translation and replication has benefited greatly from half a century of research on the closely related polioviruses. Like poliovirus, coxsackievirus gene expression is controlled largely at the translation level and coxsackievirus infection results in profound changes in the profile of mRNAs with access to the protein synthesis machinery of the host cell. This review chronicles the advances in understanding translational control by the enteroviruses, primarily in poliovirus and clarified by related viruses, and highlights areas where coxsackievirus conforms to or differs from the aggregate model. Basic IRES structure and function, proteins involved in cap-dependent and viral translation, viral modification of translation factors to achieve host translation shutoff and promotion of viral translation are discussed. The translational bases for neurovirulent phenotypes and tissue specificity are also addressed.

R. E. Lloyd
Department of Molecular Virology and Microbiology, Baylor College of Medicine,
One Baylor Plaza, 77030, Houston, TX
rlloyd@bcm.tmc.edu

# 1 Introduction

The encapsidated enteroviral genome is a single-stranded plus sense RNA approximately 7,500 nt in length that contains a 3' poly(A) tail varying from roughly 65 to 100 nt in length. The virus capsid serves to deliver the genomic RNA to the cytoplasm of the cell where the very first step in viral replication is translation of the viral RNA to produce all viral proteins. Thus, the viral genome is at first a functional messenger RNA and must attract ribosomes. Enterovirus RNAs do not contain m7GpppG cap structures at their 5' termini; rather, they are covalently bound by a small VPg peptide. In infected cells, poliovirus (PV) RNA found on polyribosomes does not contain VPg, which is removed by an unidentified cellular unlinking activity (Ambros and Baltimore 1980; Nomoto et al. 1977). Despite this finding, viral RNA containing VPg translates as well as RNA lacking it and no direct function for VPg in translation has been reported.

Most cellular RNA translates in a cap-dependent translation mechanism utilizing a complex of protein initiation factors that assemble at the 5'-cap structure. In contrast, enteroviral RNAs bind ribosomes internally to a large RNA motif known as an internal ribosome entry sequence (IRES). Some canonical and noncanonical initiation factors bind this structure to facilitate ribosome binding and translation initiation. The viral protein translation products include two proteinases that cleave the viral polyprotein and also target key host cell proteins. Cleavage of host proteins disrupts several cellular processes, including a profound inhibition of cellular translation in favor of promotion of viral translation. This translation switch occurs within 2 h of infection; however, at later time points (e.g., 5 h postinfection), viral translation is also inhibited.

Most of these events were originally characterized in poliovirus-infected cells but have been observed in other members of the Picornaviridae, especially coxsackie B viruses (CVB) and rhinoviruses (HRV). As the critical molecular events in this phenomenon were elucidated, the knowledge gained from experiments from poliovirus has been applied to related viruses to reinforce similarities and discover some differences in modes of action. This chapter will describe features of viral RNA and its protein products that regulate viral and cellular translation in poliovirus and coxsackievirus infection.

# 2 Poliovirus and Coxsackievirus IRES Structure and Function

The major features of PV and CVB RNA that regulate translation are the IRES and poly(A) tail. The uncapped enterovirus genome is protected from exonuclease degradation at the 5' end by a small cloverleaf structure that provides a binding site for poly(rC)-binding protein 2 (PCBP2) (Murray et al. 2001). Translation is driven by an approximately 490-nt internal ribosome entry site (IRES) within the 743-nt 5' untranslated region of the viral RNA. Most picornaviral IRESs have been divided into four classifications based on homology, secondary structure, and other criteria. Type I IRESs include those of poliovirus, rhinovirus, coxsackievirus, and other enteroviruses. Type II IRESs include those of foot-and-mouth disease virus (FMDV), cardioviruses such as encephalomyocarditis virus (EMCV), paraechoviruses, and kobuvirus. Type III IRESs include hepatitis A virus. Newly classified teschovirus IRES is most similar to hepatitis C virus, which is not a picornavirus, and thus may represent an ancient recombination event between an enterovirus and hepacivirus or pestivirus (Pisarev et al. 2004). Overall, there is very little similarity in sequence or secondary structure between these types of IRESs, though there is high conservation within the type designation. Although tertiary structures of large RNA molecules are difficult to predict or solve, the secondary structure of poliovirus 5′ UTR has been mapped by chemical means, genetic mapping, and in silico modeling (Le and Zuker 1990; Pilipenko et al. 1989; Rivera et al. 1988; Skinner et al. 1989). Additional evidence for the veracity of the modeled structure and stem loop assignments comes from sequences of independent virus isolates, revealing mutations and genetic drift that maintains structure in stem-loops and sequence divergence in linking regions between domains (Poyry et al. 1992).

The consensus type I IRES structure (based on PV) contains five major stem-loop regions (Fig. 1). IRES boundaries have been determined by deletion analyses to extend roughly from nt 130(5′) to nt 600 (3′). The 5′ terminal cloverleaf structure

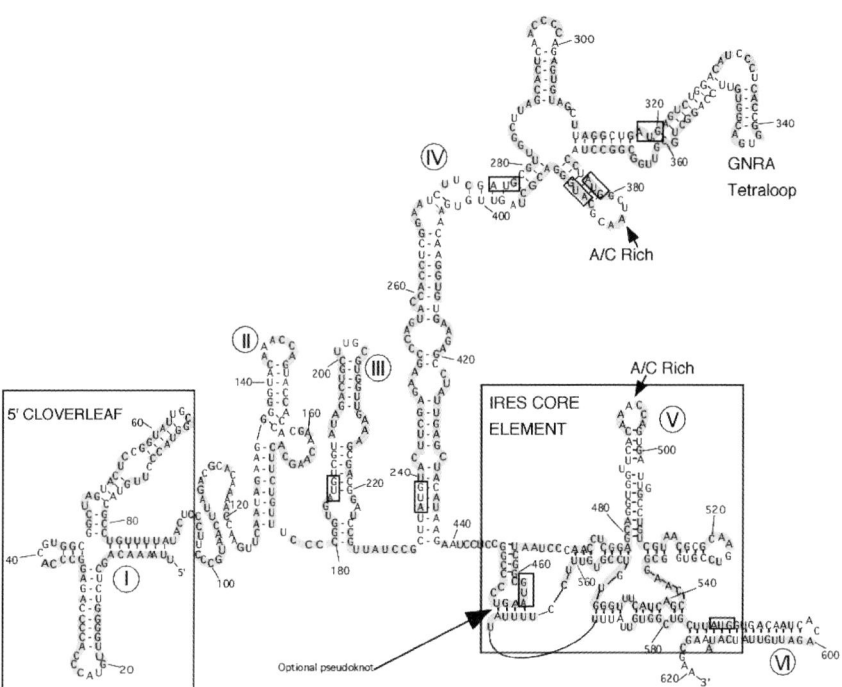

**Fig. 1** Diagram of 5′ UTR of PV type 1 Mahoney strain. The 5′ cloverleaf structure and IRES core elements are boxed. Major IRES stem-loop domain structures I-VI are denoted in circles. Nucleotides that are shaded are conserved in CVB3 Nancy strain. Note that the 3′ unstructured 120 nt of the 5′ UTR are not shown (nt 621-743). The initiator AUG is at nucleotide 743

(domain I) has been shown to be an independent functional unit critical for RNA replication that can be separated from most translation functions. However, one report suggested that the cloverleaf may influence translation, since a mutation that did not affect its RNA stabilizing function attenuated viral translation (Simoes and Sarnow 1991).

Deletion analysis showed domains IV and V to be essential for viral translation, and that domain VI is important for efficient viral translation (Dildine and Semler 1989; Percy et al. 1992). Deletions from the 5' and 3' ends of the UTR also implicate stem-loop II as being necessary for translation (Nicholson et al. 1991). In contrast, deletion of domain III is tolerated in mutant virus (Dildine and Semler 1989) and deletion of both domain III and domain VI results in only a 70% decrease in translation, remarkable in that the mutant IRES is missing 30% of its total length (Haller et al. 1993).

In addition to the conserved structural stem-loop domains, some common sequence motifs have been observed within type I and II IRES elements that have been linked to function. The most notable, in the region linking domains V and VI, is a pyrimidine-rich sequence followed by a 16- to 25-nt spacer that ends in an invariant AUG codon (AUG 586 in PV). (This region lies within the IRES core element shown in Fig. 1.) In type I IRES elements, this is not the initiator codon; however, the next downstream AUG is the initiator. In type II IRESs, the AUG within the core serves as the initiator. This combined Py-xx-AUG region has been proposed to act analogously to a Shine-Dalgarno-like sequence that functions in direct ribosome binding. In this case, paired interaction of the invariant AUG with the anticodon loop of met-tRNA$_1$ and 18S-rRNA pairing with the pyrimidine-rich region are likely to occur (Meerovitch et al. 1991; Pilipenko et al. 1992). In PV, mutational analysis and viral reversion demonstrated that the distance between the pyrimidine box and AUG must not vary beyond 20-23 nt (Gmyl et al. 1993; Pilipenko et al. 1992).

Other IRES motifs include GNRA tetra-loops that may mediate long-range RNA-RNA interactions and A/C rich loops in domains IV and V that stimulate IRES activity (Nicholson et al. 1991) and may participate in a pseudoknot with domain V (Le et al. 1992). A similar GNRA tetra-loop is also found in type II IRES elements (Robertson et al. 1999).

A combination of phylogenetic analysis and computer folding simulation revealed a common tertiary core structure found at the 3' end of all picornavirus IRES elements. This approximately 150-nt region involves stem-loop V and part of stem-loop VI, and the pyrimidine tract adjacent to the invariant AUG 586, which folds into a conserved double pseudoknot (Le and Maizel 1998) (Fig. 1; IRES core element). An additional optional pseudoknot was noted in group A and C picornaviruses. Interestingly, this common core structure is also maintained in hepatitis C virus and pestivirus IRESs (Le et al. 1996).

Comparatively little analysis has been performed on CVB IRES elements. However, most, if not all major aspects of the IRES that allow interaction with initiation factors and ribosomes are expected to be similar to PV because of the high sequence identity and phylogenetic conservation of structural motifs. Surprisingly, one report of deletion-based boundary mapping of the CVB3 IRES in full length virus constructs suggested a much shorter minimal functional IRES existed compared to PV IRES, and suggested that domain IV-V deletion mutants retained lower IRES expression. The reported minimal IRES encompassed mostly domains V and VI, though no direct translation assays were performed in this study (Liu et al. 1999). A subsequent study reported a base-pairing interaction between the oligopyrimidine stretch (~nt 558-578) and 18S rRNA that was crucial for CVB3 translation (Yang et al. 2003), similar to earlier reports with PV.

For more detailed information and other perspectives about IRES structure and function, refer to the chapters by P. Sean and B.L. Semler, R. Feuer and J.L. Whitton, this volume and (Ehrenfeld and Tererina 2002; Hellen and Sarnow 2001; Jackson 2005).

## 3 Proteins Required for Enteroviral IRES-Mediated Translation

There is a growing list of cellular polypeptides that have been reported to bind to viral RNA and facilitate viral translation. How many of these are absolutely required for viral translation vs being stimulatory is a somewhat unresolved problem. Early efforts to define proteins involved in cap-independent translation on the various picornaviral IRESs revealed a number of proteins that were identified by UV-crosslinking assays, yet some do not play a role in translation. Thus, the use of proper translation functional assays is critical to discern the important RNA-binding factors. The most stringent and defined functional assays with picornavirus IRESs have analyzed the EMCV IRES with toeprint assays reconstituted with completely purified protein factors and components. These assays revealed a minimal set of canonical and noncanonical factors that were required to bind ribosomes to IRES RNA and complete 48S complexes assembled on the initiation codon. The discussion below summarizes results of these and other assays.

### *3.1 Canonical Translation Factors*

In the normal model of translation of capped cellular mRNAs, 40S ribosomal subunits cannot bind directly to mRNA. Rather, even though two translation initiation factors, eIF3 and eIF2, are prebound to the 40S ribosome, the eIF4 family of initiation factors must also bind to the mRNA in order to successfully deliver mRNA to the ribosome. Both cap-dependent translation and picornavirus IRES-dependent translation also require the initiation factors eIF1, eIF1a, eIF5, and eIF5b, which bind ribosomes and aid in codon-anticodon interactions and 60S ribosomal subunit joining. These four factors will not be discussed further. EMCV and PV IRES elements are similar to capped mRNA in maintaining a requirement for some initiation factors to bind to primed 40S subunits and to the RNA itself. These canonical factors have been determined to be eIF4G and eIF4B that bind the viral RNA, and eIF3 and eIF2 that must prebind the 40S subunit. In addition, poly(A)-binding protein (PABP) significantly enhances IRES-mediated translation of EMCV, and especially of the PV IRES. This rather high requirement of picornaviral IRESs for canonical translation factors is in contrast to IRES elements of other viruses such as hepatitis C virus and cricket paralysis virus, which require fewer or no translation factors to bind ribosomes (Hellen and Sarnow 2001).

Eukaryotic translation initiation factor 4G (eIF4GI, initially called p220) is a large scaffolding protein that was originally identified as a novel polypeptide cleaved during poliovirus infection (Etchison et al. 1982). It is a key component of the mRNA cap-binding complex (eIF4F) and performs a scaffolding function in assembling the eIF4F complex (Grifo et al. 1983; Lamphear and Panniers 1990; Prevot et al. 2003). Other members of eIF4F include the cap-binding protein eIF4E and the DEAD-box ATP-dependent helicase eIF4A (Grifo et al. 1983). In addition to these high-affinity interactions, eIF4G also binds to poly(A)-binding protein (PABP) (Imataka et al. 1998), the eIF3E subunit of eIF3 (Lamphear et al. 1995; Lefebvre et al. 2006) and Mnk1 kinase, which phosphorylates eIF4E (Pyronnet et al. 1999; Waskiewicz et al. 1999). eIF4GI consists of five isoforms, designated a-e, with different N-termini that are derived from alternate promoter usage, alternate splicing, and alternate AUG selection during translation initiation (Byrd et al. 2005). In addition to these multiple isoforms of eIF4GI, a functional paralog, eIF-4GII, (Gradi et al. 1998) has been identified in mammalian systems.

eIF4B is a poorly understood RNA-binding protein that promotes recruitment of ribosomes to mRNA and activates ATPase-dependent helicase activity of eIF4A (Abramson et al. 1987). eIF4B also binds 18S rRNA and ribosome-bound eIF3 (Methot et al. 1996a, 1996b). Recent reports suggest that eIF4B is involved in ordered phosphorylation-dependent organization of the mRNA with eIF4F and PABP and eIF3 (Cheng and Gallie 2006; Holz et al. 2005; Shahbazian et al. 2006).

Cytoplasmic PABP was originally defined as protecting the 3' end of mRNAs against nucleolytic degradation, but is now known to be a crucial determinant for cap-poly(A) synergy in increased translational efficiency (Preiss and Hentze 1998; Preiss et al. 1998; Tarun et al. 1997). Its structure consists of four unequal RNA recognition motifs (Kuhn and Pieler 1996; Nietfeld et al. 1990) and a globular C-terminal protein interaction domain (Kozlov et al. 2001, 2004). PABP interacts with multiple translation-associated factors, including eIF4G (Imataka et al. 1998), eIF4B (Bushell et al. 2001), ribosome release factor 3 (eRF3) (Hoshino et al. 1999; Uchida et al. 2002), the deleted in azoospermia-like (DAZL) proteins (Collier et al. 2005), and the PABP-interacting proteins (PAIPs), PAIP1 (Craig et al. 1998), Paip2A (Khaleghpour et al. 2001), and Paip2B (Berlanga et al. 2006). PABP is not required for de novo initiation of translation on any capped mRNA or picornavirus IRES. However, PABP is now considered to be a canonical initiation factor since interaction between PABP and eIF4G strengthens cap-binding affinity and synergistically activates translation in a number of yeast, plant, and mammalian systems (Kahvejian et al. 2005). PABP also stimulates PV IRES-mediated translation (Bergamini et al. 2000).

The EMCV IRES, which is a type 2 IRES, is the most highly studied picornavirus IRES and has emerged as the paradigm that likely describes the basic features of how type 1 picornaviral IRESs interact with the ribosome. Pestova and colleagues performed what are now classic RNA toeprinting and sucrose gradient analyses with highly purified factors and EMCV IRES RNA that defined the minimal factors required for IRES to assemble a 48S ribosomal complex at the correct initiator codon. These studies revealed that eIF2, eIF3, eIF4A, and either complete eIF4F or the middle one-third domain of eIF4GI was required to bind ribosomes properly (Pestova et al. 1996a, 1996b). In this system, eIF4B was not required but enhanced complex formation. Further, direct binding of the central domain of eIF4G was demonstrated near the domain (J-K) of the IRES containing the initiator AUG on the 3' side of the IRES sequence (Kolupaeva et al. 1998). This region is within the type II IRES core region, and is analogous to the type I IRES core. It is likely that binding of eIF4F to the J-K loop places the 40S subunit very near the initiator AUG via the bridging function of eIF4G (mRNA/eIF4G/eIF3/40S).

So if the ribosome enters internally and is placed near the initiator AUG of the IRES directly, is there a need to scan? One would presume not. However, even though eIF4F contains eIF4A, additional eIF4A and ATP hydrolysis was required for efficient 48S complex formation on the initiator AUG in vitro (Pestova et al. 1996a, 1996b). This requirement was supported by the observation that a dominant negative inhibitor of eIF4A blocked EMCV translation in reticulocyte lysates (Pause et al. 1994b). However, it is generally thought that scanning does not occur and that the ATP requirement may instead reflect RNA melting required for binding in conjunction with eIF4B.

An important feature of viral IRES translation is that eIF4E is not required at all, reflecting the lack of a cap structure on viral RNA. The addition of eIF4E-binding protein (4EBP), a negative regulator of translation that competes with eIF4G for binding eIF4E, does not affect EMCV IRES translation (Pause et al. 1994a). Similarly, addition of m7GTP cap analog does not affect EMCV IRES function in vitro. Thus, no direct function for eIF4E in EMCV IRES translation has been found.

Although similar bottom-up reconstitution experiments with purified initiation factors have not been done with type I IRESs, it is currently thought that most key features are similar to the EMCV paradigm. That is, there should be a requirement for eIF2, eIF3, eIF4A, and eIF4F or minimally the central third domain of eIF4G, for the IRES to bind ribosomes. Intact eIF4G is absolutely not required, since viral 2A protease (2A$^{pro}$) and cellular proteases cleave eIF4G between the binding domains for eIF4E and eIF3 during the early phase of infection before the bulk of viral protein synthesis takes place (Bovee et al. 1998; Etchison et al. 1982; Lamphear et al. 1995; Zamora et al. 2002). Instead, numerous reports demonstrate that cleaved eIF4GI functions more effectively in PV IRES translation than intact eIF4GI (Borman et al. 1997; Liebig et al. 1993; Macejak and Sarnow 1991). This p100 fragment contains binding sites for eIF4A and eIF3, and it is postulated that this eIF4G fragment binds directly to the IRES RNA through its central HEAT domain (Lamphear et al. 1995; Gross et al. 2003; Oberer et al. 2005).

The actual binding site for eIF4G on the IRES has not been precisely determined but would be expected near the IRES core element encompassing stem-loop structures V and VI. Thus, supporting evidence from UV crosslinking suggests PV Sabin attenuation mutants bearing point mutations in stem-loop V bind eIF4G with lower affinity (Ochs et al. 2003). In addition, UV crosslinking experiments with wild type and mutant IRESs have been used to show that eIF4B binds to PV IRES stem-loop V and may interact with stem-loop VI (Ochs et al. 2002). The polypyrimidine tract is located between these structures. Competition assays demonstrated that eIF4B interacted more avidly with PV IRES than capped mRNA and was independent of PTB binding (Ochs et al. 2002).

What about the ribosome itself? It is likely that 40S subunits, in concert with bound canonical initiation factors, interact first with the completely conserved AUG on stem loop VI (nt 586 in PV1 and nt 592 in CVB3) and elements of the nearby polypyrimidine tract (Pilipenko et al. 1992) within the IRES core element. Despite its importance in ribosome binding, this AUG is not used as an initiator codon. Rather, all type 1 IRESs actually initiate at the very next AUG downstream, which is about 40 (HRV) or approximately 160 (PV, CVB) nt further along. Whether ribosomes shift to the next AUG via scanning or shunting is not clear. Finally, similarly to EMCV, PV IRES translation was blocked by eIF4A dominant negative inhibitors (Pause et al. 1994b), reflecting a requirement for eIF4A ATPase activity. It is not known if this reflects energy required to melt IRES structure during initial entry or subsequent scanning to the downstream AUG initiator.

## 3.2 Noncanonical Translation Factors

IRES *trans*-activating factors (ITAFs) are cellular proteins that do not have functions in normal cap-dependent translation but facilitate instances of cap-independent translation. Different viral IRESs have different ITAF requirements, though many are shared. Known ITAFs for the poliovirus IRES include polypyrimidine tract-binding protein (PTB), poly(rC)-binding protein 2 (PCBP2), upstream of N-ras (Unr), and lupus autoantigen (La). All ITAFs are RNA-binding proteins. ITAFs are not known to interact with initiation factors directly, though this may happen. It is proposed that the function of ITAFs is to serve as IRES chaperones, binding to RNA across multiple domains and stabilizing the entire IRES in a configuration suitable for binding canonical translation factors, and ultimately ribosomes (Pilipenko et al. 2000). This image complements our understanding of IRES RNA tertiary structure as constantly breathing and shifting with alternate base-pairing. ITAF protein binding may stabilize discrete local structures or drive long-range interactions that mold domains into functional units in space. Thus, it is not surprising that our first glimpses of IRESs by cryo-electron microscopy has revealed mostly condensed L- or F-shaped structures (Beales et al. 2003).

Lupus autoantigen (La) is a 52-kDa, predominantly nuclear protein. It was originally identified as an ITAF via its binding to a fragment of the poliovirus IRES spanning nucleotides 559-624 (stem-loop VI) (Meerovitch et al. 1989). La stimulates poliovirus IRES-mediated translation in rabbit reticulocyte lysates (Meerovitch et al. 1993) and discourages the aberrant translation of the poliovirus genome that is often seen in the reticulocyte system (Meerovitch et al. 1993; Svitkin et al. 1994). Truncation analysis of La has revealed an RNA interaction domain in its N-terminal 214 amino acids (Svitkin et al. 1994) and a dimerization domain in its C-terminus (Craig et al. 1997). RNA-binding activity alone is not sufficient for the stimulatory effect of La (Craig et al. 1997; Svitkin et al. 1994); ITAF stimulation is only achieved when bound La is able to form homodimers (Craig et al. 1997). More recently, RNAi depletion of La from cells resulted in decreased IRES translation and a dominant-negative La-inhibited 40S binding by PV IRES in vitro (Costa-Mattioli et al. 2004).

During PV infection, La begins to redistribute to the cytoplasm by 3 h postinfection (p.i.) (Meerovitch et al. 1993; Shiroki et al. 1999). This redistribution is concurrent with the appearance of viral 3C proteinase ($3C^{pro}$) in infected cells and is caused by a $3C^{pro}$-mediated cleavage event. This cleavage removes a nuclear localization signal on the extreme C-terminus of La, but retains the dimerization domain. The truncated La is still able to effectively stimulate translation but is effectively relocalized to the cytoplasm during the rise of viral protein synthesis (Shiroki et al. 1999).

La protein also binds to the CVB3 IRES and stimulates viral translation in a dose-dependent manner in rabbit reticulocyte lysate (Cheung et al. 2002; Ray and Das 2002). The CVB3 5'-UTR has been reported to contain multiple binding sites for La, with the strongest site at nt 210-529, a moderate site at 530-630 that overlaps with the binding site reported for PV, and a weak site at nt 1-209 (Cheung et al. 2002). Further analyses demonstrated that the conserved GAGA loop on domain VI in coxsackie B viruses is essential for La interaction with the 3'-most binding site on the IRES (Bhattacharyya and Das 2005).

It has been hypothesized that La does not directly stimulate enteroviral IRES-mediated translation, but recruits a larger stimulatory complex whose components are limiting in reticulocyte (Meerovitch et al. 1993) and whose components are easily lost during purification of the stimulatory complex (Toyoda et al. 1994). At this time, members of a putative La stimulation complex have not been identified.

Polypyrimidine tract-binding protein (PTB) is a 57-kDa mRNA splicing factor that UV-crosslinks to the PV IRES (Hellen et al. 1993; Pestova et al. 1991) and is required for viral-mediated translation (Pestova et al. 1991). Although the protein has four RNA recognition motifs (RRMs), the principal RNA-binding activity of the protein is concentrated in the two C-terminal RRMs (Kaminski et al. 1995; Oh et al. 1998; Perez et al. 1997). Unlike La, PTB binds to RNA as a monomer (Monie et al. 2005; Song et al. 2005). Full-length PTB stimulates PV-IRES activity in vitro (Back et al. 2002) and in vivo (Gosert et al. 2000; Guest et al. 2004). Of all the ITAFs, PTB has been shown most clearly to function as an RNA chaperone, stabilizing the IRES in an active conformation. Experiments directly supporting this chaperone role have been reported with type II IRESs such as EMCV (Kolupaeva et al. 1996) and FMDV (Kolupaeva et al. 1996; Song et al. 2005), but not yet with type I IRESs. The mechanism by which PTB augments type I IRES translation is currently unknown but is likely to be similar. Supporting this notion, multiple sites of PTB interaction with the PV IRES have been mapped, including nucleotides 40-288, 443-539, and 630-730 (Hellen et al. 1994).

Poly(rC)-binding protein 2 (PCBP2) is another factor required for poliovirus translation (Blyn et al. 1997) and was discovered via its interaction with stem-loop IV of the poliovirus IRES (Blyn et al. 1996). The ITAF function of PCBP2 may be as important as PTB, since extracts depleted of PCBP2 are severely inhibited in PV translation activity. An additional PCBP-binding site was identified on the cloverleaf structure (Gamarnik and Andino 1997); however, this 5′-terminal site appears to play a role in both viral translation and RNA replication, as a mutated cloverleaf unable to bind PCBP2 has defects in both activities (Parsley et al. 1997). Binding of PCBP2 to the cloverleaf is coupled with binding of 3CD$^{pro}$ to the viral RNA (Gamarnik and Andino 1997). However, the RNA-binding domain of PCBP2 alone (K homology domain 1) is insufficient to support wild type translation, and the presence of exogenous KH1 domain has a dominant-negative effect on viral translation despite possessing a dissociation constant equivalent to the full-length protein (Silvera et al. 1999). This result likely reflects a dimerization requirement for PCBP2 (Bedard et al. 2004).

An additional ITAF for type I IRESs is known as upstream of N-ras (Unr). Unr is a cytoplasmic protein that contains five cold-shock domains. Depletion experiments in reticulocyte lysates determined that Unr was required for translation of rhinovirus IRES, another type I IRES (Hunt et al. 1999). This initial study did not find that Unr stimulated PV IRES similar to HRV. Subsequently, another group reported that both HRV and PV IRES translation was severely impaired in unr($^{-/-}$) murine embryonic stem cells. Translation was restored by transient expression of Unr in unr($^{-/-}$) cells, thus revealing that in certain cell backgrounds, Unr requirements are evident for both viral IRES elements (Boussadia et al. 2003).

## 4  5′-3′ Interactions in Viral Translation

Approximately 10 years ago, a paradigm shift swept the translation research community with the revelation that most eukaryotic mRNAs are organized into pseudocircular structures via 5′-3′ interactions. Interactions in this case are not usually mediated by direct RNA base-pairing or kissing interactions; instead, translation factors that bind 5′ and 3′ structures interact with each other. Specifically, eIF4GI and eIF4GII contain N-terminal-binding motifs that bind to PABP on the dorsal side of its second RNA-recognition motif (RRM) (Deo et al. 1999; Gradi et al. 1998; Imataka et al. 1998). Additional interactions have also been described between the eIF4B N-terminal and PABP C-terminal domains (Bushell et al. 2001); thus, 5′-3′ interactions take place on more than one level, and others may be defined in the future. The result of many observations in plant, yeast, and mammalian systems indicates that interaction of PABP and eIF4GI in a so-called Closed Loop Model stimulates cap-dependent translation up to tenfold (Kahvejian et al. 2001).

But does this 5′-3′ stimulation apply to picornaviral RNAs that do not have cap structures? The expectation would be yes, since the factors that mediate 5′-3′ interactions on capped mRNAs are also present on viral RNAs, namely eIF4G, PABP, and eIF4B. Recently, UNR was also revealed to be a PABP-interacting protein, thus potentially providing another mechanism for 5′-3′ interactions that has not been explored in the viral context (Chang et al. 2004). However, eIF4GI is cleaved during virus infection, releasing the PABP-binding domain from the larger C-terminal fragment that binds the IRES core element. This might impair 5′-3′ translational synergy.

To experimentally demonstrate 5′-3′ translation synergy in vitro, two types of modified translation systems must be used: reticulocyte lysates partly depleted of ribosomes and HeLa lysates that retain the full complement of endogenous competitor mRNA. When these systems were employed, addition of poly(A) tails to reporter translation RNAs significantly stimulated translation driven by PV and EMCV IRESs. Disruption of the eIF4G-PABP interaction or cleavage of eIF4G abolished or severely reduced poly(A) tail-mediated stimulation of picornavirus IRES-driven translation (Michel et al. 2001). However, the cleavage of eIF4G by 2A$^{pro}$ has a net stimulatory effect for viral translation due to enhanced IRES function of the p100 eIF4G cleavage fragment (discussed further below). Similarly, poly(A) tails on reporter RNAs stimulated PV IRES-mediated translation in HeLa cell lysates (Bergamini et al. 2000).

Thus, the pseudocircular paradigm of cellular mRNA translation seems to apply to picornaviral mRNAs as well. It is unclear if cleavage of eIF4GI during infection interrupts 5′-3′ interactions on viral polysomes. Interactions between PABP and eIF4B or other factors may sustain translation once initiated, and cleavage of eIF4GI actually stimulates viral translation in most systems. On a different level, there is much evidence that 5′-3′ RNA interactions mediated by PABP allow prepositioning of PV 3CD polymerase precursor, which binds the 5′ cloverleaf, adjacent to the poly(A) template that must be copied in the first round of negative-strand RNA synthesis (Barton et al. 2001; Gamarnik and Andino 1998). Based on this model and the extensive experimental data that support it, it is likely that viral RNA remains organized in 5′-3′ circular fashion, perhaps modified, throughout the period of viral translation, despite cleavage of eIF4G.

## 5 Shutoff of Cellular Translation

A hallmark of enteroviral infections is host cell translation shutoff, a dramatic redeployment of the translational machinery toward viral translation and away from host messages. The viral IRES is intrinsically weaker at driving translation than a 7-methylguanosine cap and the virus must create an environment where the host messages can no longer compete for ribosomes. PV and CVB3 accomplish this task by cleavage of at least three key initiation factors during infection.

Cleavage of eIF4GI was discovered first and has dominated the related literature. There is no doubt that cleavage of eIF4GI is critical for translation shutoff

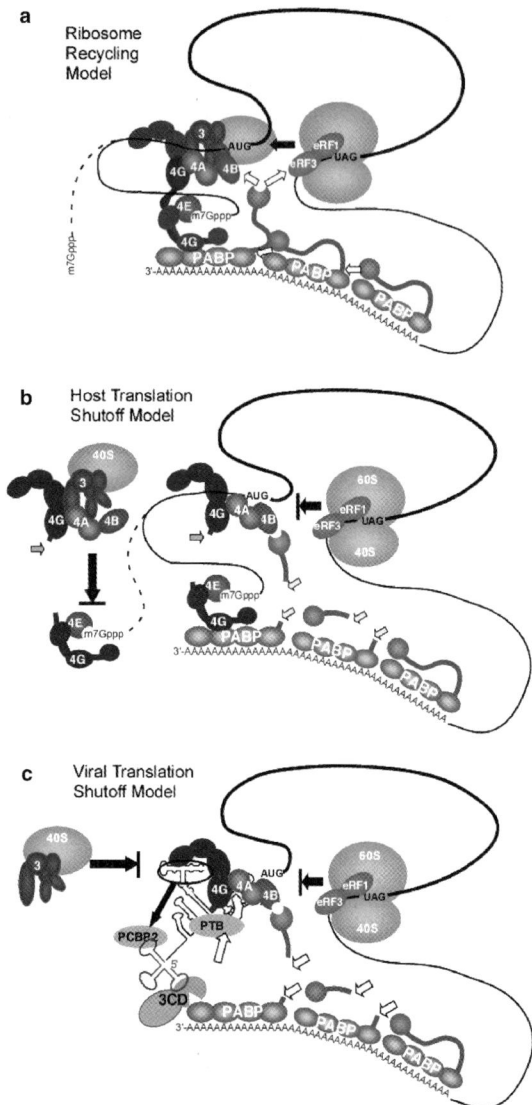

**Fig. 2** Models for the dual inhibition of de novo initiation and 3′-5′ ribosome recycling in virus-infected cells. **a** Conceptual model for ribosome recycling on capped mRNAs. The 40S subunit is depicted on the AUG codon after scanning; however, the cap structure may not be released by eIF4F during scanning (*dashed line*), thus, the 5′ UTR may be looped out (*solid line*). Ribosomes reaching stop codons bind eRF3, a binding partner of the PABP C-terminal protein-interaction domain (CTD) that may facilitate recycling of 60S subunits to waiting 40S subunits. Alternatively, both subunits may recycle. PABP-CTD may alternately bind eIF4B and eRF3 in this process. **b** Model for translation shutoff of capped mRNA. 2Apro cleavage of eIF4G (*arrows*) prohibits de novo binding of new 40S subunits to the cap structure. Recycling can continue with cleaved eIF4G but is interrupted by 3Cpro cleavage of PABP, releasing the CTD. **c** Model for shutoff of viral translation. Movement of PCBP2 from IRES to cloverleaf coupled with 3Cpro cleavage of PTB interrupts de novo binding of 40S subunits to the IRES. Recycling may continue, mediated by the complex of eIF4G/eIF4A/eIF4B that may remain in place. Recycling ribosomes are interrupted by 3Cpro cleavage of PABP. Note: La is omitted for clarity

phenotype; however, cleavage is clearly insufficient to cause drastic host translation shutoff alone. eIF4GI is cleaved by viral 2A$^{pro}$ and cellular proteinases at nearby cleavage sites early during infection (Lamphear et al. 1993; Zamora et al. 2002). The 2A proteinases from PV, CVB, and rhinoviruses all cleave eIF4G at the same site, severing amino acids 681/682, thus separating the eIF4GI fragment that binds eIF4E and PABP from the larger eIF3/eIF4A binding fragment of the protein (p100). The severed N-terminal fragment of eIF4GI can still technically circularize the genome by bridging PABP and eIF4E, but is incapable of targeting this fragmented mRNP complex to recruit the 40S ribosome (see Fig. 2). Thus, eIF4G cleavage at this site is a very efficient method to prevent de novo initiation on a capped mRNA, and has been demonstrated experimentally by many reports. If a capped reporter is added to an in vitro translation system pretreated with 2A$^{pro}$, it cannot be translated effectively (Haghighat et al. 1996; Kuyumcu-Martinez et al. 2004b).

However, when 2A$^{pro}$ is introduced into systems where translation is ongoing and polysomes are most likely circularized, the effect of eIF4GI cleavage is less dramatic and unclear. In vivo, cleavage of eIF4G occurs before host translation shutoff and the two events do not exactly coincide (Etchison et al. 1982; Pérez and Carrasco 1992). When a cleavage-resistant mutant of eIF4GI was expressed in cells together with 2A$^{pro}$, a partial rescue of translation was observed (Zhao et al. 2003); however, only a partial decrease in cellular translation was observed from 2A$^{pro}$ expression. eIF4GI cleavage and translation shutoff can be mostly unlinked if inhibitors of viral RNA replication such as 2 mM guanidine-HCl are included in infections (Bonneau and Sonenberg 1987; Irurzun et al. 1995). Further, inhibitors of viral RNA replication allow rapid and complete cleavage of eIF4GI in PV or HRV-infected cells, yet translation is diminished only approximately twofold. However, cleavage of the eIF4G paralog eIF4GII and especially PABP are also required for drastic inhibition of cap-dependent translation, and neither of these proteins are efficiently cleaved under conditions where viral RNA replication is blocked (Gradi et al. 1998; Joachims et al. 1999). eIF4GII is cleaved more slowly during infection than eIF4GI at a site similar to the eIF4GI cleavage site, particularly when viral RNA replication is blocked, and in the presence of RNA replication inhibitors PABP is not cleaved at all.

PABP is cleaved by poliovirus 3C$^{pro}$ and 2A$^{pro}$ (Joachims et al. 1999), CVB3 2A$^{pro}$ (Joachims et al. 1999; Kerekatte et al. 1999), and feline calicivirus and human norovirus 3C-like proteases (Kuyumcu-Martinez et al. 2004a). Remarkably, all of these cleavage events target the flexible linker region between the four RRMs of PABP and its major protein-protein interaction domain, and all of these cleavage events have a negative effect on translation. In PV and CVB3 infections, one N-terminal 2A$^{pro}$ cleavage product and two N-terminal 3C$^{pro}$ cleavage products of PABP accumulate (Kuyumcu-Martinez et al. 2002). Unlike eIF4GI, PABP is never cleaved to completion during infections. However, PABP concentrations in cells are much higher than eIF4GI and 3C$^{pro}$ preferentially targets polysome-PABP associated rather than the substantial pool of PABP that is not ribosome-associated (Kuyumcu-Martinez et al. 2002). This is possibly due to altered substrate conformations when PABP is in different subcellular compartments and demonstrates an economy of viral action when attacking translation machinery.

Although 2A$^{pro}$-induced cleavage of eIF4G to p100 blocks ribosomes from initiating on a capped mRNA de novo, it apparently does little to eject ribosomes already engaged on polysomes. If an in vitro translation extract with a capped reporter is allowed to form full polysomes before 2A$^{pro}$ is added, translation continues at nearly control rates for some time after eIF4G is cleaved in the system (Kuyumcu-Martinez et al. 2004b). How can translation persist if no intact eIF4G is present? Since translation continues for much longer than the period required for ribosomes to transit the ORF, it is likely that ribosomes are recycling from the 3′ stop codon region to the 5′ start codon in a mechanism that does not require intact eIF4G. In contrast, cleavage of PABP by 3C$^{pro}$ separates the RNA-binding domain of the protein from the globular protein-interaction domain, effectively abolishing 5′-3′ interactions involving the C-terminal protein-interaction motif of PABP. The resulting opening of the closed loop likely causes ribosomes present on the polysome to escape, or prevent ribosomes from recycling from the 3′ to 5′ ends of the mRNA. Consistent with this model, 3C$^{pro}$ causes inhibition of cap-dependent translation regardless of whether it is added before or after polysome formation (Kuyumcu-Martinez et al. 2004b). However, translation is never completely abrogated by cleavage of PABP alone, likely due to intact eIF4G directing de novo translation initiation. Complete translation inhibition of active polysomes required both 2A$^{pro}$ and 3C$^{pro}$ (Kuyumcu-Martinez et al. 2004b).

The data obtained to this point invoke a two-step mechanism for host-cell translation shutoff: early in infection, eIF4GI and eIF4GII cleavage prevents de novo initiation of ribosomes on cellular mRNA, and cleavage of PABP with slower kinetics completes the destruction of cellular polysomes, resulting in cellular mRNP complexes that are translationally incompetent (Fig. 2). This results in the drastic host translation shutoff observed in enteroviral infections.

## 6 Shutoff of Viral Translation

Enteroviruses shut down host-cell translation in order to use the cell's full translational potential for its own purposes. This means that the enteroviral genome must translate robustly under conditions where host-cell translation is repressed. While 2A$^{pro}$-induced cleavage of eIF4G hampers cellular translation, it significantly enhances viral IRES-mediated translation, though the mechanism remains unclear (Hambidge and Sarnow 1992). The p100 fragment of eIF4G has been shown to bind directly to EMCV IRES and likely interacts physically with the poliovirus and coxsackievirus IRES elements near the core element (Ochs et al. 2003; Pestova et al. 1996a, 1996b). The 40S ribosomal subunit can thus be recruited to the viral genome via the eIF3-p100 interaction. Additionally, cleavage of eIF4G ensures that the viral genome does not have to compete with host mRNA for translation initiation machinery.

This robust, p100-stimulated viral translation cannot continue indefinitely. The positive-strand RNA genome of an enterovirus is a direct template for translation but must also be replicated. Translation proceeds 5′ to 3′ and negative-strand replication proceeds 3′ to 5′, so the two cannot occur simultaneously. Moreover, translation of the template dominates replication in that the 3.5-MDa ribosome prevents the progression of the 53-kDa 3D$^{pol}$ if the two encounter each other (Barton et al. 1999; Gamarnik and Andino 1998). Therefore, viral translation must be shut off before RNA replication can occur. This is observed in enteroviral infections, where viral translation is inhibited late in infection after a few hours of rapid protein production. The exact mechanism for this translation-to-replication switch is unknown, but several models can be proposed.

Extrapolating from the host translation shutoff model, for viral translation shutoff to be efficient, both de novo translation initiation and 5′-3′ recycling must be abrogated. Cleavage of eIF4G satisfies this initial requirement for inhibition of cap-dependent translation, but this only strengthens translation of the viral IRES, and p100 is not further processed at late phases of infection. How then is viral de novo translation initiation blocked during infection? In this case, multiple mechanisms have been proposed. Gamarnik and Andino proposed a model where PCBP2 was recruited from its initial translation-supporting binding site in SL IV of the IRES to an alternate site on the cloverleaf, thus vacating the IRES ITAF role. Production or addition of 3CD was found to strongly inhibit viral translation in their experiments. The cloverleaf binding site was activated only after 3CD was produced and bound the cloverleaf, thus stimulating recruitment of PCBP2 to the adjacent cloverleaf stem-loop (Gamarnik and Andino 1998, 2000). This is an attractive model that incorporates a viable mechanism for viral regulation that is coordinated with the initial formation of a replicase complex on the cloverleaf.

Subsequently, other reports have indicated that several ITAFs required for viral translation (La, PTB and PCBP2) are cleaved by 3C$^{pro}$. Cleavage of La near its C-terminus has not been associated with viral translation defects; rather, it was shown to increase the cytoplasmic distribution of La, which could stimulate IRES translation later in infection (Shiroki et al. 1999). In contrast, PTB is cleaved during poliovirus infection in a manner that separates RRM1 and RRM2 from RRM3 and RRM4, resulting in inhibition of PV-IRES-mediated translation (Back et al. 2002). This cleavage separates the RNA-interacting RRMs from the protein-protein interaction RRMs (Oh et al. 1998), ostensibly preventing the cleaved protein from nucleating a complex on the IRES and sequestering sites on the IRES from intact PTB molecules. New evidence indicates that PCBP2 cleavage by 3Cpro partially represses IRES function (Perera et al. 2007). Thus, cleavage of certain key ITAFs may be a regulated step to prevent viral de novo translation initiation mid-to-late in infection, just as cleavage of eIF4G prevents de novo cellular translation initiation early in infection.

The second step of the host translation shutoff model mandates a block in the ribosome 5′-3′ recycling to release ribosomes already present on the polysome. In this case, recycling must be blocked, but translation termination allowed to proceed. It is unknown if cleavage and movement of ITAFs will also inhibit ribosome recycling, as no relevant kinetics experiments have been conducted. Also, the 5′-3′ interactions that circularize the enteroviral genome are unclear, but several candidate interactions are possible. Both PCBP2 and 3CD have been reported to interact with PABP, thus bridging the cloverleaf and poly(A) tail. However, both PCBP2 and PABP are substrates of 3C$^{pro}$ (3CD is an active protease) and pulldown experiments were conducted with mutant 3CD that did not autoprocess itself (Herold and Andino 2001). Whether fully catalytically active 3CD protease would bind similarly or merely cleave the targets in this context is not known. Though the eIF4G-p100 fragment that binds the IRES does not interact with PABP, it is likely that the p100/eIF4A/eIF4B complex stably binds on the IRES core structure. eIF4B interacts with the C-terminal binding domain of PABP that also binds eRF3 during translation termination (Bushell et al. 2001). There is evidence this minimal p100 eIF4G fragment can sustain ribosome recycling on capped messages (Kuyumcu-Martinez et al. 2004b). We have also shown that PABP cleavage by 3C$^{pro}$ inhibits PV translation in vitro and that expression of 3C$^{pro}$-cleavage resistant PABP in vivo rescues translation of viral mRNAs late in infection (Bonderoff and Lloyd, unpublished observations).

Thus, cleavage of PABP may block ribosomes from recycling from 3' to 5' ends of the viral mRNA, just as suspected for capped mRNAs. Viral translation may be able to persist longer in cells than host translation because early cleavage of eIF4G inhibits host translation but stimulates viral translation. All the known cleavages of factors involved with viral translation are catalyzed by 3C$^{pro}$, and all have much slower cleavage kinetics than eIF4G cleavage in vivo. Alternately, the ITAF Unr also interacts specifically with PABP (Chang et al. 2004), but cleavage of PABP in this case would not be expected to unlink 5'-3' interactions as Unr binds to RRM3 of PABP and cleavage of the CTD should have no effect on this interaction.

## 7 Cell-Specific Restriction of Translation

The requirements for trans-acting factors differ between related picornavirus IRESs and can account for cell type-specific variations in IRES function. Thus, ITAFs have become recognized as important determinants of viral pathology since various cell types and tissues express differing levels of ITAFs, potentially regulating levels of viral translation and ultimately replication in those locales. The issue was first raised in the context of the basis for PV neurovirulence when it was shown that attenuated type 3 Sabin and virulent type 3 Leon viruses translated equivalently in HeLa cells but attenuated Sabin virus was translation-restricted in neuroblastoma cells (Svitkin et al. 1985; La Monica and Racaniello 1989). The translation defect was found to be caused by the C472 to U mutation in the IRES (Svitkin et al. 1990). Similar findings of neuronal-cell growth defects and impaired translation were reported with type 1 Mahoney PV when mutations were introduced around the major attenuation determinant at nt 490 (Haller et al. 1996). Related experiments showed recombinant chimeric PV containing the HRV 2 IRES was growth-defective in neuroblastoma cells and attenuated for neurovirulence in PV-receptor transgenic mice (Gromeier et al. 1996). More recently, the three attenuating Sabin point mutations were introduced into the same genetic background (PV type 1) and only the Sabin type 3 mutation caused significant reductions in viral growth and reduced translation in vitro (Malnou et al. 2004). The three PV-attenuation mutations have now been introduced into CVB3 to test if the attenuation phenotype can be transferred to CVB3. Of the point mutations in the equivalent positions at CVB3 nt 484, 485, and 473, only the 473 (type 3) mutation reduced viral growth and translation, demonstrating a partial validation of the hypothesis (Ben M'hadheb-Gharbi et al. 2006).

However, the strict correlation between IRES mutations and neural attenuation phenotype was recently challenged by use of adenoviruses to deliver PV, CVB, or HCV IRES reporter constructs to different mouse tissues. It was found that the PV IRES was functional in many tissues, including those that do not support virus replication, and the IRES containing the C472U mutation in was translation-restricted in all mouse tissues tested. In the same study, recombinant polioviruses containing substitute IRESs were still highly tropic for brain and spinal cord, suggesting that important neurovirulence determinants lie outside the IRES region (Kauder and Racaniello 2004). A related follow-up study showed an age-related restriction of HRV 2 IRES function as HRV/PV chimeric virus replicated in the CNS of neonatal but not adult mice. The IRES function was defective in adult mouse tissue and thus correlated as a determinant of virulence (Kauder et al. 2006).

Thus it would seem well established that point mutations in the IRES core of all three PV serotypes are strong attenuating mutations in Sabin vaccine viruses and define a virulence hotspot that may ultimately be based on viral translation efficiency in target tissues. However, the results do not hold up in all experimental systems, serving as a reminder that important differences may exist between explanted tissue-derived cells and the parental cells in situ. We have seen above that canonical translation factors bind in this IRES region, yet reports have also shown that unique ITAFs, $2A^{pro}$, and possibly another cellular protein may also interact in this region to support translation in a cell-specific manner.

Comparison of PV Leon serotype 3 and Sabin serotype 3 IRESs revealed that PTB and a novel neural-cell-specific homolog of PTB (nPTB) bound adjacent to the attenuation mutation in domain V, but binding was less efficient on the Sabin IRES. The Sabin IRES demonstrated a translation deficit in chick neurons that was rescued by increased PTB expression in the CNS. Thus, attenuation was linked to limited PTB expression in CNS cells coupled with reduced binding by Sabin virus IRES (Guest et al. 2004).

A different series of attenuation mutations in domain V of the PV IRES is responsible for a temperature-sensitive phenotype in vitro. After passage of these mutant viruses in monkey kidney cells, many phenotypic revertants were isolated in which a second site or compensatory mutations mapped to a series of coding changes in $2A^{pro}$. These $2A^{pro}$ mutations were found throughout the protease with no obvious pattern or trend, but were not in the catalytic site (Macadam et al. 1994). Interestingly, these mutations produce cell-type-specific revertants that did not rescue in some cells, and the progeny virions exhibit mutant phenotypes when introduced to new cell types (Rowe et al. 2000). How $2A^{pro}$ may aid IRES function in this context is unknown, but involvement of a cellular protein in a complex is possible.

No related work with CVB ITAFs in target tissues has been reported, but it is possible that significant virus tropism for cardiomyocytes or pancreatic cells may be partly defined by expression of certain ITAFs. These may be shared with PV or some may be unique for CVB. For instance, PTB and ITAF(45) are required by the type II FMDV IRES for 48S complex assembly, but only PTB is required for the related Theiler's encephalomyelitis virus IRES (Pilipenko et al. 2000). CVB replication in cardiomyocytes and pancreatic cells may be restricted by limiting levels of certain ITAFs in these target tissues that correspond with initiation of limiting or persistent infections.

## 8 Concluding Remarks

Where to go from here? An explosion of new information about enterovirus translation regulation has occurred in the last 20 years that have made a lasting imprint on the both the picornavirus and translation fields. Yet much remains unclear. One of the biggest unsolved mysteries is how viruses transition virion RNA from the initial translation-competent state to a translation-incompetent, replication-competent status required to support RNA replication. In connection with this, it is important to produce formal proof that ribosomes recycle and assays for such recycling that can be utilized to formally test the hypothesis that $3C^{pro}$ cleavage of PABP interrupts 5′-3′ recycling of ribosomes. The molecular mechanism that sustains such recycling and what factors are involved is unknown. Further, more work is needed to better define molecular determinants of tissue-specific viral virulence that function at the translation level, particularly for CVB viruses, and the RNA-protein structures that promote this virulence.

On a broader level, translation regulation is rapidly extending into the overlapping mechanisms that stimulate polysome formation vs mRNA decay and stress-activated or microRNA-mediated translation-silencing mechanisms. Regulation of RNA stability is coupled to translation processes mechanistically and many translation factors are involved in both processes. In this vein, catalytically active $2A^{pro}$ was recently shown to stabilize PV RNA in vitro and is the only nonstructural viral protein required to stabilize the RNA (Jurgens et al. 2006). There is no information yet whether PV or CVB are targets of cellular microRNAs that can lead to translation silencing or RNAi-mediated degradation of viral RNA. Alternatively, it is unknown if enteroviruses interact with miRNA and RNAi mechanisms to modulate their function. The future will likely bring many new surprises.

## References

Abramson RD, Dever TE, Lawson TG, Ray BK, Thach RE, Merrick WC (1987) The ATP-dependent interaction of eukaryotic initiation factors with mRNA. J Biol Chem 262:3826-3832

Ambros V, Baltimore D (1980) Purification and properties of a HeLa cell enzyme able to remove the 5′ terminal protein from poliovirus RNA. J Biol Chem 255:6739-6744

Back SH, Kim YK, Kim WJ, Cho S, Oh HR, Kim JE, Jang SK (2002) Translation of polioviral mRNA is inhibited by cleavage of polypyrimidine tract-binding proteins executed by polioviral 3C(pro). J Virol 76:2529-2542

Barton DJ, Morasco BJ, Flanegan JB (1999) Translating ribosomes inhibit poliovirus negative-strand RNA synthesis. J Virol 73:10104-10112

Barton DJ, O'Donnell BJ, Flanegan JB (2001) 5′ cloverleaf in poliovirus RNA is a cis-acting replication element required for negative-strand synthesis. EMBO J 20:1439-1448

Beales LP, Holzenburg A, Rowlands DJ (2003) Viral internal ribosome entry site structures segregate into two distinct morphologies. J Virol 77:6574-6579

Bedard KM, Walter BL, Semler BL (2004) Multimerization of poly(rC) binding protein 2 is required for translation initiation mediated by a viral IRES. RNA 10:1266-1276

Ben M'hadheb-Gharbi M, Gharbi J, Paulous S, Brocard M, Komaromva A, Aouni M, Kean KM (2006) Effects of the Sabin-like mutations in domain V of the internal ribosome entry segment on translational efficiency of the Coxsackievirus B3. Mol Genet Genomics 276:402-412

Bergamini G, Preiss T, Hentze MW (2000) Picornavirus IRESes and the poly(A) tail jointly promote cap-independent translation in a mammalian cell-free system. RNA 6:1781-1790

Berlanga JJ, Baass A, Sonenberg N (2006) Regulation of poly(A) binding protein function in translation: characterization of the Paip2 homolog Paip2B. RNA 12:1556-1568

Bhattacharyya S, Das S (2005) Mapping of secondary structure of the spacer region within the 5′-untranslated region of the coxsackievirus B3 RNA: possible role of an apical GAGA loop in binding La protein and influencing internal initiation of translation. Virus Res 108:89-100

Blyn LB, Swiderek KM, Richards O, Stahl DC, Semler BL, Ehrenfeld E (1996) Poly(rC) binding protein 2 binds to stem-loop IV of the poliovirus RNA 5′ noncoding region: identification by automated liquid chromatography-tandem mass spectrometry. Proc Natl Acad Sci U S A 93:11115-11120

Blyn LB, Towner JS, Semler BL, Ehrenfeld E (1997) Requirement of poly(rC) binding protein 2 for translation of poliovirus RNA. J Virol 71:6243-6246

Bonneau A-M, Sonenberg N (1987) Proteolysis of the p220 component of the cap-binding protein complex is not sufficient for complete inhibition of host cell protein synthesis after poliovirus infection. J Virol 61:986-991

Borman AM, Kirchweger R, Ziegler E, Rhoads RE, Skern T, Kean KM (1997) eIF4G and its proteolytic cleavage products: effect on initiation of protein synthesis from capped, uncapped, and IRES-containing mRNAs. RNA 3:186-196

Boussadia O, Niepmann M, Creancier L, Prats AC, Dautry F, Jacquemin-Sablon H (2003) Unr is required in vivo for efficient initiation of translation from the internal ribosome entry sites of both rhinovirus and poliovirus. J Virol 77:3353-3359

Bovee ML, Marissen WE, Zamora M, Lloyd RE (1998) The predominant eIF4G-specific cleavage activity in poliovirus-infected HeLa cells is distinct from 2A protease. Virology 245:229-240

Bushell M, Wood W, Carpenter G, Pain VM, Morley SJ, Clemens MJ (2001) Disruption of the interaction of mammalian protein synthesis eukaryotic initiation factor 4B with the poly(A)-binding protein by caspase- and viral protease-mediated cleavages. J Biol Chem 276: 23922-23928

Byrd MP, Zamora M, Lloyd RE (2005) Translation of eIF4GI proceeds from multiple mRNAs containing a novel cap-dependent IRES that is active during poliovirus infection. J Biol Chem 280:18610-18622

Chang TC, Yamashita A, Chen CY, Yamashita Y, Zhu W, Durdan S, Kahvejian A, Sonenberg N, Shyu AB (2004) UNR, a new partner of poly(A)-binding protein, plays a key role in translationally coupled mRNA turnover mediated by the c-fos major coding-region determinant. Genes Dev 18:2010-2023

Cheng S, Gallie DR (2006) Wheat eukaryotic initiation factor 4B organizes assembly of RNA and eIFiso4G, eIF4A, and Poly(A)-binding protein. J Biol Chem 281:24351-2464

Cheung P, Zhang M, Yuan J, Chau D, Yanagawa B, McManus B, Yang D (2002) Specific interactions of HeLa cell proteins with Coxsackievirus B3 RNA: La autoantigen binds differentially to multiple sites within the 5′ untranslated region. Virus Res 90:23-36

Collier B, Gorgoni B, Loveridge C, Cooke H, Gray N (2005) The DAZL family proteins are PABP-binding proteins that regulate translation in germ cells. EMBO J 24:2656-2666

Costa-Mattioli M, Svitkin Y, Sonenberg N (2004) La autoantigen is necessary for optimal function of the poliovirus and hepatitis C virus internal ribosome entry site in vivo and in vitro. Mol Cell Biol 24:6861-6870

Craig AW, Svitkin YV, Lee HS, Belsham GJ, Sonenberg N (1997) The La autoantigen contains a dimerization domain that is essential for enhancing translation. Mol Cell Biol 17:163-169

Craig AWB, Haghighat A, Yu ATK, Sonenberg N (1998) Interaction of polyadenylate-binding protein with the eIF4G homologue PAIP enhances translation. Nature 392(6675): 520-523

Deo RC, Bonanno JB, Sonenberg N, Burley SK (1999) Recognition of polyadenylate RNA by the poly(A)-binding protein. Cell 98:835-845

Dildine SL, Semler BL (1989) The deletion of 41 proximal nucleotides reverts a poliovirus mutant containing a temperature-sensitive lesion in the 5′ noncoding region of genomic RNA. J Virol 63:847-862

Ehrenfeld E, Tererina NL (2002) Initiation of translation of picornavirus RNAs: structure and function of the internal ribosome entry site. In: Wimmer E (ed) Molecular biology of picornaviruses. ASM Press, Washington DC, pp 159-170

Etchison D, Milburn SC, Edery I, Sonenberg N, Hershey JWB (1982) Inhibition of HeLa cell protein synthesis following poliovirus infection correlates with the proteolysis of a 220,000-dalton polypeptide associated with eukaryotic initiation factor 3 and a cap binding protein complex. J Biol Chem 257:14806-14810

Gamarnik AV, Andino R (1997) Two functional complexes formed by KH domain containing proteins with the 5′ noncoding region of poliovirus RNA. RNA 3:882-892

Gamarnik A, Andino R (1998) Switch from translation to RNA replication in a positive-stranded RNA virus. Genes Dev 12:2293-2304

Gamarnik A, Andino R (2000) Interactions of viral protein 3CD and poly(rC)-binding protein with the 5′ untranslated region of the poliovirus genome. J Virol 74:22219-22226

Gmyl AP, Pilipenko EV, Maslova SV, Belov GA, Agol VI (1993) Functional and genetic plasticities of the poliovirus genome - quasi-infectious RNAs modified in the 5′-untranslated region yield a variety of pseudorevertants. J Virol 67:6309-6316

Gosert R, Chang KH, Rijnbrand R, Yi M, Sangar DV, Lemon SM (2000) Transient expression of cellular polypyrimidine-tract binding protein stimulates cap-independent translation directed by both picornaviral and flaviviral internal ribosome entry sites In vivo. Mol Cell Biol 20:1583-1595

Gradi A, Imataka H, Svitkin YV, Rom E, Raught B, Morino S, Sonenberg N (1998) A novel functional human eukaryotic translation initiation factor 4G. Mol Cell Biol 18:334-342

Grifo JA, Tahara SM, Morgan MA, Shatkin AJ, Merrick WC (1983) New initiation factor activity required for globin mRNA translation. J Biol Chem 258:5804-1580

Gromeier M, Alexander L, Wimmer E (1996) Internal ribosomal entry site substitution eliminates neurovirulence in intergeneric poliovirus recombinants. Proc Natl Acad Sci U S A 93:2370-2375

Gross JD, Moerke NJ, von der Haar T, Lugovskoy AA, Sachs AB, McCarthy JE, Wagner G (2003) Ribosome loading onto the mRNA cap is driven by conformational coupling between eIF4G and eIF4E. Cell 115:739-750

Guest S, Pilipenko E, Sharma K, Chumakov K, Roos RP (2004) Molecular mechanisms of attenuation of the Sabin strain of poliovirus type 3. J Virol 78:11097-11107

Haghighat A, Svitkin Y, Novoa I, Kuechler E, Skern T, Sonenberg N (1996) The eIF4G-eIF4E complex is the target for direct cleavage by the rhinovirus 2A proteinase. J Virol 70:8444-8450

Haller AA, Nguyen JH, Semler BL (1993) Minimum internal ribosome entry site required for poliovirus infectivity. J Virol 67:7461-7471

Haller AA, Stewart SR, Semler BL (1996) Attenuation stem-loop lesions in the 5′ noncoding region of poliovirus RNA: neuronal cell-specific translation defects. J Virol 70:1467-1474

Hambidge SJ, Sarnow P (1992) Translational enhancement of the poliovirus 5′ noncoding region mediated by virus-encoded polypeptide-2A Proc Natl Acad Sci U S A 89:10272-10276

Hellen CU, Sarnow P (2001) Internal ribosome entry sites in eukaryotic mRNA molecules. Genes Dev 15:1593-1612

Hellen CUT, Witherell GW, Schmid M, Shin SH, Pestova TV, Gil A, Wimmer E (1993) A cytoplasmic 57-kDa protein that is required for translation of picornavirus RNA by internal ribosomal entry is identical to the nuclear pyrimidine tract-binding protein. Proc Natl Acad Sci U S A 90:7642-7646

Hellen CUT, Pestova TV, Litterst M, Wimmer E (1994) The cellular polypeptide p57 (pyrimidine tract-binding protein) binds to multiple sites in the poliovirus 5′ nontranslated region. J Virol 68:941-950

Herold J, Andino R (2001) Poliovirus RNA replication requires genome circularization through a protein-protein bridge. Molecular Cell 7:581-591

Holz MK, Ballif BA, Gygi SP, Blenis J (2005) mTOR and S6K1 mediate assembly of the translation preinitiation complex through dynamic protein interchange and ordered phosphorylation events. Cell 123:569-580

Hoshino S, Imai M, Kobayashi T, Uchida N, Katada T (1999) The eukaryotic polypeptide chain releasing factor (eRF3/GSPT) carrying the translation termination signal to the 3′-Poly(A) tail of mRNA Direct association of erf3/GSPT with polyadenylate-binding protein. J Biol Chem 274:16677-16680

Hunt SL, Hsuan JJ, Totty N, Jackson RJ (1999) unr, a cellular cytoplasmic RNA-binding protein with five cold-shock domains, is required for internal initiation of translation of human rhinovirus RNA. Genes Dev 13:437-448

Imataka H, Gradi A, Sonenberg N (1998) A newly identified N-terminal amino acid sequence of human eIF4G binds poly(A)-binding protein and functions in poly(A)-dependent translation. EMBO J 17:7480-7489

Irurzun A, Sanchez-Palomino S, Novoa I, Carrasco L (1995) Monensin and nigericin prevent the inhibition of host translation by poliovirus, without affecting p220 cleavage. J Virol 69:7453-7460

Jackson RJ (2005) Alternate mechanisms of initiating translation of mammalian mRNAs. Biochem Soc Trans 33:1231-1241

Joachims M, van Breugel PC, Lloyd RE (1999) Cleavage of poly(A)-binding protein by enterovirus proteases concurrent with inhibition of translation in vitro. J Virol 73:718-727

Jurgens CK, Barton DJ, Sharma N, Morasco BJ, Ogram SA, Flanegan JB (2006) 2Apro is a multifunctional protein that regulates the stability, translation and replication of poliovirus RNA. Virology 345:346-357

Kahvejian A, Roy G, Sonenberg N (2001) The mRNA closed-loop model: the function of PABP and PABP-interacting proteins in mRNA translation. Cold Spring Harb Symp Quant Biol 66:293-300

Kahvejian A, Svitkin YV, Sukarieh R, M'Boutchou MN, Sonenberg N (2005) Mammalian poly(A)-binding protein is a eukaryotic translation initiation factor, which acts via multiple mechanisms. Genes Dev 19:104-113

Kaminski A, Hunt SL, Patton JG, Jackson RJ (1995) Direct evidence that polypyrimidine tract binding protein (PTB) is essential for internal initiation of translation of encephalomyocarditis virus RNA. RNA 1:924-938

Kauder SE, Racaniello VR (2004) Poliovirus tropism and attenuation are determined after internal ribosome entry. J Clin Invest 113:1743-1753

Kauder S, Kan S, Racaniello VR (2006) Age-dependent poliovirus replication in the mouse central nervous system is determined by internal ribosome entry site-mediated translation. J Virol 80:2589-2595

Kerekatte V, Keiper BD, Bradorff C, Cai A, Knowlton KU, Rhoads RE (1999) Cleavage of poly(A)-binding protein by Coxsackievirus 2A protease in vitro and in vivo: another mechanism for host protein synthesis shutoff? J Virol 73:709-717

Khaleghpour K, Svitkin YV, Craig AW, DeMaria CT, Deo RC, Burley SK, Sonenberg N (2001) Translational repression by a novel partner of human poly(A) binding protein Paip2. Mol Cell 7:205-216

Kolupaeva VG, Hellen CUT, Shatsky IN (1996) Structural analysis of the interaction of the pyrimidine tract-binding protein with the internal ribosomal entry site of encephalomyocarditis virus and foot-and-mouth disease virus RNAs. RNA 2:1199-1212

Kolupaeva VG, Pestova TV, Hellen CU, Shatsky IN (1998) Translation eukaryotic initiation factor 4G recognizes a specific structural element within the internal ribosome entry site of encephalomyocarditis virus RNA. J Biol Chem 273:18599-18604

Kozlov G, Trempe J-F, Khaleghpour K, Kahvejian A, Ekiel I, Gehring K (2001) Structure and function of the C-terminal PABC domain of human poly(A)-binding protein. Proc Natl Acad Sci U S A 98:4409-4413

Kozlov G, De Crescenzo G, Lim NS, Siddiqui N, Fantus D, Kahvejian A, Trempe JF, Elias D, Ekiel I, Sonenberg N, O'Connor-McCourt M, Gehring K (2004) Structural basis of ligand recognition by PABC, a highly specific peptide-binding domain found in poly(A)-binding protein and a HECT ubiquitin ligase. EMBO J 23:272-281

Kuhn U, Pieler T (1996) Xenopus poly(A) binding protein: functional domains in RNA binding and protein-protein interaction. J Mol Biol 256:20-30

Kuyumcu-Martinez NM, Joachims M, Lloyd RE (2002) Efficient cleavage of ribosome-associated poly(A)-binding protein by enterovirus 3C protease. J Virol 76:2062-2074

Kuyumcu-Martinez M, Belliot G, Sosnovtsev SV, Chang KO, Green KY, Lloyd RE (2004a) Calicivirus 3C-like proteinase inhibits cellular translation by cleavage of poly(A)-binding protein. J Virol 78:8172-8182

Kuyumcu-Martinez NM, Van Eden ME, Younan P, Lloyd RE (2004b) Cleavage of poly(A)-binding protein by poliovirus 3C protease inhibits host cell translation: a novel mechanism for host translation shutoff. Mol Cell Biol 24:1779-1790

La Monica N, Racaniello VR (1989) Differences in replication of attenuated and neurovirulent polioviruses in human neuroblastoma cell line SH-SY5Y. J Virol 63:2357-2360

Lamphear BJ, Panniers R (1990) Cap binding protein complex that restores protein synthesis in heat-shocked Ehrlich cell lysates contains highly phosphorylated eIF-4E. J Biol Chem 265:5333-5336

Lamphear BJ, Yan RQ, Yang F, Waters D, Liebig HD, Klump H, Kuechler E, Skern T, Rhoads RE (1993) Mapping the cleavage site in protein synthesis initiation factor-eIF-4 g of the 2A proteases from human coxsackievirus and rhinovirus. J Biol Chem 268:19200-19203

Lamphear BJ, Kirchweger R, Skern T, Rhoads RE (1995) Mapping of functional domains in eukaryotic protein synthesis initiation factor 4G (eIF4G) with picornaviral proteases - implications for cap-dependent and cap-independent translational initiation. J Biol Chem 270:21975-21983

Le SY, Maizel JV (1998) Evolution of a common structural core in the internal ribosome entry sites of picornavirus. Virus Genes 16:25-38

Le SY, Zuker M (1990) Common structures of the 5' non-coding RNA in enteroviruses and rhinoviruses - thermodynamical stability and statistical significance. J Mol Biol 216:729-741

Le SY, Chen JH, Sonenberg N, Maizel JV (1992) Conserved tertiary structure elements in the 5' untranslated region of human enteroviruses and rhinoviruses. Virology 191:858-866

Le SY, Siddiqui A, Maizel JV (1996) A common structural core in the internal ribosome entry sites of picornavirus, hepatitis C virus, and pestivirus. Virus Genes 12:135-147

Lefebvre AK, Korneeva NL, Trutschl M, Cvek U, Duzan RD, Bradley CA, Hershey JW, Rhoads RE (2006) Translation initiation factor eIF4G-1 binds to eIF3 through the eIF3e subunit. J Biol Chem 281:22917-22932

Liebig HD, Ziegler E, Yan R, Hartmuth K, Klump H, Kowalski H, Blaas D, Sommergruber W, Frasel L, Lamphear B, Rhoads R, Kuechler E, Skern T (1993) Purification of two picornaviral 2A proteinases - interaction with eIF-4g and influence on in vitro translation. Biochemistry 32:7581-7588

Liu Z, Carthy CM, Cheung C, Bohunek L, Wilson JE, McManus BM, Yang D (1999) Structural and functional analysis of the 5' untranslated region of coxsackievirus B3 RNA: In vivo translational and infectivity studies of full-length mutants. Virology 265:;206-217

Macadam AJ, Ferguson G, Fleming T, Stone DM, Almond JW, Minor PD (1994) Role for poliovirus protease 2A in cap independent translation. EMBO J 13:924-927

Macejak DG, Sarnow P (1991) Translation of uncapped RNA molecules is stimulated in poliovirus infected cells. 1991 Annual Meeting of the American Society for Virology

Malnou CE, Werner A, Borman AM, Westhof E, Kean KM (2004) Effects of vaccine strain mutations in domain V of the internal ribosome entry segment compared in the wild type poliovirus type 1 context. J Biol Chem 279:10261-10269

Meerovitch K, Pelletier J, Sonenberg N (1989) A cellular protein that binds to the 5'- noncoding region of poliovirus RNA: implications for internal translation initiation. Genes Dev 3:1026-1034

Meerovitch K, Nicholson R, Sonenberg N (1991) In vitro mutational analysis of cis-Acting RNA translational elements within the poliovirus type-2:5' untranslated region. J Virol 65: 5895-5901

Meerovitch K, YV Svitkin Lee HS, Lejbkowicz F, Kenan DJ, Chan EKL, Agol VI, Keene JD, Sonenberg N (1993) La autoantigen enhances and corrects aberrant translation of poliovirus RNA in reticulocyte lysate. J Virol 67:3798-3807

Methot N, Pickett G, Keene JD, Sonenberg N (1996a) In vitro RNA slection identifed RNA ligands that specifically bind to eukaryotic translation initiation factor 4B: the role of the RNA recognition motif. RNA 2:38-50

Methot N, Song MS, Sonenberg N (1996b) A region rich in aspartic acid, arginine, tyrosine, and glycine (DRYG) mediates eukaryotic initiation factor 4B (eIF4B) self-association and interaction with eIF3. Mol Cell Biol 16:5328-5334

Michel YM, Borman AM, Paulous S: Kean KM (2001) Eukaryotic initiation factor 4G-poly(A) binding protein interaction is required for poly(A) tail-mediated stimulation of picornavirus internal ribosome entry segment-driven translation but not for X-mediated stimulation of hepatitis C virus translation. Mol Cell Biol 21:4097-109

Monie TP, Hernandez H, Robinson CV, Simpson P, Matthews S, Curry S (2005) The polypyrimidine tract binding protein is a monomer. RNA 11:1803-1808

Murray KE, Roberts AW, Barton DJ (2001) Poly(rC) binding proteins mediate poliovirus mRNA stability. RNA 7:1126-1141

Nicholson R, Pelletier J, Le SY, Sonenberg N (1991) Structural and functional analysis of the ribosome landing pad of poliovirus type-2 - in vivo translation studies. J Virol 65: 5886-5894

Nietfeld W, Mentzel H, Pieler T (1990) The Xenopus laevis poly(A) binding protein is composed of multiple functionally independent RNA binding domains. EMBO J 9:3699-3705

Nomoto A, Kitamura N, Golini F, Wimmer E (1977) The 5'-terminal structures of poliovirion RNA and poliovirus mRNA differ only in the genome-linked protein VPg. Proc Natl Acad Sci U S A 74:5345-5349

Oberer M, Marintchev A, Wagner G (2005) Structural basis for the enhancement of eIF4A helicase activity by eIF4G. Genes Dev 19:2212-23

Ochs K, Saleh L, Bassili G, Sonntag VH, Zeller A, Niepmann M (2002) Interaction of translation initiation factor eIF4B with the poliovirus internal ribosome entry site. J Virol 76:2113-2122

Ochs K, Zeller A, Saleh L, Bassili G, Song Y, Sonntag A, Niepmann M (2003) Impaired binding of standard initiation factors mediates poliovirus translation attenuation. J Virol 77:115-122

Oh YL, Hahm B, Kim YK, Lee HK, Lee JW, Song O, Tsukiyama-Kohara K, Kohara M, Nomoto A, Jang SK (1998) Determination of functional domains in polypyrimidine-tract-binding protein. Biochem J 331:169-175

Parsley TB, Towner JS, Blyn LB, Ehrenfeld E, Semler BL (1997) Poly (rC) binding protein 2 forms a ternary complex with the 5'-terminal sequences of poliovirus RNA and the viral 3CD proteinase. RNA 3:1124-1134

Pause A, Belsham GJ, Gingras AC, Donze O, Lin TA, Lawrence JC Jr, Sonenberg N (1994a) Insulin-dependent stimulation of protein synthesis by phosphorylation of a regulator of 5'-cap function. Nature 371:762-767

Pause A, Methot N, Svitkin Y, Merrick WC, Sonenberg N (1994b) Dominant negative mutants of mammalian translation initiation factor eIF-4A define a critical role for eIF-4F in cap-dependent and cap-independent initiation of translation. EMBO J 13:1205-1215

Percy N, Barclay WS, Sullivan M, Almond JW (1992) A poliovirus replicon containing the chloramphenicol acetyltransferase gene can be used to study the replication and encapsidation of poliovirus RNA. J Virol 66:5040-5046

Perera R, Daijogo S, Walter BL, Nguyen JHC, Selmer BL (2007) Cellular protein modification by poliovirus: the two faces of poly(rC)-binding protein. J Virol 81:8919–8932

Perez I, McAfee JG, Patton JG (1997) Multiple RRMs contribute to RNA binding specificity and affinity for polypyrimidine tract binding protein. Biochemistry 36:11881-11890

Pérez L, Carrasco L (1992) Lack of direct correlation between p220 cleavage and the shut-off of host translation after poliovirus infection. Virology 189:178-186

Pestova TV, Hellen CUT, Wimmer E (1991) Translation of poliovirus RNA - role of an essential cis-acting oligopyrimidine element within the 5' nontranslated region and involvement of a cellular 57-kilodalton protein. J Virol 65:6194-6204

Pestova TV, Hellen CUT, Shatsky IN (1996a) Canonical eukaryotic initiation factors determine initiation of translation by internal ribosomal entry. Mol Cell Biol 16:6859-6869

Pestova TV, Shatsky IN, Hellen CUT (1996b) Functional dissection of eukaryotic initiation factor 4F: the 4A subunit and the central domain of the 4G subunit are sufficient to mediate internal entry of 43S preinitiation complexes. Mol Cell Biol 16:70-78

Pilipenko EV, Blinov VM, Romanova LI, Sinyakov AN, Maslova SV, Agol VI (1989) Conserved structural domains in the 5'-ultranslated region of picronaviral genomes: an analysis of the segment controlling translation and neurovirulence. Virology 168:201-209

Pilipenko EV, Gmyl AP, Maslova SV, Svitkin YV, Sinyakov AN, Agol VI (1992) Prokaryotic-like cis elements in the cap-independent internal initiation of translation on picornavirus RNA. Cell 68:119-131

Pilipenko EV, Pestova TV, Kolupaeva VG, Khitrina EV, Poperechnaya AN, Agol VI, Hellen CU (2000) A cell cycle-dependent protein serves as a template-specific translation initiation factor. Genes Dev 14:2028-2045

Pisarev AV, Chard LS, Kaku Y, Johns HL, Shatsky IN, Belsham GJ (2004) Functional and structural similarities between the internal ribosome entry sites of hepatitis C virus and porcine teschovirus, a picornavirus. J Virol 78:4487-4497

Poyry T, Kinnunen L, Hovi T (1992) Genetic variation in vivo and proposed functional domains of the 5' noncoding region of poliovirus RNA. J Virol 66:5313-5319

Preiss T, Hentze MW (1998) Dual function of the messenger RNA cap structure in poly(A)-tail-promoted translation in yeast. Nature 392:516-520

Preiss T, Muckenthaler M, Hentze MW (1998) Poly(A)-tail-promoted translation in yeast: implications for translational control. RNA 4:1321-1331

Prevot D, Darlix JL, Ohlmann T (2003) Conducting the initiation of protein synthesis: the role of eIF4G Biol Cell 95:141-156

Pyronnet S, Imataka H, Gingras AC, Fukunaga R, Hunter T, Sonenberg N (1999) Human eukaryotic translation initiation factor 4G (eIF4G) recruits mnk1 to phosphorylate eIF4E. EMBO J 18:270-279

Ray PS, Das S (2002) La autoantigen is required for the internal ribosome entry site-mediated translation of Coxsackievirus B3 RNA. Nucleic Acids Res 30:4500-4508

Rivera VM, Welsh JD, Maizel JV (1988) Comparative sequence analysis of the 5' noncoding region of the enteroviruses and rhinoviruses. Virology 165:42-50

Robertson ME, Seamons RA, Belsham GJ (1999) A selection system for functional internal ribosome entry site (IRES) elements: analysis of the requirement for a conserved GNRA tetraloop in the encephalomyocarditis virus IRES. RNA 5:1167-1179

Rowe A, Ferguson GL, Minor PD, Macadam AJ (2000) Coding changes in the poliovirus protease 2A compensate for 5' NCR domain V disruptions in a cell-specific manner. Virology 269:284-293

Shahbazian D, Roux PP, Mieulet V, Cohen MS, Raught B, Taunton J, Hershey JW, Blenis J, Pende M, Sonenberg N (2006) The mTOR/PI3K and MAPK pathways converge on eIF4B to control its phosphorylation and activity. EMBO J 25:2781-2791

Shiroki K, Isoyama T, Kuge S, Ishii T, Ohmi S, Hata S, Suzuki K, Takasaki Y, Nomoto A (1999) Intracellular redistribution of truncated La protein produced by poliovirus 3Cpro-mediated cleavage. J Virol 73:2193-2200

Silvera D, Gamarnik AV, Andino R (1999) The N-terminal K homology domain of the poly(rC)-binding protein is a major determinant for binding to the poliovirus 5'-untranslated region and acts as an inhibitor of viral translation. J Biol Chem 274:38163-38170

Simoes EAF, Sarnow P (1991) An RNA hairpin at the extreme 5' end of the poliovirus RNA genome modulates viral translation in human cells. J Virol 65:913-921

Skinner MA, Racaniello VR, Dunn G, Cooper J, Minor PD, Almond JW (1989) New model for the secondary structure of the 5' non-coding RNA of poliovirus is supported by biochemical and genetic data that also show that RNA secondary structure is important in neurovirulence. J Mol Biol 207:379-392

Song Y, Tzima E, Ochs K, Bassili G, Trusheim H, Linder M, Preissner KT, Niepmann M (2005) Evidence for an RNA chaperone function of polypyrimidine tract-binding protein in picornavirus translation. RNA 11:1809-1824

Svitkin YV, Maslova SV, Agol VI (1985) The genomes of attenuated and virulent poliovirus strains differ in their in vitro translation efficiencies. Virology 147:243-252

Svitkin YV, Cammack N, Minor PD, Almond JW (1990) Translation deficiency of the Sabin type 3 poliovirus genome: association with an attenuating mutation C472AeU. Virology 175:103-109

Svitkin YV, Meerovitch K, Lee HS, Dholakia JN, Kenan DJ, Agol VI, Sonenberg N (1994) Internal translation initiation on poliovirus RNA - further characterization of Ia function in poliovirus translation in vitro. J Virol 68:1544-1550

Tarun SZ Jr, Wells SE, Deardorff JA, Sachs AB (1997) Translation initiation factor eIF4G mediates in vitro poly(A) tail-dependent translation. Proc Natl Acad Sci U S A 94:9046-9051

Toyoda H, Koide N, Kamiyama M, Tobita K, Mizumoto K, Imura N (1994) Host factors required for internal initiation of translation on poliovirus RNA. Arch Virol 138:1-15

Uchida N, Hoshino S, Imataka H, Sonenberg N, Katada T (2002) A novel role of the mammalian GSPT/eRF3 associating with poly(A)-binding protein in Cap/Poly(A)-dependent translation. J Biol Chem 277:50286-50292

Waskiewicz AJ, Johnson JC, Penn B, Mahalingam M, Kimball SR, Cooper JA (1999) Phosphorylation of the cap-binding protein eukaryotic translation initiation factor 4E by protein kinase Mnk1 in vivo. Mol Cell Biol 19:1871-1880

Yang D, Cheung P, Sun Y, Yuan J, Zhang H, Carthy CM, Anderson DR, Bohunek L, Wilson JE, McManus BM (2003) A shine-dalgarno-like sequence mediates in vitro ribosomal internal entry and subsequent scanning for translation initiation of coxsackievirus B3 RNA. Virology 305:31-43

Zamora M, Marissen WE, Lloyd RE (2002) Multiple eIF4GI-specific protease activities present in uninfected and poliovirus-infeced cells. J Virol 76:165-177

Zhao X, Lamphear B, Xiong D, Knowlton K, Rhoads R (2003) Protection of cap-dependent protein synthesis in vivo and in vitro with an eIF4G-1 variant highly resistant to cleavage by Coxsackievirus 2A protease. J Biol Chem 278:4449-4457

# Preferential Coxsackievirus Replication in Proliferating/Activated Cells: Implications for Virus Tropism, Persistence, and Pathogenesis

R. Feuer and J. L. Whitton(✉)

1  Introduction .................................................................................................................. 150
2  Enhanced Viral Replication in Proliferating Cells: A Recurrent Theme
   Among Viruses ............................................................................................................ 151
3  Cell Cycle Effects on Coxsackievirus Infection: Evidence from Tissue
   Culture Studies ............................................................................................................ 152
   3.1  Cell Cycle Effects Are Observed for Viruses in Several
        Different Picornavirus Genera ............................................................................. 152
   3.2  Several Lines of Evidence Show that Coxsackieviruses Interact
        with the Host Cell Cycle in Tissue Culture ........................................................ 152
4  In Vivo Evidence that the Cell Cycle Affects the Outcome of CVB3 Infection ......... 154
   4.1  CVB Infection of Lymphoid Cells May Be Regulated by the Cells'
        Activation State ................................................................................................... 154
   4.2  Coxsackievirus Selectively Infects Neural Stem Cells In Vivo .......................... 155
5  Possible Molecular Mechanisms by Which the Host Cell Status Affects
   the Outcome of Coxsackievirus Infection .................................................................. 156
   5.1  Cell Cycle Effects on Virus Receptor ................................................................. 157
   5.2  Cell Cycle Effects on Virus Entry ...................................................................... 157
   5.3  Cell Cycle Effects on Virus Gene Expression: The Internal Ribosomal Entry Site ... 157
   5.4  Cell Cycle Effects on Virus Replication ............................................................ 160
   5.5  Cell Cycle Effects on Virus Maturation and Release ......................................... 162
6  The Biological Implications of the Relationship Between the Virus and the Cell Cycle .... 163
   6.1  Implications for Cell Tropism ............................................................................. 163
   6.2  Implications for Transmission ............................................................................ 163
   6.3  Facilitation of Persistent/Latent Infection .......................................................... 164
   6.4  Implications for Pathogenesis ............................................................................. 165
References ........................................................................................................................ 167

**Abstract** Coxsackieviruses cause substantial human morbidity and mortality, but the underlying molecular mechanisms of disease remain obscure. Here, we review the effects that the cell status—both cellular activation, and the cell

J. L. Whitton
Molecular and Integrative Neurosciences Dept., SP30-2110,The Scripps Research Institute,
10550 N. Torrey Pines Rd., La Jolla, CA 92037, USA
lwhitton@scripps.edu

cycle—may have on the outcome of virus infection. We propose that these viruses have evolved to undergo productive infection in cells at the $G_1/S$ stage of the cell cycle, and to preferentially establish persistence/latent infection in quiescent cells, and we provide possible explanations for these outcomes. Finally, we consider the implications of these interactions for virus transmission and host pathology.

# 1 Introduction

Viruses parasitize host cellular machinery and biochemical resources to maximize their survival advantage. Although each virus family utilizes a number of different strategies to exploit the resources of the host cell, one common theme consists of the enhanced vulnerability of cells to infection during cellular activation or proliferation. The prototypical eukaryotic cell cycle is shown diagrammatically in Fig. 1. The figure includes a conceptual representation of the possible interactions between the virus and the host cell, and the factors that determine the outcome of infection. As indicated, in some cases the virus may play the determining role, using any of the several strategies shown, and in other cases, the cell may dominate, thereby controlling the viral life cycle. In addition, one can readily conceive of scenarios in which the ascendancy oscillates between virus and cell, for example in response to changes in exogenous conditions. Furthermore, for any given virus, the nature of the cell-cycle interaction may vary depending on the cell type that it infects; a striking example of this is the species dichotomy that exists for dengue virus, which shows enhanced viral production in S-phase mosquito cells but not in S-phase human cells (Helt and Harris 2005).

It is our remit, in this article, to focus mainly on the cell-dominant aspect of the virus-cell interaction. Therefore, in this chapter we shall review published data, from our laboratory and others, which strongly suggest that coxsackieviruses (CV), and many other picornaviruses, productively infect activated or proliferating cells;

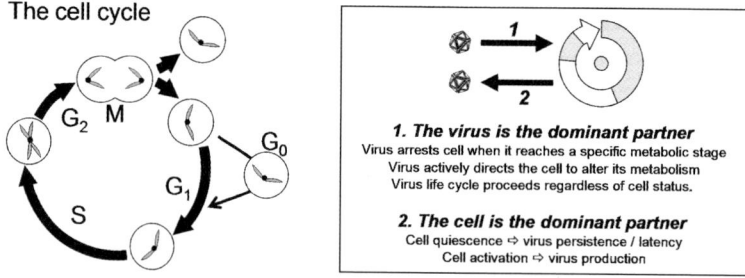

**Fig. 1** Summary of the cell cycle, and the strategies by which virus and cell may interact. *Left* a diagrammatic representation of the various stages of the eukaryotic cell cycle. *Right* a summary of the factors that may determine the outcome of virus infection, and of the possible interactions between virus and host cell

and that, in contrast, persistence/latency is a likely outcome if the infected cell is quiescent. These findings indicate that the cell status at the time of infection may be important, i.e., the cell dominates. However, we shall also review data showing that, under some circumstances, the virus dominates: virus entry and early viral gene expression may stimulate cellular activation/proliferation or inhibit the cell cycle at $G_1/S$ phase, thereby making the host cell a more hospitable environment for maximal virus production; or, conversely, may selectively arrest cell division without preventing cellular migration or differentiation. These and other effects of the virus on the biology of the host cell—for example, regulating cellular apoptosis (Carthy et al. 1998; Henke et al. 2000)—will be discussed only briefly, because they are the topic of another chapter in this volume.

## 2 Enhanced Viral Replication in Proliferating Cells: A Recurrent Theme Among Viruses

The activation state of the cell is a common theme that determines susceptibility and support for many different virus families. Indeed, the importance of an appropriate host cell environment may explain why some viruses contain homologs of cellular genes that are involved in cell-cycle control (Verschuren et al. 2004) one consequence of which may be the occurrence of virus-related tumorigenesis (Talbot and Crawford 2004); and why so many viruses encode proteins that interact with the cellular machinery that regulates cell activation, cycling, and death.

Early studies on retroviruses identified cell division as a critical factor for efficient virus transcription and replication (Fritsch and Temin 1977; Humphries and Temin 1974). A general feature for retroviruses is the need for access into the nucleus for integration into the host DNA, and this usually can only occur following the dissolution of the nucleus during cellular mitosis; but some lentiviruses have active mechanisms in certain cell types that shuttle the preintegration complex into the nucleus without nuclear dissolution (Bukrinsky 2004). Parvoviruses are well known for targeting actively dividing cells, including proliferating erythroid progenitor cells (Brown et al. 1993), lysis of which can cause severe anemia (Young and Brown 2004). Parvovirus infection also may cause myocarditis in humans, which suggests that there may be a small pool of proliferating cells in the heart that can support active parvoviral replication (Srivastava and Ivey 2006). The need for proliferating cells during robust infection by the flavivirus hepatitis C virus (HCV) has been previously described (Nelson and Tang 2006; Scholle et al. 2004); it has been proposed that nucleotide pools found in high concentrations during cellular proliferation and lower concentrations during quiescence may account for the inhibition of HCV replication during cell confluence (Nelson and Tang 2006), although the viral internal ribosome entry site (IRES) also may play a role (see Sect. 5.3 below, discussing the possible mechanisms by which the cell may regulate viral gene expression).

## 3 Cell Cycle Effects on Coxsackievirus Infection: Evidence from Tissue Culture Studies

### 3.1 Cell Cycle Effects Are Observed for Viruses in Several Different Picornavirus Genera

Judging by their omission from recent reviews of virus-cell cycle interactions (Op De Beeck and Caillet-Fauquet 1997; Swanton and Jones 2001), the interactions between the cell cycle and picornaviruses appear not to be as widely appreciated as those between DNA viruses and their cellular hosts, despite having been first suggested approximately 30 years ago (Eremenko et al. 1972b; Lake et al. 1970; Mallucci et al. 1985; Suarez et al. 1975). Early studies on poliovirus (Eremenko et al. 1972a; Eremenko et al. 1972b; Koch et al. 1974; Marcus and Robbins 1963) and mengovirus (a strain of encephalomyocarditis virus [EMCV]) (Lake et al. 1970) implicated the cell cycle in regulating the replication and cytopathicity of these members of the enterovirus and cardiovirus genera. Soon thereafter, this hypothesis was extended to include the enterovirus coxsackievirus B1 (CVB1); it was found that virus yield was higher from synchronized cells infected during S phase than from nonsynchronized populations (Suarez et al. 1975). Subsequent analyses confirmed the effects on cardioviruses: prototypical EMCV is affected by the cell cycle (Mallucci et al. 1985), and a more recent study revealed that the activation state of macrophages influenced their susceptibility to infection by Theiler's murine encephalomyelitis virus (TMEV) (Shaw-Jackson and Michiels 1997). Furthermore, as discussed below, viruses in the *Hepatovirus* and *Aphthovirus* genera also are affected. Thus, it appears that, for many members of the family Picornaviridae, the outcome of infection is inextricably linked to the status of the host cell.

### 3.2 Several Lines of Evidence Show that Coxsackieviruses Interact with the Host Cell Cycle in Tissue Culture

Recent studies by our laboratory and others confirm and extend these studies, and suggest a mechanism that may explain the ability of CV (and, perhaps, of other picornaviruses) to be influenced by the cell cycle and to persist in vivo.

#### 3.2.1 CVB3 Gene Expression and Virus Production Is Highest in Cells Arrested at the $G_1/S$ Phase

We have previously shown that the greatest levels of coxsackievirus B3 (CVB3) protein expression were observed in HeLa cells arrested at the $G_1/S$ phase of the cell cycle (Feuer et al. 2002). As one might expect, the highest levels of infectious virus also were produced during $G_1/S$, and much lower levels of virus were observed at $G_2/M$ phase.

### 3.2.2 Cells Infected When Quiescent Show Minimal Viral Protein Expression, But Harbor Infectious RNA

Even more noteworthy, quiescent cells infected with CVB3 expressed virtually no viral proteins and very limited infectious virus 24 hours (h) after infection. Within 48 h, no infectious virus could be detected in quiescent cell cultures, although inactive CVB3 RNA remained within quiescent cells. Furthermore, inactive viral RNA isolated from these quiescent cultures and transfected into proliferating HeLa cells could, once again, produce infectious virus (Feuer et al. 2002). This suggested that inactive CVB3 RNA from quiescent HeLa cell cultures was fully capable of giving rise to infectious virus once transferred into a receptive cellular environment, and that the persistent viral genome was, therefore, likely to be full-length in nature. We proposed that, by remaining within quiescent cells in the form of inactive RNA, CVB3—thought of as a highly cytolytic virus—could, in fact, exhibit the cardinal properties of viral latency.

### 3.2.3 Wounding a Confluent Culture of Cells: Increased Viral Protein Expression at Wound Margins

From these data, we hypothesized that CV targets rapidly proliferating cells and that viral RNA may persist within differentiated cells undergoing limited cellular proliferation. We attempted to test this phenomenon in a more natural in vitro setting using HeLa cells prohibited from dividing, not by serum removal, but instead by contact inhibition (Feuer et al. 2004; Feuer et al. 2002). HeLa cells were grown to high confluency in a chamber slide and then were wounded by scraping the cell surface with a sterile pipette tip. A recombinant CVB3 expressing eGFP (eGFP-CVB3) was applied at high MOI immediately after wounding, and 1 h later the inoculum was replaced with an agar overlay. eGFP expression was followed over time using fluorescence microscopy (Fig. 2). Within 12 h, low levels of viral protein expression were observed in cells adjacent to the site of the wound. By 16 h postinfection (PI), the number of infected cells and the intensity of eGFP expression increased within wounded areas of the culture. By 24 h PI and beyond, viral protein expression increased and remained localized to cells bordering the wound. These results suggested to us that only in those cells that had migrated and proliferated into the open spaces of the wounded area were able to support robust virus infection. In contrast, contact-inhibited HeLa cells visualized beyond the wound area supported little or no viral protein expression.

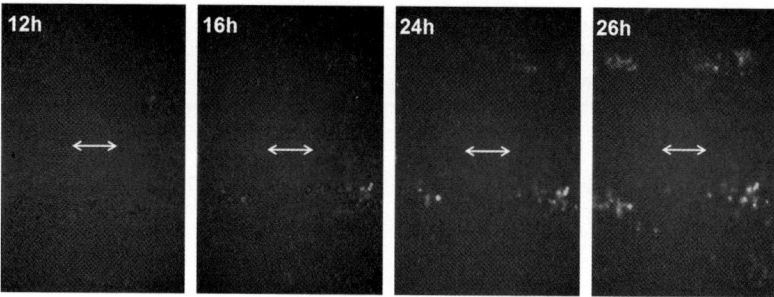

**Fig. 2** Cells at a wound border preferentially support virus protein expression. HeLa cells were grown to confluency on a chamber slide and wounded by drawing a sterile pipette tip across the surface of the slide. The culture was immediately infected with eGFP-CVB3 for 1 h and covered with an agar plug. Fluorescent images were taken to visualize viral protein production (eGFP) over time in the cultures. *Arrows* represent the direction of the wound. Within 12 h, viral protein expression was detected in a few cells near the border of the wound. By 16 h, the number and the intensity of eGFP+ cells increased. After 24 h, eGFP expression was localized to the immediate wound area, indicating that only those HeLa cells that were dividing/proliferating near or within the open area of the chamber slide could support CVB3 protein expression

## 4 In Vivo Evidence that the Cell Cycle Affects the Outcome of CVB3 Infection

The evidence that the cell cycle may affect the outcome of infection by CV (and by other picornaviruses) is not limited to tissue culture data.

### 4.1 CVB Infection of Lymphoid Cells May Be Regulated by the Cells' Activation State

Following infection of immunocompetent mice, CVB3 RNA can be identified in various tissues, including the spleen and lymph nodes, and a detailed evaluation using in situ hybridization and immunohistochemistry showed that the majority of the viral materials were present in pre-B or B cells, and that signals could also be found, albeit at lower frequencies, in CD4+ T cells and macrophages (Klingel et al. 1996); virus products were not detected in CD8+ T cells. Subsequent work from our laboratory showed that, between days 2 and 3 postinfection, a burst of CVB3 replication occurs within the germinal centers of the spleen (Mena et al. 1999), a site of active B cell proliferation and maturation (Shapiro-Shelef and Calame 2005). Others have determined that CVB3 replication in T cells is dependent upon the T cell activation enzyme, p56$^{lck}$ (Liu et al. 2000). Strains of mice (such as C57BL/6 mice) having less inherent MAP kinase activity in cardiac tissue are also somewhat less susceptible to CVB3-induced myocarditis (Opavsky et al. 2002). Intriguingly, a recent study has suggested that activated T cells may be targeted by foot-and-mouth disease virus (FMDV), and that such targeting may generate partial immunosuppression in host animals (Díaz-San Segundo et al. 2006). In this example, picornavirus targeting of activated/proliferating cells may not only maximize virus replication, but also may specifically eliminate those virus-specific T cells that would normally have limited the systemic dissemination of virus. As one might expect, immunosuppression by targeting immune effector cells is a common viral strategy (Brenchley et al. 2006; Naniche and Oldstone 2000; Rall 2003).

## 4.2 Coxsackievirus Selectively Infects Neural Stem Cells In Vivo

It is well known that enterovirus infections cause more severe disease in newborns than in adults. The favored explanation for this phenomenon has been that newborns can mount only very limited immune responses, leading to increased titers (and/or wider dissemination) of virus. However, recent studies have questioned the dogma of neonatal immunoincompetence; it has been shown that neonates, of several species, mount rather robust immune responses (Hassett et al. 1997; Martinez et al. 1997; Ridge et al. 1996; Watts et al. 1999). Therefore, we have proposed an alternative model to explain the susceptibility of neonates to enteroviral diseases: the high quantity of proliferating cells observed during normal development might increase the number of CVB3-receptive cells during the course of infection. In this scenario, greater pathology may be expected in neonates, similar to what is observed with other viruses that target proliferating cells during development (Ramirez et al. 1996; Young and Brown 2004). Our CVB3 model of neonatal infection in the central nervous system (CNS) indicates that proliferating neural stem cells, which are found within distinct regions of the developing brain, preferentially support viral infection (Feuer et al. 2003). Proliferating cells can be identified by staining for the nuclear protein Ki67 and, as shown in Fig. 3a, these cells are located in the subventricular zone, immediately adjacent to the lateral ventricle of the brain. This also is where CVB3 infection of the CNS begins (green cells; the virus expresses eGFP). These virus-infected (green) cells display phenotypic markers, such as nestin, that are characteristic of stem cells (Fig. 3b, c). We propose that the proliferative status of stem cells may make them favored target cells for infection (Feuer et al. 2005); the high quantity of proliferating neural stem cells in the neonatal CNS may contribute to the well-recognized age-related difference in susceptibility to enteroviral meningitis and encephalitis, and to the vulnerability of the fetal CNS to enteroviral disease (Daley et al. 1998; Euscher et al. 2001; Feuer et al. 2003; Hsueh et al. 2000; Kaplan et al. 1983; Modlin 1988; Modlin and Bowman 1987; Ratzan 1985; Sauerbrei et al. 2000). However, the interaction between CVB3 and neural stem cells may be more complex than the above would imply. At first blush, one would predict that infection of such highly active progenitor cells would be productive, and several lines of evidence suggest that this may be the case; many stem cells express high levels of virus protein (Fig. 3b, c), and virus RNA is readily

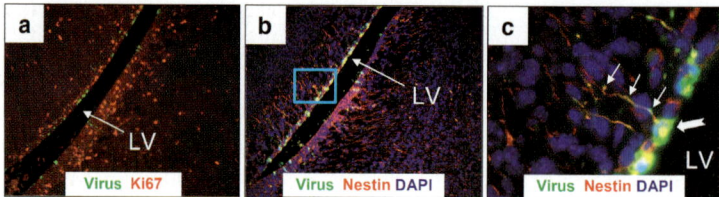

**Fig. 3** Proliferating neural stem cells are targets for CVB3 infection. One-day-old pups were infected with eGFP-CVB3 and, 24 h later, brains were harvested and fixed in 10% formalin. Paraffin-embedded sections were immunostained using antibodies against Ki67 (proliferation antigen) and nestin (marker for neural stem cells). The area around the lateral ventricle of the infected brain was visualized by fluorescence microscopy. **a** Neural stem cells are found in high numbers within the subventricular zone (SVZ) located adjacent to the lateral ventricle. The location and proliferative status of neural stem cells was revealed by Ki67 (*red*). Many infected cells (eGFP⁺) were observed primarily in the proliferative regions of the SVZ at this time point. **b** A parallel section was immunostained with nestin (*red*) revealing neural stem cells in the SVZ and their cellular processes stretching deep within the parenchyma. Nearly all infected cells expressed high levels of nestin, as determined by co-localization of signal (*yellow*). **c** Higher magnification illustrated the morphology of infected cell bodies in the SVZ expressing high levels nestin (*notched arrow*). Many infected cells also exhibited extended cellular processes, and viral protein expression co-localized with nestin signal within these processes (*arrows*)

detected around the SVZ and increases in abundance in the days following infection (Feuer et al. 2005). However, we also have found that the great majority of virus-infected (eGFP⁺) cells exiting the SVZ are Ki67⁻ (i.e., presumably nonproliferative) (Feuer et al. 2005), leading us to propose that CVB may infect highly proliferative neural stem cells but then may inhibit the cells' subsequent division, without ablating their capacity to migrate and differentiate. Thus, the outcome of infection at the single-cell level—virus production, or virus persistence—depends not only on the cell's status at the moment of infection, but also on the virus's ability to alter the cell's behavior, rendering it receptive to viral persistence. To date, we have been unable to confidently distinguish between stem cells that are productively infected and those that are persistently infected; it is difficult to directly measure virus production by single infected stem cells in vivo.

## 5 Possible Molecular Mechanisms by Which the Host Cell Status Affects the Outcome of Coxsackievirus Infection

The preceding discussion leads us to the broad concept that the outcome of CVB3 infection of a single cell-productive infection or persistence/latency-is exquisitely dependent upon the cell's metabolic/mitotic status. Infection of a quiescent cell favors persistence/latency of the virus (or of viral components), while infection of a highly active cell tips the balance toward productive infection. The former is well represented by infection of postmitotic cells such as myocytes and neurons; the latter is exemplified by CVB3 infection of pancreatic acinar cells. By what means might the cell cycle status, or other aspects of cellular metabolism, alter the outcome of virus infection? The full cycle of productive virus infection may be considered to have (at least) five stages: (1) binding to a cell surface receptor, (2) entry into the cell/uncoating, (3) viral gene expression (transcription of viral RNAs, translation of virus proteins), (4) viral genome replication, and (5) virion formation and egress. Cellular components may be involved in all of the above and, therefore, all are potentially vulnerable to cell-cycle effects.

## 5.1 Cell Cycle Effects on Virus Receptor

The amount of virus receptor expressed on the cell surface may fluctuate during the cell cycle or depending on the activation state of the cell. Epstein-Barr virus receptor expression is correlated to the cell-cycle phase (Wells et al. 1981), and the HTLV-1 receptor [the ubiquitous vertebrate glucose transporter GLUT-1 (Manel et al. 2003a)] is absent from resting T cells but is quickly upregulated upon T cell activation (Manel et al. 2003a). Most importantly, at least for the purpose of this chapter, CAR expression appears to vary only modestly (approximately two-fold) over the course of the cell cycle, and is slightly increased during the $G_2/M$ phase (Seidman et al. 2001). This suggests that something other than CAR expression is responsible for the reported preference of CVB3 for the $G_1/S$ phase of the cell cycle.

## 5.2 Cell Cycle Effects on Virus Entry

Virus entry into the cell is a complex process that is dependent on both viral and cellular factors. The precise details of enteroviral entry into the cell remain to be resolved (Hogle 2002; Tuthill et al. 2006), but endocytosis has long been considered important for poliovirus entry (Madshus et al. 1984) (as well as for entry of many other viruses). Cellular endocytic activity is abruptly reduced at the onset of mitosis (Illinger et al. 1993), suggesting that cells may be less susceptible to de novo infection at this stage of the cell cycle.

## 5.3 Cell Cycle Effects on Virus Gene Expression: The Internal Ribosomal Entry Site

Although binding to receptor, and entry, are required for successful virus infection, they are not sufficient to guarantee it. Viruses are obligate intracellular parasites and, to produce progeny, they invariably exploit the host cell, by consuming its nutrient and energy stores and usurping its macromolecules. This offers another avenue through which the cell cycle may exert it effects on the virus. The interactions between viral and host components within an infected cell are enormously complex. They involve not only interactions between viral and host proteins, but also between viral and host nucleic acids. For example, virus-encoded micro RNAs can alter host gene expression (Nair and Zavolan 2006), and organ-specific host micro-RNAs can alter viral genome abundance (Jopling et al. 2005). In addition, of course, host proteins can act on virus nucleic acids to regulate gene expression, and several regions of the picornaviral genome have been implicated in host protein interactions. It is important to note, at this point, that for CV, as for many viruses, it is difficult to clearly separate gene expression from genome replication; the two are inextricably intertwined. For example, interrupting CV RNA replication causes a dramatic decrease in the abundance of virus protein, presumably because the number of positive-strand templates available for translation is greatly reduced in the absence of RNA replication. Conversely, a severe constraint on viral translation will decrease the abundance of viral proteins needed to support robust RNA replication. Nevertheless, it is clear that genome replication of many RNA and DNA viruses is cell-cycle-associated (Muszynski et al. 2000; Naniche et al. 1999; Oleksiewicz and Alexandersen 1997; Poggioli et al. 2000; Poon et al. 1998), and some picornavirus RNA motifs appear to predominantly affect replication; thus, we have assigned this topic its own subsection (Sect. 5.4 below), and in this segment we focus solely on the picornaviral internal ribosomal entry site (IRES), which has long been considered a likely candidate for interactions with cellular proteins. The structure and detailed functional analysis of the IRES will not be presented here, because another chapter in this volume is devoted to a discussion of enterovirus translation.

### 5.3.1 Host-Cell IRESs and Their Relationship to the Cell Cycle

Several lines of evidence suggest that some eukaryotic proteins can act on IRESs in a highly-specific manner, and that some of these interactions may vary with the cell cycle.

1. Certain cellular mRNAs contain an IRES. This indicates that some host proteins must be able to interact with IRES motifs and suggests (but, obviously, does not prove) that viral IRESs may be open to regulation by cellular proteins.
2. The activity of some cellular IRESs is highly dependent on cell type (Creancier et al. 2000), indicating that the function of an individual IRES sequence can change in response to alterations in the intracellular microenvironment.
3. Cap-dependent translation of cellular proteins is most robust during the $G_1$ phase of the cell cycle, and certain cellular IRESs are most active during the $G_2/M$ phase, when cap-dependent translation is diminished (Cornelis et al. 2000; Pyronnet et al. 2000; Qin and Sarnow 2004).

4. Several of these IRES-driven cellular mRNAs encode proteins that regulate the cell cycle (Giraud et al. 2001; Pyronnet and Sonenberg 2001), possibly indicating a degree of co-evolution between IRESs and the cell cycle. Several other IRES-containing cellular mRNAs function very efficiently under conditions of cell stress and after the initiation of apoptosis. This is intriguing, because such conditions commonly occur during virus infection.
5. Host proteins that bind to cellular IRESs (IRES transacting factors; ITAFs) have been identified; one of these, ITAF-45, is a cell-cycle related, proliferation-dependent, protein (Pilipenko et al. 2000), further supporting an association between cellular IRESs and the cell cycle.

Thus, cellular IRESs may be seen as motifs that allow cellular proteins to be translated when cells are in the process of dividing, or are under stress; and the presence of IRESs in picornaviruses may explain their predilection for proliferating cells, and their ability to survive in cells that—because of the infection—are severely stressed, and/or are undergoing apoptosis. So much for cellular IRESs. What is the evidence that picornaviral IRESs interact with cellular proteins, and that their activities vary with the cell's status?

### 5.3.2 Interactions Between Picornavirus IRESs and Host Proteins

The contention that cellular proteins can modify the activity of viral IRESs is supported by both indirect and direct evidence.

1. Mutations in the IRES of wild type poliovirus result in altered cellular host range (Shiroki et al. 1997).
2. The Sabin vaccine strains of poliovirus contain IRES mutations that attenuate their growth in cells of neuronal origin (Evans et al. 1985), suggesting the possibility that a translational initiation factor whose IRES interaction is weakened by the mutations may be limiting in neurons. However, a recent study (Kauder and Racaniello 2004) showed that the Sabin vaccine IRES mutations may impair translation not only in neuronal cells, but also in the intestine, where the virus undergoes primary replication; thus, the CNS attenuation may result from lower virus titers, as has been previously suggested (Minor and Dunn 1988).
3. Several cellular ITAFs have been identified that modify the activities of a variety of picornaviral IRESs (reviewed in (Hellen and Sarnow 2001)).
4. The CVB IRES has been implicated as a key determinant of cardiovirulence (Dunn et al. 2000).

### 5.3.3 Picornaviral IRESs and the Cell Cycle: Is the $G_1$ Phase Favored, and If So, Why?

Because most cellular IRESs are most active during the $G_2/M$ phase (Cornelis et al. 2000; Pyronnet et al. 2000; Qin and Sarnow 2004), one might expect that picornaviral IRESs also may be very active during this stage of the cell cycle, and early data were consistent with this expectation, showing that poliovirus translation and RNA levels were maintained in mitotic cells (Bonneau and Sonenberg 1987). However, we have shown that, contrary to that finding, CVB3 protein expression is dramatically reduced in cells arrested in the $G_2/M$ phase, and is most abundant during the $G_1/S$ period (Feuer et al. 2002). For what reasons might evolutionary pressures have tuned the CVB IRES to work so well during the $G_1$ phase? We propose that at least two significant translational benefits might accrue. First, in order to achieve the maximal reduction in the translation of host-cell proteins, CVB must attack cap-dependent translation at its peak; to do so, the responsible virus proteins—the translation of which requires viral IRES function—must be efficiently expressed during the $G_1$ phase. Second, the dramatic decrease in cap-dependent translation in the $G_1$ phase will lead to a concomitant increase in the available amino acid pool, allowing these precious precursors to be commandeered by the viral translational apparatus. These proposed evolutionary benefits may not be restricted to the coxsackieviruses. The EMCV IRES, like the CVB3 IRES, may have greatest activity during the $G_1/S$ stages (Venkatesan et al. 2003). Indeed, the concept may extend beyond the picornavirus family. Recent data suggest that the HCV IRES activity is highest in the $G_1/S$ phase (Venkatesan et al. 2003) and that it is strongest in actively proliferating cells and weakest in quiescent cells (Honda et al. 2000; Scholle et al. 2004).

## 5.4 Cell Cycle Effects on Virus Replication

The virus sequences upon which host cell proteins might act to directly affect RNA replication include the 5' cloverleaf, the *cis*-acting element, and the poly(A) tail. As stated above, effects on virus replication will, almost inevitably, have consequent effects on viral protein abundance.

### 5.4.1 The 5' Cloverleaf Interacts with Host (and Viral) Proteins

The 5' terminus of picornavirus RNAs contains a structure whose predicted folding resembles a cloverleaf. A similar structure is present in several enterovirus and rhinovirus 5' UTRs, and forms a ribonucleoprotein complex with viral proteins, as well as the host poly r(C) binding protein (PCBP) (Andino et al. 1993). The formation of a nucleoprotein complex at the cloverleaf is an absolute requirement for poliovirus RNA replication, suggesting that it may be important for CVB, and work from our laboratory has validated this prediction (Hunziker et al. 2006). However, a recent study indicates that CVB infection in vivo gives rise to viable 5' UTR deletion mutants that lack the prototypical cloverleaf; these naturally occurring mutants appear to replicate at low levels and contain a high proportion of negative-strand

RNA (Kim et al. 2005). One could postulate that the attenuating effects of such deletions, and the resulting viral persistence, might be related to changes in the capacity of the host protein(s) to interact with the foreshortened viral motif; but, to our knowledge, this has not been demonstrated.

### 5.4.2 The *cis*-Acting Element Within the Coding Region

A *cis*-acting replication element (CRE) has been identified within the coding region of several picornaviral genomes. This structure, a hairpin with a terminal loop, has been implicated in RNA replication. The precise role of this sequence in viral replication is somewhat controversial, especially in regard to its relative importance for synthesis of the positive and negative RNA strands, but a very recent study found that, for CVB3, the sequence is involved in the initiation of synthesis of both senses of RNA (Van Ooij et al. 2006b). The role of host proteins in CRE function remains uncertain, but one study reported that a host protein bound to the poliovirus CRE element (Yin et al. 2003).

### 5.4.3 The Picornaviral Poly(A) Tail

Many positive-strand RNA viruses contain poly(A) tails at their 3' ends, and CVB is no exception. The poly(A) sequence is 80-90 nt in length, and appears to play a role in replication (Van Ooij et al. 2006a). A role for the poly(A) tail has recently been described for hepatitis A (HAV) (Kusov et al. 2005). A tailless 3' poly(A) HAV genome was unable to replicate in quiescent cells ($G_0$) or cells arrested at the $G_2$/M phase. However, the 3' poly(A) tail could be restored, functionally and physically, by cellular/viral activity regulated by the cell cycle. The authors further demonstrated that the HAV genome replication was dependent upon cell cycle status and favored cell division. Either proliferating cells (grown in media with 10% fetal bovine serum) or cells blocked at $G_1$/S or $G_1$ gave rise to the highest amount of HAV protein levels, results matching our own published results with CVB3 (Feuer et al. 2002). All these studies indicate that picornaviruses prefer proliferating cells, and that persistence may occur in quiescent cells.

### 5.4.4 Size Matters: The Materiel Requirements for RNA Replication

In Sect. 5.3.3, we presented a hypothesis to explain why the picornaviral IRES might have evolved to operate optimally during the $G_1$ phase of the cell cycle. The proposed advantages were based on protein synthesis, but we suggest that, in addition, viral replication might be enhanced during the $G_1$ phase. During the S phase, as a eukaryotic cell prepares to replicate its DNA, ribonucleotide reductase is induced. This enzyme transforms ribonucleotides into their deoxy equivalents, leading to a consequent decrease in the size of the pool of ribonucleotide precursors (Bjursell and Skoog 1980). Thus, it would be to the virus's advantage to replicate its RNA during the $G_1$ phase, before these events took place. Conversely, a cell infected during mitosis will have a small ribonucleotide pool, a situation that might be better suited to the establishment of RNA virus persistence. We have shown that cells treated with hydroxyurea-a chemical that inhibits ribonucleotide reductase, thereby preventing cellular DNA synthesis and locking cells at the $G_1/S$ phase (Elford 1972)-show vastly increased CVB3 protein expression and infectious virus production (Feuer et al. 2002). Therefore, the preference of CVB for the $G_1$ phase of the cell cycle may have evolved for both replicative and translational reasons.

## 5.5 Cell Cycle Effects on Virus Maturation and Release

Cellular proteins and pathways are involved in the release of a number of viruses. In particular, Tsg101 and the ubiquitin pathway are involved in the budding of HIV (Patnaik et al. 2000; reviewed in Mazze and Degreve 2006) arenaviruses (Urata et al. 2006), orbiviruses (Wirblich et al. 2006), and filoviruses (Harty et al. 2000), all of which are lipid-enveloped. Picornaviruses are nonenveloped, and the mechanism of release from infected cells is unknown (reviewed in Hogle 2002). Virus assembly precedes release and may proceed in a fashion that is relatively independent of cellular assistance. Following polyprotein cleavage, three capsid components (VP1 and VP3 and the myristoylated immature capsid protein VP0 [mVP0]) self-assemble into a pentameric configuration; these pentamers then spontaneously assemble into empty capsids that contain 60 copies each of mVP0, VP3, and VP1. RNA replication is required for subsequent encapsidation (Molla et al. 1991) that results in formation of a provirion containing the RNA and 60 copies each of mVP0, VP3, and VP1. Encapsidation also is associated with cleavage of mVP0 to yield the mature products, mVP4 and VP2. This cleavage, which stabilizes the virion and increases its infectivity, appears to be autocatalytic. The apparent lack of cell dependence suggests that this final phase of infection, once initiated, is unlikely to be affected by cell status, although it is possible that cellular apoptosis may play some part in the process.

Thus it appears that, in the early stage of infection, the virus benefits from closely associating itself with cellular functions. Not only does this allow the virus to exploit the cellular machinery to its own ends, the association also permits the virus to modulate the cell's status. As the infection proceeds, at least two outcomes are possible: persistence/latency or productive infection. In the former case, it behooves the virus to continue to interact with the cell but, if a productive/lytic outcome becomes inevitable, the virus can abandon its interactions and proceed remorselessly with the process of self-assembly and release.

## 6 The Biological Implications of the Relationship Between the Virus and the Cell Cycle

Regardless of the exact mechanism(s), the fact that coxsackieviruses respond to, and can modify, the host cell has many implications.

### 6.1 Implications for Cell Tropism

Highly active cells, such as pancreatic acinar cells and stem cells, may be especially susceptible to infection, as discussed above, although this is only one of many factors that regulate tropism. However, it is important to note that there are many examples of highly metabolically active cells that remain relatively unscathed during CVB infection [for example, hepatocytes, which not only are biosynthetically active, but also are capable of rapid division if required; type I IFNs play a key part in protecting these cells from the ravages of CVB infection (Wessely et al. 2001)].

### 6.2 Implications for Transmission

CVB is transmitted mainly by the fecal-oral route, but the site of its primary replication remains uncertain; this enterovirus does not appear to replicate extensively in intestinal epithelial cells, instead being found in mucosal lymphocytes as soon as 2 h following oral inoculation (Harrath et al. 2004). Interestingly, it has been shown that oral inoculation leads to pancreatic infection, but not to detectable pancreatic disease (Bopegamage et al. 2005). It is of interest that the titers of CVB3 in the pancreas following oral inoculation were extremely low [peaking at $\sim 10^2$-$10^3$ pfu (Bopegamage et al. 2005)]. This contrasts dramatically with the outcome of idiopathic pancreatitis infection, in which high pancreatic titers are reached ($\sim 10^9$/g), and severe exocrine pancreatitis is the rule. It may be relevant to human infection, which usually occurs via the fecal-oral route, which rarely causes symptomatic pancreatitis. The authors proposed a plausible explanation: that oral inoculation may lead to a stronger innate immune response, which cannot prevent pancreatic infection, but is sufficient to ameliorate disease. We speculate that an additional factor may be at play. Activity of the exocrine pancreas is highly cyclical, and is regulated by hormonal signals (e.g., cholecystokinin, CCK) some of which are triggered by food intake. We propose that the signals triggered by food ingestion (which is the time at which the host is most likely to encounter this enterovirus) change the metabolic status of pancreatic acinar cells, causing these cells to support a low level of productive infection that is sufficient to ensure viral shedding in the stool (thereby maintaining the virus in the host population) but is insufficient to cause florid pancreatitis. If true, this would indicate that—as in many other virus/host relationships-CVB and its host may have reached a position of evolutionary equilibrium. This hypothesis could be tested by administering a CCK receptor antagonist such as loxiglumide, which has been shown to have beneficial effects in several nonviral models of acute pancreatitis (Satake et al. 1999).

## 6.3   Facilitation of Persistent/Latent Infection

Picornaviruses sometimes are cited as the prototypical lytic virus; many of these agents (including poliovirus and CVB) rapidly shut down host transcriptional and translational machinery, and infectious progeny can be released in as few as 4-6 h. However, as recently reviewed in some detail (Colbere-Garapin et al. 2002), it is becoming clear that in some cases picornaviruses—or, at least, viral components—may persist. Both the cardiovirus TMEV and the aphthovirus foot-and-mouth disease virus (FMDV) can result in a carrier state, in which the infected host continuously sheds infectious particles for several years (Rodriguez et al. 1987; Salt 1998). The situation is different for enteroviruses, for which infectious particles are extraordinarily difficult to detect after the acute infection has been resolved [except in immunocompromised individuals, who may shed infectious materials for up to 20 years (Martin et al. 2004)]. However, both for poliovirus and CVB, viral RNA often can be detected in host tissues many years after initial infection. Persistence of poliovirus RNA has frequently been reported in patients suffering from postpolio syndrome (Julien et al. 1999; Leon-Monzon and Dalakas 1995; Leparc-Goffart et al. 1996; Muir et al. 1995), which occurs in approximately 50% of victims of paralytic polio, usually appearing some 30 years after the acute disease (Dalakas et al. 1986). Persistent RNA also has been frequently found in several tissues following CVB infection. Slot blot hybridization studies have shown positive signal for coxsackie virus RNA in myocardial biopsy specimens of approximately 45% of patients with myocarditis or its serious sequela, dilated cardiomyopathy (DCM), compared with none of the controls (Martino et al. 1995). Interestingly, some 43% of patients with healed myocarditis or DCM remained positive for CVB signal (Archard et al. 1991). High levels of CVB-specific neutralizing antibodies are found in about 50% of patients, and serial antibody studies show a fourfold or greater change in paired sera in approximately half of patients (Martino et al. 1995); both observations suggest that the host immune system may be sporadically re-encountering CVB antigen. The long-term prognosis following acute myocarditis is substantially worse if the acute disease was CVB-related. In one 15-year follow-up study of myocarditis patients, 25% of patients with serological evidence of CVB infection died from subsequent chronic myocarditis or cardiomyopathy, while none of the 26 patients with negative viral serology died (Levi et al. 1988). Although this finding is open to various interpretations, it is possible that the long-term morbidity and mortality results from the retention of CVB materials. Indeed, the presence of enteroviral RNA in heart tissue is associated with a poorer prognosis (Archard et al. 1991; Bowles et al. 1989). CVB RNA also has been detected in skeletal muscle, in mouse models of polymyositis (Tam et al. 1991, 1994), and in humans suffering from idiopathic inflammatory myopathy (Bowles et al. 1987). Because the topic of virus persistence is covered elsewhere in this volume, we shall provide here only our hypotheses regarding the possible link(s) between persistence and the cell cycle.

### 6.3.1 How Might Picornavirus RNA Persist in Target Tissues?

The high frequency with which viral RNA is detected, along with the difficulty in isolating infectious virus, indicate that these viruses may be carried within the host in latent form. A careful in vivo study of CVB infection reported that wild-type sequences were retained in the form of stable double-stranded RNA (Tam and Messner 1999); perhaps the double-stranded RNAs are the picornaviral equivalent of the better-characterized latent genomes of several DNA viruses. A recent, intriguing paper reported the identification of low levels of persistent CVB3 in a mouse model; these viruses had various deletions from their 5' termini, but remained viable (if only at low level) and transmissible (Kim et al. 2005). To our knowledge, this is the first paper showing the long-term in vivo persistence of infectious CVB in an immunocompetent host. It will be interesting to determine if cell cycle constraints and/or specific tissue microenvironments select for the emergence of such attenuated viruses with 5' terminal deletions.

## 6.4 Implications for Pathogenesis

Taken together, the data and hypotheses presented above have several implications for CVB pathogenesis. First, a correlation between cell division and productive infection may, in part, explain why neonates (human and mouse) are much more susceptible to CVB-associated morbidity and mortality. However, the mere requirement for organismal growth (and thus cell division) cannot completely explain the correlation, because susceptibility wanes very rapidly (Feuer et al. 2003), long before the individual has reached adulthood. Second, the correlation between cell quiescence ($G_0$ cells) and virus (or viral RNA) persistence may explain why CVB RNA is found in postmitotic cells such as skeletal and cardiac myocytes and why poliovirus RNA persists in the human CNS. Third, viral persistence/latency could lead to chronic or recurrent activation of the immune system. It has been hypothesized that a persistent, unidentified enteroviral infection may be the root of chronic human diseases such as schizophrenia (Rantakallio et al. 1997), amyotrophic lateral sclerosis (Woodall and Graham 2004; Woodall et al. 1994), Sjögren's disease (Metskula et al. 2006; Triantafyllopoulou et al. 2004), and chronic fatigue syndrome (Douche-Aourik et al. 2003; Galbraith et al. 1995). One can imagine at least two routes by which the persistence of viral materials might lead to long-term immunopathological disease.

1. Even in the absence of viral protein expression, persistent viral RNA might chronically activate the innate immune response. As noted above, some studies suggest that CVB RNA may persist in double-stranded form, and dsRNA is a strong inducer of type I interferons. Furthermore, single-stranded CVB3 RNA co-localizes with TLR-7 and TLR-8 in endosomes, activates NF-κB, and induces cytokine synthesis (Triantafilou et al. 2005). These inflammatory responses would not be specific for CVB antigens and may be diagnosed as autoimmune in nature [the possible role of autoimmunity in CVB disease has been recently reviewed by Huber (2006) and Tam (2006)].

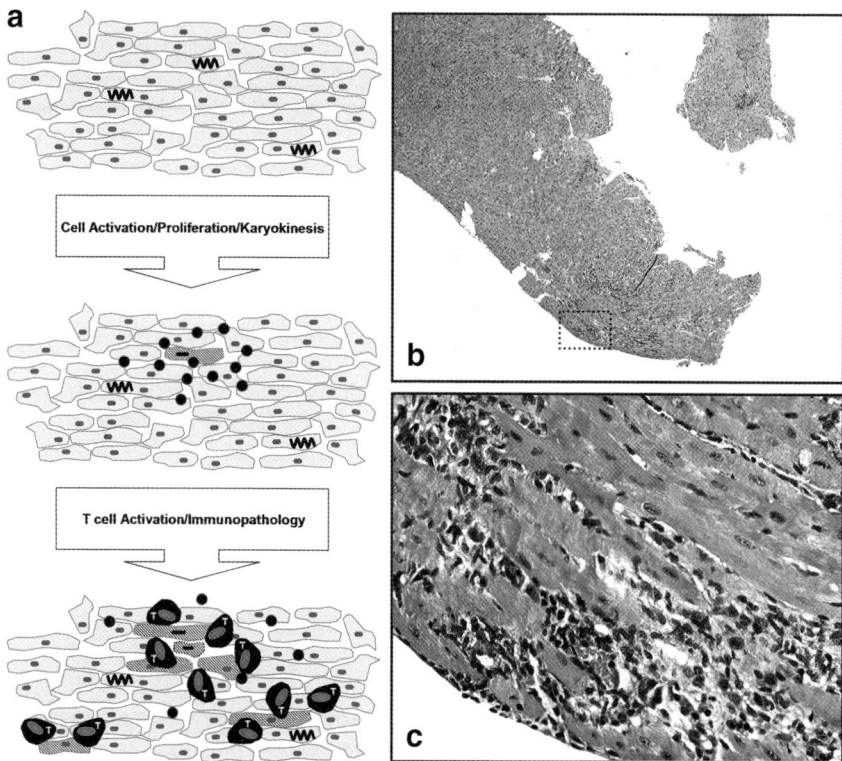

**Fig. 4** Coxsackievirus latency and reactivation: implications for chronic human disease. **a** Coxsackieviral RNA (*squiggly lines*) may persist in a latent state within quiescent, differentiated cells of the heart. Following cellular activation, proliferation, or karyokinesis, reactivation of viral RNA may lead to viral protein expression and sporadic infectious virus production (*black circles*). Latently infected cells triggering virus reactivation and newly infected adjacent cells presenting viral antigen may be targeted by virus-specific effector T cells (*black cells, white T*). The outcome over the long term of continuous virus reactivation and chronic T cell inflammation in patients suffering from myocarditis may be virus-mediated immunopathology with clinical features similar to autoimmune disease. **b** H&E staining of the heart isolated from an adult C57 BL/6 mice sacrificed 10 days after infection with CVB3. The region enclosed by a *dotted rectangle* is shown in higher magnification in **c**, and reveals the presence of inflammatory cells within the cardiac tissue

2. Ongoing, or sporadic, CVB protein expression could activate virus-specific adaptive immune responses (T cells and/or antibodies) with immunopathological consequences. This is represented diagrammatically in Fig. 4a. Examples of virus-specific immunopathology, during acute infection, are shown in Fig. 4b and c.

In conclusion, much remains to be learned about the complex relationship between coxsackieviruses and the host cell. New approaches such a nucleic acid arrays, proteomics, and systems biology are only now being applied to many virus-cell interactions and these will, hopefully, yield new concepts of compelling interest to both the academician and the clinician.

**Acknowledgements** We are grateful to Annette Lord for excellent secretarial support, and to the various members of the laboratory who have contributed to the coxsackievirus work over the years. JLW's coxsackievirus research is supported by an NIH R-01 award AI-42314, and RF was supported by an Advanced Postdoctoral Fellowship from the National Multiple Sclerosis Society (FA 1551-A-1). This is manuscript number 18427 from the Scripps Research Institute.

# References

Andino R, Rieckhof GE, Achacoso PL, Baltimore D (1993) Poliovirus RNA synthesis utilizes an RNP complex formed around the 5′-end of viral RNA. EMBO J 12:3587-3598

Archard LC, Bowles NE, Cunningham L, Freeke CA, Olsen EG, Rose ML, Meany B, Why HJ, Richardson PJ (1991) Molecular probes for detection of persisting enterovirus infection of human heart and their prognostic value. Eur Heart J 12 [Suppl D]:56-59

Bjursell G, Skoog L (1980) Control of nucleotide pools in mammalian cells. Antibiot Chemother 28:78-85

Bonneau AM, Sonenberg N (1987) Involvement of the 24-kDa cap-binding protein in regulation of protein synthesis in mitosis. J Biol Chem 262:11134-11139

Bopegamage S, Kovacova J, Vargova A, Motusova J, Petrovicova A, Benkovicova M, Gomolcak P, Bakkers J, van KF, Melchers WJ, Galama JM (2005) Coxsackie B virus infection of mice: inoculation by the oral route protects the pancreas from damage, but not from infection. J Gen Virol 86:3271-3280

Bowles NE, Dubowitz V, Sewry CA, Archard LC (1987) Dermatomyositis, polymyositis, and Coxsackie-B-virus infection. Lancet 1:1004-1007

Bowles NE, Rose ML, Taylor P, Banner NR, Morgan-Capner P, Cunningham L, Archard LC, Yacoub MH (1989) End-stage dilated cardiomyopathy. Persistence of enterovirus RNA in myocardium at cardiac transplantation and lack of immune response. Circulation 80:1128-1136

Brenchley JM, Price DA, Douek DC (2006) HIV disease: fallout from a mucosal catastrophe? Nat Immunol 7:235-239

Brown KE, Anderson SM, Young NS (1993) Erythrocyte P antigen: cellular receptor for B19 parvovirus. Science 262:114-117

Bukrinsky M (2004) A hard way to the nucleus. Mol Med 10:1-5

Carthy CM, Granville DJ, Watson KA, Anderson DR, Wilson JE, Yang D, Hunt DW, McManus BM (1998) Caspase activation and specific cleavage of substrates after coxsackievirus B3-induced cytopathic effect in HeLa cells. J Virol 72:7669-7675

Colbere-Garapin F, Pelletier I, Ouzilou L (2002) Persistent infections by picornaviruses. In: Semler BL, Wimmer E (eds) Molecular biology of picornaviruses. ASM Press, Washington DC, pp 437-448

Cornelis S, Bruynooghe Y, Denecker G, van Huffel S, Tinton S, Beyaert R (2000) Identification and characterization of a novel cell cycle-regulated internal ribosome entry site. Mol Cell 5:597-605

Creancier L, Morello D, Mercier P, Prats AC (2000) Fibroblast growth factor 2 internal ribosome entry site (IRES) activity ex vivo and in transgenic mice reveals a stringent tissue-specific regulation. J Cell Biol 150:275-281

Dalakas MC, Elder G, Hallett M, Ravits J, Baker M, Papadopoulos N, Albrecht P, Sever J (1986) A long-term follow-up study of patients with post-poliomyelitis neuromuscular symptoms. N Engl J Med 314:959-963

Daley AJ, Isaacs D, Dwyer DE, Gilbert GL (1998) A cluster of cases of neonatal coxsackievirus B meningitis and myocarditis. J Paediatr Child Health 34:196-198

Díaz-San Segundo F, Salguero FJ, de Avila A, de Marco MM, Sanchez-Martin MA, Sevilla N (2006) Selective lymphocyte depletion during the early stage of the immune response to foot-and-mouth disease virus infection in swine. J Virol 80:2369-2379

Douche-Aourik F, Berlier W, Feasson L, Bourlet T, Harrath R, Omar S, Grattard F, Denis C, Pozzetto B (2003) Detection of enterovirus in human skeletal muscle from patients with chronic inflammatory muscle disease or fibromyalgia and healthy subjects. J Med Virol 71:540-547

Dunn JJ, Chapman NM, Tracy S, Romero JR (2000) Genomic determinants of cardiovirulence in coxsackievirus B3 clinical isolates: localization to the 5' nontranslated region. J Virol 74:4787-4794

Elford HL (1972) Functional regulation of mammalian ribonucleotide reductase. Adv Enzyme Regul 10:19-38

Eremenko T, Benedetto A, Volpe P (1972a) Poliovirus replication during HeLa cell life cycle. Nat New Biol 237:114-116

Eremenko T, Benedetto A, Volpe P (1972b) Virus infection as a function of the host cell life cycle: replication of poliovirus RNA. J Gen Virol 16:61-68

Euscher E, Davis J, Holzman I, Nuovo GJ (2001) Coxsackie virus infection of the placenta associated with neurodevelopmental delays in the newborn. Obstet Gynecol 98:1019-1026

Evans DM, Dunn G, Minor PD, Schild GC, Cann AJ, Stanway G, Almond JW, Currey K, Maizel JV Jr (1985) Increased neurovirulence associated with a single nucleotide change in a noncoding region of the Sabin type 3 polio vaccine genome. Nature 314:548-550

Feuer R, Mena I, Pagarigan RR, Slifka MK, Whitton JL (2002) Cell cycle status affects coxsackievirus replication, persistence, and reactivation in vitro. J Virol 76:4430-4440

Feuer R, Mena I, Pagarigan RR, Harkins S, Hassett DE, Whitton JL (2003) Coxsackievirus B3 and the neonatal CNS: the roles of stem cells, developing neurons, and apoptosis in infection, viral dissemination, and disease. Am J Pathol 163:1379-1393

Feuer R, Mena I, Pagarigan RR, Hassett DE, Whitton JL (2004) Coxsackievirus replication and the cell cycle: a potential regulatory mechanism for viral persistence/latency. Med Microbiol Immunol (Berl) 193:83-90

Feuer R, Pagarigan RR, Harkins S, Liu F, Hunziker IP, Whitton JL (2005) Coxsackievirus targets proliferating neuronal progenitor cells in the neonatal CNS. J Neurosci 25:2434-2444

Fritsch EF, Temin HM (1977) Inhibition of viral DNA synthesis in stationary chicken embryo fibroblasts infected with avian retroviruses. J Virol 24:461-469

Galbraith DN, Nairn C, Clements GB (1995) Phylogenetic analysis of short enteroviral sequences from patients with chronic fatigue syndrome. J Gen Virol 76:1701-1707

Giraud S, Greco A, Brink M, Diaz JJ, Delafontaine P (2001) Translation initiation of the insulin-like growth factor I receptor mRNA is mediated by an internal ribosome entry site. J Biol Chem 276:5668-5675

Harrath R, Bourlet T, Delezay O, Douche-Aourik F, Omar S, Aouni M, Pozzetto B (2004) Coxsackievirus B3 replication and persistence in intestinal cells from mice infected orally and in the human CaCo-2 cell line. J Med Virol 74:283-290

Harty RN, Brown ME, Wang G, Huibregtse J, Hayes FP (2000) A PPxY motif within the VP40 protein of Ebola virus interacts physically and functionally with a ubiquitin ligase: implications for filovirus budding. Proc Natl Acad Sci U S A 97:13871-13876

Hassett DE, Zhang J, Whitton JL (1997) Neonatal DNA immunization with an internal viral protein is effective in the presence of maternal antibodies and protects against subsequent viral challenge. J Virol 71:7881-7888

Hellen CU, Sarnow P (2001) Internal ribosome entry sites in eukaryotic mRNA molecules. Genes Dev 15:1593-1612

Helt AM, Harris E (2005) S-phase-dependent enhancement of dengue virus 2 replication in mosquito cells, but not in human cells. J Virol 79:13218-13230

Henke A, Launhardt H, Klement K, Stelzner A, Zell R, Munder T (2000) Apoptosis in coxsackievirus B3-caused diseases: interaction between the capsid protein VP2 and the proapoptotic protein siva. J Virol 74:4284-4290

Hogle JM (2002) Poliovirus cell entry: common structural themes in viral cell entry pathways. Annu Rev Microbiol 56:677-702

Honda M, Kaneko S, Matsushita E, Kobayashi K, Abell GA, Lemon SM (2000) Cell cycle regulation of hepatitis C virus internal ribosomal entry site-directed translation. Gastroenterology 118:152-162

Hsueh C, Jung SM, Shih SR, Kuo TT, Shieh WJ, Zaki S, Lin TY, Chang LY, Ning HC, Yen DC (2000) Acute encephalomyelitis during an outbreak of enterovirus type 71 infection in Taiwan: report of an autopsy case with pathologic, immunofluorescence, and molecular studies. Mod Pathol 13:1200-1205

Huber SA (2006) Autoimmunity in coxsackievirus B3 induced myocarditis. Autoimmunity 39:55-61

Humphries EH, Temin HM (1974) Requirement for cell division for initiation of transcription of Rous sarcoma virus RNA. J Virol 14:531-546

Hunziker IP, Cornell CT, Whitton JL (2006) Deletions within the 5′UTR of coxsackievirus B3: consequences for virus translation and replication. Virology 360:120-128

Illinger D, Italiano L, Beck JP, Waltzinger C, Kuhry JG (1993) Comparative evolution of endocytosis levels and of the cell surface area during the L929 cell cycle: a fluorescence study with TMA-DPH. Biol Cell 79:265-268

Jopling CL, Yi M, Lancaster AM, Lemon SM, Sarnow P (2005) Modulation of hepatitis C virus RNA abundance by a liver-specific MicroRNA. Science 309:1577-1581

Julien J, Leparc-Goffart I, Lina B, Fuchs F, Foray S, Janatova I, Aymard M, Kopecka H (1999) Postpolio syndrome: poliovirus persistence is involved in the pathogenesis. J Neurol 246:472-476

Kaplan MH, Klein SW, McPhee J, Harper RG (1983) Group B coxsackievirus infections in infants younger than three months of age: a serious childhood illness. Rev Infect Dis 5:1019-1032

Kauder SE, Racaniello VR (2004) Poliovirus tropism and attenuation are determined after internal ribosome entry. J Clin Invest 113:1743-1753

Kim KS, Tracy S, Tapprich W, Bailey J, Lee CK, Kim K, Barry WH, Chapman NM (2005) 5′-terminal deletions occur in coxsackievirus B3 during replication in murine hearts and cardiac myocyte cultures and correlate with encapsidation of negative-strand viral RNA. J Virol 79:7024-7041

Klingel K, Stephan S, Sauter M, Zell R, McManus BM, Bultmann B, Kandolf R (1996) Pathogenesis of murine enterovirus myocarditis: virus dissemination and immune cell targets. J Virol 70:8888-8895

Koch AS, Eremenko T, Benedetto A, Volpe P (1974) A guanidine-sensitive step of the poliovirus RNA replication cycle. Intervirology 4:221-225

Kusov YY, Gosert R, Gauss-Muller V (2005) Replication and in vivo repair of the hepatitis A virus genome lacking the poly(A) tail. J Gen Virol 86:1363-1368

Lake RS, Winkler DC, Ludwig EH (1970) Delay of mengovirus-induced cytopathology in mitotic L-cells. J Virol 5:262-263

Leon-Monzon ME, Dalakas MC (1995) Detection of poliovirus antibodies and poliovirus genome in patients with the post-polio syndrome. Ann N Y Acad Sci 753:208-218

Leparc-Goffart I, Julien J, Fuchs F, Janatova I, Aymard M, Kopecka H (1996) Evidence of presence of poliovirus genomic sequences in cerebrospinal fluid from patients with postpolio syndrome. J Clin Microbiol 34:2023-2026

Levi G, Scalvini S, Volterrani M, Marangoni S, Arosio G, Quadri A (1988) Coxsackie virus heart disease: 15 years after. Eur Heart J 9:1303-1307

Liu P, Aitken K, Kong YY, Opavsky MA, Martino T, Dawood F, Wen WH, Kozieradzki I, Bachmaier K, Straus D, Mak TW, Penninger JM (2000) The tyrosine kinase p56lck is essential in coxsackievirus B3-mediated heart disease. Nat Med 6:429-434

Madshus IH, Olsnes S, Sandvig K (1984) Requirements for entry of poliovirus RNA into cells at low pH. EMBO J 3:1945-1950

Mallucci L, Wells V, Beare D (1985) Cell cycle position and expression of encephalomyocarditis virus in mouse embryo fibroblasts. J Gen Virol 66:1501-1506

Manel N, Kim FJ, Kinet S, Taylor N, Sitbon M, Battini JL (2003a) The ubiquitous glucose transporter GLUT-1 is a receptor for HTLV. Cell 115:449-459

Manel N, Kinet S, Battini JL, Kim FJ, Taylor N, Sitbon M (2003a) The HTLV receptor is an early T-cell activation marker whose expression requires de novo protein synthesis. Blood 101:1913-1918

Marcus PI, Robbins E (1963) Viral inhibition in the metaphase-arrest cell. Proc Natl Acad Sci U S A 50:1156-1164

Martin J, Odoom K, Tuite G, Dunn G, Hopewell N, Cooper G, Fitzharris C, Butler K, Hall WW, Minor PD (2004) Long-term excretion of vaccine-derived poliovirus by a healthy child. J Virol 78:13839-13847

Martinez X, Brandt C, Saddallah F, Tougne C, Barrios C, Wild F, Dougan G, Lambert PH, Siegrist CA (1997) DNA immunization circumvents deficient induction of T helper type 1 and cytotoxic T lymphocyte responses in neonates and during early life. Proc Natl Acad Sci U S A 94:8726-8731

Martino TA, Liu P, Petric M, Sole MJ (1995) Enteroviral myocarditis and dilated cardiomyopathy: a review of clinical and experimental studies. 291-351

Mazze FM, Degreve L (2006) The role of viral and cellular proteins in the budding of human immunodeficiency virus. Acta Virol 50:75-85

Mena I, Perry CM, Harkins S, Rodriguez E, Gebhard JR, Whitton JL (1999) The role of B lymphocytes in coxsackievirus B3 infection. Am J Pathol 155:1205-1215

Metskula K, Salur L, Mandel M, Uibo R (2006) Demonstration of high prevalence of SS-A antibodies in a general population: association with HLA-DR and enterovirus antibodies. Immunol Lett 106:14-18

Minor PD, Dunn G (1988) The effect of sequences in the 5′ non-coding region on the replication of polioviruses in the human gut. J Gen Virol 69:1091-1096

Modlin JF (1988) Perinatal echovirus and group B coxsackievirus infections. Clin Perinatol 15:233-246

Modlin JF, Bowman M (1987) Perinatal transmission of coxsackievirus B3 in mice. J Infect Dis 156:21-25

Molla A, Paul AV, Wimmer E (1991) Cell-free, de novo synthesis of poliovirus. Science 254:1647-1651

Muir P, Nicholson F, Sharief MK, Thompson EJ, Cairns NJ, Lantos P, Spencer GT, Kaminski HJ, Banatvala JE (1995) Evidence for persistent enterovirus infection of the central nervous system in patients with previous paralytic poliomyelitis. Ann N Y Acad Sci 753:219-232

Muszynski KW, Thompson D, Hanson C, Lyons R, Spadaccini A, Ruscetti SK (2000) Growth factor-independent proliferation of erythroid cells infected with Friend spleen focus-forming virus is protein kinase C dependent but does not require Ras-GTP. J Virol 74:8444-8451

Nair V, Zavolan M (2006) Virus-encoded microRNAs: novel regulators of gene expression. Trends Microbiol 14:169-175

Naniche D, Oldstone MB (2000) Generalized immunosuppression: how viruses undermine the immune response. Cell Mol Life Sci 57:1399-1407

Naniche D, Reed SI, Oldstone MBA (1999) Cell cycle arrest during measles virus infection: a G0-like block leads to suppression of retinoblastoma protein expression. J Virol 73:1894-1901

Nelson HB, Tang H (2006) Effect of cell growth on hepatitis C virus (HCV) replication and a mechanism of cell confluence-based inhibition of HCV RNA and protein expression. J Virol 80:1181-1190

Oleksiewicz MB, Alexandersen S (1997) S-phase-dependent cell cycle disturbances caused by Aleutian mink disease parvovirus. J Virol 71:1386-1396

Op De Beeck A, Caillet-Fauquet P (1997) Viruses and the cell cycle. Prog Cell Cycle Res 3:1-19

Opavsky MA, Martino T, Rabinovitch M, Penninger J, Richardson C, Petric M, Trinidad C, Butcher L, Chan J, Liu PP (2002) Enhanced ERK-1/2 activation in mice susceptible to coxsackievirus-induced myocarditis. J Clin Invest 109:1561-1569

Patnaik A, Chau V, Wills JW (2000) Ubiquitin is part of the retrovirus budding machinery. Proc Natl Acad Sci U S A 97:13069-13074

Pilipenko EV, Pestova TV, Kolupaeva VG, Khitrina EV, Poperechnaya AN, Agol VI, Hellen CU (2000) A cell cycle-dependent protein serves as a template-specific translation initiation factor. Genes Dev 14:2028-2045

Poggioli GJ, Keefer C, Connolly JL, Dermody TS, Tyler KL (2000) Reovirus-induced G2/M cell cycle arrest requires sigma1s and occurs in the absence of apoptosis. J Virol 74:9562-9570

Poon B, Grovit-Ferbas K, Stewart SA, Chen IS (1998) Cell cycle arrest by Vpr in HIV-1 virions and insensitivity to antiretroviral agents. Science 281:266-269

Pyronnet S, Sonenberg N (2001) Cell-cycle-dependent translational control. Curr Opin Genet Dev 11:13-18

Pyronnet S, Pradayrol L, Sonenberg N (2000) A cell cycle-dependent internal ribosome entry site. Mol Cell 5:607-616

Qin X, Sarnow P (2004) Preferential translation of internal ribosome entry site-containing mRNAs during the mitotic cycle in mammalian cells. J Biol Chem 279:13721-13728

Rall GF (2003) Measles virus 1998-2002: progress and controversy. Annu Rev Microbiol 57:343-367

Ramirez JC, Fairen A, Almendral JM (1996) Parvovirus minute virus of mice strain I multiplication and pathogenesis in the newborn mouse brain are restricted to proliferative areas and to migratory cerebellar young neurons. J Virol 70:8109-8116

Rantakallio P, Jones P, Moring J, von Wendt L (1997) Association between central nervous system infections during childhood and adult onset schizophrenia and other psychoses: a 28-year follow-up. Int J Epidemiol 26:837-843

Ratzan KR (1985) Viral meningitis. Med Clin North Am 69:399-413

Ridge JP, Fuchs EJ, Matzinger P (1996) Neonatal tolerance revisited: turning on newborn T cells with dendritic cells. Science 271:1723-1726

Rodriguez M, Oleszak E, Leibowitz J (1987) Theiler's murine encephalomyelitis: a model of demyelination and persistence of virus. Crit Rev Immunol 7:325-365

Salt JS (1998) Persistent infection with foot-and-mouth disease virus. Top Trop Virol 1:77-129

Satake K, Kimura K, Saito T (1999) Therapeutic effects of loxiglumide on experimental acute pancreatitis using various models. Digestion 60 [Suppl 1]:64-68

Sauerbrei A, Gluck B, Jung K, Bittrich H, Wutzler P (2000) Congenital skin lesions caused by intrauterine infection with coxsackievirus B3. Infection 28:326-328

Scholle F, Li K, Bodola F, Ikeda M, Luxon BA, Lemon SM (2004) Virus-host cell interactions during hepatitis C virus RNA replication: impact of polyprotein expression on the cellular transcriptome and cell cycle association with viral RNA synthesis. J Virol 78:1513-1524

Seidman MA, Hogan SM, Wendland RL, Worgall S, Crystal RG, Leopold PL (2001) Variation in adenovirus receptor expression and adenovirus vector-mediated transgene expression at defined stages of the cell cycle. Mol Ther 4:13-21

Shapiro-Shelef M, Calame K (2005) Regulation of plasma-cell development. Nat Rev Immunol 5:230-242

Shaw-Jackson C, Michiels T (1997) Infection of macrophages by Theiler's murine encephalomyelitis virus is highly dependent on their activation or differentiation state. J Virol 71:8864-8867

Shiroki K, Ishii T, Aoki T, Ota Y, Yang WX, Komatsu T, Ami Y, Arita M, Abe S, Hashizume S, Nomoto A (1997) Host range phenotype induced by mutations in the internal ribosomal entry site of poliovirus RNA. J Virol 71:1-8

Srivastava D, Ivey KN (2006) Potential of stem-cell-based therapies for heart disease. Nature 441:1097-1099

Suarez M, Contreras G, Fridlender B (1975) Multiplication of Coxsackie B1 virus in synchronized HeLa cells. J Virol 16:1337-1339

Swanton C, Jones N (2001) Strategies in subversion: de-regulation of the mammalian cell cycle by viral gene products. Int J Exp Pathol 82:3-13

Talbot SJ, Crawford DH (2004) Viruses and tumours-an update. Eur J Cancer 40:1998-2005

Tam PE (2006) Coxsackievirus myocarditis: interplay between virus and host in the pathogenesis of heart disease. Viral Immunol 19:133-146

Tam PE, Messner RP (1999) Molecular mechanisms of coxsackievirus persistence in chronic inflammatory myopathy: viral RNA persists through formation of a double-stranded complex without associated genomic mutations or evolution. J Virol 73:10113-10121

Tam PE, Schmidt AM, Ytterberg SR, Messner RP (1991) Viral persistence during the developmental phase of Coxsackievirus B1-induced murine polymyositis. J Virol 65:6654-6660

Tam PE, Schmidt AM, Ytterberg SR, Messner RP (1994) Duration of virus persistence and its relationship to inflammation in the chronic phase of coxsackievirus B1-induced murine polymyositis. J Lab Clin Med 123:346-356

Triantafilou K, Orthopoulos G, Vakakis E, Ahmed MA, Golenbock DT, Lepper PM, Triantafilou M (2005) Human cardiac inflammatory responses triggered by Coxsackie B viruses are mainly Toll-like receptor (TLR) 8-dependent. Cell Microbiol 7:1117-1126

Triantafyllopoulou A, Tapinos N, Moutsopoulos HM (2004) Evidence for coxsackievirus infection in primary Sjögren's syndrome. Arthritis Rheum 50:2897-2902

Tuthill TJ, Bubeck D, Rowlands DJ, Hogle JM (2006) Characterization of early steps in the poliovirus infection process: receptor-decorated liposomes induce conversion of the virus to membrane-anchored entry-intermediate particles. J Virol 80:172-180

Urata S, Noda T, Kawaoka Y, Yokosawa H, Yasuda J (2006) Cellular factors required for Lassa virus budding. J Virol 80:4191-4195

Van Ooij MJ, Polacek C, Glaudemans DH, Kuijpers J, van Kuppeveld FJ, Andino R, Agol VI, Melchers WJ (2006a) Polyadenylation of genomic RNA and initiation of antigenomic RNA in a positive-strand RNA virus are controlled by the same cis-element. Nucleic Acids Res 34:2953-2965

Van Ooij MJ, Vogt DA, Paul A, Castro C, Kuijpers J, van Kuppeveld FJ, Cameron CE, Wimmer E, Andino R, Melchers WJ (2006b) Structural and functional characterization of the coxsackievirus B3 CRE(2C): role of CRE(2C) in negative- and positive-strand RNA synthesis. J Gen Virol 87:103-113

Venkatesan A, Sharma R, Dasgupta A (2003) Cell cycle regulation of hepatitis C and encephalomyocarditis virus internal ribosome entry site-mediated translation in human embryonic kidney 293 cells. Virus Res 94:85-95

Verschuren EW, Jones N, Evan GI (2004) The cell cycle and how it is steered by Kaposi's sarcoma-associated herpesvirus cyclin. J Gen Virol 85:1347-1361

Watts AM, Stanley JR, Shearer MH, Hefty PS, Kennedy RC (1999) Fetal immunization of baboons induces a fetal-specific antibody response. Nat Med 5:427-430

Wells A, Steen HB, Godal T, Klein G (1981) Epstein-Barr virus receptor expression is correlated to cell cycle phase. J Recept Res 2:285-298

Wessely R, Klingel K, Knowlton KU, Kandolf R (2001) Cardioselective infection with coxsackievirus B3 requires intact type I interferon signaling: implications for mortality and early viral replication. Circulation 103:756-761

Wirblich C, Bhattacharya B, Roy P (2006) Nonstructural protein 3 of bluetongue virus assists virus release by recruiting ESCRT-I protein Tsg101. J Virol 80:460-473

Woodall CJ, Graham DI (2004) Evidence for neuronal localisation of enteroviral sequences in motor neurone disease/amyotrophic lateral sclerosis by in situ hybridization. Eur J Histochem 48:129-134

Woodall CJ, Riding MH, Graham DI, Clements GB (1994) Sequences specific for enterovirus detected in spinal cord from patients with motor neurone disease. BMJ 308:1541-1543

Yin J, Paul AV, Wimmer E, Rieder E (2003) Functional dissection of a poliovirus cis-acting replication element [PV-cre(2C)]: analysis of single- and dual-cre viral genomes and proteins that bind specifically to PV-cre RNA. J Virol 77:5152-5166

Young NS, Brown KE (2004) Parvovirus B19. N Engl J Med 350:586-597

# Section III
# Host-Virus Interaction

# The Impact of CVB3 Infection on Host Cell Biology

D. Marchant, X. Si, H. Luo, B. McManus(✉), and D. Yang

| | | |
|---|---|---|
| 1 | Coxsackie Adenovirus Receptor and Decay Accelerating Factor and Host Cell Signalling............................................................................................... | 178 |
| 2 | Host-Cell-Protein Interactions with the 5′ Untranslated Region of Coxsackievirus B3.. | 180 |
| 3 | The Inflammatory Response to CVB3 Infection and Host Susceptibility ...................... | 181 |
| 4 | Coxsackievirus B3 is a Primary Pathogen for Viral Myocarditis .................................. | 182 |
| | 4.1 Early Direct Injury of Myocardium by Coxsackievirus B3 Infection .................... | 182 |
| 5 | CVB3 Infection and Host-Cell Survival ........................................................................ | 183 |
| | 5.1 CVB3-Induced Cell Death Pathways...................................................................... | 183 |
| | 5.2 Cardiac Remodelling: The Matrix Metalloproteases............................................. | 183 |
| 6 | Coxsackievirus B3 Infection and the Web of Host Intracellular Signalling Pathways.... | 185 |
| | 6.1 The Src Family Kinases and ERK .......................................................................... | 186 |
| | 6.2 Akt ........................................................................................................................... | 187 |
| | 6.3 JNK, JAK, and STAT ............................................................................................. | 188 |
| | 6.4 The Ubiquitin-Proteasome System ......................................................................... | 188 |
| 7 | CVB3 Infection and Infectomics ................................................................................... | 189 |
| | 7.1 Differential RNA Display of Murine Myocarditis.................................................. | 189 |
| | 7.2 Microarray Infectomics ........................................................................................... | 190 |
| 8 | Therapeutic Potential ..................................................................................................... | 190 |
| | 8.1 Receptor Analogues ................................................................................................ | 191 |
| | 8.2 Antisense Oligonucleotide Therapy........................................................................ | 192 |
| 9 | Conclusion ..................................................................................................................... | 194 |
| References............................................................................................................................ | | 194 |

**Abstract** CVB3 myocarditis can lead to dilated cardiomyopath (DCM). DCM is one of the leading causes of the need for heart transplantation, so it is important to understand the life cycle of CVB3 and its interactions with the host cell. Infection causes rapid death of host cardiomyocytes by altering normal cellular homeostasis for the efficient release of progeny virion. In this chapter, we will examine the impact that CVB3 replication has on host cell biology, from events that take place at receptor ligation to progeny virus release. The primary focus will be on the myriad of signalling pathways that are activated

B. McManus
The James Hogg iCAPTURE centre, University of British Columbia,
Providence Health Care, St. Pauls Hospital, Vancouver, BC, Canada
BMcManus@mrl.ubc.ca

at all stages of virus replication and their downstream effects. We will also discuss some of the extracellular effects of infection as well as immune and matrixmetalloprotease activation. Interactions of host cell proteins with the 5′ untranslated region (UTR) are required for translation and replication of CVB3. These interactions do not always benefit the virus since the interactions of a 28-kDa host protein with the 5′ UTR are thought to be responsible for inhibitory activity against CVB3. Finally, we will discuss how the elucidation of the different stages of replication has provided the opportunity to develop novel strategies for combating CVB3 infection.

## 1 Coxsackie Adenovirus Receptor and Decay Accelerating Factor and Host Cell Signalling

Like all the other picornaviruses, the replication of Coxsackievirus type B (CVB) takes place entirely within the cytoplasm. To enter the cell, CVB binds the coxsackie adenovirus receptor (CAR) and, in some settings, decay accelerating factor [DAF (CD55); Fig. 1a]. There are virus strains that use CAR only; thus, the role that DAF plays in the life cycle is apparently dispensable for productive replication (Fig. 1b). It has been suggested that DAF plays a chaperone-like role in polarised gut epithelial cells during entry, to direct the virion within cell-cell tight junctions where the major CAR entry receptor resides (Coyne and Bergelson 2006; Fig. 1a). In these cells, the virus then enters the cell via a caveolae-mediated route.

DAF binding and clustering by CVB3 triggers Abl kinase activation, on the surface of CaCo cells, which in turn induces Rac-mediated actin rearrangement and CVB3 movement to cell-cell tight junctions (Coyne and Bergelson 2006; Fig. 1a). Upon arriving at the tight junction, CVB3 interacts with CAR, which induces conformational changes that result in formation of the 135 S A-particle and release of the viral genomic RNA into the cell cytoplasm. Another consequence of DAF binding is Fyn activation and subsequent caveolin phosphorylation, which leads to uptake of CVB3 into caveosomes (Coyne and Bergelson 2006).

Isolates that do not use DAF as a co-receptor may utilize a clathrin- and pH-dependent route of entry, such as in nonpolarised HeLa cells (Fig. 1b; Zautner et al. 2006; Chung et al. 2005). Here, infection is sensitive to lysosomotropic agents and monensin, suggesting a requirement for an acidified compartment for entry. By using a dominant negative inhibitor of dynamin, a large GTPase required for many pathways of endocytosis, and electron microscopy, it has been suggested that a clathrin- and dynamin-dependent pathway is utilised for entry. In light of the studies discussed here, it still remains to be seen what effect these different pathways of entry have on cell tropism and the pathology of CVB3 infection in vivo.

Some CVB3 strains do not use DAF; however, they are still capable of causing myocarditis in the mouse model (Selinka et al. 2002). Thus, the use of DAF by CVB3 isolates for entry does not appear to be associated with CVB3 pathogenesis. Though it has been observed that a hemagglutinating strain of CVB3 that uses DAF

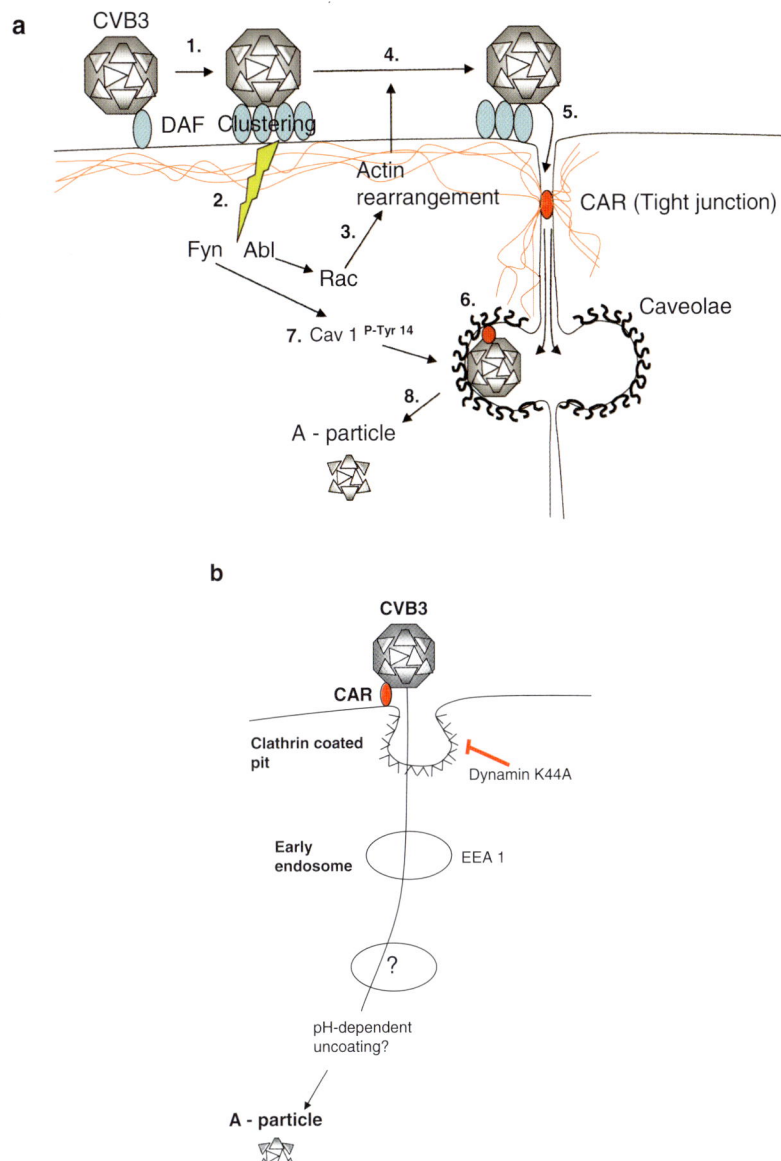

**Fig. 1** Route of entry taken by coxsackie virus B3 into (**a**) polarised CaCo cells and (**b**) unpolarised HeLa cells. **a** 1, CVB3-induced DAF clustering leads to 2, activation of Abl and Fyn kinases. Abl activation results in 3, Rac mediated 4, actin rearrangement, and 5, movement of CVB3 into the tight junction where CVB3 binds its primary receptor, CAR. 6, CVB3 enters CaCo cells via caveolae where 7, tyrosine phosphorylation at position 14 of caveolin-1 results leads to 8, A - particle formation (Coyne and Bergelson 2006). **b** CVB3 enters HeLa cells by a clathrin mediated, EEA1 positive route (Chung et al. 2005)

for entry showed significantly less acute infection of myocardium than a nonhaemagglutinating strain of CVB3 that could not use DAF for entry (Selinka et al. 2002). However, Martino et al. (1998) demonstrated that isolates which cause the most severe myocarditis in the mouse model had more pronounced interactions with DAF upon entry. So it is not yet clear what role DAF plays in the pathogenesis of CVB3-induced myocarditis.

## 2 Host-Cell-Protein Interactions with the 5′ Untranslated Region of Coxsackievirus B3

After entry and achieving transport to the site of productive replication, the virus must interact with host cell factors in order to initiate efficient expression of progeny virion. These factors mediate translation of the polyprotein from the plus sense CVB3 RNA genome. Since *Picornaviruses* carry a plus sense RNA genome in the virion, they are able to begin translation directly from that transcript, immediately after entry, or produce more plus sense RNA from the virus-encoded RNA-dependent RNA polymerase. The interaction of translation initiation factors with the 5′ UTR of poliovirus (Dildine and Semler 1992), and CVB3 (Cheung et al. 2002) in HeLa cells has been described. We reported the interaction of proteins required for translation initiation in the 5′ UTR, identifying the La autoantigen in the process (Cheung et al. 2002). This 52-kDa protein, found in patients with systemic lupus erythematosus, was one of the first host-cell proteins shown to interact with the IRES of polio and hepatitis C viruses (Dasgupta et al. 2004). The majority of the La protein is localized in the nucleus and appears to be involved in small RNA biogenesis including pre-tRNA maturation, stabilisation of nascent RNAs, nuclear retention of nascent transcripts, and RNA pol III transcription termination. However, cellular stress such as poliovirus infection, and possibly CVB3 as well, causes redistribution of the La protein from the nucleus to the cytoplasm, possibly by removal of the C-terminal nuclear localisation signal by a viral protease. So it is conceivable that the La autoantigen, in concert with other proteins, may be involved in resolving the secondary structure of CVB3 RNA, an action that is required before the eIF4G/F complex can properly bind the UTR and initiate translation. Proteins of sizes that are congruent with what is known about the translation initiation complex, like eIF4A, 4B, G and F, were also found to interact with the CVB3 5'UTR in this study (Cheung et al. 2002).

Despite being the primary entry receptor, the expression of CAR in tissues does not always correlate with susceptibility to infection by CVB3. Host intracellular factors can also affect virus-cell tropism. In CVB3 susceptible A/J mice there is an apparent inverse association between tissue-organ tropism of the virus and the presence of a 28-kDa protein (Cheung et al. 2005). This 28-kDa protein binds CVB3 RNA in the 5′ UTR region at nucleotides 210-529 of the antisense transcript only, suggesting that binding is sequence specific (Fig. 2). The role of this binding may be to inhibit transcription of the CVB3 antisense RNA to sense RNA because it was shown in 4- and 10-week-old mice that the interaction of the 28-kDa protein with the UTR region of CVB3 is associated with low viral load within the respective tissue. Despite the

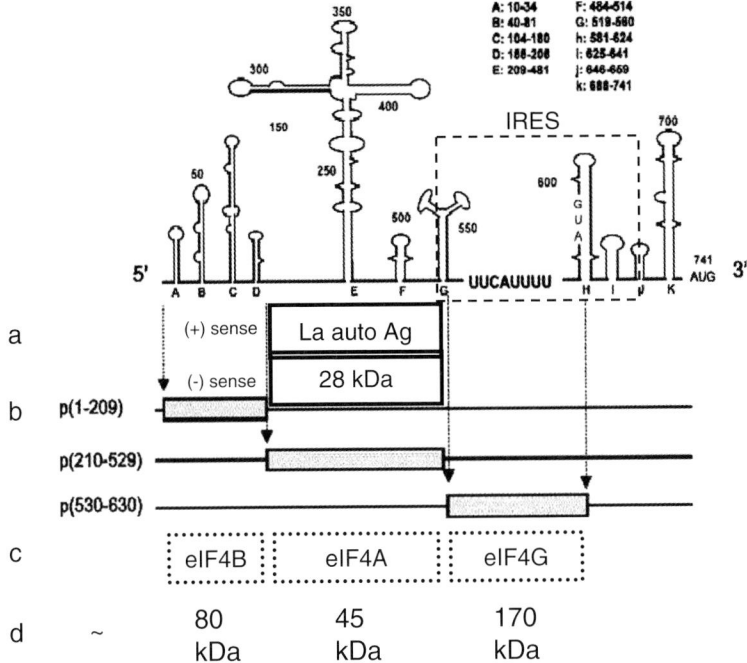

**Fig. 2** Host cell protein interactions within the coxsackie virus B3 5' untranslated region and location of 28-kDa protein binding. (**a, c**) Interacting host cell proteins and regions of binding within the 5' UTR. *Thick solid boxes* denote confirmed protein interactions and the dotted boxes denote a putative interaction. (**b**) Regions of interest in the 5' UTR. (**c**) Putative translation initiation and elongation protein binding sites within the UTR and (**d**) their molecular weights. The location of the core sequence of the internal ribosomal entry site (IRES) is indicated. Derived from Cheung et al. (2002)

interaction of a myriad of proteins with the 5' UTR, the 28-kDa protein was the most common interacting protein among tissue types and age of the mice. The kidneys have the lowest viral load when compared to heart, liver, brain, and pancreas and demonstrate the strongest interaction of the 28-kDa protein with the CVB3 negative sense 5' UTR RNA. In the other organs tested, the interaction of the 28-kDa protein with the 5' UTR was more pronounced in the 10-week-old mice than the 4-week-old mice, also suggesting that this protein is expressed in an age-controlled manner.

## 3 The Inflammatory Response to CVB3 Infection and Host Susceptibility

With successful viral entry and replication comes inflammation as the host attempts to clear itself of the invader. Myocarditis is clinically defined as inflammation and injury of the heart muscle, which can be brought on by an assortment of viruses and autoimmune processes (Kim et al. 2001). In children, acute forms of viral myocarditis are more common, often accompanying fulminant multiorgan disease, whereas chronic forms present more commonly in adults (Martin et al. 1994; McCarthy et al. 2000; reviewed in Kim et al. 2001). Myocarditis predominates in adolescent and adult males, and the majority of female patients are women in their third trimester of pregnancy. The BALB/c mouse model of CVB3 infection closely mirrors this observation, and, as is the case with polio virus infection, a strong Th2 immune response is associated with control of viral replication (Huber et al. 1982; Huber and Lodge 1984; Huber and Pfaeffle 1994). Huber et al. (Huber and Pfaeffle 1994) observed that male BALB/c mice had a strong Th1 response, whereas female mice developed a strong Th2 response and exhibited less morbidity and mortality. Disease suppression could also be abrogated by treatment of female mice with monoclonal anti-IL 4 and anti-CD3 and CD4 antibodies, which may also suggest that a predominantly Th1 response allows virus to replicate unchecked during CVB3 infection of myocardium.

## 4 Coxsackievirus B3 is a Primary Pathogen for Viral Myocarditis

### 4.1 Early Direct Injury of Myocardium by Coxsackievirus B3 Infection

Immune infiltration was once thought to be the major, if not the only important cause of CVB3-induced pathology; however, there is mounting evidence that suggests direct infection may be the primary cause of CVB3-induced injury and the attendant myocarditis.

Regardless of the immunological capacity of the mice, myocarditic lesions can be seen by histochemistry and immunohistochemistry as early as 2 days after infection. These lesions are spatially coincident with positivity by in-situ hybridization for CVB3 genome in the myocardium; viral replication is correlated with the number of myocarditic lesions present within each of the murine hosts (Arola et al. 1995; McManus et al. 1993; Kandolf et al. 1987). Immune-deficient C3H/HeSnJ SCID mice develop overwhelming myocardial damage and die approximately 2 weeks after infection; however, the nature of this disease course is similar to that seen in immune-competent animals except that pathology is more severe and prolonged in the SCID cases, suggesting that direct infection causes pathology regardless of immune status. Further, artificial chemical induction of immunodeficiency aggravates disease in CVB3-infected, and otherwise immune-competent, A/J mice. In these instances, enhanced and prolonged disease was shown at an early timepoint, 7 days after infection, and a late time-point, 30 days after infection.

Other less obvious examples of cytopathic effects (CPE) due to direct CVB3 infection is that mice that are innately resistant to infection by CVB3 experience less pronounced myocarditis (McManus et al. 1993). By controlling viral load with

CVB3 receptor analogues, without modulation of any arm of the immune system, one can abrogate cardiac injury in the mouse model (Yuan et al. 2004). Thus we can see that with loss of immunological control of virus infection or in the absence of any immune control, the result is severe pathology due to virus replication, and pathology can be abrogated by controlling virus replication.

## 5  CVB3 Infection and Host-Cell Survival

### *5.1  CVB3-Induced Cell Death Pathways*

CVB3 induces overwhelming cytopathology in order for progeny virion to escape the infected host cell. Thus, the *Picornaviruses* have evolved mechanisms for effecting cell death and CPE in the host. There is extensive caspase cleavage that occurs shortly after CVB3 infection of cardiomyocytes (Carthy et al. 1998, 2003; Yuan et al. 2003, 2005a); however, it is still unclear whether this cleavage truly represents typical apoptotic cell death alone or a combined apoptosis and necrosis (Fig. 3). We, and others, have reported that caspases 2, 3, 6, 7, 8, and 9 are cleaved during CVB3 infection, and that this cleavage is associated with significant cytopathic effects. Caspase 3 has even been reported to be cleaved by CVB3 protease 2A and possibly 3C as well (Fig. 3; Chau et al. 2007; Calandria et al. 2004). However, we have also noted that inhibition of cleavage of various caspases with inhibitors and overexpression of Bcl-2 and Bcl-xL only retrieves about 60% of viability and also only slows the ultimate loss of cell viability due to infection. The possibility that other pathways of cell death are also being activated during infection is high in light of these results. Indeed, inhibition of glycogen synthase kinase 3β (GSK-3b) activity suppressed cytopathic effects and virus replication via a β-catenin pathway, whereas the general caspase inhibitor zVAD.fmk had no effect on virus-induced CPE (Yuan et al. 2005a). Here it was reported that CVB3 stimulates GSK 3β activity, which results in decreased β-catenin expression; this activity assists in the onset of cell death. Apoptotic cell death, which is programmed and step-wise, ultimately leads to the shrinkage of the cell and cellular constituents within a secure vimentin cage. So this type of cell death could serve to trap virion within, leading to phagocytosis and destruction by phagocytic cells. So from the studies discussed above, CVB3 may trigger a state of overwhelming cellular distress, causing chaotic necrosis, making the cell burst, and releasing maximum amounts of progeny virion.

### *5.2  Cardiac Remodelling: The Matrix Metalloproteases*

Matrix metalloproteinases (MMPs), collectively called matrixins, are proteinases that participate in extracellular matrix (ECM) degradation (Visse and Nagase 2003).

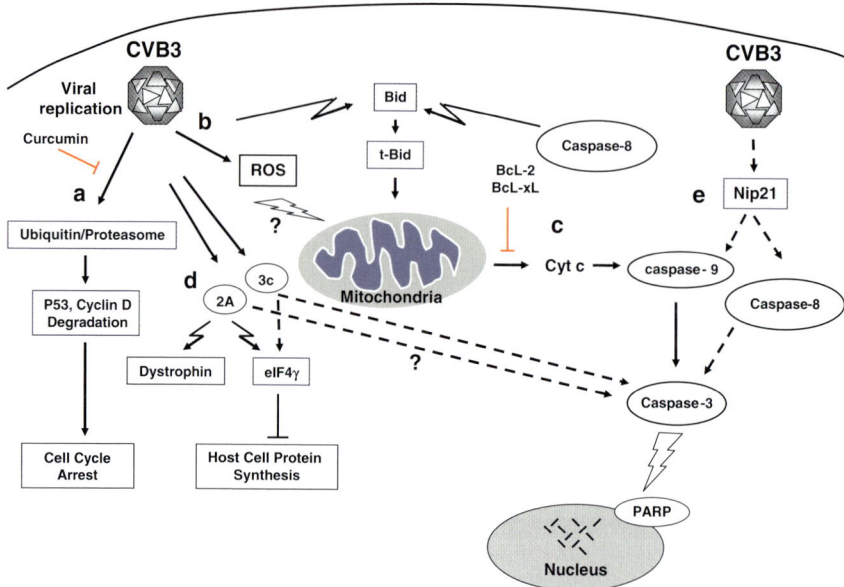

**Fig. 3** CVB3-induced cell death events in infected cells. (**a**) CVB3 replication activates the ubiquitin-proteasome system that leads to the degradation of p53, cyclin D and cell cycle arrest. (**b**) Our preliminary data indicates that CVB3 enhances reactive oxygen species (ROS) production within the cytoplasm of infected cells, which may lead to an increase in the permeability of the mitochondrial outer membrane. (**c**) CVB3 replication leads to the release of cytochrome c (cyt c) from mitochondria. The release of Cyt c results in activation of members of the caspase family, eventually leading to DNA fragmentation within the nucleus, the irreversible stage of cell death. Overexpression of the pro-life members of the BcL-2 family, BcL-2, and BcL-xL can block cyt c release from mitochondria, underlining the important protective role of these proteins. (**d**) It has been suggested that virus 2A and 3C proteases can also directly cleave caspase-3 through a Cyt c-independent pathway. 2A also cleaves eIF4 gamma and dystrophin, which results in the blockage of host protein synthesis and induction of cytopathic effects (CPE) in host cells, respectively. (**e**) Nip21, is a mouse homologue of the human Nip2 gene. Overexpression of this gene induces apoptosis at an accelerated rate in infected cells via caspase 3 cleavage. The fate of infected host cells is determined by the balance between the virus attempting to overcome host-cell defenses and the host cell trying to control pathogen replication

Shortly after the heart is insulted by CVB3 infection and viral load is controlled, extracellular remodelling must take place to help clear infection and damaged cells and repair tissue damage. The MMPs are key proteins that mediate these events. The MMPs are a family of over 24 vertebrate genes, within the superfamily of zinc endoproteinases, whose activity was first described in tail remodelling of metamorphosing tadpoles (Parks et al. 2004; Elkington and Friedland 2006). They can be produced by almost any cell type and are produced as inactive proenzymes that are activated by proteolytic cleavage (Visse and Nagase 2003). The MMPs are mediators of inflammatory-cell migration because they can degrade all components of the ECM, clearing the path for migrating cells of the immune and repair responses.

MMPs remodel the ECM after infection is cleared. Studies using BALB/c H-$2^d$ mice demonstrate upregulation of MMPs 1, 3, 8, and 9 at the protein level in the heart after infection with cardiotropic CVB3 (Pauschinger et al. 2005; Li et al. 2002). Our laboratory has shown that MMPs 2, 9, and 12 are upregulated in male A/J mice 9 days after infection, and further, that MMPs 2 and 9 demonstrate enhanced activity at this time-point (Cheung et al. 2006). In keeping with the gelatinase and elastase specificity of MMP 2 and 9 and MMP 12, respectively, there was increased collagen at the sites of myocardial infection. Any discrepancy between our data and the data from other laboratories may be explained by sampling at different time-points, but, most importantly, because of the use of different mouse strains. However, it is interesting that between the three studies, MMP 8 and 9 are the common denominators among the MMPs, expressed and active during CVB3 myocarditis regardless of mouse strain (Pauschinger et al. 2005; Cheung, unpublished observations).

The tissue inhibitors of metalloproteinases (TIMPs) bind the active site of the MMPs in a molar ratio of 1:1, blocking substrate binding. Thus far, only four TIMPs have been cloned; however, these four inhibit all MMPs tested (Visse and Nagase 2003). In keeping with the above results, TIMP 3 and 4 were found to be downregulated at the corresponding timepoints during CVB3-induced myocarditis in A/J and BALB/c mice (Pauschinger et al. 2005; Cheung et al. 2006), and the expression levels of TIMP 1 and 2 remained the same throughout infection (Cheung et al. 2006).

As their name suggests, the MMPs are responsible for the degradation and turnover of the ECM but they are also responsible for modulating the immune response (Parks et al. 2004). Many cytokines and chemokines are directly cleaved by MMPs, thereby resulting in enhancement, inactivation, or antagonism of chemokine and cytokine activities. Thus, the MMPs may also help to direct inflammation to the site of pathogen insult, but the exact role that MMPs play in most inflammation reactions remains to be elucidated (Parks et al. 2004). Heymans et al. (2006) showed that decreased expression and activity of urokinase-type plasminogen activator and MMP 9 resulted in less inflammation at early time-points after infection in the mouse model. As we discussed earlier, there was increased expression and activity of MMP 9 in both the A/J and BALB/c mouse models. So it will be interesting to see whether remodelling and MMP mediated inflammatory signals act in concert or abrogate one another (Fig. 4).

## 6 Coxsackievirus B3 Infection and the Web of Host Intracellular Signalling Pathways

Thus far we have discussed how CVB3 alteration of host-cell biology affects the steps of virus replication and host-cell fate. We will now focus on the myriad of data that have been obtained by our laboratory and others in the realm

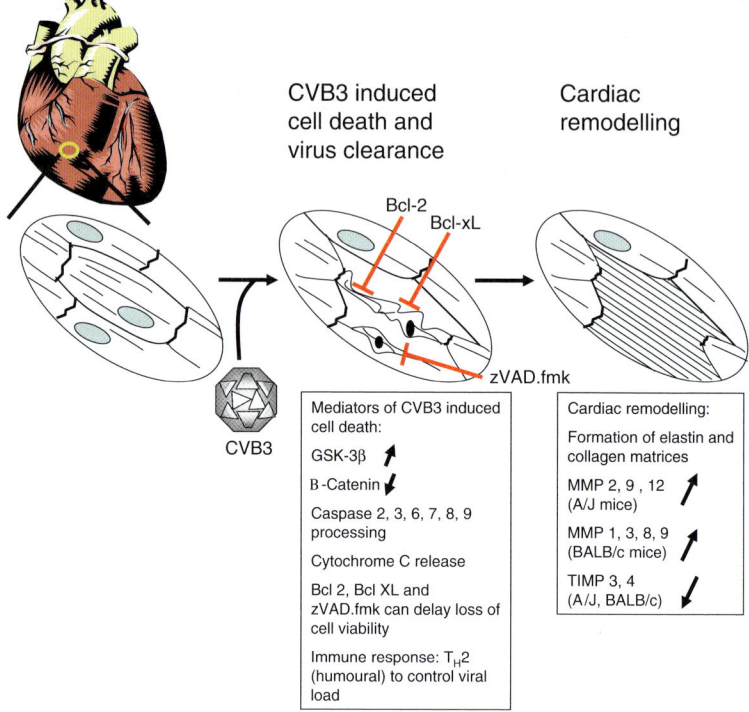

**Fig. 4** Relationships among cell death, virus clearance and cardiac remodelling during coxsackie virus B3 infection. *Up and down arrows* denote increased or decreased expression, respectively

of virus-induced intracellular signalling. In this section, we will attempt to make sense of the current understanding of the CVB3 web of signalling pathways in the context of virus replication and cell fate.

## 6.1 The Src Family Kinases and ERK

CVB3 infection results in activation of multiple simultaneous intracellular signalling pathways, although the temporal and spatial activation and relative roles of these pathways remain to be elucidated. During CVB3 infection of host cells, several cellular proteins become tyrosine phosphorylated; there is tyrosine phosphorylation of two proteins with molecular masses of 48 and 200 kDa in CVB3 infected HeLa and Vero cells, respectively (Huber et al. 1999; Huber et al. 1997). Further subcellular fractionation experiments revealed that the 48-kDa CVB3-induced phosphoprotein is localized to the cytosol, whereas the 200-kDa phosphoprotein is membrane-associated. Interestingly, the Src family kinase inhibitor Herbimycin A attenuates such tyrosine phosphorylation events and reduces production of CVB3 progeny virions, suggesting that Src family kinase-mediated signalling pathways are required for efficient replication of CVB3, possibly at the point of particle assembly. Indeed, the Src family kinase, $p56^{lck}$, is required for viral replication in T cells and probably cardiac myocytes and plays essential roles in CVB3 pathogenesis (Liu et al. 2000). Mice deficient for the $p56^{lck}$ gene are completely protected from CVB3-induced myocarditis and subsequent dilated cardiomyopathy. Remember that, as described in Sect. 1 of this chapter, the CVB3 co-receptor DAF-associated Fyn kinase, another member of the Src family kinases, has been shown to regulate CVB3 entry through epithelial tight junctions in polarized cells (Coyne and Bergelson 2006).

Another component of this story is the bi-phasic phosphorylation of ERK during CVB3 infection in non-polarised cells (Cunningham et al. 2003; Luo et al. 2002; Opavsky et al. 2002). We and others have found that the early transient activation of ERK possibly results from the engagement of CVB3 with receptors CAR and DAF (Luo et al. 2002; Opavsky et al. 2002; Lim et al. 2005). Meanwhile, the late activation of ERK requires replication of CVB3, possibly mediated by cleavage of RasGAP by CVB3 $3C^{pro}$ that results in activation of the Ras/Raf/MAPK pathway. Src family kinases are likely involved in the early activation of ERK, as specific src inhibitors reduce ERK phosphorylation following receptor engagement. The importance of the ERK pathway is established, as specific inhibition of ERK leads to the attenuation of viral replication, decreased virus cleavage of host proteins, and attenuation of host cell death (Luo et al. 2002).

## *6.2 Akt*

The critical role of the phophatidyl-3-kinase (PI3K)/Akt pathway in supporting CVB3 replication has also been investigated by our laboratory (Esfandiarei et al. 2004). We have found that CVB3 infection leads to the activation of the PI3K/Akt pathway, and inhibition of this pathway using specific pharmacological inhibitors or dominant-negative mutants significantly impairs virus production yet increases apoptosis of infected cells. We postulate that integrin-linked kinase (ILK) is involved in the activation of Akt following CVB3 infection, and our recent studies clearly demonstrate that kinase activity of ILK is required for efficient CVB3 replication in both HeLa cells and HL-1 murine cardiomyocytes. We have found that the activity of GSK3β, a downstream kinase of the PI3K/Akt pathway, is increased following CVB3 infection and depends on tyrosine phosphorylation (Yuan et al. 2005a). In fact, inhibition of GSK3β increases accumulation and nuclear translocation of β-catenin and significantly reduces CPE and apoptosis of infected cells, as discussed earlier. CVB3 infection stimulates GSK3β activity via a tyrosine kinase-dependent mechanism, which contributes to CVB3-induced CPE.

## 6.3 JNK, JAK, and STAT

We and others have also investigated the relative roles of the stress-activated protein kinase (SAPK) pathway and CVB3 (Kim et al. 2004; Si et al. 2005a). Both the p38 MAPK and the JNK pathways are activated during CVB3 infection (Si et al. 2005a). Although it is reported that activation of JNK is required for the upregulation of cyr61 and induction of apoptosis of infected cells (Kim et al. 2004), our results suggest that JNK activation plays a role in host defence through the activation of activating transcription factor 2 (ATF-2) rather than just by inducing cell death (Si et al. 2005a). Another protein, reported to play a role in host antiviral defense, is the Janus kinase (JAK)-signal transducers and activators of transcription (STAT) pathway in CVB3 infection (Irie-Sasaki et al. 2001; Metcalf et al. 2000; Yasukawa et al. 2003). Our studies have also found that p38 MAPK activation is necessary for CVB3-induced caspase-3 activation and CVB3 viral progeny release (Kim et al. 2004; Si et al. 2005a). This caspase-3 activation leads to prominent production of reactive oxygen species (ROS) in infected cells (Si et al. 2005a). Consequently, it is possible that activation of the p38 MAPK pathway, production of ROS, and subsequent activation of NFκB, play a major signalling role in the regulation of expression and secretion of pro-inflammatory cytokines in the progression of CVB3-induced pathogenesis.

## 6.4 The Ubiquitin-Proteasome System

The ubiquitin-proteasome system (UPS) is a major protein degradation system in mammalian cells. Degradation of a protein via the ubiquitin-proteasome pathway involves two successive steps: (1) conjugation of multiple ubiquitin moieties to the substrate and (2) degradation of the tagged protein by the downstream 26S proteasome complex (Glickman and Ciechanover 2002). We have previously shown that CVB3 infection resulted in degradation of several host proteins, such as cell-cycle regulator cyclin D1 and transcription factors p53 and β-catenin (Fig. 3; Luo et al. 2003a, b). Degradation of these proteins during CVB3 infection is likely through the UPS because proteasome inhibitors can reverse CVB3-induced degradation of these proteins. Further, CVB3 likely utilizes this system for its replication, as CVB3 RNA-dependent RNA polymerase 3D is ubiquitinated in infected cells (unpublished result). We have also shown that dysregulation of the UPS by proteasome inhibitors, pyrrolidine dithiocarbamate (PDTC) or curcumin effectively reduces synthesis of CVB3 viral proteins and release of CVB3 progeny virions (Si et al. 2005a; Fig. 5).

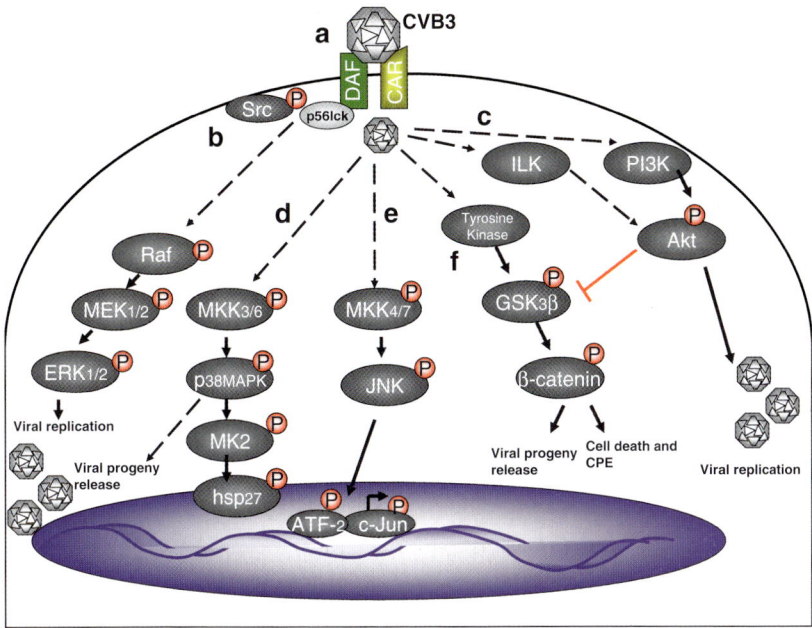

**Fig. 5** CVB3-induced intracellular signalling pathways and networks in infected cells. (**a**) Attachment of CVB3 to its receptors, CAR and DAF, (**b**) induces a rapid and transient activation of src-family kinases and the ERK pathway. (**c**) CVB3 infection also leads to activation of PI3 kinase and integrin-linked kinase (*ILK*) that leads to activation of Akt. CVB3-induced anti-apoptotic survival signalling early in the lifecycle may be mediated by the activation of Akt. In later stages of the CVB3 lifecycle, stress-activated protein kinases including (**d**) p38 MAPK and (**e**) JNK are activated. Activation of p38 MAPK plays a role in CVB3 progeny virion release, while activation of JNK leads to phosphorylation of transcription factors ATF-2 and c-Jun. In addition, (**f**) GSK3β is also activated through tyrosine phosphorylation, and such activation plays a role in CVB3-induced CPE and apoptosis of infected cells for viral progeny release

## 7 CVB3 Infection and Infectomics

### 7.1 *Differential RNA Display of Murine Myocarditis*

The transcriptional profile from two or more sources can be compared by differential RNA display (Stein and Liang 2002). It is especially useful in comparing the transcriptome profiles between virus-infected and -uninfected cells and can quickly and easily identify candidate host genes that are necessary for virus replication. Such experiments serve as effective screening and fishing expeditions, particularly when used as a tool to screen the alteration of host-cell biology by intracellular pathogens. Our laboratory used differential display to identify which host genes are down- or upregulated during CVB3 replication in mouse hearts (Yang et al. 1999). Total RNA was extracted from hearts of 4- and 10-week-old A/J (H-$2^a$) mice at day 4 after

CVB3 infection, and mRNAs were detected by reverse transcriptase-PCR and subsequently analysed. Twenty-eight upregulated or downregulated bands were selected from the sequencing gels; among these, two upregulated and three downregulated cDNA fragments were confirmed by Northern hybridisation. DNA sequence analysis and GenBank searching have determined that four of the five candidate genes are homologous to genes encoding *Mus musculus* inducible GTPase. The remaining candidate gene matches an unpublished cDNA clone, *M musculus* Nip21 mRNA (GenBank accession number, AF035207), which is homologous to human Nip2, a Bcl-2 binding protein.

We demonstrated that Nip21 expression could induce apoptosis via caspase-dependent mitochondria activation (Fig. 3; Zhang et al. 2002). We found that activation of caspase-3 (probably via capases 8 and 9) and cleavage of poly-(ADP-ribose) polymerase occurred 2 h earlier than in vector-transfected control cells, suggesting that Nip21 expression enhances CVB3-induced apoptosis. We also demonstrated a significant decrease in HeLa cell and H9c2 cell viability with Nip21 overexpression.

## 7.2 Microarray Infectomics

While differential RNA display can reveal the transcriptional profile of a few dozen genes, in a few days of work, the advent of microarrays permits the simultaneous transcriptional profiling of over 10,000 genes in half a day. We used cDNA microarrays to probe differential gene expression in the myocardium following virus infection (McManus et al. 2002). Following virus infection, there are global decreases in metabolic and mitochondrial genes, increases in signalling genes and distinctive patterns in other functional groups. To establish early gene expression profiles in infected cells by themselves, we also used oligonucleotide arrays in an in vitro model of CVB3 infection. Notably, we have found increased expression of transcription factors c-fos and c-jun downstream of extracellular signal-related kinase, a pathway which is crucial for virus replication and pathogenesis.

## 8 Therapeutic Potential

There have been major advances in antiviral strategy because of the response to the growing hepatitis C virus (HCV) and human immunodeficiency virus (HIV) pandemics and the threat of an avian influenza pandemic (Fauci 2006) (Morens et al. 2004). However, despite these advances, more research is required for the treatment of CVB3 infection since the broad spectrum anti-Picornaviridae drug, Pleconaril (Florea et al. 2003), the only specific treatment of CVB3 infection, is not effective against all strains of this virus. One of the prototypic CVB3 strains, the Nancy strain, shown to be resistant to Pleconaril treatment, is dependent on only a single mutation at AA 1092 of CVB3 (Schmidtke et al. 2005). Therefore, further research into developing a robust and specific therapeutic regime is required to combat CVB3 infection in adult and especially paediatric patients.

Ribavirin is a broad-spectrum nucleoside-analogue virustatic drug that leads to increased mutation rates in RNA viruses, resulting in terminated virus replication; it has been shown to be effective against RSV bronchiolitis, measles, and Lassa fever, in vivo (Gilbert and Knight 1986). It was also demonstrated early to be effective against enteroviruses (Sidwell et al. 1979). The efficacy of Ribavirin on CVB3 infection, specifically, has been reported in human myocardial fibroblasts (Heim et al. 1997) and in a mouse model of myocarditis (Kishimoto et al. 1988). However, a common side effect of Ribavirin treatment is haemolytic anaemia (Nomura et al. 2004), so Ribavirin may be contraindicative in patients with acute and chronic myocarditis. Therefore, the current strategies for combating CVB3 infection certainly allow room for novel and safer drugs.

## 8.1 Receptor Analogues

Inhibiting the binding of viruses, and thus their replication, by treatment with soluble receptor analogues, ligands, and proteoglycans has been demonstrated on both enveloped (HIV) (McKnight and Weiss 2003) and unenveloped viruses (papillomaviruses) alike (De Clercq 2002; Buck et al. 2006). The interruption of CVB3 infection by IgG antibody-CAR and DAF fusion proteins has proven to be relatively successful, in vitro, which is not surprising since a strong humoral immune response is associated with clearance of enterovirus infections (Fig. 6a) (MacLennan et al. 2004; Sawyer 2002; Rotbart 2002). Our laboratory has demonstrated efficient CVB3 neutralisation using CAR and DAF analogues, in vitro, fused to the Fc region of IgG (Yanagawa et al. 2003; Yanagawa et al. 2004; Lim et al. 2006). The antibody Fc region promotes the solubilisation of the receptor, systemically, and for recognition by Fc receptors present on antigen-presenting cells such as macrophages so that, shortly after the virus is bound by the soluble receptor-Fc peptide, the virus is cleared by phagocytic cells. It was later found that the Fc-fused CAR receptor binds to CVB3 with 5,000- to 10,000-fold more affinity than soluble Fc-DAF fusion proteins (Goodfellow et al. 2005). Thus, the incorporation of the Fc-fused receptor analogue blocks virus infection, sterically, and labels the virus for clearance from the host.

Proof of the in vitro principle was demonstrated in an in vivo mouse model for the Fc-CAR (Yanagawa et al. 2004), Fc-DAF fusions (Yanagawa et al. 2003), and for an extended Fc-CAR:DAF fusion protein (Lim et al. 2006). Immune competent A/J mice were treated with the soluble Fc-CAR or DAF, either before, during or after infection, and the treatment groups consequentive presented with significantly lower cardiac viral loads. Since infectious myocarditis tends to occur in the already immunocompromised, it would be interesting to see the effect of these soluble

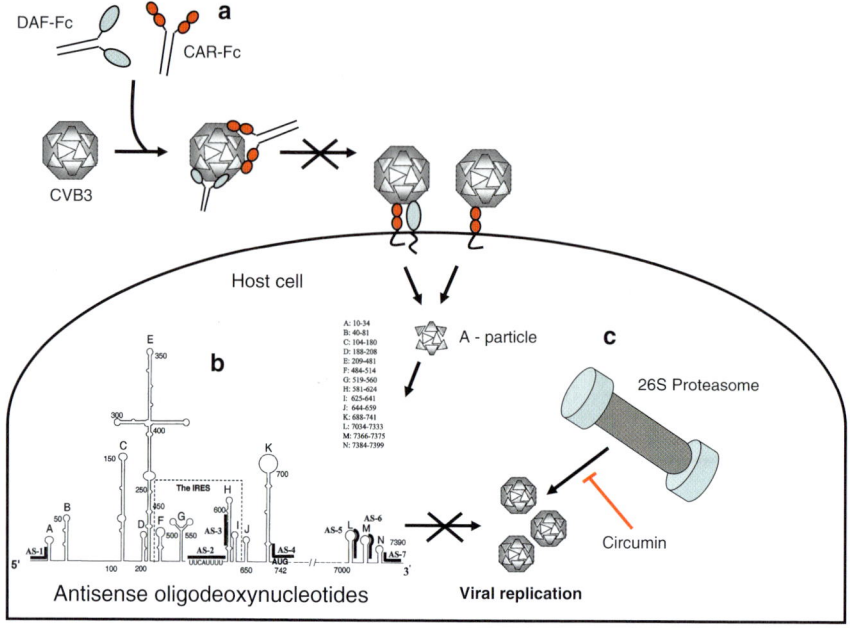

**Fig. 6** Therapeutic strategies for combating CVB3 infection. (**a**) Receptor analogue - Fc fusion proteins have proven effective in blocking virus entry, in vitro and controlling viral load in vivo. (**b**) Established infection and virus turnover can be controlled by treatment with antisense oligodeoxynucleotides. Phosphorothioate nucleotides improve the half-life of the oligodeoxynucleotides systemically and are readily taken up by host cells. (**c**) CVB3 replication is dependent upon the ubiquitin proteasome network for efficient replication. Curcumin, found in turmeric, inhibits CVB3 replication by inhibition of the 26S proteasome

receptor analogues on CVB3-infected immune-deficient animals, such as C3H/HeSnJ SCID mice. This novel method of treatment may serve to supplement the humoral response that infectious myocarditis patients require to clear CVB3 infection.

## 8.2 Antisense Oligonucleotide Therapy

The only specific treatment for CVB3 infection is Pleconaril, and some virus strains are resistant to this treatment (Schmidtke et al. 2005), so our laboratory has investigated the use of oligonucleotides for the specific, and tailored, inhibition of CVB3 infection (Yuan et al. 2004, 2005b; Yang et al. 1997; Wang et al. 2001).

Antisense oligodeoxynucleotides (AS-ODNs) targeting the untranslated regions are promising therapeutic reagents for the treatment of CVB3-induced myocarditis, shown in vitro in both HeLa cells and HL1 cardiomyocytes (Yuan et al. 2004; Yang et al. 1997; Yuan et al. 2005b; Wang et al. 2001), and in vivo in a mouse model (Yuan et al. 2004). The replacement of one of the nonbridging oxygen atoms at the phosphorus, with sulphur, lends greater stability to the phosphorothioate compounds as compared to native oligodeoxynucleotides, because of lower substrate specificity to nucleases. These oligos are apparently well distributed systemically after intravenous administration and are even taken-up, in the absence of a carrier by cells, as shown in an in vivo mouse model (Zhao et al. 1998). Our laboratory designed AS-ODNs directed against the 5' and 3' UTRs of CVB3 (Fig. 6b) (Wang et al. 2001). Out of the seven that were designed, the AS-ODN directed against the 3' terminal UTR demonstrated the greatest inhibitory activity of CVB3 replication. AS7, directed against the 3' UTR proximal terminus, decreased viral load in HL-1 cardiomyocytes and, most importantly, in A/J mice by approximately fivefold and significantly reduced cardiac injury caused by CVB3 infection (Yuan et al. 2004). It is not surprising then, with demonstration of inhibition of HIV-1 (Lisziewicz et al. 1992), hepatitis B (Offensperger et al. 1993), and foot-and-mouth disease viruses (Rosas et al. 2003) that antisense deoxynucleotides have made it to clinical trials and the anticytomegalovirus drug, Vitravene has been approved by the American Food and Drug Administration.

CVB3 is an RNA virus, like HIV, and as such it will mutate and evolve at a faster rate than a DNA virus (Holmes 2003). It is for this reason that we propose that any treatment of CVB3 infection be with more than one drug, such as the triple drug therapy regime used to treat HIV infection. The purpose of this strategy is to stall the emergence of single drug-resistance mutants, giving the host time to control infection. Individual treatments administered in succession to clear persistent enterovirus infections have proved incapable of clearing these infections (MacLennan et al. 2004). When individuals with agammaglobulinaemia are vaccinated with live attenuated polio virus, they lack the ability to control their infection and can become life-long shedders of polio, thus acting as a potential reservoir for future polio outbreaks (MacLennan et al. 2004). Attempts have been made to help clear these people of their infection and consecutive monotherapies proved to work initially, on their own, only to have infection re-emerge, perhaps because of the outgrowth of resistance mutants. Therefore, treatment of CVB3 infection, particularly chronic, may be better tackled with multiple simultaneous therapeutic approaches, removing any opportunity for the virus to outgrow resistance mutants and giving the host every opportunity to clear infection.

The discovery of the receptors for CVB3 provided the opportunity to design therapeutic receptor analogues that effectively lower viral load and abrogate myocarditis in the mouse model. Research into virus-dependent signalling pathways, which become activated later in the CVB3 life cycle, have raised the possibility of using curcumin, a chemical that is present in turmeric, as a drug for treating CVB3-induced myocarditis (Fig. 6c). These examples underline the importance of basic science in drug discovery; high-throughput genomics technology and the understanding of the viral life cycle provide obvious opportunities for developing novel therapeutic regimens to infectious disease.

## 9 Conclusion

CVB3 alters normal cellular function in order to optimise progeny virion assembly and to promote cell death for release of virion. It is now clear that CVB3 induces overwhelming cytopathology that is responsible for the majority of cell death and CPE during CVB3 infection. From studying host-cell signalling during CVB3 infection, we learned that CVB3 entry, replication and CPE are highly dependent on host cell src signalling and we are only starting to understand the roles of other signalling pathways during CVB3 infection: receptor sequestration during entry to host-cell antiviral signalling, cell death, and progeny virion release. So we can see that by elucidating virus-induced alterations of host-cell biology, we can develop novel therapeutics that target these alterations and block CVB3 replication and pathogenesis.

## References

Arola A, Kalimo H, Ruuskanen O, Hyypia T (1995) Experimental myocarditis induced by two different coxsackievirus B3 variants: aspects of pathogenesis and comparison of diagnostic methods. J Med Virol 47:251-259

Buck CB, Thompson CD, Roberts JN, et al (2006) Carrageenan is a potent inhibitor of papillomavirus infection. PLoS Pathog 2:e69

Calandria C, Irurzun A, Barco A, Carrasco L (2004) Individual expression of poliovirus 2Apro and 3Cpro induces activation of caspase-3 and PARP cleavage in HeLa cells. Virus Res 104:39-49

Carthy CM, Granville DJ, Watson KA, Anderson DR, Wilson JE, Yang D, Hunt DW, McManus BM (1998) Caspase activation and specific cleavage of substrates after coxsackievirus B3-induced cytopathic effect in HeLa cells. J Virol 72:7669-7675

Carthy CM, Yanagawa B, Luo H, Granville DJ, Yang D, Cheung P, Cheung C, Esfandiarei M, Rudin CM, Thompson CB, Hunt DW, McManus BM (2003) Bcl-2 and Bcl-xL overexpression inhibits cytochrome c release, activation of multiple caspases, and virus release following coxsackievirus B3 infection. Virology 313:147-157

Chau D, Yuan J, Zhang H, et al (2007) Coxsackievirus B3 proteases 2A and 3C induce apoptotic cell death through mitochondrial injury and cleavage of eIF4GI but not DAP5/p97/NAT1. Apoptosis 12:513-524

Cheung C, Luo H, Yanagawa B, Leong HS, Samarasekera D, Lai JC, Suarez A, Zhang J, McManus BM (2006) Matrix metalloproteinases and tissue inhibitors of metalloproteinases in coxsackievirus-induced myocarditis. Cardiovasc Pathol 15:63-74

Cheung P, Zhang M, Yuan J, Chau D, Yanagawa B, McManus B, Yang D (2002) Specific interactions of HeLa cell proteins with Coxsackievirus B3 RNA: La autoantigen binds differentially to multiple sites within the 5′ untranslated region. Virus Res 90:23-36

Cheung PK, Yuan J, Zhang HM, Chau D, Yanagawa B, Suarez A, McManus B, Yang D (2005) Specific interactions of mouse organ proteins with the 5′ untranslated region of coxsackievirus B3: potential determinants of viral tissue tropism. J Med Virol 77:414-424

Chung SK, Kim JY, Kim IB, Park SI, Paek KH, Nam JH (2005) Internalization and trafficking mechanisms of coxsackievirus B3 in HeLa cells. Virology 333:31-40

Coyne CB, Bergelson JM (2006) Virus-induced Abl and Fyn kinase signals permit coxsackievirus entry through epithelial tight junctions. Cell 124:119-131

Cunningham KA, Chapman NM, Carson SD (2003) Caspase-3 activation and ERK phosphorylation during CVB3 infection of cells: influence of the coxsackievirus and adenovirus receptor and engineered variants. Virus Res 92:179-186

Dasgupta A, Das S, Izumi R, Venkatesan A, Barat B (2004) Targeting internal ribosome entry site (IRES)-mediated translation to block hepatitis C and other RNA viruses. FEMS Microbiol Lett 234:189-199

De Clercq E (2002) Strategies in the design of antiviral drugs. Nat Rev Drug Discov 1:13-25

Dildine SL, Semler BL (1992) Conservation of RNA-protein interactions among picornaviruses. J Virol 66:4364-4376

Elkington PT, Friedland JS (2006) Matrix metalloproteinases in destructive pulmonary pathology. Thorax 61:259-266

Esfandiarei M, Luo H, Yanagawa B, Suarez A, Dabiri D, Zhang J, McManus BM (2004) Protein kinase B/Akt regulates coxsackievirus B3 replication through a mechanism which is not caspase dependent. J Virol 78:4289-4298

Fauci AS (2006) Emerging and re-emerging infectious diseases: influenza as a prototype of the host-pathogen balancing act. Cell 124:665-670

Florea NR, Maglio D, Nicolau DP (2003) Pleconaril, a novel antipicornaviral agent. Pharmacotherapy 23:339-348

Gilbert BE, Knight V (1986) Biochemistry and clinical applications of ribavirin. Antimicrob Agents Chemother 30:201-205

Glickman MH, Ciechanover A (2002) The ubiquitin-proteasome proteolytic pathway: destruction for the sake of construction. Physiol Rev 82:373-428

Goodfellow IG, Evans DJ, Blom AM, Kerrigan D, Miners JS, Morgan BP, Spiller OB (2005) Inhibition of coxsackie B virus infection by soluble forms of its receptors: binding affinities, altered particle formation, and competition with cellular receptors. J Virol 79:12016-12024

Heim A, Grumbach I, Pring-Akerblom P, Stille-Siegener M, Müller G, Kandolf R, Figulla HR (1997) Inhibition of coxsackievirus B3 carrier state infection of cultured human myocardial fibroblasts by ribavirin and human natural interferon-alpha. Antiviral Res 34:101-111

Heymans S, Pauschinger M, De Palma A, Kallwellis-Opara A, Rutschow S, Swinnen M, Vanhoutte D, Gao F, Torpai R, Baker AH, Padalko E, Neyts J, Schultheiss HP, Van de Werf F, Carmeliet P, Pinto YM (2006) Inhibition of urokinase-type plasminogen activator or matrix metalloproteinases prevents cardiac injury and dysfunction during viral myocarditis. Circulation 114:565-573

Holmes EC (2003) Molecular clocks and the puzzle of RNA virus origins. J Virol 77:3893-3897

Huber M, Selinka HC, Kandolf R (1997) Tyrosine phosphorylation events during coxsackievirus B3 replication. J Virol 71:595-600

Huber M, Watson KA, Selinka HC, Carthy CM, Klingel K, McManus BM, Kandolf R (1999) Cleavage of RasGAP and phosphorylation of mitogen-activated protein kinase in the course of coxsackievirus B3 replication. J Virol 73:3587-3594

Huber SA, Lodge PA (1984) Coxsackievirus B-3 myocarditis in Balb/c mice. Evidence for autoimmunity to myocyte antigens. Am J Pathol 116:21-29

Huber SA, Pfaeffle B (1994) Differential Th1 and Th2 cell responses in male and female BALB/c mice infected with coxsackievirus group B type 3. J Virol 68:5126-5132

Huber SA, Job LP, Auld KR (1982) Influence of sex hormones on Coxsackie B-3 virus infection in Balb/c mice. Cell Immunol 67:173-179

Irie-Sasaki J, Sasaki T, Matsumoto W, Opavsky A, Cheng M, Welstead G, Griffiths E, Krawczyk C, Richardson CD, Aitken K, Iscove N, Koretzky G, Johnson P, Liu P, Rothstein DM, Penninger JM (2001) CD45 is a JAK phosphatase and negatively regulates cytokine receptor signalling. Nature 409:349-354

Kandolf R, Ameis D, Kirschner P, Canu A, Hofschneider PH (1987) In situ detection of enteroviral genomes in myocardial cells by nucleic acid hybridization: an approach to the diagnosis of viral heart disease. Proc Natl Acad Sci USA 84:6272-6276

Kim KS, Hufnagel G, Chapman NM, Tracy S (2001) The group B coxsackieviruses and myocarditis. Rev Med Virol 11:355-368

Kim SM, Park JH, Chung SK, Kim JY, Hwang HY, Chung KC, Jo I, Park SI, Nam JH (2004) Coxsackievirus B3 infection induces cyr61 activation via JNK to mediate cell death. J Virol 78:13479-13488

Kishimoto C, Crumpacker CS, Abelmann WH (1988) Ribavirin treatment of murine coxsackievirus B3 myocarditis with analyses of lymphocyte subsets. J Am Coll Cardiol 12:1334-1341

Li J, Schwimmbeck PL, Tschope C, Leschka S, Husmann L, Rutschow S, Reichenbach F, Noutsias M, Kobalz U, Poller W, Spillmann F, Zeichhardt H, Schultheiss HP, Pauschinger M (2002) Collagen degradation in a murine myocarditis model: relevance of matrix metalloproteinase in association with inflammatory induction. Cardiovasc Res 56:235-247

Lim BK, Nam JH, Gil CO, Yun SH, Choi JH, Kim DK, Jeon ES (2005) Coxsackievirus B3 replication is related to activation of the late extracellular signal-regulated kinase (ERK) signal. Virus Res 113:153-157

Lim BK, Choi JH, Nam JH, Gil CO, Shin JO, Yun SH, Kim DK, Jeon ES (2006) Virus receptor trap neutralizes coxsackievirus in experimental murine viral myocarditis. Cardiovasc Res 71:517-526

Lisziewicz J, Sun D, Klotman M, Agrawal S, Zamecnik P, Gallo R (1992) Specific inhibition of human immunodeficiency virus type 1 replication by antisense oligonucleotides: an in vitro model for treatment. Proc Natl Acad Sci U S A 89:11209-11213

Liu P, Aitken K, Kong YY, Opavsky MA, Martino T, Dawood F, Wen WH, Kozieradzki I, Bachmaier K, Straus D, Mak TW, Penninger JM (2000) The tyrosine kinase p56lck is essential in coxsackievirus B3-mediated heart disease. Nat Med 6:429-434

Luo H, Yanagawa B, Zhang J, Luo Z, Zhang M, Esfandiarei M, Carthy C, Wilson JE, Yang D, McManus BM (2002) Coxsackievirus B3 replication is reduced by inhibition of the extracellular signal-regulated kinase (ERK) signaling pathway. J Virol 76:3365-3373

Luo H, Zhang J, Dastvan F, Yanagawa B, Reidy MA, Zhang HM, Yang D, Wilson JE, McManus BM (2003a) Ubiquitin-dependent proteolysis of cyclin D1 is associated with coxsackievirus-induced cell growth arrest. J Virol 77:1-9

Luo H, Zhang J, Cheung C, Suarez A, McManus BM, Yang D (2003b) Proteasome inhibition reduces coxsackievirus B3 replication in murine cardiomyocytes. Am J Pathol 163:381-385

MacLennan C, Dunn G, Huissoon AP, Kumararatne DS, Martin J, O'Leary P, Thompson RA, Osman H, Wood P, Minor P, Wood DJ, Pillay D (2004) Failure to clear persistent vaccine-derived neurovirulent poliovirus infection in an immunodeficient man. Lancet 363:1509-1513

Martin AB, Webber S, Fricker FJ, Jaffe R, Demmler G, Kearney D, Zhang YH, Bodurtha J, Gelb B, Ni J, et al (1994) Acute myocarditis. Rapid diagnosis by PCR in children. Circulation 90:330-339

Martino TA, Petric M, Brown M, Aitken K, Gauntt CJ, Richardson CD, Chow LH, Liu PP (1998) Cardiovirulent coxsackieviruses and the decay-accelerating factor (CD55) receptor. Virology 244:302-314

McCarthy RE 3rd, Boehmer JP, Hruban RH, Hutchins GM, Kasper EK, Hare JM, Baughman KL (2000) Long-term outcome of fulminant myocarditis as compared with acute (nonfulminant) myocarditis. N Engl J Med 342:690-695

McKnight A, Weiss RA (2003) Blocking the docking of HIV-1. Proc Natl Acad Sci U S A 100:10581-10582

McManus BM, Chow LH, Wilson JE, Anderson DR, Gulizia JM, Gauntt CJ, Klingel KE, Beisel KW, Kandolf R (1993) Direct myocardial injury by enterovirus: a central role in the evolution of murine myocarditis. Clin Immunol Immunopathol 68:159-169

McManus BM, Yanagawa B, Rezai N, Luo H, Taylor L, Zhang M, Yuan J, Buckley J, Triche T, Schreiner G, Yang D (2002) Genetic determinants of coxsackievirus B3 pathogenesis. Ann N Y Acad Sci 975:169-179

Metcalf D, Greenhalgh CJ, Viney E, Willson TA, Starr R, Nicola NA, Hilton DJ, Alexander WS (2000) Gigantism in mice lacking suppressor of cytokine signalling-2. Nature 405:1069-1073

Morens DM, Folkers GK, Fauci AS (2004) The challenge of emerging and re-emerging infectious diseases. Nature 430:242-249

Nomura H, Tanimoto H, Kajiwara E, Shimono J, Maruyama T, Yamashita N, Nagano M, Higashi M, Mukai T, Matsui Y, Hayashi J, Kashiwagi S, Ishibashi H (2004) Factors contributing to ribavirin-induced anemia. J Gastroenterol Hepatol 19:1312-1317

Offensperger WB, Offensperger S, Walter E, Teubner K, Igloi G, Blum HE, Gerok W (1993) In vivo inhibition of duck hepatitis B virus replication and gene expression by phosphorothioate modified antisense oligodeoxynucleotides. EMBO J 12:1257-1262

Opavsky MA, Martino T, Rabinovitch M, Penninger J, Richardson C, Petric M, Trinidad C, Butcher L, Chan J, Liu PP (2002) Enhanced ERK-1/2 activation in mice susceptible to coxsackievirus-induced myocarditis. J Clin Invest 109:1561-1569

Parks WC, Wilson CL, Lopez-Boado YS (2004) Matrix metalloproteinases as modulators of inflammation and innate immunity. Nat Rev Immunol 4:617-629

Pauschinger M, Rutschow S, Chandrasekharan K, Westermann D, Weitz A, Peter Schwimmbeck L, Zeichhardt H, Poller W, Noutsias M, Li J, Schultheiss HP, Tschope C (2005) Carvedilol improves left ventricular function in murine coxsackievirus-induced acute myocarditis association with reduced myocardial interleukin-1beta and MMP-8 expression and a modulated immune response. Eur J Heart Fail 7:444-452

Rosas MF, Martinez-Salas E, Sobrino F (2003) Stable expression of antisense RNAs targeted to the 5' non-coding region confers heterotypic inhibition to foot-and-mouth disease virus infection. J Gen Virol 84:393-402

Rotbart HA (2002) Treatment of picornavirus infections. Antiviral Res 53:83-98

Sawyer MH (2002) Enterovirus infections: diagnosis and treatment. Semin Pediatr Infect Dis 13:40-47

Schmidtke M, Hammerschmidt E, Schüler S, Zell R, Birch-Hirschfeld E, Makarov VA, Riabova OB, Wutzler P (2005) Susceptibility of coxsackievirus B3 laboratory strains and clinical isolates to the capsid function inhibitor pleconaril: antiviral studies with virus chimeras demonstrate the crucial role of amino acid 1092 in treatment. J Antimicrob Chemother 56:648-656

Selinka HC, Wolde A, Pasch A, Klingel K, Schnorr JJ, Küpper JH, Lindberg AM, Kandolf R (2002) Comparative analysis of two coxsackievirus B3 strains: putative influence of virus-receptor interactions on pathogenesis. J Med Virol 67:224-233

Si X, Luo H, Morgan A, Zhang J, Wong J, Yuan J, Esfandiarei M, Gao G, Cheung C, McManus BM (2005a) Stress-activated protein kinases are involved in coxsackievirus B3 viral progeny release. J Virol 79:13875-13881

Si X, McManus BM, Zhang J, Yuan J, Cheung C, Esfandiarei M, Suarez A, Morgan A, Luo H (2005a) Pyrrolidine dithiocarbamate reduces coxsackievirus B3 replication through inhibition of the ubiquitin-proteasome pathway. J Virol 79:8014-8023

Sidwell RW, Robins RK, Hillyard IW (1979) Ribavirin: an antiviral agent. Pharmacol Ther 6:123-146

Stein J, Liang P (2002) Differential display technology: a general guide. Cell Mol Life Sci 59:1235-1240

Visse R, Nagase H (2003) Matrix metalloproteinases and tissue inhibitors of metalloproteinases: structure, function, and biochemistry. Circ Res 92:827-839

Wang A, Cheung PK, Zhang H, Carthy CM, Bohunek L, Wilson JE, McManus BM, Yang D (2001) Specific inhibition of coxsackievirus B3 translation and replication by phosphorothioate antisense oligodeoxynucleotides. Antimicrob Agents Chemother 45:1043-1052

Yanagawa B, Spiller OB, Choy J, Luo H, Cheung P, Zhang HM, Goodfellow IG, Evans DJ, Suarez A, Yang D, McManus BM (2003) Coxsackievirus B3-associated myocardial pathology and viral load reduced by recombinant soluble human decay-accelerating factor in mice. Lab Invest 83:75-85

Yanagawa B, Spiller OB, Proctor DG, Choy J, Luo H, Zhang HM, Suarez A, Yang D, McManus BM (2004) Soluble recombinant coxsackievirus and adenovirus receptor abrogates coxsackievirus b3-mediated pancreatitis and myocarditis in mice. J Infect Dis 189:1431-1439

Yang D, Wilson JE, Anderson DR, Bohunek L, Cordeiro C, Kandolf R, McManus BM (1997) In vitro mutational and inhibitory analysis of the cis-acting translational elements within the 5' untranslated region of coxsackievirus B3: potential targets for antiviral action of antisense oligomers. Virology 228:63-73

Yang D, Yu J, Luo Z, Carthy CM, Wilson JE, Liu Z, McManus BM (1999) Viral myocarditis: identification of five differentially expressed genes in coxsackievirus B3-infected mouse heart. Circ Res 84:704-712

Yasukawa H, Yajima T, Duplain H, Iwatate M, Kido M, Hoshijima M, Weitzman MD, Nakamura T, Woodard S, Xiong D, Yoshimura A, Chien KR, Knowlton KU (2003) The suppressor of cytokine signaling-1 (SOCS1) is a novel therapeutic target for enterovirus-induced cardiac injury. J Clin Invest 111:469-478

Yuan J, Cheung PK, Zhang H, Chau D, Yanagawa B, Cheung C, Luo H, Wang Y, Suarez A, McManus BM, Yang D (2004) A phosphorothioate antisense oligodeoxynucleotide specifically inhibits coxsackievirus B3 replication in cardiomyocytes and mouse hearts. Lab Invest 84:703-714

Yuan J, Zhang J, Wong BW, Si X, Wong J, Yang D, Luo H (2005a) Inhibition of glycogen synthase kinase 3beta suppresses coxsackievirus-induced cytopathic effect and apoptosis via stabilization of beta-catenin. Cell Death Differ 12:1097-1106

Yuan J, Cheung PK, Zhang HM, Chau D, Yang D (2005b) Inhibition of coxsackievirus B3 replication by small interfering RNAs requires perfect sequence match in the central region of the viral positive strand. J Virol 79:2151-2159

Yuan JP, Zhao W, Wang HT, Wu KY, Li T, Guo XK, Tong SQ (2003) Coxsackievirus B3-induced apoptosis and caspase-3. Cell Res 13:203-209

Zautner AE, Jahn B, Hammerschmidt E, Wutzler P, Schmidtke M (2006) N- and 6-O-sulfated heparan sulfates mediate internalization of coxsackievirus B3 variant PD into CHO-K1 cells. J Virol 80:6629-6636

Zhang HM, Yanagawa B, Cheung P, Luo H, Yuan J, Chau D, Wang A, Bohunek L, Wilson JE, McManus BM, Yang D (2002) Nip21 gene expression reduces coxsackievirus B3 replication by promoting apoptotic cell death via a mitochondria-dependent pathway. Circ Res 90:1251-1258

Zhao Q, Zhou R, Temsamani J, Zhang Z, Roskey A, Agrawal S (1998) Cellular distribution of phosphorothioate oligonucleotide following intravenous administration in mice. Antisense Nucleic Acid Drug Dev 8:451-458

# Host Immune Responses to Coxsackievirus B3

S. Huber

1 Introduction ............................................................................................................ 200
2 Innate Immunity .................................................................................................... 200
　2.1 Virus-Cell Interactions as Host Response to Infection ............................................ 200
　2.2 Innate Effector Mechanisms ................................................................................ 206
3 Adaptive Immunity ................................................................................................ 210
　3.1 Virus-Specific Immunity ..................................................................................... 210
　3.2 Autoimmunity ................................................................................................... 212
4 Summary ............................................................................................................... 213
References .................................................................................................................. 214

**Abstract** Group B coxsackieviruses are members of the picornavirus family of small nonenveloped RNA viruses and have been associated with diseases of multiple organs including the heart, acinar and islet pancreas, liver, skeletal muscle, central nervous system, and testes. Damage to tissues occurs not only from the direct virus replication and infection of cells, but also from the host response to infection. However, without host immunity and response, the viruses are not appropriately cleared and chronic infection occurs. The host response to coxsackieviruses is diverse and complex. Also, the host response both benefits and is detrimental to the virus. This review discusses the major aspects of the host response to coxsackieviruses and attempts to demonstrate the interplay between the virus and the cell, which ultimately determines both the type and strength of the adaptive immune response as well as whether autoimmunity will follow the infection.

S. Huber
University of Vermont, USA 05405 Burlington, VT
Sally.Huber@uvm.edu

# 1 Introduction

Coxsackieviruses are members of the picornavirus family of small non-enveloped RNA viruses which replicate in the cell cytoplasm and are usually considered to be released from infected cells through cell lysis (Rueckert 1996). As with nearly all microbial infections, the host response is both diverse and complex. Furthermore, host response to the virus, in one form or another, may be essential to the virus for replication. As discussed below, studies have found that coxsackieviruses can only successfully replicate in cells during the G1/S phase of the cycle because of requirements for virus RNA translation (Feuer et al. 2003). For only cells already in cycle to be able to support virus replication would substantially increase the virus inoculum necessary to establish an infection. It is far more likely that the virus can itself cause the cells it binds to enter the cell cycle and/or become activated. There are several different mechanisms by which such activation can occur. These include virus cross-linking of cellular molecules used as the virus receptor and signal transduction through this cross-linking; and signaling through toll-like receptor (TLR) recognition of viral molecules, most notably single-stranded (ssRNA) and double-stranded (dsRNA) viral RNA. Virus-induced cellular activation is also the first step in the host response to the infection since TLR signaling is a potent inducer of immune cell proliferation and cytokine/chemokine expression. As a rule, the adaptive immune response is usually required for complete virus clearance and individuals with seriously compromised adaptive immunity may develop chronic infections. Since that microbe-specific immune response takes a substantial amount of time (7-14 days for the primary immune response) to develop, the innate immune response is needed to help keep the microbial infection in check until the adaptive immune response is in full gear. The innate immune response also plays an essential role in directing the developing adaptive immune response. This review discusses various aspects of the host response to coxsackievirus infections that are necessary both for controlling the virus and that also may contribute to tissue injury during infection.

# 2 Innate Immunity

## 2.1 Virus-Cell Interactions as Host Response to Infection

### 2.1.1 Virus-Virus Receptor Interactions

The initial step in virus replication is virus attachment to one or more cell surface receptors, which leads to internalization of the virus within the cell. Generally, receptor expression is a major determinant in viral tropism and species specificity of virus infection (Schneider-Schaulies 2000). Two cellular receptors are best known for coxsackieviruses. These are decay accelerating factor (DAF, CD55) and coxsackievirus-adenovirus receptor (CAR). CAR is a member of the immunoglobulin superfamily and is located in tight junctions (junction-associated molecule; JAM) (Bazzoni 2003). These molecules are not only necessary for the structural formation and stabilization of tight junctions, primarily in epithelial and endothelial cells, but also function in promoting leukocyte transmigration. CAR binds immunoglobulin and B lymphocytes and therefore might be involved in transmigration of these lymphocytes (Carson 2001; Carson and Chapman 2001). All six coxsackievirus B serotypes bind CAR (Martino et al. 2000) and the extracellular domain of the molecule is adequate to promote virus entry (Wang and Bergelson 1999). Although CAR is normally present in tight junctions, exposure to TNFα and IFNγ can cause dispersal from the tight junctions and increase JAM accessibility to either leukocytes or viruses (Ozaki et al. 1999). The same cytokines also reportedly suppress CAR expression in endothelial cells (Vincent et al. 2004). These two effects could promote coxsackievirus-induced disease both by making residual CAR more available to virus and weakening the tight junction integrity, which could promote leukocyte transmigration. The other receptor, DAF, is more widely distributed in tissues and cells than CAR, and its true function is to protect cells against complement mediated lysis (Medof et al. 1987). Unlike CAR, DAF is expressed in nonattached cells such as leukocytes and may be responsible for the infection and replication of coxsackievirus B3 in B cells, dendritic cells, and activated T cells (Anderson et al. 1996; Liu et al. 2000a), although this has not been shown. Chimeric fusion proteins (DAF-Fc and CAR-Fc) are highly effective in inhibiting coxsackievirus infection in the heart but CAR-Fc alone prevents virus infection of the pancreas (Goodfellow et al. 2005; Yanagawa et al. 2003, 2004). Although coxsackievirus interactions with both CAR and DAF lead to optimal infection (Selinka et al. 2004), CAR appears to be the dominant receptor and can lead to infection with little or no DAF involvement (Pasch et al. 1999).

Although CAR and DAF are the best known of the coxsackievirus B3 receptors, it seems probable that other receptors/co-receptors must also exist (Schmidtke et al. 2000). The CVB3 PD variant binds more effectively to DAF than to CAR, but is also able to bind to and infect CAR and DAF deficient cell lines. This interaction occurs through heparin sulfates (Zautner et al. 2006). Virus internalization was substantially slower when heparin sulfate was used rather than CAR, but the total time required for completion of the virus replication cycle was unchanged by receptor utilization. Other receptors are also likely to exist. CAR, DAF, and heparin sulfates are very widely distributed within the body, yet coxsackievirus does not infect equally all tissues positive for these molecules. In an experimental model, coxsackievirus B3 infection of DBA/2 mice results in high virus titers in heart, liver, pancreas and brain, but little replication in other tissues. Furthermore, when virus replicating in the heart was isolated and used to reinfect naïve mice, the heart virus isolates showed substantially better tropism to the heart than to other tissues such as liver, while isolates from the liver were more tropic to the liver of naïve mice than to the heart (Huber et al. 1990). This variation in virus isolate tropism depended upon the ability of the viruses to bind to tissue-specific endothelial cells. There is no evidence that endothelial cells isolated from different tissues vary substantially in CAR/DAF or heparin sulfate expression.

Virus-cellular receptor interactions also form the initial host response to infection. Most virus receptors belong to the immunoglobulin, integrin, or other superfamilies with important biological functions. Virus cross-linking the receptor may in many cases mimic the natural ligation of these molecules and result in highly effective transduction of signals in cells. Examples include:

1. HIV, which uses CD4 and co-receptors CCR5 and CXCR4, and activates various kinase pathways leading to increased cytokine and MMP9 induction (Collman et al. 2000; Del Corno et al. 2001; Liu et al. 2000b; Misse et al. 2001),
2. Epstein-Barr virus gp350, which cross-links CD21 and results in B cell proliferation and cytokine induction (D'Addario et al. 1999, 2000, 2001),
3. Herpesvirus, which uses herpesvirus entry mediator protein, a member of the TNF receptor superfamily, which can suppress T cell proliferation and activation (Cheung et al. 2005).

Similar signal transductions probably occur with DAF, one of the coxsackievirus receptors. DAF is a glycosylphosphatidylinositol (GPI)-anchored membrane glycoprotein (Brodbeck et al. 2000; Miwa et al. 2001, 2002; Song et al. 1996), and signaling through this molecule can induce calcium flux (Peiffer et al. 1998). Antibody cross-linking of DAF molecules activates p56lck and p59fyn and promotes T cell proliferation and cytokine induction (Davis et al. 1988; Shenoy-Scaria et al. 1992). If antibody cross-linking of DAF can induce signal transduction, it is reasonable that coxsackieviruses with high binding avidity to DAF could also cross-link and signal through these molecules. Activation of p56lck is necessary for coxsackievirus replication in T cells, multiorgan pathogenicity, and viral persistence in vivo (Liu et al. 2000a). Therefore, viruses having high avidity DAF-binding and enhanced p56lck activation should show increased in vivo virulence. However, signaling through DAF is also directly immunomodulatory. $DAF^{-/-}$ mice immunized with ovalbumin have a substantially increased T cell response compared to immunized $DAF^{+/+}$ animals (Liu et al. 2005), and animals lacking DAF also develop aggravated experimental allergic encephalomyelitis. Complement is crucial to this effect since $C3^{-/-}DAF^{-/-}$ double knockout mice showed normal T cell responsiveness. Similar studies found that antigen presenting cells lacking DAF-induced enhanced T cell responses (Heeger et al. 2005).

### 2.1.2 Toll-Like Receptors

Pattern recognition receptors (PRR) are cellular molecules that recognize pathogen-associated molecular patterns (PAMPs) to activate host innate and adaptive immunity and stimulate production of cytokines. PAMPs are conserved microbial structures that are found in broad categories of infectious agents, are usually important in either microbial infection or replication, and are not often mutated (Abreu and Arditi 2004). Recognition of these conserved ligands allows very rapid host responses to infection without requiring specific antigen activation that would take 7-14 days for maximum effect. There are 11 known mammalian Toll-like receptors (TLR) reacting to different molecular species(Abreu and Arditi 2004; Hasan et al. 2005; Lauw et al. 2005; Netea et al. 2004; O'Neill 2004) (Table 1). Additional complexity of PAMP recognition is provided by the formation of heterodimers between different TLR, such as between TLR1/TLR6, TLR1/TLR2, or TLR2/TLR6 (Schroder et al. 2005). TLR are transmembrane proteins that have conserved intracellular domains and leucine-rich extracellular domains (Abreu and Arditi 2004; Means et al. 2000). Studies by Zarember and Godowski (Zarember and Godowski 2002) found that nearly all tissues express at least low levels of mRNA for at least one TLR, and several cells and tissues express multiple TLRs. Heart expresses modest levels of TLR5 and has minimal expression of TLR3 and TLR4, but negligible expression of the other TLRs. In contrast, lymphoid cells express very high levels of all TLRs but the distinct populations of mononuclear cells can have substantial variation in TLR expression. $CD8^+$ cells express high levels of TLR3 but very low levels of TLR2 and TLR4, while B cells express high levels of TLR9 and TLR10. The ability of lymphoid cells to express TLRs, and especially the distinct patterns of expression, may result in different effects of PAMPs on the immune response (reviewed in Pulendran 2005; Netea et al. 2005). Generally, TLRs 3, 4, 5, and 9 are considered to preferentially activate Th1 responses, while TLR2 preferentially induces Th2 immunity (Netea et al. 2005). In the resting stage, plasmacytoid dendritic cells express TLR7 and TLR9 but not TLR3, while monocyte-derived dendritic cells express TLR3 but not TLRs7 and 9, suggesting that the different dendritic cell subsets might have distinct responses to pathogens. This is important since different dendritic cell subsets preferentially activate Th1 or Th2 adaptive immunity and also have different abilities to make specific host response molecules such as type I interferons under selected circumstances. The differential ability of TLR signaling to bias adaptive immunity toward a Th1 or Th2 phenotype may also depend not only on the type of TLR activated, but upon the intensity of the activation. Thus, low-dose LPS stimulation of TLR4 promotes Th2 response, while higher doses of ligand promote Th1 responses. Care needs to be taken when

**Table 1** TLRs and their legends

| | |
|---|---|
| TLR 1 | Triacyl lipopeptides |
| TLR 2 | Glycosylphosphatidiyl inositol–linked proteins, lipoproteins, zymosan |
| TLR 3 | Double–stranded RNA |
| TLR 4 | LPS, heat shock proteins |
| TLR 5 | Flagellin |
| TLR 6 | Diacylated lipoprotein and peptidoglycan |
| TLR 7/TLR 8 | Single–stranded RNA |
| TLR 9 | Bactein CpG motifs, herpes viral DNA |
| TLR 10 | Unknown |
| TLR 11 | Profilin–like protein (Toxoplasma) |

evaluating TLR expression through mRNA levels, however. Studies indicate that naïve CD4$^+$ cells express TLR2 and TLR4 mRNA and intracellular protein but not cell-surface protein. After activation, CD4$^+$ cells rapidly express cell-surface TLR2 and TLR4 (Liew et al. 2004). Stimulation of activated CD4$^+$ cells with the TLR2 ligand (bacterial lipoprotein) substantially enhances IFNγ, IL-2, and TNFα expression by the T cells but addition of the TLR4 ligand, LPS, has minimal effect. TLR2 is also constitutively expressed on memory T cells, indicating that microbial PAMPs might directly augment a memory response.

The TLR intracellular domain contains toll-interleukin (IL)-1 response (TIR) domains, which have homology to the IL-1/IL-18 receptors and act as a platform for the recruitment of kinases (Abreu and Arditi 2004; Vogel et al. 2003). Different TLRs activate distinct signal pathways depending upon the adaptor molecules recruited to the TIR. The most common adaptor molecule is MyD88, which interacts with TLRs 2, 4, 5, 7, and 9. MyD88 recruits interleukin receptor-associated kinases (IRAKs), which activate TRAF6 and IKK. This phosphorylates IkB, leading to NFkB activation and increased expression of pro-inflammatory cytokine (IL-1β, TNFα, and IL-6) expression (Michelsen et al. 2004). MyD88 signaling also increases inducible nitric oxide synthetase expression (Hosoi et al. 2004), which has known antiviral activity in coxsackievirus B3 infections (Lowenstein et al. 1996). MyD88 signaling enhances CD80 and CD86 costimulatory molecule expression that binds CD28 on T lymphocytes during antigen presentation (Horng et al. 2002). Another adaptor molecule is Toll-IL-1 receptor domain-containing adaptor protein (TIRAP), which signals through TLR1, TLR2, TLR4, and TLR6 but not through TLR5, TLR7, TLR9, IL-1, or IL-18 (Horng et al. 2002). TIRAP activates pro-inflammatory cytokine responses, but does not upregulate co-stimulatory molecule expression. Two other adaptors are Toll-IL-1 receptor domain-containing adaptor-inducing interferon-beta (TRIF) and TRIF-related adaptor molecule (TRAM). TRIF binds to TLR3 and TLR4 (O'Neill 2004; Vogel et al. 2003). TRAM interacts with TRIF, but only in TLR4 signal pathways. The TRIF/TRAM pathways act independently of MyD88 and activate interferon response factors (IRF3 and IRF7), leading to increased expression of type I interferons, RANTES, and IP-10. Thus, different patterns of adaptor molecule usage by TLR can result in complex signaling (Vogel et al. 2003) (Fig. 1).

Various studies have been conducted investigating the role of TLRs in coxsackievirus infections and associated diseases. MyD88$^{-/-}$ mice reduce expression of various cytokines and molecules known to be important in either host response to coxsackievirus infection (pro-inflammatory cytokines including TNFα and IFNγ) or in viral infection itself (CAR and p56lck) (Fuse et al. 2005; Liu et al. 2000a). However, type I interferon expression is increased in infected MyD88$^{-/-}$ mice and virus titers are reduced presumably because of this interferon response (Lutton and Gauntt 1985). Studies by Triantafilou et al. show that cardiac cells express a minimal amount of TLR1-8 prior to coxsackievirus infection but after infection show increased expression of TLR7 and TLR8 (Triantafilou et al. 2005), and interaction of single-stranded coxsackievirus RNA with TLR7 and TLR8 in endosomes recruits MyD88 and results in NFkB dependent pro-inflammatory cytokine

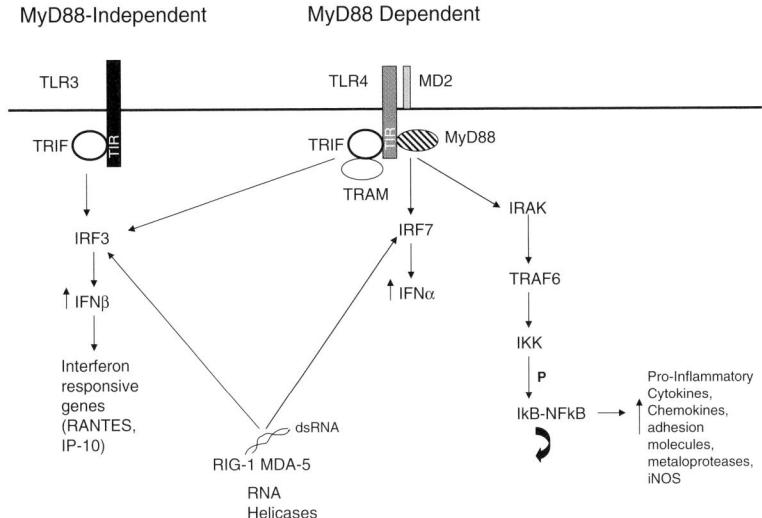

**Fig. 1**

expression. Surprisingly, coxsackieviruses also upregulate TLR4, which theoretically should not be upregulated by viruses (Fairweather et al. 2005). It is not known why coxsackievirus infection upregulates TLR4, but studies indicate that IFNγ promotes and IL-4 supressesTLR4 and TLR3 expression (Mueller et al. 2006). Coxsackievirus B3 infections usually cause strong IFNγ responses in mice (Huber et al. 1996). Upregulation of TLR4 may be a by-product of the host response to the virus. However, in the absence of a ligand for TLR4, one would not have expected the upregulated TLR4 to be functional. Since it clearly is, this must mean that either the virus or some other molecule present during coxsackievirus infections must be a viable TLR4 ligand.

### 2.1.3 Type I Interferons

Picornaviruses make double-stranded RNA (dsRNA) as part of their replicative cycle, and this dsRNA initiates profound alterations in cell physiology (Gibaudo et al. 1991). The dsRNA induces type 1 interferons through both TLR-dependent and -independent mechanisms (Conzelmann 2005). As indicated above, TLR3 and TLR4 activate IRF3, which binds to the interferon β promoter, leading to expression of this gene. Signaling through TLR3, 4, 7, and 9 also activates IRF-7, which binds to the interferon α promoter, leading to production of this mediator. In addition to the TLR-dependent mechanisms, two DEXD/H box RNA helicases have been identified that can directly bind cytosolic virus. These helicases, RIG-1 and

MDA-5, contain two caspase recruiting domains (CARDs) in the N terminal portion of the molecule and presumably activate IRF-3 through TBK1 (Andrejeva et al. 2004; Yoneyama et al. 1998, 2004). There are eight classes of type I interferons (Oritani and Kanakura 2005; Pestka et al. 2004). These are IFNα, IFNβ, IFNξ, IFNδ, IFNτ, IFNε, IFNκ, and IFNω, although not all of these are found in all species. For example, IFNs α,ε,κ, and ω are found in humans, while IFNs δ,τ, and ξ are not. Furthermore, there are subtypes of IFNs (14 genes for IFNα in humans and 12-13 proteins) (Pestka et al. 2004). Some are probably tissue-restricted because human keratinocytes make IFNκ, while IFNε promoters have reproductive endocrine sites (Hardy et al. 2004; LaFleur et al. 2001; Pestka et al. 2004). All type I interferons, except possibly IFNω, use the IFN-αR1 and IFN-αR2c cellular receptors and activate various signal pathways, the best known of which are the JAK1 (associated with the IFN-αR1 intracellular domain) or Tyk2 (associated with the IFN-αR2 intracellular domain) and Stat1 and Stat2 pathways activated by IFNα and IFNβ. Stat1/Stat2 transcription factors bind to ISRE recognition sequence in gene promoters to induce gene transcription. Two major proteins activated by IFNs are RNA-dependent protein kinase (PKR) and 2'5'-oligoadenylate synthetase (OAS), which in turn activate eIF2α and RNase L, respectively (Oritani and Kanakura 2005). The former molecule inhibits viral protein synthesis and activates the NFkB pathway, while RNase L can cleave viral RNA. Both PKR and RNase L promote cellular apoptosis, which can eliminate infected cells before progeny virions can be produced. An additional antiviral effect of IFNα/β is its ability to induce expression of $NOS_2$ (Diefenbach et al. 1998), and nitric oxide is an important host defense against coxsackievirus infections by inhibiting viral RNA and protein synthesis (Saura et al. 1999; Zaragoza et al. 1997, 1998).

In addition to their direct antiviral activities, type I interferons have potent biological activities, especially on immunological functions, such as upregulation of major histocompatibility complex class I and class II antigens, cell cycle arrest, stimulation of T cell and natural killer cell activity, activation of pro-apoptotic factors, and inhibition of anti-apoptotic factors. Many types of cells can produce type I IFNs but at least some studies indicate that plasmacytoid dendritic cells are the dominant natural source of type I IFNs in viral infections (Asselin-Paturel et al. 2001, 2005). Type I IFNs are at least partially responsible for inducing dendritic cell maturation and may be important in dendritic cell migration to T cell-rich areas in the spleen (Asselin-Paturel et al. 2005).

## 2.2 Innate Effector Mechanisms

### 2.2.1 Alternate Complement Activation Pathway

Coxsackievirus B3 directly binds to the third component of complement (C3) and while not activating the classical pathway of complement, does activate the alternative pathway as shown by the production of C3 degradation products (Anderson et al. 1997). Virus-C3 interaction is necessary for virus retention in germinal centers where B cells may act as early targets of infection and a source of virus dissemination (Mena et al. 1999). The mechanism(s) by which C3 is important in virus localization are not clearly understood, but since DAF is both a receptor for coxsackieviruses and binds C3, it is possible that virus-C3 interactions might provide an additional mechanism for DAF-virus binding for normally non-DAF-binding viral variants (Orthopoulos et al. 2004). Similarly, virus-C3 complexes could infect macrophage and other cells expressing C3 receptors. Whether coxsackievirus binding to and activation of complement also causes neutralization of the virus, as has been reported for other virus systems (Ikeda et al. 1998), is not known. Complement activation can be highly injurious to the host since C3a and C5a degradation products act as anaphylatoxins, which can recruit leukocytes to the site of infection. Thus, virus-induced activation of complement could substantially promote early inflammation without requiring stimulation of the adaptive immune system.

### 2.2.2 Mast Cells

Several studies have correlated mast cells and/or mast cell degranulation with myocarditis and cardiac fibrosis. Mast cells are increased in hearts of patients with lymphocytic and giant cell myocarditis and are more closely associated with areas of fibrosis in the heart than with areas of active inflammation (Turlington and Edwards 1988). In a mouse model of coxsackievirus-induced myocarditis, degranulating mast cells were observed within 6 h of infection and mast cell numbers were increased in mouse strains susceptible to virus-induced heart disease compared to resistant mouse strains (Fairweather et al. 2004). This provides circumstantial evidence that early mast cell degranulation and activation may determine disease susceptibility. It is not clear why coxsackieviruses would cause rapid mast cell degranulation. The usual mechanism of degranulation involves an initial immune response to an antigen (or virus) with the production of IgE antibodies that bind to Fc receptors on mast cells, thus arming them. At subsequent antigen (virus) exposure, antigen binding to the Fab portion of the IgE triggers mast cell degranulation (Dakhama et al. 2004). However, evidence also indicates that dsRNA may directly activate cytokine production by mast cell using TLR3 (Kulka et al. 2004). Mast cells produce a wide variety of pharmacologic factors, which can be divided into those that are preformed (histamine, heparin, and serine proteases) and those that are induced with mast cell activation (cytokines, leukotrienes, prostaglandins, and thromboxanes). These factors are released through either slow exocytic degranulation or rapid anaphylactic degranulation, which occurs within minutes (Abraham and Malaviya 1997). TNF$\alpha$ is one pro-inflammatory cytokine that is preformed and stored in mast cells for immediate release with activation (Gordon and Galli 1991). Furthermore, studies have shown that exogenous administration of TNF$\alpha$ is sufficient to convert coxsackievirus B3-induced myocarditis-resistant mice into disease-susceptible animals (Huber 2004; Lane et al. 1992). Thus, it is reasonable that mouse strains naturally having increased numbers of mast cells in the heart would show increased viral myocarditis susceptibility since virus-induced mast cell degranulation would release high levels of TNFα needed for pathogenesis. Many other important factors are made by mast cells including IL-1, IL-6, IL-8, IL-10, IL-12, IL-13, GM-CSF, MCP-1, MIP-1α, and RANTES, all of which may promote inflammation (Abraham and Malaviya 1997). Histamine is also a major modulator of immune responsiveness and can be necessary for regulation (promotion or suppression) of autoimmunity, depending upon the histamine receptor used (Jutel et al. 2006; Kunzmann et al. 2003; Ma et al. 2002; Musio et al. 2006). There are four histamine receptors (H1R, H2R, H3R, and H4R), which are hepta-helical G-protein-coupled receptors (Jutel et al. 2002). Signaling through H1R promotes both Th1 (IFNγ$^+$) and Th2 (IL-4$^+$) responses, while signaling through H2R suppresses both Th1/2 responses. H2R and H3R suppress IL-12 and enhance IL-10 expression by dendritic cells. H1R-deficient mice produce higher levels of IgG1 and IgE antibodies, while H2R-deficient mice showed reduced IgG3 and IgE responses, demonstrating that histamine regulates B cell as well as T cell responses. Thus, histamine can have diverse effects on host responses to infection.

### 2.2.3 Natural Killer, Natural Killer T cells, and γδ TCR$^+$ T cells

Innate immunity represents a constitutive or rapidly induced host response to infection, which reacts broadly to many different microbial agents. Innate effectors include natural killer, natural killer T, and γδ$^+$ T cells. Natural killer (NK) cells are non-T cells lacking the T cell receptor and CD3 (Biron et al. 1999). Type I interferons increase NK cell cytotoxicity, which is mediated through perforin and results in infected cell death prior to completion of the virus replication cycle. NK cells express receptors that can either increase or suppress their cytotoxic activities. NK cells recognize MHC class I molecules through killer immunoglobulin-like (KIR) receptors, which suppress NK cytotoxicity. This is one mechanism for NK cell differentiation between infected/transformed and normal cells, as transformation or virus infection can cause downregulation of MHC molecules (Moretta and Moretta 2004; Vossen et al. 2002). In contrast to the KIRs, natural cytotoxic receptors (NCR) activate NK cells, making them more cytolytic (Moretta and Moretta 2004). Evidence suggests that NK cells are important in controlling coxsackievirus B infections in vivo (Gauntt et al. 1988, 1989; Vella and Festenstein 1992) as depletion of these cells substantially increases virus titers in the heart or pancreas. NK cells directly interact with both dendritic cells and activated T cells, causing maturation of the dendritic cells and increased activation of the T cells (Zingoni et al. 2005). Interactions occur through upregulation of OX40L on the NK cells and OX40 on activated CD4$^+$ lymphocytes. A subpopulation of NK cells are NKT cells. Unlike NK cells, these NKT cells express a T cell receptor comprised of an invariant alpha chain (Vα14-Jα18 mice; Vα24-Jα18 humans) and one of several different beta chains (Vβ2, Vβ7, and Vβ8.2) (Godfrey and Kronenberg 2004). Although initially described as CD4$^+$, it is now known that NKT cells can also be CD4$^-$. As with NK cells, NKT cells are cytolytic to cells and also modulate developing adaptive immune responses either through interactions with T cells and dendritic cells or through production of high levels of cytokines. Subpopulations of NKT cells can either produce Th1-associated cytokines (IFNγ) or Th2 cytokines (IL-4). NKT cells are protective in several bacterial, viral, and parasitic infections (Godfrey and Kronenberg 2004), but it is questionable whether these cells are important in coxsackievirus B3 infections, at least in the heart. Coxsackievirus B3 infection of Jα18-deficient mice had no effect on either coxsackievirus-induced myocarditis or cardiac virus titers (Huber et al. 2003). However, studies by Exley et al. (2001) found that NKT cells protect against encephalomyocarditis virus (EMCV)-induced diabetes, and that in the absence of NKT cell function, virus titers in the islets are substantially increased. EMCV is a picornavirus, as is coxsackievirus B3. Whether these results indicate that NKT cells are protective only in selected picornavirus infections or whether there are differences in NKT cell migration and function in different picornavirus infected tissues is not known.

Many NKT cells recognize nonclassical MHC class I antigens on target cells. The most common implicated molecules belong to the CD1 family. CD1 comprises two groups of molecules: CD1 group I and group II. Group I consists of CD1a, CD1b, and CD1c, while Group II consists of CD1d (Stanic et al. 2003). Mice express only the CD1d molecules. CD1d presents self- and possibly microbial glycolipid antigens (Bendelac et al. 1997; Burdin and Kronenberg 1999). Although cardiac myocytes are normally CD1d-negative, these cells can rapidly upregulate CD1d expression when infected with coxsackievirus B3 (Huber et al. 2003). Thus, infected myocytes will express self-glycolipid antigens, which are not expressed on uninfected cells. Restricting CD1d upregulation only to infected somatic cells should have the advantage of restricting CD1d-dependent innate immunity killing only to the infected cells while sparing adjacent uninfected cells. Both infection and exposure to TNFα are required for CD1d induction (Huber and Sartini 2005a). TNFα signaling requires TNFR1 but is independent of TNFR2. What the viral signal is in CD1d expression is not currently known, but it seems probable that TLRs would be the most likely candidates.

A subpopulation of T cells expressing the γδ TcR also recognize CD1d (Huber et al. 2003). The γδ$^+$ T cells often accumulate at sites of inflammation because of either infection (Carding 1990; Carding et al. 1990; Carding and Egan 2000; O'Brien et al. 1989; Sandor et al. 1995) or autoimmunity (Mukasa et al. 1997; Olive 1995; Peterman et al. 1993). The γδ$^+$ T cells are delineated by the variable (V) genes (γ- and δ-chains) they express, and distinct subpopulations of γδ$^+$ cells are often found in different diseases and can have either beneficial or detrimental effects (Mukasa et al. 1997, 1998; Olive 1995; Roark et al. 1996; Sandor et al. 1995). In coxsackievirus-induced myocarditis, Vγ4$^+$ T cells are highly pathogenic and promote both CD4$^+$ Th1 and autoimmune cytolytic CD8$^+$ T cell responses (Huber and Lodge 1984; Huber et al. 2002a, 2002b; Huber 2004). In contrast, Vγ1$^+$ cells promote CD4$^+$Th2 cell and virus neutralizing antibody responses (Huber et al. 2000; Huber and Sartini 2005b). Therefore, while Vγ4$^+$ cells are selectively pathogenic by initiating autoimmunity induction in coxsackievirus B3 infection, Vγ1$^+$ cells in the same animals can be protective through their effect on antiviral humoral immunity. Co-segregation of Vγ haplotype and function has been observed in other disease models (O'Brien et al. 2000, 2001) and has led to the hypothesis that whenever a specific Vγ subpopulation is activated in a disease, it will always have the same effect on developing adaptive immune responses (O'Brien et al. 2001).

# 3 Adaptive Immunity

## 3.1 *Virus-Specific Immunity*

Evidence suggests that both humoral and cellular immunity are important in coxsackievirus clearance (Tam 2006). Coxsackievirus infection of B cell-deficient mice results in chronic, high-titer viral infections in heart, liver, kidney, lung, pancreas, and spleen for the life of the mice (Mena et al. 1999). Transfer of naïve T or B cells to these animals had no effect on persistent infection, but transfer of immune T or B cells at least transiently suppressed virus infection. These results are similar to those showing that infection of nude (athymic) mice also resulted in persistent infection (Sato et al. 1994; Schnurr et al. 1984). However, other reports suggest that neither neutralizing antibody nor T cells control virus infection, but that macrophages are the dominant factor controlling viral clearance (Woodruff and Woodruff 1974; Woodruff 1979). Why different investigators reach opposite conclusions on the mechanisms of virus control is not clear but might suggest that either distinct immune mechanisms affect viral replication using different coxsackievirus variants or that different mouse strains might control virus replication through distinct means. For example, different strains of athymic mice were used by the investigators varying in their conclusion that T cells promote virus clearance (Schnurr et al. 1984; Woodruff and Woodruff 1974). T cells are important mediators of pathogenicity in coxsackievirus infections (Huber and Lodge 1986; Huber 1997; Opavsky et al. 1999; Woodruff and Woodruff 1974). However, the same immunopathogenic mechanisms are not important in all infected mice even when the animals are infected with the identical coxsackievirus B3 variant. Thus, infection of BALB/c mice results in induction of autoimmune $CD8^+$ effectors without autoimmune antibody generation or humoral pathogenicity, while infection of DBA/2 mice results in an exclusively $CD4^+$ cell-dependent autoantibody mediated myocarditis (Guthrie et al. 1984; Huber and Lodge 1984, 1986). In A/J animals, both $CD4^+$ and $CD8^+$ cells are pathogenic with both humoral and cellular immunopathogenic mechanisms involved (Lodge et al. 1987). Although it has never been proven, it is highly likely that clinical viral myocarditis will be even more complex than the experiment murine models and involve even additional pathogenic mechanisms. The importance of this is that the same therapies may not prove beneficial to all forms of myocarditis. Certainly, murine studies have shown that while the humoral autoimmunity in coxsackievirus-infected mice responds favorably to immunosuppression, $CD8^+$ cell-dependent myocarditis is actually made worse by immunosuppression (Estrin et al. 1986; Herzum et al. 1991; Huber et al. 1988). Since it is not clear what the dominant pathogenic mechanisms are in clinical myocarditis, use of a single therapy might miss beneficial results in a subpopulation of patients while concluding that the treatment is ineffective in the myocarditis population as a whole (Mason et al. 1995). The murine models of viral myocarditis are not likely to be useful in clinical disease until a thorough understanding of what pathogenic mechanisms dominate in patients and whether specific mechanisms are responsible for the poor outcome in approximately 10% of clinical cases.

$CD8^+$ effector cells are frequently mediators of virus clearance in vivo (Kanto and Hayashi 2006). However, to date, no $CD8^+$ T cell epitopes have been identified in coxsackievirus B3 infection. The only $CD8^+$ effectors identified to date are autoimmune against heart antigens and fail to recognize or kill infected myocytes (Huber and Lodge 1984; Slifka et al. 2001). Studies by Slifka et al. (2001) introduced a known $CD8^+$ CTL epitope from LCMV virus into a recombinant coxsackievirus B3 and showed that mice infected with the recombinant virus failed to generate a strong $CD8^+$ LCMV-specific response. This result raises the question of whether coxsackieviruses use immune evasion to prevent $CD8^+$ virus-specific responses (Vossen et al. 2002). Many viruses use distinct mechanisms for downregulating virus-specific immunity. These include inhibiting proteosome degradation of viral peptides, peptide transport into the ER where the peptides can be bound to MHC class I molecules, transcription or transport of MHC class I molecules, and expression of CD4 (see Vossen et al. 2002). Viruses may additionally promote MHC molecule degradation. The question is whether coxsackieviruses use any of these or other unknown mechanisms to inhibit virus-specific $CD8^+$ T cell responses. Ubiquitination of proteins will result in transport and degradation of those proteins by proteosomes (Monaco 1995). Certain viral proteins, such as the MIR-1, MIR-2, and two E3 ubiquitin ligases of Karposi's sarcoma-associated herpesvirus, ubiquinate and downregulate MHC class I molecules (Cadwell and Coscoy 2005). Although no specific ubiquinating viral proteins have been identified for coxsackieviruses or other picornaviruses, it is clear that coxsackieviruses interact with the ubiquitin/proteosome pathway during their replication (Luo et al. 2003a) and that coxsackievirus facilitates ubiquitination and degradation of at least some cellular proteins such as cyclin D1 (Luo et al. 2003b). However, MHC class I expression is increased in coxsackievirus B3-infected hearts (Seko et al. 1990, 1996). While this might suggest that facilitated ubiquitination by coxsackieviruses does not alter MHC class I expression by infected cells, in truth the published studies do not specifically co-localize MHC expression with virus in the same cells. IFNγ can upregulate MHC expression (Seko et al. 1990), and it is possible that the enhanced MHC class I in the heart is on uninfected myocardial cells. Certainly, this would promote susceptibility of the uninfected cells to autoimmune lysis through antigen presentation of self-peptides by the induced MHC molecules, but downregulation of MHC on infected cells would be protective from virus-specific effectors.

Although a second method of inhibiting virus peptide presentation on infected cells might be through the general inhibition of protein synthesis found during infection, it seems clear that total protein synthesis is not suppressed in infected myocytes. As discussed above, infected myocytes were evaluated for CD1d expression in the hearts of infected mice by confocal microscopy and demonstrated that infected (virus protein-positive) myocytes upregulate CD1d expression (Huber et al. 2003). Since no virus-negative myocytes were CD1d-positive, these infected myocytes must have been able to synthesize de novo the CD1d molecules. Why CD1d expression would be upregulated while MHC class I molecule expression might be inhibited is not clear, except that antigen loading of CD1d molecules occurs in the endosome pathway, not in the ER, as is the case with MHC class I molecules (Chiu et al. 2002). It is possible that different localization of CD1d and MHC class I molecules within the cell affect ubiquitination and degradation.

## 3.2 Autoimmunity

Autoimmunity is the reaction of the immune system to self-antigens. Three criteria need to be met for autoimmune disease. There needs to be a pool of autoreactive lymphocytes, these autoreactive lymphocytes must undergo activation and migration to the relevant tissue, and the normal immunoregulatory mechanisms that should inhibit autoimmunity must fail (Horwitz and Sarvetnick 1999). Although deletion of self-reactive T cell clones should occur in the thymus during T cell ontogeny, this requires the presence of the self-antigen in the thymus. Thus, sequestered self-molecules may not result in clonal deletion of cells reactive to them (Huber 2005; Theofilopoulos 1995a 1995b). An alternative approach for suppressing autoimmunity is through anergy of self-reactive T cells by inappropriate antigen presentation in the periphery. This usually means presentation of antigen by nonprofessional antigen-presenting cells that lack the appropriate co-stimulatory factors needed for optimal naïve T cell activation. However, many self-reactive T cell clones may remain because the antigens they recognize are not available to the immune system and are in cells that normally do not express MHC antigens. Autoimmunity in this case may occur during virus infection when the virus kills cells, thus releasing substantial amounts of self-antigens, and also recruits professional antigen-presenting cells to the site of infection as part of the antiviral immune response. The release of cellular components during infection may also lead to epitope spreading. Epitope spreading follows an initial immune response to a specific epitope or antigen by making more antigens available to the immune system and promoting responses to these new epitopes or antigens (Vanderlugt and Miller 2002). Epitope spreading is found in several viral diseases such as Theiler's murine encephalomyelitis virus infections, where virus infection causes tissue damage in the central nervous system, leading to endogenous antigen release at the time when activated antigen-presenting cells are available to process the self-antigens (Miller et al. 2001). Coxsackievirus B3 infection first results in antiviral antibodies, but ultimately to autoantibodies to a wide variety of cardiac antigens (Latif et al. 1999; Schwimmbeck et al. 2004). Similar observations occur in clinical myocarditis and dilated cardiomyopathy patients who often have autoantibodies to multiple cardiac antigens (Caforio et al. 2001, 2002; Pankuweit et al. 1997). The concurrent presence of multiple autoantibodies is more consistent with epitope spreading than with antigenic mimicry.

Other mechanisms for induction of autoimmunity are also known. The classical method is antigenic mimicry between an infectious agent and self-molecules. It has been clearly shown that antigenic mimicry exists between the M5 protein of group A streptococcus (Cunningham et al. 1986). Interestingly, antibodies to streptococcal protein and cardiac myosin also neutralize coxsackievirus B3, indicating that bacteria that induce autoimmune heart disease and a virus that induces autoimmune heart disease share a common cross-reactive epitope with cardiac myosin (Cunningham et al. 1994). The significance of this observation is that multiple sequential infections with different microbes sharing the same cross-reactive epitope to cardiac myosin could activate memory autoimmune T cells and result in ever-increasing autoimmune responses.

A final mechanism for autoimmunity induction is through the release of cryptic epitopes. Cryptic epitopes are peptides that are not normally generated during antigen processing or are generated at subthreshold concentration to induce an immune response (Sercarz et al. 1993). Under some circumstances, these cryptic epitopes become available in sufficient quantities to initiate a pathogenic immune response. Cryptic epitopes in human cardiac myosin can induce autoimmune myocarditis in rats (Li et al. 2004), demonstrating the biological potential for this mechanism in this disease. The question is how cryptic epitopes could be produced during coxsackievirus infections. There seem to be two major probabilities. First, coxsackieviruses have virus-specific proteases (2A, 3C, 3CD) that are required for processing the large polypeptide translated from the open reading frame of the viral RNA into the 11 virus proteins (Rueckert 1996). However, the 2A protease also cleaves myocyte contractile proteins such as dystrophin (Badorff et al. 1999, 2000). It is highly likely that the cleavage sites for the viral protease would differ than those for cellular proteases, thus providing the opportunity for novel (cryptic) epitope formation. A second possibility is that virus-induced ubiquitination of self-proteins in infected cells can lead to increased self-molecule degradation and release of increased amounts of naturally occurring cryptic epitopes (Caporossi et al. 1998).

# 4 Summary

Coxsackieviruses, like all microbes, induce complex host responses to infection. These are basically divided into innate or immediate responses that are either non-specific to the particular pathogen or identify broadly expressed molecules such as pathogen-associated molecular patterns and adaptive immunity, which is specific to the initiating microbial infection. Neither the innate or adaptive host responses are truly independent, however, since there is substantial cross-talk between the two pathways. Clearly, the innate response is a major factor in determining both the type and strength of the subsequent adaptive immunity. Secondly, factors made by the adaptive immune effectors can influence innate mediators, such as macrophage, dendritic cells, NK/NKT cells, and $\gamma\delta^+$ T cells. Also, the virus itself may require activation of the host response for optimal replication. Coxsackieviruses need cells in G1/S to support viral replication (Feuer et al. 2003). While it is possible that this means that only virus-binding cells already in cycle will successfully replicate, a more likely scenario is that the virus itself causes cells it binds to enter the cycle in order to support the virus replication. Although signaling through TLRs might provide the necessary proliferative signal, this seems unlikely in the initial infection when both viral dsRNA and ssRNA are minimal. However, the ability of the virus-virus receptor (DAF) interaction to directly signal cell proliferation would allow cells to enter cycle without TLR signals.

**Acknowledgements** The work from this laboratory has been supported by NIH grants HL58583, HL 80594, and AI 45666. The authors wish to thank Mr. Kevin Kolinich for help with the manuscript.

# References

Abraham SN, Malaviya R (1997) Mast cells in infection and immunity. Infect Immun 65:3501-3508

Abreu MT, Arditi M (2004) Innate immunity and toll-like receptors: clinical implications of basic science research. J Pediatr 144:421-429

Anderson D, Wilson J, Carthy C, Yang D, Kandolf R, McManus B (1996) Direct interactions of coxsackievirus B3 with immune cells in the splenic compartment of mice susceptible or resistant to myocarditis. J Virol 70:4632-4645

Anderson D, Carthy C, Wilson J, Yang D, Devine D, McManus B (1997) Complement component 3 interactions with coxsackievirus B3 capsid proteins: innate immunity and the rapid formation of splenic antiviral germinal centers. J Virol 71:8841-8845

Andrejeva J, Childs KS, Young DF, Carlos TS, Stock N, Goodbourn S, Randall RE (2004) The V proteins of paramyxoviruses bind the IFN-inducible RNA helicase, mda-5, and inhibit its activation of the IFN-beta promoter. Proc Natl Acad Sci U S A 101:17264-17269

Asselin-Paturel C, Boonstra A, Dalod M et al (2001) Mouse type I IFN-producing cells are immature APCs with plasmacytoid morphology. Nat Immunol 2:1144-1150

Asselin-Paturel C, Brizard G, Chemin K, Boonstra A, O'Garra A, Vicari A, Trinchieri G (2005) Type I interferon dependence of plasmacytoid dendritic cell activation and migration. J Exp Med 201:1157-1167

Badorff C, Lee GH, Lamphear BJ, Martone ME, Campbell KP, Rhoads RE, Knowlton KU (1999) Enteroviral protease 2A cleaves dystrophin: evidence of cytoskeletal disruption in an acquired cardiomyopathy. Nat Med 5:320-326

Badorff C, Lee GH, Knowlton KU (2000) Enteroviral cardiomyopathy: bad news for the dystrophin-glycoprotein complex. Herz 25:227-232

Bazzoni G (2003) The JAM family of junctional adhesion molecules. Curr Opin Cell Biol 15:525-530

Bendelac A, Rivera MN, Park SH, Roark JH (1997) Mouse CD1-specific NK1 T cells: development, specificity, and function. Annu Rev Immunol 15:535-562

Biron CA, Nguyen KB, Pien GC, Cousens LP, Salazar-Mather TP (1999) Natural killer cells in antiviral defense: function and regulation by innate cytokines. Annu Rev Immunol 17:189-220

Brodbeck WG, Kuttner-Kondo L, Mold C, Medof ME (2000) Structure/function studies of human decay-accelerating factor. Immunology 101:104-111

Burdin N, Kronenberg M (1999) CD1-mediated immune responses to glycolipids. Curr Opin Immunol 11:326-331

Cadwell K, Coscoy L (2005) Ubiquitination on nonlysine residues by a viral E3 ubiquitin ligase. Science 309:127-130

Caforio AL, Mahon NJ, McKenna WJ (2001) Cardiac autoantibodies to myosin and other heart-specific autoantigens in myocarditis and dilated cardiomyopathy. Autoimmunity 34:199-204

Caforio AL, Mahon NJ, Tona F, McKenna WJ (2002) Circulating cardiac autoantibodies in dilated cardiomyopathy and myocarditis: pathogenetic and clinical significance. Eur J Heart Fail 4:411-417

Caporossi AP, Bruno G, Salemi S et al (1998) Autoimmune T-cell response to the CD4 molecule in HIV-infected patients. Viral Immunol 11:9-17

Carding SR (1990) A role for gamma/delta T cells in the primary immune response to influenza virus. Res Immunol 141:603-606

Carding SR, Egan PJ (2000) The importance of gamma delta T cells in the resolution of pathogen-induced inflammatory immune responses. Immunol Rev 173:98-108

Carding SR, Allan W, Kyes S, Hayday A, Bottomly K, Doherty PC (1990) Late dominance of the inflammatory process in murine influenza by gamma/delta + T cells. J Exp Med 172:1225-1231

Carson SD (2001) Receptor for the group B coxsackieviruses and adenoviruses: CAR. Rev Med Virol 11:219-226

Carson SD, Chapman NM (2001) Coxsackievirus and adenovirus receptor (CAR) binds immunoglobulins. Biochemistry 40:14324-14329

Cheung TC, Humphreys IR, Potter KG et al (2005) Evolutionarily divergent herpesviruses modulate T cell activation by targeting the herpesvirus entry mediator cosignaling pathway. Proc Natl Acad Sci U S A 102:13218-13223

Chiu YH, Park SH, Benlagha K, Forestier C, Jayawardena-Wolf J, Savage PB, Teyton L, Bendelac A (2002) Multiple defects in antigen presentation and T cell development by mice expressing cytoplasmic tail-truncated CD1d. Nat Immunol 3:55-60

Collman RG, Yi Y, Liu QH, Freedman BD (2000) Chemokine signaling and HIV-1 fusion mediated by macrophage CXCR4: implications for target cell tropism. J Leukoc Biol 68:318-323

Conzelmann KK (2005) Transcriptional activation of alpha/beta interferon genes: interference by nonsegmented negative-strand RNA viruses. J Virol 79:5241-5248

Cunningham M, Hall N, Krisher K, Spanier A (1986) A study of anti-group A streptococcal monoclonal antibodies crossreactive with myosin. J Immunol 136:293

Cunningham M, Antone S, Gulizia J, McManus B, Fishetti V, Gauntt C (1994) Cytotoxic and viral neutralizing antibodies cross-react with streptococcal M protein, enteroviruses and human cardiac myosin. Proc Natl Acad Sci U S A 91:5543-5547

D'Addario M, Ahmad A, Xu JW, Menezes J (1999) Epstein-Barr virus envelope glycoprotein gp350 induces NF-kappaB activation and IL-1beta synthesis in human monocytes-macrophages involving PKC and PI3-K. FASEB J 13:2203-2213

D'Addario M, Ahmad A, Morgan A, Menezes J (2000) Binding of the Epstein-Barr virus major envelope glycoprotein gp350 results in the upregulation of the TNF-alpha gene expression in monocytic cells via NF-kappaB involving PKC, PI3-K and tyrosine kinases. J Mol Biol 298:765-778

D'Addario M, Libermann TA, Xu J, Ahmad A, Menezes J (2001) Epstein-Barr virus and its glycoprotein-350 upregulate IL-6 in human B-lymphocytes via CD21, involving activation of NF-kappaB and different signaling pathways. J Mol Biol 308:501-514

Dakhama A, Park JW, Taube C et al (2004) The role of virus-specific immunoglobulin E in airway hyperresponsiveness. Am J Respir Crit Care Med 170:952-959

Davis LS, Patel SS, Atkinson JP, Lipsky PE (1988) Decay-accelerating factor functions as a signal transducing molecule for human T cells. J Immunol 141:2246-2252

Del Corno M, Liu QH, Schols D, de Clercq E, Gessani S, Freedman BD, Collman RG (2001) HIV-1 gp120 and chemokine activation of Pyk2 and mitogen-activated protein kinases in primary macrophages mediated by calcium-dependent, pertussis toxin-insensitive chemokine receptor signaling. Blood 98:2909-2916

Diefenbach A, Schindler H, Donhauser N, Lorenz E, Laskay T, MacMicking J, Rollinghoff M, Gresser I, Bogdan C (1998) Type 1 interferon (IFNalpha/beta) and type 2 nitric oxide synthase regulate the innate immune response to a protozoan parasite. Immunity 8:77-87

Estrin M, Smith C, Huber S (1986) Coxsackievirus B-3 myocarditis T cell autoimmunity to heart antigens is resistant to cyclosporin A treatment. Am J Pathol 125:18-25

Exley M, Bigley N, Cheng O et al (2001) CD1d-reactive T-cell activation leads to amelioration of disease caused by diabetogenic encephalomyocarditis virus. J Leukoc Biol 69:713-718

Fairweather D, Frisancho-Kiss S, Gatewood S, Njoku D, Steele R, Barrett M, Rose NR (2004) Mast cells and innate cytokines are associated with susceptibility to autoimmune heart disease following coxsackievirus B3 infection. Autoimmunity 37:131-145

Fairweather D, Frisancho-Kiss S, Rose NR (2005) Viruses as adjuvants for autoimmunity: evidence from Coxsackievirus-induced myocarditis. Rev Med Virol 15:17-27

Feuer R, Mena I, Pagarigan RR, Hassett DE, Whitton JL (2003) Coxsackievirus replication and the cell cycle: a potential regulatory mechanism for viral persistence/latency. Med Microbiol Immunol (Berl) 193:83-90

Fuse K, Chan G, Liu Y, Gudgeon P, Husain M, Chen M, Yeh WC, Akira S, Liu PP (2005) Myeloid differentiation factor-88 plays a crucial role in the pathogenesis of Coxsackievirus B3-induced myocarditis and influences type I interferon production. Circulation 112:2276-2285

Gauntt C, Godney E, Lutton C (1988) Host factors regulating viral clearance. Pathol Immunopathol Res 7:251-265

Gauntt C, Godney E, Lutton C, Fernandes G (1989) Role of natural killer cells in experimental murine myocarditis. Srpinger Semin Immunopathol 11:51-59

Gibaudo G, Lembo D, Cavallo G, Landolf S, Lengyel P (1991) Interferon action: binding of viral RNA to the 40 kilodalton 2'-5' oligoadenylate synthetase in interferon-treated HeLa cells infected with encephalomyocarditis virus. J Virol 65:1748-1757

Godfrey DI, Kronenberg M (2004) Going both ways: immune regulation via CD1d-dependent NKT cells. J Clin Invest 114:1379-1388

Goodfellow IG, Evans DJ, Blom AM, Kerrigan D, Miners JS, Morgan BP, Spiller OB (2005) Inhibition of coxsackie B virus infection by soluble forms of its receptors: binding affinities, altered particle formation, and competition with cellular receptors. J Virol 79:12016-12024

Gordon JR, Galli SJ (1991) Release of both preformed and newly synthesized tumor necrosis factor alpha (TNF-alpha)/cachectin by mouse mast cells stimulated via the Fc epsilon RI A mechanism for the sustained action of mast cell-derived TNF-alpha during IgE-dependent biological responses. J Exp Med 174:103-107

Guthrie M, Lodge PA, Huber SA (1984) Cardiac injury in myocarditis induced by Coxsackievirus group B, type 3 in Balb/c mice is mediated by Lyt 2 + cytolytic lymphocytes. Cell Immunol 88:558-567

Hardy MP, Owczarek CM, Jermiin LS, Ejdeback M, Hertzog PJ (2004) Characterization of the type I interferon locus and identification of novel genes. Genomics 84:331-345

Hasan U, Chaffois C, Gaillard C et al (2005) Human TLR10 is a functional receptor, expressed by B cells and plasmacytoid dendritic cells, which activates gene transcription through MyD88. J Immunol 174:2942-2950

Heeger PS, Lalli PN, Lin F, Valujskikh A, Liu J, Muqim N, Xu Y, Medof ME (2005) Decay-accelerating factor modulates induction of T cell immunity. J Exp Med 201:1523-1530

Herzum M, Huber SA, Weller R, Grebe R, Maisch B (1991) Treatment of experimental murine Coxsackie B3 myocarditis. Eur Heart J 12 [Suppl D]:200-202

Horng T, Barton GM, Flavell RA, Medzhitov R (2002) The adaptor molecule TIRAP provides signalling specificity for Toll-like receptors. Nature 420:329-333

Horwitz M, Sarvetnick N (1999) Viruses, host responses, and autoimmunity. Immunol Rev 169:241-253

Hosoi T, Suzuki S, Nomura J, Ono A, Okuma Y, Akira S, Nomura Y (2004) Bacterial DNA induced iNOS expression through MyD88-p38 MAP kinase in mouse primary cultured glial cells. Brain Res Mol Brain Res 124:159-164

Huber S (1997) Coxsackievirus-induced myocarditis is dependent on distinct immunopathogenic responses in different strains of mice. Lab Invest 76:691-701

Huber S (2004) T cells in coxsackievirus induced myocarditis. Viral Immunol 17:152-164

Huber S (2005) Cellular autoimmunity in myocarditis. Heart Failure Clinic 1:321-331

Huber S, Lodge P (1984) Coxsackievirus B3 myocarditis in Balb/c mice: evidence for autoimmunity to myocyte antigens. Am J Path 116:21-29

Huber S, Lodge P (1986) Coxsackievirus B3 myocarditis: identification of different mechanisms in DBA/2 and Balb/c mice. Am J Pathol 122:284-291

Huber SA, Sartini D (2005a) Roles of tumor necrosis factor alpha (TNF-alpha) and the p55 TNF receptor in CD1d induction and coxsackievirus B3-induced myocarditis. J Virol 79:2659-2665

Huber S, Sartini D (2005b) T cells expressing the Vgamma1 T-cell receptor enhance virus-neutralizing antibody response during coxsackievirus B3 infection of BALB/c mice: differences in male and female mice. Viral Immunol 18:730-739

Huber SA, Weller A, Herzum M, Lodge PA, Estrin M, Simpson K, Guthrie M (1988) Immunopathogenic mechanisms in experimental picornavirus-induced autoimmunity. Pathol Immunopathol Res 7:279-291

Huber S, Haisch C, Lodge P (1990) Functional diversity in vascular endothelial cells: role in coxsackievirus tropism. J Virol 64:4516-4522

Huber S, Mortensen A, Moulton G (1996) Modulation of cytokine expression by CD4+ T cells during coxsackievirus B3 infections of BALB/c mice initiated by cells expressing the gammadelta+ T cell receptor. J Virol 70:3039-3045

Huber S, Graveline D, Newell M, Born W, O'Brien R (2000) Vgamma1+ T cells suppress and Vgamma4+ T Cells promote susceptibility to coxsackievirus B3-induced myocarditis in mice. J Immunol 165:4174-4181

Huber SA, Sartini D, Exley M (2002a) Vgamma4(+) T cells promote autoimmune CD8(+) cytolytic T-lymphocyte activation in coxsackievirus B3-induced myocarditis in mice: role for CD4(+) Th1 cells. J Virol 76:10785-10790

Huber S, Shi C, Budd RC (2002b) Gammadelta T cells promote a Th1 response during coxsackievirus B3 infection in vivo: role of Fas and Fas ligand. J Virol 76:6487-6494

Huber S, Sartini D, Exley M (2003) Role of CD1d in coxsackievirus B3-induced myocarditis. J Immunol 170:3147-3153

Ikeda F, Haraguchi Y, Jinno A, Iino Y, Morishita Y, Shiraki H, Hoshino H (1998) Human complement component C1q inhibits the infectivity of cell-free HTLV-I. J Immunol 161:5712-5719

Jutel M, Watanabe T, Akdis M, Blaser K, Akdis CA (2002) Immune regulation by histamine. Curr Opin Immunol 14:735-740

Jutel M, Blaser K, Akdis CA (2006) The role of histamine in regulation of immune responses. Chem Immunol Allergy 91:174-187

Kanto T, Hayashi N (2006) Immunopathogenesis of hepatitis C virus infection: multifaceted strategies subverting innate and adaptive immunity. Intern Med 45:183-191

Kulka M, Alexopoulou L, Flavell RA, Metcalfe DD (2004) Activation of mast cells by double-stranded RNA: evidence for activation through Toll-like receptor 3. J Allergy Clin Immunol 114:174-182

Kunzmann S, Mantel PY, Wohlfahrt JG, Akdis M, Blaser K, Schmidt-Weber CB (2003) Histamine enhances TGF-beta1-mediated suppression of Th2 responses. FASEB J 17:1089-1095

LaFleur DW, Nardelli B, Tsareva T et al (2001) Interferon-kappa, a novel type I interferon expressed in human keratinocytes. J Biol Chem 276:39765-39771

Lane J, Neumann D, Lafond-Walker A, Herkowitz A, Rose N (1992) Interleukin 1 or tumor necrosis factor can promote Coxsackie B3-induced myocarditis in resistant B10.A mice. J Exp Med 175:1123-1129

Latif N, Zhang H, Archard LC, Yacoub MH, Dunn MJ (1999) Characterization of anti-heart antibodies in mice after infection with coxsackie B3 virus. Clin Immunol 91:90-98

Lauw FN, Caffrey DR, Golenbock DT (2005) Of mice and man: TLR11 (finally) finds profilin. Trends Immunol 26:509-511

Li Y, Heuser JS, Kosanke SD, Hemric M, Cunningham MW (2004) Cryptic epitope identified in rat and human cardiac myosin S2 region induces myocarditis in the Lewis rat. J Immunol 172:3225-3234

Liew FY, Komai-Koma M, Xu D (2004) A toll for T cell costimulation. Ann Rheum Dis 63 Suppl 2, ii76-ii78

Liu J, Miwa T, Hilliard B, Chen Y, Lambris JD, Wells AD, Song WC (2005) The complement inhibitory protein DAF (CD55) suppresses T cell immunity in vivo. J Exp Med 201:567-577

Liu P, Aitken K, Kong YY et al (2000a) The tyrosine kinase p56lck is essential in coxsackievirus B3-mediated heart disease. Nat Med 6:429-434

Liu QH, Williams DA, McManus C et al (2000b) HIV-1 gp120 and chemokines activate ion channels in primary macrophages through CCR5 and CXCR4 stimulation. Proc Natl Acad Sci U S A 97:4832-4837

Lodge P, Herzum M, Olszewski J, Huber S (1987) Coxsackievirus B3-myocarditis: acute and chronic forms of the disease caused by different immunopathogenic mechanisms. Am J Pathol 128:455-463

Lowenstein C, Hill S, Lafond-Walker A, Wu J, Allen G, Landavere M, Rose N, Herkowitz A (1996) Nitric oxide inhibits viral replication in murine myocarditis. J Clin Invest 97:1837-1843

Luo H, Zhang J, Cheung C, Suarez A, McManus BM, Yang D (2003a) Proteasome inhibition reduces coxsackievirus B3 replication in murine cardiomyocytes. Am J Pathol 163:381-385

Luo H, Zhang J, Dastvan F, Yanagawa B, Reidy MA, Zhang HM, Yang D, Wilson JE, McManus BM (2003b) Ubiquitin-dependent proteolysis of cyclin D1 is associated with coxsackievirus-induced cell growth arrest. J Virol 77:1-9

Lutton C, Gauntt C (1985) Ameliorating effect of IFN-B and anti-IFN-B on coxsackievirus B3-induced myocarditis in mice. J Interferon Res 5:137-146

Ma RZ, Gao J, Meeker ND et al (2002) Identification of Bphs, an autoimmune disease locus, as histamine receptor H1. Science 297:620-623

Martino TA, Petric M, Weingartl H et al (2000) The coxsackie-adenovirus receptor (CAR) is used by reference strains and clinical isolates representing all six serotypes of coxsackievirus group B and by swine vesicular disease virus. Virology 271:99-108

Mason J, O'Connell JB, Herskowitz A et al (1995) A clinical trial of immunosuppressive therapy for myocarditis. The Myocarditis Treatment Trial Investigators. N Engl J Med 333:269-275

Means TK, Golenbock DT, Fenton MJ (2000) The biology of Toll-like receptors. Cytokine Growth Factor Rev 11:219-232

Medof ME, Walter EI, Rutgers JL, Knowles DM, Nussenzweig V (1987) Identification of the complement decay-accelerating factor (DAF) on epithelium and glandular cells and in body fluids. J Exp Med 165:848-864

Mena I, Perry CM, Harkins S, Rodriguez F, Gebhard J, Whitton JL (1999) The role of B lymphocytes in coxsackievirus B3 infection. Am J Pathol 155:1205-1215

Michelsen KS, Doherty TM, Shah PK, Arditi M (2004) TLR signaling: an emerging bridge from innate immunity to atherogenesis. J Immunol 173:5901-5907

Miller SD, Katz-Levy Y, Neville KL, Vanderlugt CL (2001) Virus-induced autoimmunity: epitope spreading to myelin autoepitopes in Theiler's virus infection of the central nervous system. Adv Virus Res 56:199-217

Misse D, Esteve PO, Renneboog B, Vidal M, Cerutti M, St Pierre Y, Yssel H, Parmentier M, Veas F (2001) HIV-1 glycoprotein 120 induces the MMP-9 cytopathogenic factor production that is abolished by inhibition of the p38 mitogen-activated protein kinase signaling pathway. Blood 98:541-547

Miwa T, Sun X, Ohta R, Okada N, Harris CL, Morgan BP, Song WC (2001) Characterization of glycosylphosphatidylinositol-anchored decay accelerating factor (GPI-DAF) and transmembrane

DAF gene expression in wild-type and GPI-DAF gene knockout mice using polyclonal and monoclonal antibodies with dual or single specificity. Immunology 104:207-214

Miwa T, Zhou L, Hilliard B, Molina H, Song WC (2002) Crry, but not CD59 and DAF, is indispensable for murine erythrocyte protection in vivo from spontaneous complement attack. Blood 99:3707-3716

Monaco JJ (1995) Pathways for the processing and presentation of antigens to T cells. J Leukoc Biol 57:543-547

Moretta L, Moretta A (2004) Unravelling natural killer cell function: triggering and inhibitory human NK receptors. EMBO J 23:255-259

Mueller T, Terada T, Rosenberg IM, Shibolet O, Podolsky DK (2006) Th2 cytokines down-regulate TLR expression and function in human intestinal epithelial cells. J Immunol 176:5805-5814

Mukasa A, Lahn M, Pflum E, Born W, O'Brien R (1997) Evidence that the same gamma delta T cells respond during infection-induced and autoimmune inflammation. J Immunol 159:5787-5794

Mukasa A, Yoshida H, Kobayashi N, Matsuzaki G, Nomoto K (1998) Gamma delta T cells in infection-induced and autoimmune-induced testicular inflammation. Immunol 95:395-401

Musio S, Gallo B, Scabeni S et al (2006) A key regulatory role for histamine in experimental autoimmune encephalomyelitis: disease exacerbation in histidine decarboxylase-deficient mice. J Immunol 176:17-26

Netea MG, Van der Meer JW, Kullberg BJ (2004) Toll-like receptors as an escape mechanism from the host defense. Trends Microbiol 12:484-488

Netea MG, Van der Meer JW, Sutmuller RP, Adema GJ, Kullberg BJ (2005) From the Th1/Th2 paradigm towards a Toll-like receptor/T-helper bias. Antimicrob Agents Chemother 49:3991-3996

O'Brien R, Happ M, Dallas A, Palmer E, Kubo R, Born W (1989) Stimulation of a major subset of lymphocytes expressing T cell receptor gamma/delta by an antigen derived from *Myocbacterium tuberculosis*. Cell 57:667

O'Brien R, Yin X, Huber S, Ikuta K, Born W (2000) Depletion of a gamma-delta T cell subset can increase host resistance to a bacterial infection. J of Immunology 165:6472-6479

O'Brien R, Lahn M, Born W, Huber S (2001) TCR and function co-segregate in gd T cell subsets. Karger, Basel

Olive C (1995) Gamma delta T cell receptor variable region usage during the development of experimental allergic encephalomyelitis. J Neuroimmunol 62:1-7

O'Neill LA (2004) TLRs: Professor Mechnikov, sit on your hat. Trends Immunol 25:687-693

Opavsky MA, Penninger J, Aitken K, Wen WH, Dawood F, Mak T, Liu P (1999) Susceptibility to myocarditis is dependent on the response of alphabeta T lymphocytes to coxsackieviral infection. Circ Res 85:551-558

Oritani K, Kanakura Y (2005) IFN-zeta/ limitin: a member of type I IFN with mild lympho-myelosuppression. J Cell Mol Med 9:244-254

Orthopoulos G, Triantafilou K, Triantafilou M (2004) Coxsackie B viruses use multiple receptors to infect human cardiac cells. J Med Virol 74:291-299

Ozaki H, Ishii K, Horiuchi H, Arai H, Kawamoto T, Okawa K, Iwamatsu A, Kita T (1999) Cutting edge: combined treatment of TNF-alpha and IFN-gamma causes redistribution of junctional adhesion molecule in human endothelial cells. J Immunol 163:553-557

Pankuweit S, Portig I, Lottspeich F, Maisch B (1997) Autoantibodies in sera of patients with myocarditis: characterization of the corresponding proteins by isoelectric focusing and N-terminal sequence analysis. J Mol Cell Cardiol 29:77-84

Pasch A, Kupper JH, Wolde A, Kandolf R, Selinka HC (1999) Comparative analysis of virus-host cell interactions of haemagglutinating and non-haemagglutinating strains of coxsackievirus B3. J Gen Virol 80:3153-3158

Peiffer I, Servin AL, Bernet-Camard MF (1998) Piracy of decay-accelerating factor (CD55) signal transduction by the diffusely adhering strain *Escherichia coli* C1845 promotes cytoskeletal F-actin rearrangements in cultured human intestinal INT407 cells. Infect Immun 66:4036-4042

Pestka S, Krause CD, Walter MR (2004) Interferons, interferon-like cytokines, and their receptors. Immunol Rev 202:8-32

Peterman GM, Spencer C, Sperling AI, Bluestone JA (1993) Role of gamma delta T cells in murine collagen-induced arthritis. J Immunol 151:6546-6558

Pulendran B (2005) Variegation of the immune response with dendritic cells and pathogen recognition receptors. J Immunol 174:2457-2465

Roark C, Vollmer M, Campbell P, Born W, O'Brien R (1996) Response of a gamma delta+ T cell receptor invariant subset during bacterial infection. J Immunol 156:2214-2220

Rueckert R (1996) Picronaviruses, 3rd edn. Lippincott-Raven, Philadelphia

Sandor M, Sperling A, Cook G, Weinstock J, Lynch R, Bluestone J (1995) Two waves of gamma delta T cells expressing different V delta genes are recruited into schistosome-induced liver granulomas. J Immunol 155:275-284

Sato S, Tsutsumi R, Burke A, Carson G, Porro V, Seko Y, Okumura K, Kawana R, Virmani S (1994) Persistence of replicating coxsackievirus B3 in the athymic murine heart is associated with development of myocarditic lesions. J Gen Virol 75:2911-2924

Saura M, Zaragoza C, McMillan A, Quick RA, Hohenadl C, Lowenstein JM, Lowenstein CJ (1999) An antiviral mechanism of nitric oxide: inhibition of a viral protease. Immunity 10:21-28

Schmidtke M, Selinka H, Heim A, Jahn B, Tonew M, Kandolf R, Stelzner A, Zell R (2000) Attachment of coxsackievirus B3 variants to various cell lines: mapping of phenotypic differences to capsid protein VP1. Virology 275:77-88

Schneider-Schaulies J (2000) Cellular receptors for viruses: links to tropism and pathogenesis. J Gen Virol 81:1413-1429

Schnurr DP, Cao Y, Schmidt NJ (1984) Coxsackievirus B3 persistence and myocarditis in N: NIH(S) II nu/nu and +/nu mice. J Gen Virol 65:1197-1201

Schroder NW, Diterich I, Zinke A et al (2005) Heterozygous Arg753Gln polymorphism of human TLR-2 impairs immune activation by *Borrelia burgdorferi* and protects from late stage Lyme disease. J Immunol 175:2534-2540

Schwimmbeck PL, Bigalke B, Schulze K, Pauschinger M, Kuhl U, Schultheiss HP (2004) The humoral immune response in viral heart disease: characterization and pathophysiological significance of antibodies. Med Microbiol Immunol (Berl) 193:115-119

Seko Y, Tsuchimochi H, Nakamura T, Okumura K, Naito S, Imataka K, Fujii J, Takaku F, Yazaki Y (1990) Expression of major histocompatibility complex class I antigen in murine ventricular myocytes infected with Coxsackievirus B3. Circ Res 67:360-367

Seko Y, Yagita H, Okumura K, Yazaki Y (1996) Expression of vascular cell adhesion molecule-1 in murine hearts with acute myocarditis caused by coxsackievirus B3. 180:450-454

Selinka HC, Wolde A, Sauter M, Kandolf R, Klingel K (2004) Virus-receptor interactions of coxsackie B viruses and their putative influence on cardiotropism. Med Microbiol Immunol (Berl) 193:127-131

Sercarz E, Lehmann P, Ametani A, Benichou G, Miller A, Moudgil K (1993) Dominance and crypticity of T cell antigenic determinants. Ann Rev Immunol 11:729-766

Shenoy-Scaria AM, Kwong J, Fujita T, Olszowy MW, Shaw AS, Lublin DM (1992) Signal transduction through decay-accelerating factor. Interaction of glycosyl-phosphatidylinositol anchor and protein tyrosine kinases p56lck and p59fyn 1. J Immunol 149:3535-3541

Slifka MK, Pagarigan R, Mena I, Feuer R, Whitton JL (2001) Using recombinant coxsackievirus B3 to evaluate the induction and protective efficacy of CD8+ T cells during picornavirus infection. J Virol 75:2377-2387

Song WC, Deng C, Raszmann K, Moore R, Newbold R, McLachlan JA, Negishi M (1996) Mouse decay-accelerating factor: selective and tissue-specific induction by estrogen of the gene encoding the glycosylphosphatidylinositol-anchored form. J Immunol 157:4166-4172

Stanic AK, Park JJ, Joyce S (2003) Innate self recognition by an invariant, rearranged T-cell receptor and its immune consequences. Immunology 109:171-184

Tam PE (2006) Coxsackievirus myocarditis: interplay between virus and host in the pathogenesis of heart disease. Viral Immunol 19:133-146

Theofilopoulos A (1995a) The basis for autoimmunity: Part I, mechanisms of aberrant self-recognition. Immunol Today 16:90-98

Theofilopoulos A (1995b) The basis for autoimmunity: Part II, genetic predisposition. Immunol Today 16:150-159

Triantafilou K, Orthopoulos G, Vakakis E, Ahmed MA, Golenbock DT, Lepper PM, Triantafilou M (2005) Human cardiac inflammatory responses triggered by Coxsackie B viruses are mainly Toll-like receptor (TLR) 8-dependent. Cell Microbiol 7:1117-1126

Turlington BS, Edwards WD (1988) Quantitation of mast cells in 100 normal and 92 diseased human hearts. Implications for interpretation of endomyocardial biopsy specimens. Am J Cardiovasc Pathol 2:151-157

Vanderlugt CL, Miller SD (2002) Epitope spreading in immune-mediated diseases: implications for immunotherapy. Nat Rev Immunol 2:85-95

Vella C, Festenstein H (1992) Coxsackievirus B4 infection of the mouse pancreas: the role of natural killer cells in the control of virus replication and resistance to infection. J Gen Virol 73:1379-1386

Vincent T, Pettersson RF, Crystal RG, Leopold PL (2004) Cytokine-mediated downregulation of coxsackievirus-adenovirus receptor in endothelial cells. J Virol 78:8047-8058

Vogel SN, Fitzgerald KA, Fenton MJ (2003) TLRs: differential adapter utilization by toll-like receptors mediates TLR-specific patterns of gene expression. Mol Interv 3:466-477

Vossen MT, Westerhout EM, Soderberg-Naucler C, Wiertz EJ (2002) Viral immune evasion: a masterpiece of evolution. Immunogenetics 54:527-542

Wang X, Bergelson JM (1999) Coxsackievirus and adenovirus receptor cytoplasmic and transmembrane domains are not essential for coxsackievirus and adenovirus infection. J Virol 73:2559-2562

Woodruff J (1979) Lack of correlation between neutralizing antibody production and suppression of coxsackievirus B-3 replication in target organs: evidence for involvement of mononuclear inflammatory cells in defense. J Immunol 123:31-36

Woodruff J, Woodruff J (1974) Involvement of T lymphocytes in the pathogenesis of coxsackievirus B3 heart disease. J Immunol 113:1726-1734

Yanagawa B, Spiller OB, Choy J et al (2003) Coxsackievirus B3-associated myocardial pathology and viral load reduced by recombinant soluble human decay-accelerating factor in mice. Lab Invest 83:75-85

Yanagawa B, Spiller OB, Proctor DG, Choy J, Luo H, Zhang HM, Suarez A, Yang D and McManus BM (2004) Soluble recombinant coxsackievirus and adenovirus receptor abrogates coxsackievirus b3-mediated pancreatitis and myocarditis in mice. J Infect Dis 189:1431-1439

Yoneyama M, Suhara W, Fukuhara Y, Fukuda M, Nishida E, Fujita T (1998) Direct triggering of the type I interferon system by virus infection: activation of a transcription factor complex containing IRF-3 and CBP/p300. EMBO J 17:1087-1095

Yoneyama M, Kikuchi M, Natsukawa T, Shinobu N, Imaizumi T, Miyagishi M, Taira K, Akira S, Fujita T (2004) The RNA helicase RIG-I has an essential function in double-stranded RNA-induced innate antiviral responses. Nat Immunol 5:730-737

Zaragoza C, Ocampo CJ, Saura M, McMillan A, Lowenstein CJ (1997) Nitric oxide inhibition of coxsackievirus replication in vitro. J Clin Invest 100:1760-1767

Zaragoza C, Ocampo C, Saura M, Leppo M, Wei XQ, Quick R, Moncada S, Liew FY, Lowenstein CJ (1998) The role of inducible nitric oxide synthase in the host response to Coxsackievirus myocarditis. Proc Natl Acad Sci U S A 95:2469-2474

Zarember KA, Godowski PJ (2002) Tissue expression of human Toll-like receptors and differential regulation of Toll-like receptor mRNAs in leukocytes in response to microbes, their products, and cytokines. J Immunol 168:554-561

Zautner AE, Jahn B, Hammerschmidt E, Wutzler P, Schmidtke M (2006) N- and 6-O-sulfated heparan sulfates mediate internalization of coxsackievirus B3 variant PD into CHO-K1 cells. J Virol 80:6629-6636

Zingoni A, Sornasse T, Cocks BG, Tanaka Y, Santoni A, Lanier LL (2005) NK cell regulation of T cell-mediated responses. Mol Immunol 42:451-454

# Pediatric Group B Coxsackievirus Infections

J. R. Romero

| | | |
|---|---|---|
| 1 | Epidemiology of Group B Coxsackieviruses | 223 |
| 2 | Clinical Syndromes Associated with Group B Coxsackievirus Infections | 225 |
| | 2.1 Asymptomatic Infections | 225 |
| | 2.2 Febrile Illness Without an Apparent Focus | 225 |
| | 2.3 Dermatologic Manifestations | 226 |
| | 2.4 Respiratory Syndromes | 227 |
| | 2.5 Central Nervous System Syndromes | 228 |
| | 2.6 Musculoskeletal Syndromes | 230 |
| | 2.7 Severe Neonatal Infections | 230 |
| 3 | Diagnosis | 232 |
| 4 | Treatment | 233 |
| 5 | Prevention | 234 |
| References | | 235 |

**Abstract** The CVB have long been recognized as significant pathogens of infants and children. Although the major route for transmission of the CVB is fecal-oral, vertical transmission from mother to infant is also possible. This review will focus on the more common or clinically relevant CVB-related syndromes, their diagnosis, treatment, and prevention.

## 1 Epidemiology of Group B Coxsackieviruses

The importance of the group B coxsackieviruses (CVBs) as pediatric pathogens is documented in a recent Centers for Disease Control and Prevention (CDC) Enteroviral Surveillance Report spanning 36 years. Five of the six CVB serotypes,

J. R. Romero
University of Nebraska Medical Center, 986495 Nebraska Medical,
Omaha, NE 68198-6495, USA
jrromero@unmc.edu

CVB 1-5, were among the 15 most commonly isolated enterovirus serotypes from 1970 to 2005 (CDC 2006). The ranking of CVB 1-5 was 13th, 6th, 11th, 9th, and 4th, respectively. These five CVB serotypes accounted for nearly 25% of the 49,637 known enteroviral serotypes reported to the CDC during that period of time.

In temperate regions of the world, like that of the United States, CVB infections occur predominantly during the summer and fall (i.e., June-October) (Moore et al. 1984). This pattern of seasonality is particularly pronounced for CVB 4 and 5. However, while approximately 80% of CVB infections occur during the summer and fall, the advent of molecular diagnostic methods has shown that they can also be associated with infections during the winter months (Byington et al. 1999; Chambon et al. 2001) The patterns of annual circulation of the individual CVB serotypes vary (CDC 2006). Coxsackieviruses B 1, 3, and 4 exhibit an epidemic pattern: periods of increased activity that occur at irregular intervals and may last from 1 to last several years. Coxsackievirus B2 has an endemic pattern of circulation that exhibits annual variability. Community-wide outbreaks of CVB 2 are uncommon. However, reports of focal outbreaks in summer camps and football teams have been reported (Alexander et al. 1993; Schiff 1979). Coxsackievirus B5 is unique among the six serotypes in that it has epidemic pattern that cycles at 3- to 6-year intervals and last for 1 year. Coxsackievirus B6 is an uncommonly reported serotype in the United States; of nearly 50,000 isolates reported to the CDC from 1970 to 2005 only 68 were CVB6.

The CVB have long been recognized as significant pathogens of infants and children (Berlin et al. 1993; Kaplan et al. 1983). Ninety percent of reported CVB infections occur in infants, children, and adolescents (CDC 2006). Even more significant is that approximately 50% of all CVB infections occur in infants younger than 1 year of age. Over two-thirds of CVB1 isolates reported to the CDC were for specimens obtained from infants less than 1 year of age. Approximately, 63% of CVB 2, 3, and 4 isolates from 1970 to 2005 were from the same age group.

The major route for transmission of the CVB is fecal-oral. The highest incidence (75% of cases) infections occurs in infants less than 1 year and toddlers 1-4 years of age (CDC 2006). The lack of adequate personal hygiene exhibited by infants and children, in general, may help to explain the high incidence of CVB infections among these groups.

In addition to fecal-oral transmission, vertical transmission from mother to infant is also possible (Kaplan et al. 1983; Modlin and Rotbart 1997). Evidence for intrauterine (Bates 1970; Bendig et al. 2003; Burch et al. 1968; Kaplan et al. 1983; Ouellet et al. 2004,) and peripartum (Kaplan et al. 1983; Krajden et al. 1983) transmission exists. Based on the presence of viremia or onset clinical symptoms, Abzug has calculated that approximately 22% of fatal neonatal CVB infections are the result of intrauterine infection (Abzug 2004).

Group B coxsackievirus infections may involve nearly every organ system (Cherry 2004), resulting in myriad clinical syndromes. It is important to keep in mind that no single CVB serotype is uniquely linked to a single specific clinical syndrome or disease. Thus, any member of the species may cause any of a number of clinical illnesses.

## 2 Clinical Syndromes Associated with Group B Coxsackievirus Infections

This review will focus on the more common or clinically relevant CVB-related syndromes. Myocarditis and diabetes associated with CVB infections are covered extensively in other chapters in this volume. It is beyond the scope of this chapter to discuss every possible clinical manifestation of CVB infection. Readers desiring such information are directed to the single most authoritative review of enteroviral infections (Cherry 2004).

### 2.1 Asymptomatic Infections

As with other members of the genus *Enteroviridae*, many, if not the majority, of CVB infections are subclinical (i.e., asymptomatic) in nature. This holds true even for CVB infections in neonates (Hall and Miller 1969; Jenista et al. 1984). The estimated frequencies of asymptomatic infection due the CVB are shown in Table 1.

### 2.2 Febrile Illness Without an Apparent Focus

The single most common clinical manifestation of CVB infection is that of a nonspecific febrile illness in children. During the summer and early fall, the CVB are common causes of fever without an apparent focus in infants and children less than 2 years of age (Dagan et al. 1989; Leggiadro et al. 1983). In one report, the enteroviruses (including CVB) were responsible for nearly two-thirds of all hospitalizations for evaluation of possible sepsis during the summer and fall (Dagan et al. 1989).

Group B coxsackievirus infections in neonates and infants less than 3 months of age may result in fever and lethargy as their sole clinical manifestation (Jenista et al. 1984). In Kaplan et al.'s report of the CVB disease in infants less than 3 months of age, over 90% of infants had a temperature greater than or

**Table 1** Approximate frequency of asymptomatic infection with the group B coxsackieviruses (Cherry 2004)

| Serotype | Percent of asymptomatic infections |
|---|---|
| CVB2 | 11%–50% |
| CVB3 | 25%–96% |
| CVB4 | 30%–70% |
| CVB5 | 5%–40% |

equal to 38.3°C (Kaplan et al. 1983). As in older patients, the fever may be biphasic. The average duration of fever is 3 days (Kaplan et al. 1983). Other signs and symptoms may include poor feeding, lethargy, irritability, emesis, diarrhea, rash, and signs of upper respiratory tract infection (Lake et al. 1976; Rorabaugh et al. 1993).

In infants and children, the onset of this syndrome is abrupt, consisting of fever (38°-40°C) and malaise. In greater that 25% of cases, the fever may have a biphasic nature: present for a day followed by a 2- to 3-day period of euthermia preceding a recurrence of the fever (Cherry 2004; Moore et al. 1984). In patients old enough to relate it, headache is common accompaniment. Sore throat is common. Other symptoms and signs may include nausea, emesis, myalgias, mild abdominal discomfort, and rash (discussed in Sect. 2.3). Findings on physical examination are minimal and may consist of mild conjunctival and pharyngeal injection as well lymphadenopathy. The illness is generally short-lived, lasting, on average, 3-4 days.

## 2.3 Dermatologic Manifestations

Cutaneous manifestations associated with CVB infection have been well described (Table 2) (Cherry 2004). The frequency of exanthems as part of the clinical spectrum of CVB infection varies by serotype, being rare in association with CVB3 infection and relatively frequent with infections by CVB5 and CVB6. Coxsackievirus B5 and B6 infections are more commonly associated with exanthematous manifestations. Approximately 10% of children with CVB5 infection during outbreaks have rash. The rash is typically maculopapular in nature. However, petechial, roseola-like and urticarial rashes have been reported. In one report, 20% of cases during an epidemic of CVB6 had associated rash.

**Table 2** Exanthems associated with group B coxsackievirus infections (Cherry 2004)

| Serotype | Frequency of exanthem | Described exanthems |
|---|---|---|
| Coxsackievirus B1 | Occasional | Common: maculopapular<br>Other: hand–foot–mouth–like rash |
| Coxsackievirus B2 | Rare | Maculopapular, vesicular, petechial |
| Coxsackievirus B3 | Occasional | Common: maculopapular<br>Other: petechial, vesicular, hand–foot–mouth–like rash |
| Coxsackievirus B4 | Occasional | Morbilliform: petechial, urticarial |
| Coxsackievirus B5 | 10% | Common: maculopapular<br>Other: petechial, urticarial, roseola–like, hand–foot–mouth–like rash |
| Coxsackievirus B6 | 20% | Morbilliform |

## 2.4 Respiratory Syndromes

The CBVs have been linked to multiple common infectious syndromes involving the upper and lower respiratory tracts (Table 3) (Cherry 2004; Chonmaitree and Mann 1995). Fourteen percent of patients with CVB infections reported to the CDC from 1970 to 1979 had respiratory syndromes (Moore et al. 1984). It is of note that respiratory specimens were the most frequent source of isolation for CVB 1 and 3 (CDC 2006)

The summer cold associated with CVB infection differs from that of the common cold of rhinoviruses in that it is characterized by fever, nasal congestion, sneezing, rhinorrhea, and pharyngitis. (Carmichael et al. 1968; Cherry 2004; Chonmaitree and Mann 1995; Hable et al. 1970). The syndrome lasts 3-6 days.

Herpangina (Parrott et al. 1951; Zahorsky 1920; Zahorsky 1924) has been clearly linked to CVB infections. (Cherry 2004; Cherry et al. 1965; Nakayama et al. 1989) Onset is sudden with fever that may be as high as 41°C. The higher temperatures tend to be more common in younger children. The enanthem is located primarily to the anterior pillars of the tonsillar fauces. However, lesions may also be seen on the soft palate, uvula, and tonsils. Rarely, the posterior buccal surfaces and dorsal tip of the tongue may be involved. The lesions are described as papulovesicular, grayish-white in color with an areola of erytherma and measure 1-2 mm in diameter. Over 2-3 days, the lesions progress from papular to vesicular, increase in size (3-4 mm) and, ultimately, ulcerate. On average, five lesions are present. Associated complaints include sore throat, sialorrhea, anorexia, dysphagia, abdominal pain, and emesis.

In CVB-associated pharyngitis, the onset is abrupt and has as its principle complaints fever (38°-40°C) and sore throat (Cherry 2004; Chonmaitree and Mann 1995). Children may complain of anorexia, generalized malaise headache, and/or myalgia. Gastrointestinal manifestations such as nausea, emesis, and diarrhea may

**Table 3** Respiratory tract syndromes associated with group B coxsackievirus infections (Cherry 2004; Chonmaitree and Mann 1995)

| Clinical syndrome | Coxsackievirus serotypes |
|---|---|
| Upper respiratory tract | |
| Summer cold | 1, 2, 3, 4, 5, 6 (rarely) |
| Stomatitis | 2, 5 |
| Herpangina | 1, 2, 3, 4, 5 |
| Pharyngitis | 1, 2, 3, 4, 5 |
| Otitis media | 4, 5 |
| Parotitis | 3, 4 |
| Croup | 4, 5 |
| Lower respiratory tract | |
| Bronchitis | 1, 3, 4 |
| Bronchiolitis | 1, 2, 3, 4, 5 |
| Pneumonia | 1, 2, 3, 4, 5, 6 |

be present. On physical examination, the pharynx and tonsils are erythematous. In severe cases, exudates may be seen. Cervical lymphadenopathy is usually present on palpation of the neck.

The CVB have been linked to summer and fall pneumonia, bronchiolitis, and bronchitis (Eckert et al. 1967; Goldwater 1995). The clinical presentations of CVB lower respiratory infections include fever, tachypnea, cough, retractions and, in some, cyanosis (Eckert et al. 1967; Goldwater 1995). Auscultation of the chest may reveal rales, rhonchi, and wheezes. A persistent cough has been reported in association with CVB6 pneumonias. Hospitalization for as long as 6 weeks was required for some patients (Goldwater 1995). Perihilar infiltrates may be seen on chest x-ray. In two fatal cases of CVB5 pneumonia, postmortem examination revealed interstitial pneumonia that was characterized by a mononuclear cell infiltration of the alveolar walls. The alveoli contained protein-rich exudate with localized hyaline membrane formation (Flewet 1965).

## 2.5 Central Nervous System Syndromes

The CVBs are frequently associated with infections of the central nervous system (CNS), and, in particular, meningitis (CDC 2006; Moore et al. 1984). During the decade of the 1970s, meningitis, encephalitis, and paralysis comprised 56%, 15%, and 1%, respectively, of the clinical syndromes associated with CVB1-5 infections reported to the CDC (Moore 1984). One study found that approximately 50% of enteroviruses isolated from the cerebrospinal fluid (CSF) of children less than 2 years of age with aseptic meningitis were CVB (CVB2, CVB4, and CVB5) (Berlin et al. 1993). Cerebrospinal fluid was the most frequent source for the detection of CVB2, CVB4, and CVB5 from 1970 to 2005 (CDC 2006). In the case of the latter CVB serotype, nearly 50% of isolates recovered from 1970 to 2005 were from CSF.

The onset of CVB meningitis is typical of that of other enteroviruses in that it is abrupt. Fever (38°-40°C) is the most common presenting sign and may be biphasic (Cherry 2004; Moore et al. 1984; Singer et al. 1980; Wilfert et al. 1983). Signs of meningeal irritation (i.e., nucal rigidity, Brudzinski's and Kernig's signs) are seen in the majority of children older than 1 year, but is uncommon in young infants (Rorabaugh et al. 1993). Photophobia is a common clinical component of the disease. Headache, most likely as a result of increased intracranial pressure, is present in nearly all who can report it. The headache may be ameliorated by the performance of a lumbar puncture (Jaffe et al. 1989). Nonspecific findings, singly or in combinations, such as rash, malaise, sore throat, abdominal pain, nausea, vomiting, and myalgias are common. Other nonneurologic CVB-associated syndromes such as herpangina, pleurodynia, or myocarditis may occur concurrently with the meningitis. Young children and infants may be irritable or lethargic and exhibit other nonspecific signs such as poor feeding, emesis, diarrhea, and rash (Rorabaugh et al. 1993). In infants, the fontanelle may be full or bulging. In some patients, the entire illness may exhibit a biphasic course consisting of an initial period of nonspecific signs and symptoms (e.g., fever, headache, gastrointestinal symptoms, myalgia), which resolve and subsequently recrudesce accompanied by evidence of frank neurologic involvement (Cherry 2004). Potential complications of meningitis include seizures, coma, increased intracranial pressure, as well as inappropriate secretion of antidiuretic hormone, and have been described in approximately 9% of cases (Chemtob et al. 1985; Rorabaugh et al. 1993).

Typically, the evaluation of the CSF in cases of meningitis reveals a mild to moderate (100-500 cells/mm$^3$) lymphocytic pleocytosis. Pleocytosis in excess of 1,000 cells/mm$^3$ is possible, but generally the exception (Arrieta et al. 1991; Kaplan et al. 1983; Severien 1994; Singer et al. 1980). A predominantly polymorphonuclear pleocytosis may be present if the CSF is obtained early in the course of illness (Cherry 2004; Singer et al. 1980; Wilfert et al. 1983). Interestingly, CSF eosinophilia has been reported with meningitis due to CVB4 (Chesney et al. 1980). The CSF protein concentration is commonly in the range of 40-80 mg/dl and associated with a normal CSF glucose concentration in the majority of cases. However, reports of hypoglycorrhachia in association with CVB3 and CVB1 meningitis exist and may occur more frequently in young infants (Chesney et al. 1978; Severien et al. 1994; Chiou et al. 1998).

The CVBs have also been associated with encephalitis and meningoencephalitis, although with significantly less frequency than meningitis (Charney et al. 1979; Heathfield et al. 1967; Horstmann et al. 1968; Modlin et al. 1991; Moore et al. 1984). The onset is similar to that of meningitis with a abrupt appearance of fever followed by or coincident with alterations in mental status such as confusion, lethargy, somnolence, or irritability. Seizures, either generalized or focal, and coma may occur in some. In patients with meningoencephalitis, signs of meningeal irritation accompany the mental status changes. Occasionally encountered findings include signs of increased intracranial pressure (i.e., blurring of the optic discs), cranial nerve abnormalities, truncal ataxia, and apnea (Modlin and Rotbart 1997). In encephalitis, the CSF parameters may reveal only mild elevation in the protein concentration or be normal. If the patient has meningoencephalitis CSF pleocytosis may be present. Neuroimaging in patients with CVB-associated encephalitis or meningoencephalitis may reveal diffuse lesions in the brain (Charney et al. 1979).

The CVBs have also been associated with acute paralysis (i.e., poliomyelitis) alone or in association with encephalitis (encephalomyelitis) (Cherry 2004; Horstman and Yamada 1968). The presentation may mimic that of poliovirus-related paralysis with fever present at the time of onset of the paralysis and a rapid progression to flaccid paralysis. Sensory function is left intact (Dietz et al. 1995). Other less commonly associated conditions with CVB infection include Guillain-Barré syndrome, cerebellar ataxia, and transverse myelitis (Cherry 2004).

In patients with congenital impaired humoral immunity, CVB infection may result in chronic meningoencephalitis. This syndrome, designated as chronic enteroviral meningoencephalitis in agammaglobulinemia (CEMA) (McKinney et al. 1987), was originally described in patients with X-linked agammaglobulinemia. A case of CVB3 meeting the case definition, as well as a possible case involving

CVB2, have been reported. This syndrome is a constellation of neurologic symptoms that includes headache, seizures, hearing loss, lethargy/coma, weakness, ataxia, paresthesias, and loss of cognitive skills. A significant number of patients also have nonneurologic manifestations that occur as a result of chronic enteroviral infection of other organs and include a dermatomyositis-like syndrome, peripheral ligneous edema, exanthems, and hepatitis.

## 2.6 Musculoskeletal Syndromes

The CVBs have been documented to be the causal agents of epidemic and sporadic pleurodynia (Bornholm disease) (Bain et al. 1961; Cherry 20004; Curnen et al. 1949; Huebner et al. 1953; Ryder et al. 1959; Weller 1950). While CVB1, CVB2, CVB3, and CVB5 have all been linked to outbreaks of epidemic pleurodynia, the latter two serotypes are the most commonly found. All six CVB serotypes have been associated with sporadic cases of pleurodynia. The disease was fully characterized in a monograph by Sylvest, which also provided its common geographically linked name: Bornholm disease (Sylvest 1934). Pleurodynia typically occurs as an epidemic disease and is the result of virally induced inflammation of intercostal and abdominal muscles rather than of the pleura, as its name would suggest.

An incubation period of approximately 4 days precedes the onset of fever and pain, which is typically located to the chest and upper abdomen. The fever may be biphasic; resolving and reappearing in association with the pain. The pain is spasmodic in nature and may be exacerbated by deep inspiration, coughing, or sneezing. It may be so severe as to result in diaphoresis and pallor in the affected individual. During the spasms, patients tend to be tachypneic, breathing shallowly and grunting. If the pain is localized to the abdomen it may be confused with conditions associated with acute abdomen (i.e., appendicitis, peritonitis, intestinal obstruction). Abdominal pain may be more commonly seen in children. Tenderness on palpation of the involved muscles may be found on physical examination. Auscultation of the lungs may reveal a pleural friction rub. The duration of the illness is generally less than 1 week.

## 2.7 Severe Neonatal Infections

A recent report based on data from the CDC's National Enteroviral Surveillance System (NESS) has provided novel and detailed information regarding the epidemiology and outcome of CVB-related neonatal disease in the United States (Khetsuriani et al. 2006). From 1983 to 2003, the five common CVBs (B1-5) accounted for 35% of all identified enteroviral isolates from infants less than 1 month of age. These five serotypes all ranked among the ten most common isolates: CVB2-$2^{nd}$, CVB5-$3^{rd}$, CVB4-$4^{th}$, CVB3-$8^{th}$, and CVB1-$9^{th}$. Group B coxsackievirus infections of neonates may occur sporadically (Jenista et al. 1984; Kaplan et al. 1983; Lake et al. 1976) or as the result of nursery outbreaks (Brightman et al. 1966; Eilard et al. 1974; Farmer 1968; Isacsohn et al. 1994; Rantakallio et al. 1970; Reiss-Levy et al. 1986; Swender et al. 1974).

Although outcome information was available in only approximately 18% of reports from 1983 to 1998, it provides a sobering view of the mortality associated with CVB infections in neonates. Mortality associated with neonatal infections due to CVB2, CVB3, and CVB4 was 11.1%, 18.8%, and 40%, respectively (Khetsuriani 2006). The mortality rate associated with CVB4 infection in neonates was greater than twice that associated with infections due echoviruses 9 and 11, both which are also known to be major enteroviral pathogens in the neonatal period. These findings support earlier reports that pointed toward a poorer outcome in neonates with CVB infections (Lake et al. 1976; Kaplan et al. 1983). The greatest mortality rates are seen with myocarditis and hepatitis in association with disseminated intravascular coagulopathy (Kaplan et al. 1983; Abzug 2001) Myocarditis in association with hepatitis was associated with particularly poor prognosis.

Severe CVB neonatal disease consists of combinations of meningitis, encephalitis, meningoencephalitis, myocarditis, hepatitis, coagulopathy, and pneumonia. While CVB-associated neonatal sepsis is most commonly the encephalomyocarditis syndrome consisting of heart failure due to severe myocarditis in association with meningoencephalitis (Isacsohn et al. 1994; Kaplan et al. 1984; Khetsuriani et al. 2006; Kibrick et al. 1958), reports of their association with hemorrhage hepatitis syndrome exist (Chou et al. 1995; Clavell et al. 1999; Wallot et al. 2004; Wang et al. 1998), a syndrome more commonly associated with echovirus 11 neonatal infection (Khetsuriani et al. 2006).

Perinatal maternal infection with CBV has been reported in infants with severe neonatal CVB infections (Baker et al. 1980; Kaplan et al. 1983). Symptomatic infection of the mother has been reported to between 10 days prior to delivery and 5 days postpartum. Maternal clinical syndromes include nonfocal febrile illness, upper respiratory tract illness, pleurodynia, meningitis, and abdominal pain (Kaplan et al. 1983). The latter may be severe enough to mimic appendicitis, chorioamnionitis, or placental abruption.

Severe neonatal CVB disease typically presents in the first 2 weeks of life (Kaplan et al. 1983; Baker et al. 1980). As with other CVB-related syndromes, the fever and illness may have a biphasic course. The initial clinical findings may consist of hypotonia, lethargy, fever, poor feeding, and abdominal distension. Meningoencephalitis may be manifested by profound lethargy, seizures, focal neurologic abnormalities, full fontanelle, and nuchal rigidity (Kaplan et al. 1983; Kibrick et al. 1958). Evaluation of the CSF may show a predominantly lymphocytic pleocytosis, increased protein concentration, and normal glucose concentration. Hypoglycorrhachia may be present in some infants (Kaplan et al. 1983).

The clinical manifestations of myocarditis include evidence of congestive heart failure: poor perfusion, cyanosis, hypotension, metabolic acidosis, oliguria, etc. (Kaplan et al. 1983). Electrocardiographic findings include tachydysrrhythmias and evidence of myocardial infarction (Hornung et al. 1999, Kaplan et al. 1983,

Lu et al. 2005, Shah et al. 1998). The echocardiogram may demonstrate poor ventricular contractility, a reduced shortening fraction, or pericardial effusion.

In cases of hepatitis, infants have jaundice, hepatomegaly, abdominal distension, and coagulopathy (Clavell et al. 1999; Kaplan et al. 1983; Wang et al. 1998). Laboratory abnormalities include elevated transaminases, hyperbilirubinemia, thrombocytopenia, and evidence of disseminated intravascular coagulopathy.

## 3 Diagnosis

The principle means by which the majority of clinical laboratories continue to attempt detection of CVB from clinical specimens is by viral isolation using either traditional cell culture or shell vial culture combined with the use of monoclonal antibodies. These approaches suffer from significant limitations. The time to detection of the enteroviruses using cell culture methods is too long to impact on clinical management. Using traditional cell culture, the mean time for detection of the enteroviruses from the CSF of patients with enteroviral meningitis ranges from approximately 4 to 8 days (Rotbart 1995) and from 2 to 3 days with shell vial culture (Klespies et al. 1998; Van Doornum et al. 1998). Because high titers of CVB may be found in feces or upper respiratory tract secretions, cell culture detection of the presence may be faster (i.e., 1-3 days) (Romero 2006). Of major clinical importance is the lack of sensitivity of cell culture-based approaches. In 25%-35% of CSF specimens from patients with symptoms consistent with EV meningitis, cell culture is negative (Chonmaitree et al. 1988). Attempts to isolate the CVB using cell culture is limited in part by the lack of permissiveness to infection by CVB by some cell lines used (Romero 2006). Monkey kidney cell lines (e.g., rhesus, cynomolgus, and Buffalo green) and some human cell lines (e.g., HeLa, Hep-2, A549, CaCo2, and NCI-H292) have good sensitivity for detection of the CBVs (Otero 2001; Reigel 1985; Romero 2006). In contrast, HK, HELF, HK, and RD cell lines are poor substrates for isolation of the CVBs. Specific identification of the serotype is accomplished using the most sensitive and rapid means currently available for the detection of CBVs and the enteroviruses, in general, that of nucleic aid detection using traditional or real-time reverse transcription-polymerase chain reaction (RT-PCR) or nucleic acid sequence-based amplification (NASBA) (Romero 2006). Reverse transcription PCR should be considered the method of choice for detection of the CBVs from clinical specimens such as CSF, blood, or tissue. Techniques have been developed for RT-PCR detection of EV from sources such as blood, myocardium, and liver (Romero 1999). The confluence of nucleic and amplification technology and bioinformatics has resulted in a novel powerful method for molecular serotyping of the NPEV directly from clinical specimens (Nix et al. 2006; Oberste et al. 1999, 2003). The amplicons generated are rapidly sequenced using PCR and computationally compared to a

VP1 sequence database of all known EVs. This method greatly reduces the time required for identification of the EV from weeks to days.

Serologic diagnosis of CVB infections is not clinically useful. ELISA assays for the detection of CVB-specific IgM antibodies have been reported (Dorries 1983; Goldwater 1995). Assays for the detection of homotypic IgM responses are generally impractical unless there is suspicion for a specific CVB serotype. The IgM response may be nonspecific, leading to a false-positive result; the lack of sensitivity and the cross-reactivity with non-EV pathogens has not been thoroughly studied (Romero 2006).

The interpretation of cell culture and nucleic acid amplification assay results requires that the site from which the CVB was detected be taken into account. Because the CBVs are shed from the oropharynx and gastrointestinal tract for weeks to months after infection, the significance of their detection from these sites must be interpreted with caution. Detection of the CVB at these sites does not automatically establish causality in the clinical syndrome being evaluated. The detection of a CVB serotype from the throat or stool of a patient with meningitis, encephalitis, myocarditis, or febrile illness may represent viral shedding as a result of antecedent infection and have no relationship with the current clinical problem. In strong contrast, the identification of a CVB serotype from the CSF, blood, tissue, or sterilely obtained urine, strongly supports the causal role the serotype isolated. Samples from the latter sites are the ideal sources from which to diagnose CVB infections.

The exception to the discussion above is in diagnosis of neonatal CVB infections occurring within the first 2 weeks of life. Because the incubation period for the enteroviruses is 3-6 days, the identification of a CVB in the stool or oropharynx from neonates with compatible syndromes in is highly supportive of their role in disease.

## 4 Treatment

There is no currently FDA-approved antiviral for the treatment of CVB infections in humans. The search for effective drugs to treat picornaviral infections has led to the development of three compounds (Buontempo et al. 1997; Kaiser et al. 2000; Rogers et al. 1999; Romero 2001; Witherell 2000). One of these, pleconaril, has undergone phase II and III studies. Pleconaril intercalates into the hydrophobic pockets of the EV capsid and is believed to inhibit receptor binding and viral uncoating (McKinlay et al. 1992). It has a broad antipicornaviral spectrum, including the CVB. Pleconaril is orally bioavailable and well tolerated across all age groups, including neonates (Abdel-Rahman et al. 1998, 1999; Kearns et al. 2000). In all age groups, oral dosing resulted in serum levels at 12 h that exceeded those required to inhibit the replication of more than 90% of EV tested (Abdel-Rahman et al. 1998, 1999, Abzug et al. 2003; Kearns 2000). The most commonly reported adverse event has been nausea. Use in children, it has been possibly associated with headache (Sawyer et al. 1999). Pleconaril can induce cytochrome P3A4 activity in humans. Concomitant use of pleconaril and oral contraceptives has resulted in menstrual irregularities.

Anecdotal reports of the pleconaril for the therapy of severe neonatal enteroviral infections, including that of CVB, and immunocompromised children has suggested benefit in some cases (Aradottir et al. 2001; Bauer et al. 2002; Nowak-Wegrzyn et al. 2001; Rentz et al. 2006; Rotbart and Webster 2001). A multicenter, placebo-controlled randomized trial of pleconaril for the treatment of severe (i.e., hepatitis, myocarditis, coagulopathy, myocarditis) is currently being conducted by the National Institute of Allergy and Infectious Disease Collaborative Antiviral Study Group of the National Institutes of Health.

Only a single randomized, double-blind, placebo-controlled study of pleconaril for treatment of EV meningitis in infants less than 12 months of age has been published (Abzug et al. 2003). Unfortunately, the trial failed to demonstrate clinical efficacy due the small numbers of patients enrolled, lack of adequate cultures, and the short, benign nature of the clinical illness. Although pleconaril was well tolerated, drug accumulation was noted. In contrast, a post hoc evaluation of efficacy of pleconaril for the treatment of meningitis in adults did demonstrate faster resolution of headache in individuals with moderate to severe nausea (Desmond et al. 2006). However, for those patients with severe headache, resolution was significantly slower.

Attempts to use immunoglobulin preparations for the therapy of EV infections in immunocompromised hosts and neonates have met with varied outcomes. Systemically or intrathecally administered immunoglobulins delivered for the treatment of CEMA in patients with X-linked agammaglobulinemia has met with variable results in eradicating EV from infected individuals (McKinney et al. 1987; Quartier et al. 2000). Anecdotal reports of intravenous immunoglobulin for the treatment of severe neonatal EV infections, including CVB, also suggest limited success, but they are insufficient for deriving conclusions regarding it use (Johnston et al. 1989; Kimura et al. 1999; Murry 1996; Valduss et al. 1993; Wong et al. 1989). Only a single small randomized trial evaluating the efficacy of intravenously administered immunoglobulin for the treatment of neonatal enteroviral infections has been conducted (Abzug 1995). In that study, receipt of intravenous gammaglobulin containing a neutralizing titer of greater than or equal to 1:800 to the enteroviral isolate of the infected infant resulted in more rapid cessation of viremia and viruria in association with subtle clinical benefits.

# 5 Prevention

No vaccines exist for the prevention of the CVB infections. Potable water supplies, modern sewage systems, and waste treatment plants all aid in the elimination CVB outbreaks. Attention to good hygienic practices (i.e., handwashing) will help reduce exposure to and transmission of the CVB. For patients hospitalized with

CVB-related syndromes, infection control measures consist of standard precautions and, for infants and children, contact precautions for the duration of the illness. Women who are pregnant should avoid contact with individuals with probable or known CVB infection. Emergent delivery of pregnant women who have illnesses consistent with those caused by the CVB should be delayed unless there is concern for the viability of the fetus or the condition is that of an obstetrical emergency. This may permit the fetus to acquire maternally derived antibodies to the infecting CVB serotype, possibly protecting the newborn (Modlin et al. 1981).

# References

Abdel-Rahman SM, Kearns GL (1998) Single-dose pharmacokinetics of a pleconaril (VP63843) oral solution and effect of food. Antimicrob Agents Chemother 42:2706-2709
Abdel-Rahman SM, Kearns GL (1999) Single oral dose escalation pharmacokinetics of pleconaril (VP 63843) capsules in adults. J Clin Pharmacol 39:613-618
Abzug MJ (2001) Prognosis for neonates with enterovirus hepatitis and coagulopathy. Pediatr Infect Dis J 20:758-763
Abzug MJ (2004) Presentation, diagnosis, and management of enterovirus infections in neonates. Paediatr Drugs 6:1-10
Abzug MJ, Cloud G, Bradley J, Sanchez P, Romero J, Powell D, Lepow M, Mani C, Capparelli EV, Blount S, Lakeman F, Whitley RJ, Kimberlin DW, Collaborative Antiviral Study Group (2003) Double blind, placebo-controlled trial of pleconaril in infants with enterovirus meningitis. Pediatr Infect Dis J 22:335-341
Alexander JP Jr, Chapman LE, Pallansch MA, Stephenson WT, Torok TJ, Anderson LJ (1993) Coxsackievirus B2 infection and aseptic meningitis: a focal outbreak among members of a high school football team. J Infect Dis 167:1201-1205
Aradottir E, Alonso EM, Shulman ST (2001) Severe neonatal enteroviral hepatitis treated with pleconaril. Pediatr Infect Dis J 20:457-459
Arrieta AC, Stutman HR (1991) Coxsackie B2 meningitis with unusually high white blood cell count in cerebrospinal fluid. Pediatr Infect Dis J 10:250-251
Bain HW, McLean DM, Walkere SJ (1961) Epidemic pleurodynia (Bornholm disease) due to Coxsackie B-5 virus. The inter-relationship of pleurodynia, benign pericarditis and aseptic meningitis. Pediatrics 27:889-903
Baker DA, Phillips (1980) CA maternal and neonatal infection with coxsackievirus Obstet Gynecol 55(3 Suppl):12S-15S
Bates HR Jr (1970) Coxsackie virus B3 calcific pancarditis and hydrops fetalis. Am J Obstet Gynecol 106:629-630
Bauer S, Gottesman G, Sirota L, Litmanovitz I, Ashkenazi S, Levi I (2002) Severe Coxsackie virus B infection in preterm newborns treated with pleconaril. Eur J Pediatr 161:491-493
Bendig JW, Franklin OM, Hebden AK, Backhouse PJ, Clewley JP, Goldman AP, Piggott N (2003) Coxsackievirus B3 sequences in the blood of a neonate with congenital myocarditis, plus serological evidence of maternal infection. J Med Virol 70:606-609
Berlin LE, Rorabaugh ML, Heldrich F, Roberts K, Doran T, Modlin JF (1993) Aseptic meningitis in infants <2 years of age: diagnosis and etiology. J Infect Dis 168:888-892
Brightman VJ, Scott TF, Westphal M Boggs TR (1966) An outbreak of coxsackie B-5 virus infection in a newborn nursery. J Pediatr 69:179-192
Buontempo PJ, Cox S, Wright-Minogue J, DeMartino JL, Skelton AM, Ferrari E, Albin R, Rozhon EJ, Girijavallabhan V, Modlin JF, O'Connell JF (1997) SCH 48973 a potent, broad-spectrum, antienterovirus compound. Antimicrob Agents Chemother 41:1220-1225

Burch GE, Sun SC, Chu K, Sohal RS, Colcolough HL (1968) Interstitial and coxsackievirus B myocarditis in infants and children. A comparative histologic and immunofluorescent study of 50 autopsied hearts. JAMA 203:1-8

Byington CL, Taggart EW, Carroll KC, Hillyard DR (1999) A polymerase chain reaction-based epidemiologic investigation of the incidence of nonpolio enteroviral infections in febrile and afebrile infants 90 days and younger. Pediatrics 103:E27

Carmichael J, McGuckin R, Gardner PS (1968) Outbreak of Coxsackie type B2 virus in a children's home in Newcastle upon Tyne. BMJ 2:532-533

Centers for Disease Control and Prevention (2006) Enterovirus Surveillance- United States 1970-2005 MMWR 55:1-20

Chambon M, Archimbaud C, Bailly JL, Henquell C, Regagnon C, Charbonne F, Peigue-Lafeuille H (2001) Circulation of enteroviruses and persistence of meningitis cases in the winter of 1999-2000. J Med Virol 65:340-347

Charney EB, Orecchio EJ, Zimmerman RA, Berman PH (1979) Computerized tomography in infantile encephalitis. Am J Dis Child 133:803-805

Chemtob S, Reece ER, Mills EL (1985) Syndrome of inappropriate secretion of antidiuretic hormone in enteroviral meningitis. Am J Dis Child 139:292-294

Cherry JD (2004) Enteroviruses and parechoviruses. In: Feigin RD, Cherry JD, Demmler G, Kaplan SL (eds) Textbook of pediatric infectious diseases, 5th edn, vol. 2. Saunders, Philadelphia, pp 1984-2041

Cherry JD, Jahn CL (1965) Herpangina: the etiologic spectrum. Pediatrics 36:632-634

Chesney JC, Hoganson GE, Wilson MH (1980) CSF eosinophilia during an acute coxsackie B4 viral meningitis. Am J Dis Child 134:703

Chesney PJ, Quennec P, Clark C (1978) Hypoglycorrhachia and coxsackie B3 meningoencephalitis. Am J Clin Pathol 70:947-948

Chiou CC, Liu WT, Chen SJ, Soong WJ, Wu KG, Tang RB, Hwang B (1998) Coxsackievirus B1 infection in infants less than 2 months of age. Am J Perinatol 15:155-159

Chonmaitree T, Mann L (1995) Respiratory Infections. In: Rotbart HA (ed) Human enterovirus infections. American Society of Microbiology, Washington DC, pp 255-270

Chonmaitree T, Ford C, Sanders C, Lucia HL (1988) Comparison of cell cultures for rapid isolation of enteroviruses. J Clin Microbiol 26:2576-2580

Chou LL, Chang CP, Wu LC (1995) Neonatal coxsackievirus B1 infection associated with severe hepatitis: report of three cases. Zhonghua Min Guo Xiao Er Ke Yi Xue Hui Za Zhi 36:296-299

Clavell M, Barkemeyer B, Martinez B, Craver R, Correa H, Gohd R, Schmidt-Sommerfeld E (1999) Severe hepatitis in a newborn with coxsackievirus B5 infection. Clin Pediatr (Phila) 38:739-741

Curnen EC, Shaw EW, Melnick JL (1949) Disease resembling nonparalytic poliomyelitis associated with a virus pathogenic for infant mice. JAMA 141:894-901

Dagan R, Hall CB, Powell KR, Menegus MA (1989) Epidemiology and laboratory diagnosis of infection with viral and bacterial pathogens in infants hospitalized for suspected sepsis. J Pediatr 115:351-356

Desmond RA, Accortt NA, Talley L, Villano SA, Soong SJ, Whitley RJ (2006) Enteroviral meningitis: natural history and outcome of pleconaril therapy. Antimicrob Agents Chemother 50:2409-2414

Dietz V, Andrus J, Olive JM, Cochi S, de Quadros C (1995) Epidemiology and clinical characteristics of acute flaccid paralysis associated with non-polio enterovirus isolation: the experience in the Americas. Bull World Health Organ 73:597-603

Dorries R, Ter Meulen V (1983) Specificity of IgM antibodies in acute human coxsackievirus B infections, analysed by indirect solid phase enzyme immunoassay and immunoblot technique. 64:159-167

Eckert HL, Portnoy B, Salvatore MA, Ressler R (1967) Group B Coxsackie virus infection in infants with acute lower respiratory disease. Pediatrics 39:526-531

Eilard T, Kyllerman M, Wennerblom I, Eeg-Olofsson O, Lycke E (1974) An outbreak of Coxsackie virus type B2 among neonates in an obstetrical ward. Acta Paediatr Scand 63:103-107

Farmer K, Patten PT (1968) An outbreak of coxsackie B5 infection in a special care unit for newborn infants. N Z Med J 68:86-89

Flewet TH (1965) Histological study of two cases of Coxsackie B virus pneumonia in children. J Clin Path 18:743-736

Goldwater PN (1995) Immunoglobulin M capture immunoassay in investigation of coxsackievirus B5 and B6 outbreaks in south Australia. J Clin Microbiol 33:1628-1631

Hable KA, O'Connell EJ, Herrmann EC Jr (1970) Group B coxsackieviruses as respiratory viruses. Mayo Clin Proc 45:170-176

Hall CB, Miller DG (1969) The detection of a silent coxsackie B-5 virus perinatal infection. J Pediatr 75:124-127

Heathfield KW, Pilsworth R, Wall BJ, Corsellis JA (1967) Coxsackie B5 infections in Essex, 1965, with particular reference to the nervous system. Q J Med 36:579-595

Hornung TS, Bernard EJ, Howman-Giles RB, Sholler GF (1999) Myocardial infarction complicating neonatal enterovirus myocarditis. J Paediatr Child Health 35:309-312

Horstmann DM, Yamada N (1968) Enterovirus infections of the central nervous system. Res Publ Assoc Res Nerv Ment Dis 44:236-253

Huebner RJ, Risser JA, Bell JA, Beeman EA, Beigelman PM, Strong JC (1953) Epidemic pleurodynia in Texas; a study of 22 cases. N Engl J Med 248:267-274

Isacsohn M, Eidelman AI, Kaplan M, Goren A, Rudensky B, Handsher R, Barak Y (1994) Neonatal coxsackievirus group B infections: experience of a single department of neonatology. Isr J Med Sci 30:371-374

Jaffe M, Srugo I, Tirosh E, Colin AA, Tal Y (1989) The ameliorating effect of lumbar puncture in viral meningitis. Am J Dis Child 143:682-685

Jenista JA, Powell KR, Menegus MA (1984) Epidemiology of neonatal enterovirus infection. J Pediatr 104:685-690

Johnston JM, Overall JC Jr (1989) Intravenous immunoglobulin in disseminated neonatal echovirus 11 infection. Pediatr Infect Dis J 8:254-256

Kaiser L, Crump CE, Hayden FG (2000) In vitro activity of pleconaril and AG7088 against selected serotypes and clinical isolates of human rhinoviruses. Antiviral Res 47:215-220

Kaplan MH, Klein SW, McPhee J, Harper RG (1983) Group B coxsackievirus infections in infants younger than three months of age: a serious childhood illness. Rev Infect Dis 5:1019-1032

Kearns GL, Bradley JS, Jacobs RF, Capparelli EV, James LP, Johnson KM, Abdel-Rahman SM (2000) Single dose pharmacokinetics of pleconaril in neonates. Pediatr Infect Dis J 19:833-839

Khetsuriani N, Lamonte A, Oberste MS, Pallansch M (2006) Neonatal enterovirus infections reported to the national enterovirus surveillance system in the United States 1983-2003. Pediatr Infect Dis J 25:889-893

Kibrick S, Benirschke K (1958) Severe generalized disease (encephalohepatomyocarditis) occurring in the newborn period and due to infection with coxsackie virus, group B: evidence of intrauterine infection with this agent. Pediatrics 22:857-874

Kimura H, Minakami H, Harigaya A, Takeuchi H, Tachibana A, Otsuki K (1999) Treatment of neonatal infection caused by coxsackievirus B3. J Perinatol 19:388-390

Klespies SL, Cebula DE, Kelley CL, Galehouse D, Maurer CC (1996) Detection of enteroviruses from clinical specimens by spin amplification shell vial culture and monoclonal antibody assay. J Clin Microbiol 34:1465-1467

Krajden S, Middleton PJ (1983) Enterovirus infections in the neonate. Clin Pediatr (Phila) 22:87-92

Lake AM, Lauer BA, Clark JC, Wesenberg RL, McIntosh K (1976) Enterovirus infections in neonates. J Pediatr 89:787-791

Leggiadro RJ, Darras BT (1983) Viral and bacterial pathogens of suspected sepsis in young infants. Pediatr Infect Dis 2:287-289

Lu JC, Koay KW, Ramers CB, Milazzo AS (2005) Neonate with coxsackie B1 infection, cardiomyopathy and arrhythmias. JAMA 97:1028-1030

McKinney RE Jr, Katz SL, Wilfert CM (1987) Chronic enteroviral meningoencephalitis in agammaglobulinemic patients. Rev Infect Dis 9:334-356

McKinlay MA, Pevear DC, Rossmann MG (1992) Treatment of the picornavirus common cold by inhibitors of viral uncoating and attachment. Ann Rev Microbiol 46:635-654

Modlin JF, Rotbart HA (1997) Group B coxsackie disease in children. Curr Top Microbiol Immunol 223:53-80

Modlin JF, Polk BF, Horton P, Etkind P, Crane E, Spiliotes A (1981) Perinatal echovirus infection: risk of transmission during a community outbreak. N Engl J Med 305:368-371

Modlin JF, Dagan R, Berlin LE, Virshup DM, Yolken RH, Menegus M (1991) Focal encephalitis with enterovirus infections. Pediatrics 88:841-845

Moore M, Kaplan MH, McPhee J, Bregman DJ, Klein SW (1984) Epidemiologic, clinical, and laboratory features of Coxsackie B1-B5 infections in the United States, 1970-79. Public Health Rep 99:515-522

Murry DL (1996) Neonatal enterovirus infection: neutralization by intravenous immune globulin. Clin Infect Dis 22:397-398

Valduss D, Murray DL, Karna P, Lapour K, Dyke J (1993) Use of intravenous immunoglobulin in twin neonates with disseminated coxsackie B1 infection. Clin Pediatr (Phila) 32:561-563

Nakayama T, Urano T, Osano M, Hayashi Y, Sekine S, Ando T, Makinom S (1989) Outbreak of herpangina associated with Coxsackievirus B3 infection. Pediatr Infect Dis J 8: 495-498

Nix WA, Oberste MS, Pallansch MA (2006) Sensitive, seminested PCR amplification of VP1 sequences for direct identification of all enterovirus serotypes from original clinical specimens. J Clin Microbiol 44:2698-2704

Nowak-Wegrzyn A, Phipatanakul W, Winkelstein JA, Forman MS, Lederman HM (2001) Successful treatment of enterovirus infection with the use of pleconaril in 2 infants with severe combined immunodeficiency. Clin Infect Dis 32:E13-E14

Oberste MS, Maher K, Kilpatrick DR, Flemister MR, Brown BA, Pallansch MA (1999) Typing of human enteroviruses by partial sequencing of VP1. J Clin Microbiol 37:1288-1293

Oberste MS, Nix WA, Maher K, Pallansch MA (2003) Improved molecular identification of enteroviruses by RT-PCR and amplicon sequencing. J Clin Virol 26:375-377

Otero JR, Folgueira L, Trallero G, Prieto C, Maldonado S, Babiano MJ, Martinez-Alonso I (2001) A-549 is a suitable cell line for primary isolation of coxsackie B viruses. J Med Virol 65:534-536

Ouellet A, Sherlock R, Toye B, Fung KF (2004) Antenatal diagnosis of intrauterine infection with coxsackievirus B3 associated with live birth. Infect Dis Obstet Gynecol 12:23-26

Parrott RH, Ross S, Burke FG, Rice EC (1951) Herpangina; clinical studies of a specific infectious disease. N Engl J Med 245:275-280

Quartier P, Foray S, Casanova JL, Hau-Rainsard I, Blanche S, Fischer A (2000) Enteroviral meningoencephalitis in X-linked agammaglobulinemia: intensive immunoglobulin therapy and sequential viral detection in cerebrospinal fluid by polymerase chain reaction. Pediatr Infect Dis J 19:1106-1108

Rantakallio P, Lapinleimu K, Mantyjarvi R (1970) Coxsackie B 5 outbreak in a newborn nursery with 17 cases of serous meningitis. Scand J Infect Dis 2:17-23

Reigel F (1985) Isolation of human pathogenic viruses from clinical material on CaCo2 cells. J Virol Methods 12:323-327

Reiss-Levy E, Baker A, Don A, Caldwell G (1986) Two concurrent epidemics of enteroviral meningitis in an obstetric neonatal unit. Aust N Z J Med 16:365-372

Rentz AC, Libbey JE, Fujinami RS, Whitby FG, Byington CL (2006) Investigation of treatment failure in neonatal echovirus 7 infection. Pediatr Infect Dis J 25:259-262

Rogers JM, Diana GD, MCKinlay MA (1999) Pleconaril. A broad spectrum antipicornaviral agent. Adv Exp Med Biol 458:69-76

Romero JR (1999) Reverse-transcription polymerase chain reaction detection of the enteroviruses. Arch Pathol Lab Med 123:1161-1169

Romero JR (2001) Pleconaril: a novel antipicornaviral drug. Expert Opin Investig Drugs 10:369-379

Romero JR (2006) Entroviruses and parechoviruses. In: Murray PR, Baron EJ, Jorgensen JH, Landry ML, Pfaller MA (eds) Manual of clinical microbiology, 9th edn. American Society for Microbiology, Washington DC, pp 1392-1404

Rorabaugh ML, Berlin LE, Heldrich F, Roberts K, Rosenberg LA, Doran T, Modlin JF (1993) Aseptic meningitis in infants younger than 2 years of age: acute illness and neurologic complications. Pediatrics 92:206-211

Rotbart HA (1995) Enteroviral infections of the central nervous system. Clin Infect Dis 20:971-981

Rotbart HA, Webster AD (2001) Treatment of potentially life-threatening enterovirus infections with pleconaril. Clin Infect Dis 32:228-235

Ryder DE, Doane FW, Zbitnew A, Rhodes AJ (1959) Report of an outbreak of Bornholm disease, with isolation of Coxsackie B5 virus: Toronto, 1958. Can J Public Health 50:265-269

Sawyer MH, Saez-Llorenz X, Avilles CL, O'Ryan M, Romero J (1999) Oral pleconaril reduces the duration and severity of enteroviral meningitis in children. Pediatric Academic Societies' Annual Meeting, San Francisco, CA, Abstr 1012

Schiff GM (1979) Coxsackievirus B epidemic at a boys' camp. Am J Dis Children 133:782-785

Severien C, Jacobs KH, Schoenemann W (1994) Marked pleocytosis and hypoglycorrhachia in coxsackie meningitis. Pediatr Infect Dis J 13:322-323

Shah SS, Hellenbrand WE, Gallagher PG (1998) A trial flutter complicating neonatal Coxsackie B2 myocarditis. Pediatr Cardiol 19:185-186

Singer JI, Maur PR, Riley JP, Smith PB (1980) Management of central nervous system infections during an epidemic of enteroviral aseptic meningitis. J Pediatr 96:559-563

Swender PT, Shott RJ, Williams ML (1974) A community and intensive care nursery outbreak of coxsackievirus B5 meningitis. Am J Dis Child 127:42-45

Sylvest E (1934) Epidemic myalgia: Bornholm disease. Andersen H (transl). London, Oxford University Press

Valduss D, Murray DL, Karna P, Lapour K, Dyke J (1993) Use of intravenous immunoglobulin in twin neonates with disseminated coxsackie B1 infection. Clin Pediatr (Phila) 32:561-563

Van Doornum GJ, De Jong JC (1998) Rapid shell vial culture technique for detection of enteroviruses and adenoviruses in fecal specimens: comparison with conventional virus isolation method. J Clin Microbiol 36:2865-2868

Wallot MA, Metzger-Boddien C, Auth M, Kehle J, Enders G, Dirsch O, Fiedler M, Voit T (2004) Acute liver failure associated with Coxsackie virus B2 infection in a neonate. Eur J Pediatr 163:116-117

Wang SM, Liu CC, Yang YJ, Yang HB, Lin CH, Wang JR (1998) Fatal coxsackievirus B infection in early infancy characterized by fulminant hepatitis. J Infect 37:270-273

Weller TH, Enders JF, Buckingham M, Finn JJ Jr (1950) The etiology of epidemic pleurodynia: a study of two viruses isolated from a typical outbreak. J Immunol 65:337-346

Wilfert CM, Lehrman SN, Katz SL (1983) Enteroviruses and meningitis. Pediatr Infect Dis 2:333-334

Witherell G (2000) AG-7088 Pfizer. Curr Opin Investig Drugs 1:297-302

Wong SN, Tam AY, Ng TH, Ng WF, Tong CY, Tang TS (1989) Fatal Coxsackie B1 virus infection in neonates. Pediatr Infect Dis J 8:638-641

Zahorsky J (1920) Herpetic sore throat. South Med J 13:871

Zahorsky J (1924) Herpangina. Arch Pediatr 41:181

# CVB-Induced Pancreatitis and Alterations in Gene Expression

A. I. Ramsingh

| | | |
|---|---|---|
| 1 | Introduction | 242 |
| 2 | The Pancreas: Anatomy and Physiology | 242 |
| 3 | Acute Pancreatitis | 243 |
| | 3.1 Clinical Acute Pancreatitis | 243 |
| | 3.2 CVB4-Induced Mild Acute Pancreatitis | 244 |
| | 3.3 CVB4-Induced Severe Acute Pancreatitis | 245 |
| 4 | Chronic Pancreatitis | 246 |
| | 4.1 Clinical Chronic Pancreatitis | 246 |
| | 4.2 CVB4-Induced Chronic Pancreatitis | 247 |
| 5 | Gene Expression Profiles During CVB4-Induced Pancreatitis | 248 |
| | 5.1 Global Gene Expression During CVB4-P and CVB4-V Infection | 248 |
| | 5.2 Role of Macrophages in CVB4-Induced Pancreatitis | 248 |
| | 5.3 NF-κB Activation During CVB4-Induced Pancreatitis | 250 |
| 6 | Chronic Inflammation and Cancer | 251 |
| | 6.1 Chronic Pancreatitis and Pancreatic Cancer | 251 |
| | 6.2 Gene Expression Profiles in Pancreatic Cancer | 253 |
| 7 | Summary | 255 |
| | References | 256 |

**Abstract** While infections with the group B coxsackieviruses are generally asymptomatic, these viruses occasionally cause acute and chronic inflammatory diseases of the pancreas and heart. Chronic inflammatory diseases are of major clinical concern, since they predispose affected individuals to cancer. For example, chronic pancreatitis, which can develop from acute pancreatitis, is a major risk factor for pancreatic cancer, a disease with an exceedingly poor prognosis. It is crucial that we understand the mechanisms underlying the development of acute pancreatitis and the progression to chronic disease, if we are to identify new therapeutic targets to prevent the deterioration of the exocrine pancreas. We have developed a mouse model that allows us to explore the mechanisms by which CVB4 causes acute and

A. I. Ramsingh
Wadsworth Center, New York State Department of Health,
120 New Scotland Avenue, Albany, NY 12208, USA
ramsingh@wadsworth.org

chronic pancreatitis. The present review develops the idea that disease progression is a continuum, from acute inflammatory disease to chronic inflammatory disease to cancer, and that accompanying changes in gene expression are both quantitative and qualitative.

# 1 Introduction

The group B coxsackieviruses (CVBs) were discovered over 50 years ago during the search for polio-like viruses (Gifford and Dalldorf 1951; Melnick et al. 1949). In humans, most infections with the group B coxsackieviruses are asymptomatic (Huber and Ramsingh 2004; Melnick 1996). When symptomatic infections occur, the group B viruses generally cause mild diseases of the upper respiratory tract, the gastrointestinal tract, and the skin. Less frequently, the group B viruses cause inflammatory diseases of the pancreas, heart, and central nervous system. The group B viruses have been implicated in both acute inflammatory diseases (acute pancreatitis, myocarditis) and chronic inflammatory diseases (chronic pancreatitis, type 1 diabetes, dilated cardiomyopathy). While much progress has been made in our understanding of the molecular determinants of CVB virulence and of the immune response to infection, the precise mechanisms by which CVBs cause acute and chronic inflammatory diseases remain to be elucidated. We need to understand the mechanisms by which CVBs cause acute and chronic inflammatory diseases, because chronic inflammation is a risk factor for cancer. More specifically, chronic pancreatitis is a major risk factor for pancreatic cancer, which has a poor prognosis. The present review develops the idea that disease progression is a continuum, from acute inflammatory disease to chronic inflammatory disease to cancer. As a result, accompanying changes in gene expression are expected to be both quantitative and qualitative.

# 2 The Pancreas: Anatomy and Physiology

The pancreas, an organ of endodermal origin, is located near the duodenum and behind the stomach (Cotran et al. 1989; Slack 1995). The dual functions of the pancreas are reflected in its organization into endocrine and exocrine tissues. The endocrine pancreas produces hormones that are secreted into the bloodstream and that regulate glucose homeostasis. The structural units of the endocrine pancreas are the islets of Langerhans, which are spheroidal clusters embedded in the exocrine tissue. The endocrine cells consist of insulin-secreting beta cells, glucagon-secreting alpha cells, somatostatin-secreting delta cells, and pancreatic polypeptide-secreting PP or F cells. Endocrine cells account for only a small fraction of the total pancreas, approximately 4% in the adult rat (Slack 1995). Dysregulation in glucose homeostasis leads to diabetes (types 1 and 2). Although the CVBs have been implicated in the development of type 1 diabetes, the focus of the present review is CVB infection of the exocrine pancreas rather than CVB-induced endocrine disease.

The exocrine pancreas, accounting for 80%-85% of the pancreas, consists of a branched network of acinar and duct cells that produce and deliver digestive enzymes into the gastrointestinal tract (Hezel et al. 2006). Digestive enzymes produced by the exocrine pancreas include proteases, amylases, lipases, and nucleases. As a result, the exocrine pancreas is a key regulator of protein and carbohydrate digestion. The structural unit of the exocrine pancreas is the acinus, which consists of a group of acinar cells (8-12 cells) (Slack 1995). Acinar cells synthesize digestive enzymes that are packaged into secretory granules termed zymogen granules. Enzymes are secreted into intercalated ducts, which drain into larger interlobular ducts and eventually into the main pancreatic duct. The main pancreatic duct empties into the duodenum, either directly or via the common bile duct. Enzymes are generally synthesized as inactive precursors and become activated only after they enter the duodenum. Inappropriate activation of digestive enzymes within the pancreas leads to pancreatitis (Whitcomb 2006).

## 3 Acute Pancreatitis

### 3.1 Clinical Acute Pancreatitis

Pancreatitis, an inflammatory disease of the exocrine pancreas, occurs as either an acute or a chronic disease. Acute pancreatitis is a common clinical condition that is increasing in incidence, and is classified as mild or severe disease (Kingsnorth and O'Reilly 2006). Mild acute pancreatitis is generally self-limiting, while severe acute pancreatitis can lead to a systemic inflammatory response syndrome with respiratory and cardiovascular failure (Bhatia 2004; Kingsnorth and O'Reilly 2006; Whitcomb 2006). Most patients experience mild, self-limited attacks, but about 25% of patients develop severe acute pancreatitis, accompanied by significant morbidity and mortality. Pathological changes associated with acute pancreatitis include inflammation, acinar cell death, and edema of the exocrine tissues. The severity of acute pancreatitis is reflected in the extent of tissue destruction. Although gallstones and excessive alcohol consumption are important risk factors in adults, idiopathic pancreatitis represents a fairly high percentage (22%) of total cases (Lankisch et al. 1996; Lankisch et al. 1991). Idiopathic pancreatitis probably includes cases of viral etiology.

Data supporting a link between CVB infections and pancreatitis come from case reports (Kibrick and Benirschke 1958), serological studies (Arnesjo et al. 1976; Capner et al. 1975; Ozsvar et al. 1992), and animal models (Huber and Ramsingh 2004). CVB infection as a possible cause of pancreatitis was first reported in 1958 (Kibrick and Benirschke 1958). In that report, CVB4 was isolated from a child with systemic infection whose pancreas showed focal necrosis and inflammation. The clinical data is supported by mouse models of CVB-induced pancreatitis, which mimic many features of the human disease and provide excellent systems for investigating the underlying mechanisms of disease development (Huber and Ramsingh 2004).

## 3.2 CVB4-Induced Mild Acute Pancreatitis

Early studies showed that mice infected with CVBs develop pancreatitis, while the endocrine pancreas is not affected (Dalldorf and Gifford 1952; Pappenheimer et al. 1951; Ross et al. 1974). In extending these studies, we have developed unique models of CVB4-induced pancreatitis (Huber and Ramsingh 2004). Our model system uses two variants of the JVB strain of CVB4, CVB4-P, and CVB4-V. The CVB4-P variant causes a mild acute pancreatitis that resolves within 10 days of infection, while the CVB4-V variant causes a severe acute disease (Fig. 1). CVB4-P-induced disease is thus similar to the self-limiting acute pancreatitis observed in humans. Infected mice appear healthy and well-groomed and do not show visible signs of infection. CVB4-P replicates in pancreatic acinar cells, with viral replication peaking 2 days after infection (Ostrowski et al. 2004). At the peak of replication, the infected pancreas appears essentially normal. Serum amylase, a marker of pancreatitis, peaks 3 days after infection and returns to baseline levels 4 days after infection (Caggana et al. 1993). Pathological changes in the pancreas are evident 4-8 days after infection and coincide with the inflammatory response (Ostrowski et al. 2004). Pathological changes are focal and include moderate acinar cell necrosis (up to 50%), inflammation, and edema. The islets of Langerhans and pancreatic ducts appear unaffected. At 8-10 days after infection, infectious

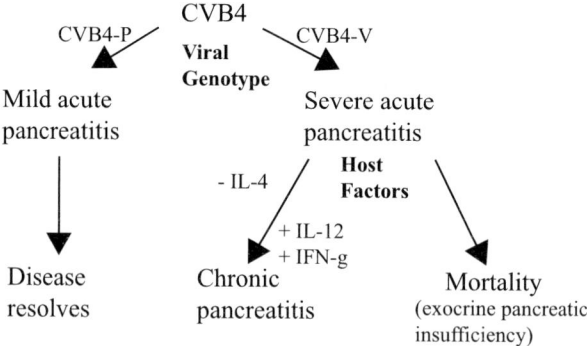

**Fig. 1** CVB4 induces mild acute pancreatitis, severe acute pancreatitis, and chronic pancreatitis. The severity of acute pancreatitis is determined by the viral genotype, while the outcome of severe acute disease is influenced by host factors including cytokines

virus is no longer detected, and tissue damage in the exocrine pancreas is resolved. Again, this is similar to clinical pancreatitis, given that a single attack of acute pancreatitis, with injury and loss of acinar cells, is generally assumed to be followed by full recovery of the structure and function of the exocrine pancreas (Elsasser et al. 1986). The pancreas thus clearly has regenerative capacity after limited injury and tissue loss. We investigated the pancreatic repair process subsequent to CVB4-P infection, and showed that there is a significant increase in the number of mitotically active acinar cells, but not in the numbers of islet or ductal cells. The data suggest that differentiated acinar cells participate in the regenerative process in this model (Ostrowski et al. 2004).

## 3.3   CVB4-Induced Severe Acute Pancreatitis

While CVB4-P-induced pancreatitis mimics the self-limiting disease observed in humans, CVB4-V infection serves as an excellent model for severe acute pancreatitis, associated with significant morbidity and mortality. CVB4-V infects acinar cells and causes a generalized infection of the exocrine pancreas (Ostrowski et al. 2004). Although the kinetics of replication for the two viruses is similar, overall replication is higher during CVB4-V infection. Maximum viral titers are observed 2 days after infection, and infectious virus is cleared in 2-3 weeks. Serum amylase peaks 2 days after infection and returns to baseline levels by 4 days after infection (Caggana et al. 1993). Pathological changes in the pancreas are evident early in infection (1-2 days) and are more severe that those observed during CVB4-P infection. Such pathological changes include widespread acinar cell necrosis, complete loss of acinar cells, inflammation, fibrosis, acinoductular metaplasia (morphological change from an acinus to a duct-like structure), and fat replacement. We have shown that mortality associated with CVB4-V infection is due to exocrine pancreatic insufficiency (Ramsingh et al. 1999).

Although gallstones, excessive alcohol consumption, and infections cause acute pancreatitis, the exact mechanisms by which these diverse factors induce tissue injury and cell death in the exocrine pancreas are not well understood. Historically, acinar cell death was thought to be due to necrosis (Bhatia 2004). During CVB4-P-induced mild acute pancreatitis, infected acinar cells display characteristics of apoptotic cells, with little necrosis, while in CVB4-V-induced severe acute pancreatitis, infected acinar cells undergo extensive degranulation and necrosis, with little apoptosis (Ostrowski et al. 2004). Biochemical and morphological studies of other experimental models of acute pancreatitis (Gukovskaya et al. 1996; Kaiser et al. 1995) support our observation of apoptotic cell death during mild acute disease and necrotic cell death during severe acute disease. The data are interesting in light of our knowledge of cell death and inflammation (Majai et al. 2006). Necrotic cell death initiates an inflammatory response in macrophages, while apoptotic cell death does not elicit pro-inflammatory responses. In the pancreas, necrotic cell death is particularly damaging, since digestive enzymes can become activated and cause additional injury. Tissue damage associated with pancreatitis is linked to inappropriate activation of trypsinogen to trypsin, a key enzyme in the activation of pancreatic zymogens, coupled with the inability to inactivate or eliminate trypsin (Whitcomb 2006).

The severity of acute pancreatitis appears to be determined by multiple factors, including the mechanism of cell death (necrosis or apoptosis), the immune response (pro-inflammatory or anti-inflammatory), and the ability to control trypsinogen activation. In our model of CVB4-induced pancreatitis, we have shown that the viral genotype determines the severity of acute disease. A single amino acid substitution in the VP1 capsid protein (met-129 → thr-129) is sufficient to induce severe, acute pancreatitis (Caggana et al. 1993). Furthermore, the single amino acid substitution in the VP1 capsid of CVB4-P alters both viral replication (A. Ramsingh, unpublished data) and immunogenicity (Halim and Ramsingh 2000; Ramsingh et al. 1997). The data suggest that both viral replication and immune responses contribute to the severity of CVB4-induced acute pancreatitis. Our working model is that the rapid replication and spread of CVB4-V throughout the exocrine pancreas are perceived as a high-danger signal that triggers necrotic cell death and inappropriate activation of digestive enzymes, resulting in an exaggerated pro-inflammatory response, all of which contribute to extensive destruction of the exocrine pancreas. Extensive tissue destruction results in the development of severe acute disease that does not resolve. In contrast, CVB4-P infection is perceived as a low-danger signal because this virus causes a focal infection of acinar cells and replicates to a lesser extent than CVB4-V. The low-danger signal triggers apoptotic cell death, less activation of digestive enzymes, and an anti-inflammatory response, all of which contribute to limited tissue destruction and the development of mild acute disease, which subsequently resolves.

## 4 Chronic Pancreatitis

### *4.1 Clinical Chronic Pancreatitis*

Chronic pancreatitis is a painful and debilitating disease, in which a progressive, destructive inflammatory process destroys the exocrine pancreas, resulting in exocrine pancreatic insufficiency (Mergener and Baillie 1997; Stevens et al. 2004). The inability of the pancreas to then produce digestive enzymes results in malabsorption of dietary nutrients, giving rise to malnutrition and weight loss. Chronic pancreatitis can develop from one episode of severe acute pancreatitis, or from recurrent episodes of acute disease. Pathological changes associated with chronic pancreatitis include inflammation, extensive loss of exocrine tissue, acinoductular metaplasia, edema, and fibrosis. While the morphological changes associated with mild acute pancreatitis are reversible, the alterations in chronic disease are irreversible, presumably due to failures in the repair process.

Chronic pancreatitis is a complex disease involving both genetic predisposition and environmental factors (Stevens et al. 2004). Environmental factors associated with chronic pancreatitis are the same as those associated with recurrent acute pancreatitis and include excessive alcohol consumption. As with acute pancreatitis, serological studies have established a link between CVB infections and chronic pancreatitis. Our model of CVB4-induced chronic pancreatitis resembles the situation in which one episode of severe acute disease leads to chronic disease.

## 4.2 CVB4-Induced Chronic Pancreatitis

In our studies of immunological factors that could influence the outcome of CVB4-V infection, we identified several cytokines that were able to reduce the morbidity and mortality associated with acute disease. Interestingly, mice that survive CVB4-V infection in these models develop chronic pancreatitis, which is characterized by prolonged inflammation in the absence of infectious virus, extensive destruction of the exocrine pancreas, acinoductular metaplasia, edema, and fibrosis (Fig. 1). The pathological features of CVB4-V-induced chronic pancreatitis are similar to those observed in clinical disease. Cytokines that have been implicated in the severity of acute pancreatitis include IL-4, IL-12, and IFN-γ. Knockout mice lacking IL-4 (Ramsingh et al. 1999) survive CVB4-V-induced acute disease and develop chronic pancreatitis. Wild type mice treated with either IL-12 or IFN-γ after CVB4-V infection also survive the acute phase of disease and develop chronic pancreatitis (Potvin et al. 2003). Histological analyses reveal that infected wild-type mice show complete destruction of the exocrine pancreas. In contrast, infected knockout mice and infected, cytokine-treated mice retain some acini. Presumably, the absence of IL-4 or the presence of IL-12 or IFN-γ decreases mortality associated with acute disease, by limiting the extent of destruction of the exocrine pancreas. Both IL-12 and IFN-γ are T helper 1 (TH1) type cytokines, while IL-4 T helper 2 (TH2) IL-4 is a T helper 2 (TH2) cytokine (Janeway et al. 2001). TH1 cytokines inhibit the production of TH2 cytokines and TH2 cytokines inhibit the production of TH1 cytokines. Since IL-12 and IFN-γ can ameliorate the severity of CVB4-V-induced acute disease, the data suggest that a predominant TH1 response participates in limiting the extent of destruction of the exocrine pancreas. A predominant TH1 response is also associated with the development of chronic disease. Data from the IL-4 knockout mice provide additional support for this idea. In the absence of IL-4, a TH1 response dominates, helping to limit the extent of pancreatic destruction in the acute phase of disease and promoting the development of chronic disease. The exact mechanism by which a TH1 response limits acinar cell destruction and promotes the development of chronic inflammatory disease is not yet understood.

## 5 Gene Expression Profiles During CVB4-Induced Pancreatitis

### 5.1 Global Gene Expression During CVB4-P and CVB4-V Infection

Since multiple factors contribute to the severity of acute pancreatitis, we sought to explore the molecular events occurring during mild and severe disease, by analyzing global gene expression in our model of CVB4-induced pancreatitis through the use of high density oligonucleotide arrays (Ostrowski et al. 2004). The experimental design consisted of three treatment groups, CVB4-P-infected, CVB4-V-infected, and mock-infected mice. The time interval chosen for analysis of pancreatic gene expression was 4-6 days after infection, because at this time point the repair process is underway in CVB4-P infection, while characteristic features of chronic disease (fibrosis, metaplasia, inflammation, tissue loss) are already apparent in CVB4-V infection. We identified 129 genes whose expression was highest in CVB4-P infection and 336 genes whose expression was highest in CVB4-V infection. The two groups of genes were sorted on the basis of function. Genes whose expression was highest in CVB4-P infection have roles in cell growth, embryogenesis, inhibiting apoptosis, angiogenesis, and immune responses. The increased expression of genes in these functional groups is certainly compatible with the repair process that is ongoing in the exocrine pancreas. While infected acinar cells undergo apoptosis at the peak of viral replication (2 days postinfection), apoptosis is already under control by 4-6 days after infection when CVB4-P titers are diminishing. For CVB4-V infection, the 336 genes were sorted into five functional groups: apoptosis, remodeling of the extracellular matrix, development of fibrosis, acinoductular metaplasia, and immune responses. Again, the increased expression of genes in these functional groups is compatible with the hallmarks of chronic pancreatitis, fibrosis, tissue loss, inflammation, and metaplasia.

### 5.2 Role of Macrophages in CVB4-Induced Pancreatitis

Given that the immune response to injury of the exocrine pancreas plays a key role in the pathogenesis of pancreatitis, we focused our attention on genes that are involved in the inflammatory response. A major player in the inflammatory response is the macrophage. Mononuclear phagocytes express specialized and polarized functional properties in response to cytokines and microbial products. In keeping with the TH1/TH2 nomenclature, polarized macrophages are referred to as M1 and M2 cells (Mantovani et al. 2005; Mantovani et al. 2002) and participate in innate and adaptive immune responses. Fully polarized M1 and M2 cells are the extremes of a continuum. As such, polarization of macrophage function constitutes an operationally useful conceptual framework to describe the plasticity of mononuclear phagocytes. In general, M1 cells (a) are efficient producers of effector molecules (reactive oxygen and nitrogen species) and inflammatory cytokines (IL-1b,

TNF-α, IL-6), (b) participate as inducers and effectors in polarized TH1 responses, and (c) mediate resistance against intracellular microbes. In contrast, M2 cells (a) produce low levels of proinflammatory cytokines, (b) promote the development of TH2 responses, and (c) participate in tissue repair. Furthermore, M2 cells are more prone to participate in immunoregulatory activities.

In evaluating the macrophage response to CVB4 infection, we used data from the microarray study (Ostrowski et al. 2004) and from multiplex cytokine studies (A. Ramsingh, unpublished data). We showed that the expression of *Ccl2*, *Ccl3*, *Ccl4*, and *Ccl5* is greater in CVB4-V than in CVB4-P infection. In addition, TNF-α and IL-6 expression is higher in CVB4-V than in CVB4-P infection. Since these cytokines and chemokines are indicative of an M1 phenotype, the data suggest that macrophage polarization is skewed toward M1 cells during CVB4-V infection. The diminished expression of these genes during CVB4-P infection indicates that macrophage polarization is skewed toward M2 cells. An additional observation is that the expression of both CCL2 and TNF-α is prolonged during CVB4-V infection but transient during CVB4-P infection, suggesting that the inflammatory response is sustained during CVB4-V infection but controlled during CVB4-P infection.

Although the data indicate that macrophage polarization is skewed toward M1 cells in CVB4-V infection and M2 cells in CVB4-P infection, it must be borne in mind that macrophage polarization is just a conceptual framework. As an example, another characteristic of M2 cells is that arginine metabolism is shifted to the production of L-ornithine, which can feed into pathways with entirely different outcomes. For example, L-ornithine can favor collagen biosynthesis, and hence fibrosis (Mills 2001). Alternatively, L-ornithine can favor the synthesis of polyamines, and hence cell proliferation. The enzyme responsible for converting L-arginine into L-ornithine is arginase, which has two isoforms, ARG1 and ARG2, both of which possess the same enzymatic activity. We observed that the expression of *Arg1* is higher during CVB4-V infection than during CVB4-P infection, while *Arg2* expression is restricted to CVB4-P infection. The assumption is that during CVB4-V infection, M1-type cells predominate, yet M2-type cells are also present. The inflammatory milieu during CVB4-V infection may cause M2 cells to participate in collagen synthesis, resulting in fibrosis. During CVB4-P infection, the assumption is that M2 type cells dominate, yet M1 type cells are also present. The inflammatory milieu during CVB4-P infection may cause M2 cells to favor polyamine synthesis, resulting in repair of the exocrine pancreas via acinar cell proliferation. The data highlight the intricate interactions between immune responses and the resolution of pancreatic injury, and between immune responses and the development of chronic inflammatory disease.

In pursuing the link between immune responses and the resolution of pancreatic injury, we focused on the expression of another gene, *Pap1* (or *Reg2*). The expression of *Pap1* is higher during CVB4-P infection than during CVB4-V infection, suggesting that *Pap1* expression is not linked to the severity of acute pancreatitis (Ostrowski et al. 2004). *Pap1* expression is upregulated in a variety of diseases, including acute pancreatitis, ulcerative colitis, human colorectal cancer, and hepatocellular carcinoma (Christa et al. 1999; Iovanna et al. 1993; Ogawa et al. 2003).

The physiological role of PAP1 in disease is unclear. Recent studies show that PAP1 has an anti-inflammatory effect. Infusion of anti-PAP1 antibodies into rats with acute pancreatitis increased the severity of disease (Vasseur et al. 2004). Pancreatitis was also exacerbated in a rat model, when an antisense PAP oligonucleotide was used to block the expression of PAP (Zhang et al. 2004). Recent work suggests that PAP1 inhibits the inflammatory response, by blocking NF-κB activation (Folch-Puy et al. 2006). PAP1 could be a major anti-inflammatory factor in epithelial cells. Given that the pancreas is prone to an exaggerated pro-inflammatory response when digestive enzymes are prematurely activated, the anti-inflammatory effect of PAP1 may function to contain the damage. The increased expression of *Pap1*, coupled with the dominance of an M2 response during CVB4-P infection, could conceivably contribute to the regulation of the inflammatory response.

## 5.3 NF-κB Activation During CVB4-Induced Pancreatitis

The scenario that is emerging from the microarray and multiplex cytokine studies is that inflammatory/immune responses during CVB4-P infection are self-limiting, while those responses are prolonged during CVB4-V infection. A key player in the inflammatory response and in the control of the inflammatory response is NF-κB. Is there a difference in NF-κB activation during CVB4-induced mild acute pancreatitis and during severe acute pancreatitis? The prediction is that NF-κB activation is greater during CVB4-V infection because of the sustained inflammatory response. NF-κB comprises a group of transcription factors that are activated by over 460 stimuli and that control the expression of more than 200 genes involved in a variety of physiological processes, particularly immunity and inflammation (Liu and Malik 2006). While this review presents only a brief overview of NF-κB activation, several recent reviews (Dolcet et al. 2005; Li et al. 2005; Liu and Malik 2006) provide in-depth coverage of NF-κB activation.

NF-κB transcription factors are dimers composed of various combinations of members of the NF-κB/Rel family (Dolcet et al. 2005; Li et al. 2005; Liu and Malik 2006). The activities of NF-κB are regulated by the IκB family of proteins. IκBs inhibit NF-κB activation via three mechanisms: sequestration of NF-κB dimers in the cytoplasm, facilitation of the dissociation of NF-κB dimers from DNA, and export of NF-κB dimers from the nucleus. A central regulator of NF-κB activity is the IκB kinase (IKK) complex. NF-κB activators activate various signal transduction pathways that ultimately lead to activation of IKK, a step that causes the phosphorylation of IκB proteins. Upon phosphorylation, the IκB proteins become ubiquinated and degraded, resulting in the release of NF-κB. NF-κB then translocates to the nucleus and activates NF-κB motif-containing promoters. NF-κB regulates the expression of myriad genes, including those encoding cytokines (IL-1b, IL-6, TNF-$\alpha$, GM-CSF), chemokines (IL-8, RANTES, MIP-1$\alpha$, MCP-1), pro-angiogenic factors (VEGF), adhesion molecules

(VCAM-1, ICAM-1, E-selectin), anti-apoptotic proteins (BCL2, BCL-X, cIAP, FLIP), and inducible enzymes (iNOS, COX-2, MMP9), and those promoting proliferation (cyclins, c-MYC).

Data from the microarray study (Ostrowski et al. 2004) show that the expression of several genes, including *Il-1b*, *Ccl5* (*RANTES*), *Ccl3* (*Mip-1α*), *Vcam-1*, *Icam-1*, and cyclin D1, is greater during CVB4-V infection than during CVB4-P infection. Data from multiplex cytokine studies (A. Ramsingh, unpublished data) show that IL-6, TNF-α, and CCL2 (MCP-1) are present at higher levels in CVB4-V infection than in CVB4-P infection. The combined data suggest that NF-κB activation is greater during severe acute disease than during mild acute disease. NF-κB also activates several negative feedback mechanisms (including increased expression of IκB) that ultimately control the inflammatory response. Failure to control the inflammatory response leads to chronic inflammation. The prolonged expression of MCP-1 and TNF-α during CVB4-V infection suggests that the development of chronic pancreatitis may be due, in part, to a failure in the control of NF-κB activation. The scenario that is emerging from these studies is that CVB4-P-induced mild acute pancreatitis is associated with a backdrop of immunoregulatory activities, involving a predominant M2 macrophage response, high expression of *Pap1*, and a controlled NF-κB response, against which tissue repair is occurring (Fig. 2). Cross-talk between controlled inflammatory/immune responses and the processes involved in cell growth and development leads to repair of the exocrine pancreas. On the other hand, CVB4-V-induced severe acute pancreatitis is associated with a backdrop of immune dysregulation, involving a predominant M1 macrophage response, low expression of *Pap1*, and sustained NF-κB activation, against which tissue damage is occurring (Fig. 2). Cross-talk between dysregulated inflammatory/immune responses and the processes involved in tissue destruction sets the stage for the development of chronic inflammatory disease. Because the inflammatory/immune response is intimately linked to the processes of tissue repair and tissue destruction, we need to understand (a) the controls that determine a well-regulated immune response and (b) the factors that lead to dysregulated immune responses. As we unravel the complex processes contributing to acute pancreatitis, the information gained will help us to design therapeutics that will prevent progression to chronic inflammatory disease and that will restore the functionality of affected tissues.

## 6 Chronic Inflammation and Cancer

### *6.1 Chronic Pancreatitis and Pancreatic Cancer*

Although the link between chronic inflammation and cancer was made more than 100 years ago by Rudolf Virchow (Balkwill and Mantovani 2001), only within the last decade has this view gained significant support. Several cancers are associated

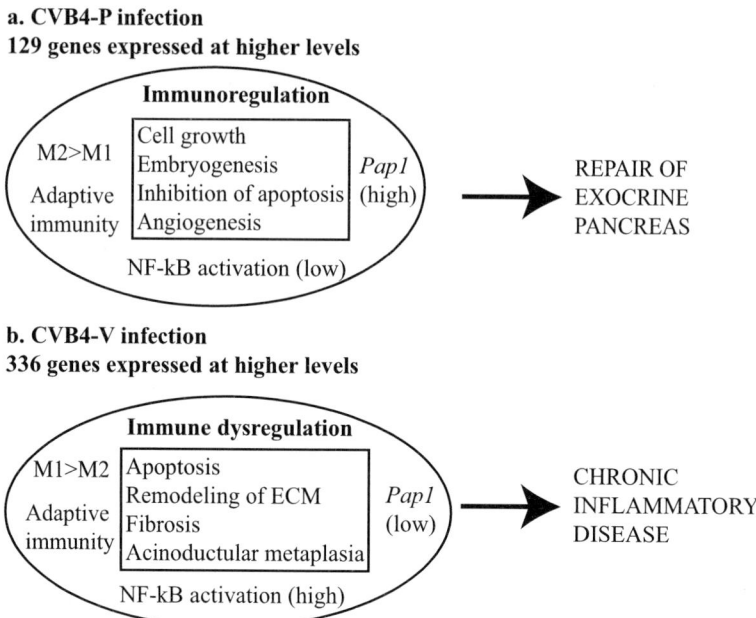

**Fig. 2** Models highlighting the cross-talk between inflammatory/immune responses and processes involved in tissue repair and tissue destruction. **a** CVB4-P infection is associated with immunoregulatory activities involving a predominant M2 macrophage response, high expression of *Pap1*, and a controlled NF-κB response against which tissue repair occurs. **b** CVB4-V infection is associated with immune dysregulation involving a predominant M1 macrophage response, low expression of *Pap1*, and sustained NF-κB activation against which tissue damage occurs

with chronic inflammation, including lung, esophageal, gastric, cervical, bladder, prostate, colorectal, and pancreatic cancers (Vakkila and Lotze 2004). Chronic pancreatitis is a consistent risk factor for pancreatic cancer, which has the worst prognosis of all cancers. The median survival is less than 6 months and the 5-year survival is only 3%-5% (Hezel et al. 2006). The pathway between inflammation and cancer involves multiple steps and converging risk factors. Pancreatic cancer is usually adenocarcinoma, a genetic disorder of pancreatic parenchymal cells in which there is progressive accumulation of mutations that parallel metaplasia and dysplasia. Mutations occur in *KRAS2*, then in *Id-1/Id-2*, *p53*, cyclin D1, and *p16/CDKN2A*, and later in *DPC4/MADH4* and *BRCA2* (Whitcomb 2004).

How does chronic inflammation accelerate the oncogenic process? A chronic inflammatory environment may accelerate the oncogenic process by promoting DNA damage and cellular proliferation (De Visser et al. 2006; Li et al. 2005). The mechanisms underlying the resolution of inflammation are not well understood. Cells from both the innate and adaptive arms of the immune system are found in the chronic inflammatory milieu and may contribute to oncogenesis. Cells involved in innate immune responses (e.g., macrophages) can generate free radicals that induce DNA damage. The accumulation of DNA damage in dormant premalignant cells may push these cells over the threshold into malignancy. Cells involved in innate immune responses can promote angiogenesis and tissue remodeling via the production of cytokines, chemokines, and matrix metalloproteinases. As a result, pro-inflammatory mediators secreted by leukocytes or stromal cells favor growth of malignant cells, resulting in tumor progression and metastasis. Failure to control the inflammatory response thus leads to chronic inflammation and a milieu that favors tumorigenesis.

## 6.2 Gene Expression Profiles in Pancreatic Cancer

Gene expression technologies are being used to identify genes that are differentially expressed between neoplastic tissues and their normal counterparts. The expectation is that differentially expressed genes will form the basis of novel diagnostic markers or therapeutic targets. An important consideration in studies of gene expression in pancreatic adenocarcinoma is the high desmoplastic reaction that occurs in these tumors (Logsdon et al. 2003; Nakamura et al. 2004). A typical tumor exists as a solid mass containing neoplastic cells, surrounded by a dense fibrous stroma that contains proliferating fibroblasts, stellate cells, small endothelial cell-lined vessels, inflammatory cells, and residual parenchymal components of the pancreas. Because of this, comparisons between adenocarcinomas and normal pancreas fail to account for the contribution of the stromal elements. Two approaches have been used to overcome this problem. One approach compares gene profiles of adenocarcinomas and chronic pancreatitis since lesions with abundant stroma are also present in chronic pancreatitis (Iacobuzio-Donahue et al. 2003; Logsdon et al. 2003). The other approach compares gene profiles of pure populations of pancreatic cancer cells, obtained by laser microbeam dissection, and normal pancreatic ductal cells (Nakamura et al. 2004). These studies identified genes that are differentially expressed in pancreatic cancer.

Since chronic pancreatitis is a risk factor for the development of pancreatic cancer and since chronic disease can develop from acute disease, we envision a continuum in which disease progresses from an acute phase to a chronic phase and eventually to cancer (given the appropriate genetic background). The prediction is that genes involved in disease progression would be expressed in all phases of disease but at varying levels. To investigate this possibility, we generated a list of genes whose increased expression was associated with pancreatic cancer, by compiling the results of several studies (Iacobuzio-Donahue et al. 2003; Logsdon et al. 2003; Nakamura et al. 2004). We then compared the compiled data with our list of genes whose expression increased during CVB4-P or CVB4-V infection (Ostrowski et al. 2004) to identify common genes (Table 1). This analysis was not intended to generate an exhaustive list of common genes, since oligonucleotide arrays probing different gene sets were used in the studies of pancreatic cancer, and since our

**Table 1** Genes whose expression is elevated in CVB4–induced acute pancreatitis, CVB4–induced chronic pancreatitis, and pancreatic cancer

| Gene | Function |
|---|---|
| **CVB4–V>CVB4–P** | |
| **ECM** | |
| Secreted protein, acidic, cysteine–rich | Inhibits proliferation, increases ECM |
| Lectin, galactoside–binding, soluble, 3 binding protein | Growth regulation, cell adhesion, metaplasia |
| A disintegrin and metalloproteinase domain 8 | Cell adhesion, signaling |
| Matrix metalloproteinase 7 | Remodeling ECM, acinoductular metaplasia |
| Cofilin 1 (non–muscle) | Actin binding, regulates actin filament turnover |
| Collagen, type I, alpha 1 | Fibrosis |
| Collagen, type I, alpha 2 | Fibrosis |
| Collagen, type III, alpha 1 | Fibrosis |
| Collagen, type VI, alpha 1 | Fibrosis |
| **Cell growth** | |
| Development and differentiation enhancing factor 1 | Cell differentiation |
| Granulin | Growth factor, mitogenic for keratinocytes and fibroblasts |
| Cyclin D1 | Cell cycle control |
| BCL2–interacting killer | Pro–apoptotic |
| Amyloid beta (A4) precursor protein | Cell proliferation |
| **Inflammatory/Immune** | |
| Heme oxygenase (decycling) 1 | Stress protein produced by inflammatory cells |
| Integrin, beta 2 | Neutrophil activation |
| Lymphocyte cytosolic protein 1 | Increased expression in IL–2–stimulated T cells |
| **Cellular processes** | |
| S100 calcium–binding protein A4 | Signal transduction |
| S100 calcium–binding protein A6 | Signal transduction |
| S100 calcium–binding protein A10 | Signal transduction |
| Phosphofructokinase, platelet | Glucose metabolism |
| Pyruvate kinase | Glucose metabolism |
| ATPase, Na$^+$/K$^+$ transporting, beta 3 polypeptide | Transport |
| **CVB4–P>CVB4–V** | |
| Claudin 3 | Tight junctions, adhesion |
| Claudin 4 | Tight junctions, adhesion |
| **CVB4–P specific** | |
| Adenylate kinase 3 | Energy metabolism |
| Proteasome subunit, beta type, 3 | Protein degradation |
| RHO GDP dissociation inhibitor beta | Signal transduction |

We complied genes whose expression increased during pancreatic cancer from several studies (Iacobuzio–Donahue et al. 2003; Logsdon et al. 2003; Nakamura et al. 2004) and then compared them to genes whose expression increased during CVB4 infection (Ostrowski et al. 2004), to identify common genes

comparison was done using data from both mouse and human oligonucleotide arrays. We identified 28 common genes: 23 of which showed higher expression in CVB4-V than in CVB4-P infection, two of which showed higher expression in CVB4-P than in CVB4-V infection, and three that were specifically upregulated in CVB4-P infection. The 23 genes whose expression is higher during CVB4-V than in CVB4-P infection function in remodeling of the extracellular matrix, development of fibrosis, metaplasia, cell growth, and inflammatory/immune responses, all processes that are associated, to varying degrees, with acute pancreatitis, chronic pancreatitis, and pancreatic cancer. The quantitative changes in gene expression raise questions regarding the control of gene expression. What controls the expression of genes common to acute pancreatitis, chronic pancreatitis, and pancreatic cancer? Which factors have roles in controlling the level of gene expression? Can gene expression be modulated? Can we identify therapeutic targets for chronic pancreatitis and pancreatic cancer?

## 7 Summary

We have developed models of CVB4-induced acute and chronic pancreatitis that resemble the human diseases and have begun to elucidate the mechanisms by which CVB4 causes acute and chronic inflammatory diseases. We have found that the viral genotype determines the severity of acute pancreatitis, while host determinants influence the outcome of infection. We have also found that severe acute pancreatitis can progress to chronic disease and that several cytokines (IL-4, IL-12, and IFN-$\gamma$) influence the development of chronic pancreatitis. The scenario that is emerging from gene profiling and multiplex cytokine studies is that CVB4-P-induced mild acute pancreatitis is associated with a backdrop of immunoregulatory activities, against which tissue repair is occurring. Cross-talk between regulated inflammatory/immune responses and the processes involved in cell growth and development leads to repair of the exocrine pancreas. On the other hand, CVB4-V-induced severe pancreatitis is associated with a backdrop of immune dysregulation, against which tissue damage is occurring. Cross-talk between dysregulated inflammatory/immune responses and the processes involved in tissue destruction sets the stage for the development of chronic inflammatory disease. The link between inflammatory/immune responses and the processes of tissue repair and tissue destruction highlights the need to understand the controls that govern well-regulated immune responses and the breakdown of such controls that leads to dysregulated immune responses. An understanding of these controls may provide new therapeutic targets for the treatment of chronic pancreatitis and pancreatic cancer.

**Acknowledgements** This work was supported, in part, by funding from the National Institutes of Health and the American Heart Association.

# References

Arnesjo B, Eden T, Ihse I, Nordenfelt E, Ursing B (1976) Enterovirus infections in acute pancreatitis-a possible etiological connection. Scand J Gastroenterol 11:645-649

Balkwill F, Mantovani A (2001) Inflammation and cancer: back to Virchow? Lancet 357:539-545

Bhatia M (2004) Apoptosis of pancreatic acinar cells in acute pancreatitis: is it good or bad? J Cell Mol Med 8:402-409

Caggana M, Chan P, Ramsingh A (1993) Identification of a single amino acid residue in the capsid protein VP1 of coxsackievirus B4 that determines the virulent phenotype. J Virol 67:4797-4803

Capner P, Lendrum R, Jeffries DJ, Walker G (1975) Viral antibody studies in pancreatic disease. Gut 16:866-870

Christa L, Simon MT, Brezault-Bonnet C, Bonte E, Carnot F, Zylberberg H, Franco D, Capron F, Roskams T, Brechot C (1999) Hepatocarcinoma-intestine-pancreas/pancreatic associated protein (HIP/PAP) is expressed and secreted by proliferating ductules as well as by hepatocarcinoma and cholangiocarcinoma cells. Am J Pathol 155:1525-1533

Cotran RS, Kumar V, Robbins SL (1989) Robbins pathologic basis of disease. WB Saunders Company, Philadelphia

Dalldorf G, Gifford R (1952) Adaptation of group B Coxsackie virus to adult mouse pancreas. J Exp Med 96:491-497

De Visser KE, Eichten A, Coussens LM (2006) Paradoxical roles of the immune system during cancer development. Nat Rev Cancer 6:24-37

Dolcet X, Llobet D, Pallares J, Matias-Guiu X (2005) NF-kB in development and progression of human cancer. Virchows Arch 446:475-482

Elsasser HP, Lutcke H, Kern HF (1986) Acinar and duct cell replication and regeneration. In: Go VLW, Gardner JD, Brooks FP, Lebenthal E, DiMagno EP, Scheele GA (eds) The exocrine pancreas: biology, pathobiology, and diseases. Raven Press, New York, pp 45-53

Folch-Puy E, Granell S, Dagorn JC, Iovanna JL, Closa D (2006) Pancreatitis-associated protein I suppresses NF-kappa B activation through a JAK/STAT-mediated mechanism in epithelial cells. J Immunol 176:3774-3779

Gifford R, Dalldorf G (1951) The morbid anatomy of experimental Coxsackie virus infection. Am J Pathol 27:1047-1063

Gukovskaya AS, Perkins P, Zaninovic V, Sandoval D, Rutherford R, Fitzsimmons T, Pandol SJ, Poucell-Hatton S (1996) Mechanisms of cell death after pancreatic duct obstruction in the opossum and the rat. Gastroenterology 110:875-884

Halim S, Ramsingh AI (2000) A point mutation in VP1 of coxsackievirus B4 alters antigenicity. Virology 269:86-94

Hezel AF, Kimmelman AC, Stanger BZ, Bardeesy N, Depinho RA (2006) Genetics and biology of pancreatic ductal adenocarcinoma. Genes Dev 20:1218-1249

Huber S, Ramsingh AI (2004) Coxsackievirus-induced pancreatitis. Viral Immunol 17:358-369

Iacobuzio-Donahue CA, Ashfaq R, Maitra A, Adsay NV, Shen-Ong GL, Berg K, Hollingsworth MA, Cameron JL, Yeo CJ, Kern SE, Goggins M, Hruban RH (2003) Highly expressed genes in pancreatic ductal adenocarcinomas: a comprehensive characterization and comparison of the transcription profiles obtained from three major technologies. Cancer Res 63:8614-8622

Iovanna JL, Keim V, Bosshard A, Orelle B, Frigerio JM, Dusetti N, Dagorn JC (1993) PAP, a pancreatic secretory protein induced during acute pancreatitis, is expressed in rat intestine. Am J Physiol 265:G611-G618

Janeway CA, Travers P, Walport M, Shlomchik MJ (2001) Immunobiology: the immune system in health and disease. Garland Publishing, New York

Kaiser AM, Saluja AK, Sengupta A, Saluja M, Steer ML (1995) Relationship between severity, necrosis, and apoptosis in five models of experimental acute pancreatitis. Am J Physiol 269: C1295-C1304

Kibrick S, Benirschke K (1958) Severe generalized disease (encephalohepatomyocarditis) occurring in the newborn period and due to infection with Coxsackie virus, group B; evidence of intrauterine infection with this agent. Pediatrics 22:857-875

Kingsnorth A, O'Reilly D (2006) Acute pancreatitis. BMJ 332:1072-1076

Lankisch PG, Schirren CA, Kunze E (1991) Undetected fatal acute pancreatitis: why is the disease so frequently overlooked? Am J Gastroenterol 86:322-326

Lankisch PG, Burchard-Reckert S, Petersen M, Lehnick D, Schirren CA, Kohler H, Stockmann F, Peiper HJ, Creutzfeldt W (1996) Morbidity and mortality in 602 patients with acute pancreatitis seen between the years 1980-1994. Z Gastroenterol 34:371-377

Li Q, Withoff S, Verma IM (2005) Inflammation-associated cancer: NF-kappaB is the lynchpin. Trends Immunol 26:318-325

Liu SF, Malik AB (2006) NF-kappa B activation as a pathological mechanism of septic shock and inflammation. Am J Physiol Lung Cell Mol Physiol 290:L622-L645

Logsdon CD, Simeone DM, Binkley C, Arumugam T, Greenson JK, Giordano TJ, Misek DE, Kuick R, Hanash S (2003) Molecular profiling of pancreatic adenocarcinoma and chronic pancreatitis identifies multiple genes differentially regulated in pancreatic cancer. Cancer Res 63:2649-2657

Majai G, Petrovski G, Fesus L (2006) Inflammation and the apopto-phagocytic system. Immunol Lett 104:94-101

Mantovani A, Sozzani S, Locati M, Allavena P, Sica A (2002) Macrophage polarization: tumor-associated macrophages as a paradigm for polarized M2 mononuclear phagocytes. Trends Immunol 23:549-555

Mantovani A, Sica A, Locati M (2005) Macrophage polarization comes of age. Immunity 23:344-346

Melnick JL (1996) Enteroviruses: polioviruses, coxsackieviruses, echoviruses, and newer enteroviruses. In: Fields BN, Knipe DM, Howley PM, et al (eds) Fields virology. Lippincott-Raven, Philadelphia, pp 655-705

Melnick JL, Shaw EW, Curnen EC (1949) A virus isolated from patients diagnosed as non-paralytic poliomyelitis or aseptic meningitis. Proc Soc Exp Biol Med 71:344-349

Mergener K, Baillie J (1997) Chronic pancreatitis. Lancet 350:1379-1385

Mills CD (2001) Macrophage arginine metabolism to ornithine/urea or nitric oxide/citrulline: a life or death issue. Crit Rev Immunol 21:399-425

Nakamura T, Furukawa Y, Nakagawa H, Tsunoda T, Ohigashi H, Murata K, Ishikawa O, Ohgaki K, Kashimura N, Miyamoto M, Hirano S, Kondo S, Katoh H, Nakamura Y, Katagiri T (2004) Genome-wide cDNA microarray analysis of gene expression profiles in pancreatic cancers using populations of tumor cells and normal ductal epithelial cells selected for purity by laser microdissection. Oncogene 23:2385-2400

Ogawa H, Fukushima K, Naito H, Funayama Y, Unno M, Takahashi K, Kitayama T, Matsuno S, Ohtani H, Takasawa S, Okamoto H, Sasaki I (2003) Increased expression of HIP/PAP and regenerating gene III in human inflammatory bowel disease and a murine bacterial reconstitution model. Inflamm Bowel Dis 9:162-170

Ostrowski SE, Reilly AA, Collins DN, Ramsingh AI (2004) Progression or resolution of coxsackievirus B4-Induced pancreatitis-a genomic analysis. J Virol 78:8229-8237

Ozsvar Z, Deak J, Pap A (1992) Possible role of Coxsackie-B virus infection in pancreatitis. Int J Pancreatol 11:105-108

Pappenheimer AM, Kunz LJ, Richardson S (1951) Passage of Coxsackie virus (Connecticut-5 strain) in adult mice with production of pancreatic disease. J Exp Med 94:45-64

Potvin DM, Metzger DW, Lee WT, Collins DN, Ramsingh AI (2003) Exogenous interleukin-12 protects against lethal infection with coxsackievirus B4. J Virol 77:8272-8279

Ramsingh AI, Lee WT, Collins DN, Armstrong LE (1997) Differential recruitment of B and T cells in coxsackievirus B4-induced pancreatitis is influenced by a capsid protein. J Virol 71:8690-8697

Ramsingh AI, Lee WT, Collins DN, Armstrong LE (1999) T cells contribute to disease severity during coxsackievirus B4 infection. J Virol 73:3080-3086

Ross ME, Hayashi K, Notkins AL (1974) Virus-induced pancreatic disease: alterations in concentration of glucose and amylase in blood. J Infect Dis 129:669-676

Slack JM (1995) Developmental biology of the pancreas. Development 121:1569-1580

Stevens T, Conwell DL, Zuccaro G (2004) Pathogenesis of chronic pancreatitis: an evidence-based review of past theories and recent developments. Am J Gastroenterol 99:2256-2270

Vakkila J, Lotze MT (2004) Inflammation and necrosis promote tumour growth. Nat Rev Immunol 4:641-648

Vasseur S, Folch-Puy E, Hlouschek V, Garcia S, Fiedler F, Lerch MM, Dagorn JC, Closa D, Iovanna JL (2004) p8 improves pancreatic response to acute pancreatitis by enhancing the expression of the anti-inflammatory protein pancreatitis-associated protein I. J Biol Chem 279:7199-7207

Whitcomb DC (2004) Inflammation and cancer V. Chronic pancreatitis and pancreatic cancer. Am J Physiol Gastrointest Liver Physiol 287:G315-G319

Whitcomb DC (2006) Clinical practice. Acute pancreatitis. N Engl J Med 354:2142-2150

Zhang H, Kandil E, Lin YY, Levi G, Zenilman ME (2004) Targeted inhibition of gene expression of pancreatitis-associated proteins exacerbates the severity of acute pancreatitis in rats. Scand J Gastroenterol 39:870-881

# The CVB and Etiology of Type 1 Diabetes

K. M. Drescher(✉) and S. M. Tracy

| | | |
|---|---|---|
| 1 | Introduction | 259 |
| 2 | Autoimmunity: Hygiene's Dirty Little Secret? | 260 |
| 3 | A Case for Viruses in T1D Onset | 261 |
| 4 | Modeling T1D in Mice | 263 |
| 5 | T1D Development in Humans: A Series of Unfortunate Events | 264 |
| | 5.1 Impact of Virus Dose on T1D Onset | 264 |
| | 5.2 Impact of Virus Strain on T1D Onset | 265 |
| | 5.3 Impact of Microenvironment on Ability of CVB to Initiate T1D in the Host | 266 |
| | 5.4 Final Thoughts | 268 |
| References | | 270 |

**Abstract** The origins of type 1 diabetes (T1D) are largely unknown. Fewer than 50% of the cases of the disease are attributable to host genetics, indicating that environmental factors are involved in disease development. The most often cited environmental agents implicated as initiators of T1D are the human enteroviruses, in particular the group B coxsackieviruses (CVB). Although the connection between the CVB and T1D has not been firmly established, significant evidence supports the role of these pathogens in T1D development.

## 1 Introduction

Type 1 diabetes (T1D; also referred to as juvenile diabetes or insulin-dependent diabetes mellitus) is caused by autoimmune destruction of insulin-producing beta cells in the pancreatic islets of Langerhans, ultimately resulting in the inability of

K. M. Drescher
Department of Medical Microbiology and Immunology, Creighton University
School of Medicine, Omaha, NE 68178, USA
kdresche@creighton.edu

the affected individual to regulate blood glucose levels. Although T1D has been recognized for centuries (Sanders 2002), the etiology of the disease remains elusive. Host genetics clearly impacts disease development, and certain HLA-DR and DQ alleles have been associated with increased risk for disease (Dorman and Bunker 2000). Despite the role of host genetics in T1D development, fewer than 50% of the cases can be attributed to genetics alone (Haller et al. 2005). Concordance rates of T1D between identical twins are 30%-40%, suggesting that nongenetic environmental factors also impact disease onset (Redondo et al. 2001; Lo et al. 1991; Barnett et al. 1981). Similar to what is observed in asthma and allergy, T1D rates are increasing worldwide (Onkamo et al. 1999; Karvonen et al. 1993, 1997; Pitkaniemi et al. 2004). Higher incidences of T1D are generally clustered in highly industrialized countries and in urban versus rural regions (Karvonen et al. 1993; Karvonen et al. 1997; Green et al. 2001). Due to the physical, financial, social, and emotional impact of T1D on the affected individuals and their families, significant effort has been made to identify factors that initiate disease development. Further, because of the increased incidence of T1D in industrialized countries, studies have focused on examining parameters that are over- or under-represented in highly vs less developed countries. Key environmental factors that have been investigated over the years include toxin exposure (Brenden et al. 2001), diet (Birgisfottir et al. 2006; Wahlberg et al. 2006; Centers for Disease Control 2006a), and infectious diseases (Otonkoski et al. 2000; Williams et al. 1900; Diaz-Horta et al. 2001; Cabrera-Rode et al. 2003; Smith et al. 1998; Yoon et al. 1979; Cabrera-Rode et al. 2005).

## 2 Autoimmunity: Hygiene's Dirty Little Secret?

In the past century, focus was increasingly placed on the biological cleanliness of society-better water quality and advances in waste processing systems are common in highly industrialized societies. Individuals are no longer as commonly exposed to the morass of pathogens that were routinely spread between individuals via the oral-fecal transmission route as in the past. Vaccination campaigns aimed at common childhood infections have further prevented regular disease outbreaks even in immunologically naïve persons through the establishment of herd immunity that depresses circulating pathogens in the population in general. Even factors such as fewer children per household have contributed to reduced exposure of individuals to infectious agents. Over several generations, these elements have resulted in alterations in the pattern of exposure to many potential pathogens in a large proportion of society, particularly in highly developed countries compared to previous generations. Whereas a person was in the past regularly exposed to certain disease agents starting at birth, the age of first exposure to many viruses may be much higher now than in the past. For example, prior to widespread measles vaccination of children in developed countries, most individuals experienced infections with the measles virus before the age of 15 (Centers for Disease Control 2006a). Now, it is not unusual to hear of outbreaks of common childhood infections such as measles or chickenpox occurring in nonimmune adults (Kumar et al. 2005; Rawson et al. 2001; Iowa Measles Response Team 2005). A similar story also developed for the emergence of the annual poliovirus-caused poliomyelitis epidemics of the twentieth century (Nathanson and Martin 1979; Paul 1971)

A lack or diminished frequencies of exposure to a wide array of infectious agents-both bacteria and viruses-was postulated to be one of the root causes of increased rates of asthma and allergy is highly developed countries and has been termed the hygiene hypothesis (Strachan 1989; Bunimovich-Mendrazitsky and Stone 2005). The hypothesis states that there is an increased rate of asthma in individuals whose immune systems are not exposed to a large number of infectious agents early in life. Data in support of the hygiene hypothesis includes the finding that children in daycare and children with larger numbers of siblings are less prone to the development of asthma than children who did not attend daycare or children without siblings (Strachan 1989, 2000; Lemanske 2004). The basis of the hypothesis lies in the concept that the adaptive immune system is shaped early in life and that in order to achieve an ideal immune status, the immune system must be stimulated by foreign agents (Strachan 1989). Failure to provide the immune system with appropriate targets upon which to act (that is, an array of bacteria, viruses, and parasites) predisposes the host to the development of inappropriate immune reactions to either self or environmental constituents (that is, autoimmune disease). Paraphrasing an old expression, clean hands are the devil's workshop. While the hygiene hypothesis was originally invoked to account for the increases in asthma and allergic diseases (Strachan 1989; Bunimovich-Mendrazitsky and Stone 2005), it has also been suggested as a mechanism underlying increased rates of T1D (Kolb and Elliott 1994), rheumatoid arthritis, multiple sclerosis and inflammatory bowel disease, indicating a common mechanism for the development of inappropriate immune responses aimed at self (Bach 2001). The hygiene hypothesis does not disregard the importance of other factors (for example, host genetics) in the development of autoimmune diseases.

## 3 A Case for Viruses in T1D Onset

Epidemiological evidence has been used to implicate infectious agents, particularly viruses, as triggers of autoimmune diseases including multiple sclerosis (Monteyne et al. 1998), rheumatoid arthritis (Nishioka et al. 1993), and systemic lupus erythematosus (James and Harley 1998). Similarly, the rationale implicating viruses as initiators of T1D is largely founded on epidemiological correlations (Szopa et al. 1993; Jenson and Rosenberg 1984; Gamble 1976). The leading candidates for an infectious trigger of T1D are the human enteroviruses (HEVs), in particular the group B coxsackieviruses (CVBs).

The human enteroviruses (HEVs) are ubiquitous and circulate commonly (Palacios and Oberste 2005; Centers for Disease Control 2000a, 2006b). The HEVs are causative agents of serious diseases such as poliomyelitis, myocarditis, and aseptic meningitis (Pallansch and Roos 2001), but despite such associations, the majority of HEVs cause mild or subclinical disease in humans that often go unnoticed or are self-limiting. Rarely do illnesses caused by HEVs proceed to the level of identification of the serotype of the virus.

Several lines of evidence support a role for HEVs in T1D onset in humans. Seasonal fluctuations in T1D onset rates in humans are reported, a situation that is predicted based on the seasonal variations in the circulation of viruses in the wild (Tani et al. 1995; Mikulecky et al. 2000; Mikulecky and Michalkova 2001). There have been case reports describing instances of sudden onset T1D in individuals following viral infections (Gamble 1976; Hyoty 2002). Certainly, not all individuals exposed to HEVs are candidates for T1D development; were that the case, T1D would be far more common given the common circulations of HEVs every year. Rather, a series of events must occur simultaneously in order to trigger T1D onset: virus dose, virus strain, host genetics, and the pancreatic microenvironment are all likely contributors to disease onset. All of these factors (and likely some additional ones) must be right in order for the host to develop T1D-it is, with all due respect to Lemony Snicket, a series of unfortunate events.

Because the pancreas is difficult to biopsy, this procedure is not performed frequently; as a result, it is difficult to directly establish whether the pancreas is infected with either HEVs or other infectious agents. A small number of studies have isolated virus following T1D onset in the patient (e.g., Otonkoski et al. 2000; Smith et al. 1998; Yoon et al. 1979). Other non-CVB HEVs have also been isolated in association with T1D onset [all echoviruses classified, like CVB, in the HEV-B species (Otonkoski et al. 2000; Williams et al. 1900; Diaz-Horta et al. 2001; Cabrera-Rode et al. 2003, 2005; Gamble 1976; Paananen et al. 2003)]. Additionally, analysis of sera from newly diagnosed diabetics has revealed increased levels of antienterovirus neutralizing antibodies as compared with controls (Frisk and Tuvemo 2004). Further supporting a role for an infectious etiology in T1D, interferon-$\alpha$, a key mediator in the antiviral immune response, was detected in one study in the plasma of 70% of newly diagnosed diabetics, suggesting that these patients recently experienced viral infections (Chehadeh et al. 2000). There is further an inverse correlation between the incidence of anti-HEV antibodies in the sera of women of childbearing age and T1D rates (Viskari et al. 2005; Atkinson and Leiter 1999). Despite the associations of the HEVs, and CVBs in particular, with T1D onset, there are also reports refuting the association of CVBs as T1D triggers (e.g., Fuchtenbusch et al. 2001; Goldacre et al. 2005; Hierholzer and Farris 1974).

Each of the T1D-linked HEV (Otonkoski et al. 2000; Williams et al. 1900; Diaz-Horta et al. 2001; Cabrera-Rode et al. 2003, 2005; Yoon et al. 1979; Paananen et al. 2003; Maria et al. 2005; Dotta et al. 2007; Andreoletti et al. 1997) is an HEV-B; no HEV belonging to species A, C, or D has been isolated to date in association with T1D onset. In fairness, it can be said that an insufficient number of isolations has occurred to clearly implicate the HEV-B as the primary, perhaps only, HEV group responsible for HEV-induced T1D onset. As well, the HEV-B are commonly identified in annual circulations (Centers for Disease Control 2000a, b, 2006b).

Even poorly conducted studies support this: a recent report (Dotta et al. 2007) from Italy described a CVB4 strain associated with several recent T1D onset cases. While the pathology results soundly indicated HEV protein detectable in patient islets, the authors' sequence data of the viral genome clearly indicated a laboratory contamination. Although the prototype CVB4 strain (JVB Benschoten) was isolated in 1951 in New York, they reported the Italian Tuscany strain shows 1% or less variation at both nucleotide and amino acid levels from the 55-year-old ancestor strain (or, 99% identity with the now extinct 1951 genome). This observation is therefore extremely suspect, for sequence space is incredibly vast and HEV isolates do not remain in stasis but evolve rapidly because of diverse pressures (Domingo et al. 2006; Domingo et al. 1997; Romero et al. 1997). This can be easily and rapidly assessed using widely available Genbank sequences: for example, two CVB2 strains, isolated in the *same* year (2004) in Korea, differ by 7% between themselves, or two strains of CVB5 isolated *just 2 years apart* (1952 and 1954) vary 19% between themselves. It is a pity from a virology standpoint that this strain was not isolated but the clinical results nonetheless indicated an HEV infection was present and perhaps of greatest interest, three of six studied diabetics were positive for HEV in situ, suggesting that-at least in this study-as many as 50% of T1D cases can be associated with an HEV infection. Clearly, a take-home lesson from these various sources is that HEV are indeed associated with T1D onset in at least some cases, but it is not clear *which* HEV are the primary instigators. One may postulate at present that (1) only HEV-B strains are involved, based on available data, or that (2) other HEVs also play a role but much less often. It must also be said that it is not clear how many clinical isolates would be required to convince the community to support one hypothesis and deny the other.

If HEV are initiators of sudden onset T1D, it would be reasonable to postulate that there would be reports of outbreaks of T1D in a defined geographic region (particularly in populations with a genetic predisposition for the disease) in humans following HEV circulations. The lack of documentation regularly describing clusters of sudden onset T1D within geographic areas experiencing HEV outbreaks could be explained by one or a combination of the following: (1) the strain(s) of CVB that induces T1D is relatively rare and does not circulate with regularity (or arise commonly) in humans, (2) the natural dose of CVB that humans are exposed to during outbreaks of HEV is insufficient to cause T1D in most humans, (3) the majority of the human population is refractory to sudden onset T1D at any given time. This review will examine each of these scenarios using data generated primarily in mouse models of type 1 diabetes.

## 4 Modeling T1D in Mice

One of the most extensively characterized animal models used to study T1D is the nonobese diabetic (NOD) mouse (Atkinson and Leiter 1999). The NOD mouse is genetically predisposed to develop T1D, and by 8 weeks of age, the islets of NOD mice become infiltrated with inflammatory cells (i.e., insulitis). Beta cell-killing insulitis becomes increasingly extensive as the animals age. Most islets are extensively infiltrated by 12-15 weeks of age, and NOD mice begin to become diabetic as defined by elevated glucose levels in the blood and urine. By about 30-35 weeks of age, 70%-100% of female NOD mice are diabetic (male NOD mice have a lower T1D incidence than female mice, and thus are seldom used to model T1D). Host genetics contributes strongly to T1D development in this mode, as mice readily develop the disease when maintained in pristine environments; approximately 20 loci have been identified as being involved in diabetes development (Anderson and Bluestone 2005).

The NOD mouse is exquisitely sensitive to perturbations of the animal's immune system. Over 100 methods have been employed to alter the incidence of T1D in the female NOD mouse (Anderson and Bluestone 2005; Shoda et al. 2005). Among the successful methods utilized to impact T1D onset rates in NOD mice include manipulating the levels of certain cytokines (Christen et al. 2001; Holz et al. 2001; Von Herrath et al. 2005; Jacob et al. 1990) and altering the availability of molecules involved in immune interactions (Anderson and Bluestone 2005; Shoda et al. 2005). The role of various infectious agents has also been examined in this model. T1D development can be suppressed in this model by numerous viruses including lymphocytic choriomeningitis virus (Oldstone 1988; Oldstone et al. 1990), mouse hepatitis virus (Wilberz et al. 1991), lactate dehydrogenase virus (Takei et al. 1992), encephalomyocarditis virus (Hermitte et al. 1990), and HEVs such as the group B coxsackieviruses (Tracy et al. 2002). However, among the large number of means that have been employed by researchers to alter T1D development in the NOD mouse model, the CVBs are unique in that they represent a natural, common infectious agent of humans that is also etiologically linked to the human disease.

## 5 T1D Development in Humans: A Series of Unfortunate Events

### 5.1 Impact of Virus Dose on T1D Onset

Certain CVB strains have been described as diabetogenic in the literature, a descriptor assigned based on the ability of the particular virus strain of to infect the islets in vivo in an animal model or caused by the isolation of the particular virus strain from the human pancreas following sudden onset diabetes (Yoon et al. 1979; Yoon et al. 1978). We hypothesized that one factor impacting the diabetic potential of any given CVB strain was related to the ability of the virus to achieve high titers relatively quickly (that is, the replication rate of the virus). If a sufficient titer of virus were attained in the pancreas prior to triggering the immune system to control the infection, then the host might be at increased risk for T1D development. Using the NOD mouse model of T1D, this hypothesis was examined. CVB3/GA is poorly pathogenic strain of CVB3 (Lee et al. 2005; Tracy et al. 2000) that rarely induces sudden onset T1D in older (12 weeks of age) female NOD mice at a dose of $5\times10^5$ $TCID_{50}$s per mouse (Drescher et al. 2004). Based on this phenotype, this virus strain would be categorized to be nondiabetogenic. CVB3/28, a virulent virus that triggers sudden onset T1D in 70% of prediabetic female NOD mice (Tracy et al. 2002) (a diabetogenic strain) was also used in these experiments. Female NOD mice were injected with $5\times10^5$ $TCID_{50}$ of either CVB3/28 or CVB3/GA and infectious virus titers in pancreas tissue determined. Mice inoculated with CVB3/28 showed that this virus strain replicated more rapidly and to 3-log higher pancreatic viral titers compared with CVB3/GA (Kanno et al. 2006). Based on these data, we hypothesized that inoculation of older NOD mice with higher doses of CVB3/GA would result in increased numbers of sudden onset T1D similar to those observed in mice injected with strains that exhibit more rapid replication. Administration of a 100-fold increase in the dose of CVB3/GA increased the level of sudden onset T1D in female NOD mice to 30% (Kanno et al. 2006). While this rate did not approach that of the CVB3/28-injected animals (70% of the animals experienced sudden onset T1D), the data suggested virus load, rather than a specific undefined diabetogenic phenotype, as the mechanism by which CVB may initiate T1D. If these changes were solely related to the dose of virus to which the animal was exposed, one would predict that administration of lower doses of CVB3/28 to NOD mice would result in lower levels of T1D onset compared to mice receiving the standard dose of $5\times10^5$ TCID50s of virus. Indeed, when lower doses of CVB3/28 were administered to 12-week-old female NOD mice, sudden onset diabetes rates dropped to those observed in the high-dose CVB3/GA-injected mice (Drescher et al. 2004). These studies strongly suggest that any strain of CVB has the potential to induce T1D if inoculated in older, prediabetic animals provided the mouse is administered a sufficient dose of CVB. Based on the animal data, one of the factors that may limit T1D incidence in humans is the dose of virus one encounters that initiates an infection.

## 5.2 Impact of Virus Strain on T1D Onset

One may postulate that the paucity of reported outbreaks of human T1D despite regular circulations of the HEVs is that CVB strains responsible for triggering T1D are rare and humans seldom encounter these particular viruses. From a basic virology standpoint, this begs the question-can the diabetogenic capacity of a particular virus be predicted based on the genetic characteristics of the virus, or are other factors involved in determining diabetogenicity? For example, domain II of the 5′ NTR of CVB has been demonstrated to be a key factor in determining a myocarditic viral phenotype (Dunn et al. 2000, 2003). To test whether a similar genetic determinant could be responsible for imparting the myocarditic phenotype, we used a poorly pathogenic strain of CVB3 mice, CVB3/GA, and a virulent strain of CVB3, CVB3/28. The 5′ NTRs of CVB3/28 and CVB3/GA were examined to determine whether elements in the NTR could account for the increased diabetogenic capacity of CVB3/28 relative to CVB3/GA. The 5' NTR of CVB3/GA was replaced with that from CVB3/28, and a chimeric CVB3/28 virus strain was also prepared, with the 5' NTR of CVB3/GA replacing that of CVB3/28 (Kanno et al. 2006). To examine the replication capacity of the chimeric viruses, single-step growth curves were generated using MIN6 cells, a mouse pancreatic beta cell line, to compare the growth of the chimeric viruses with those of the parental strains. Replacement of the 5' NTR of CVB3/28 with that of CVB3/GA reduced replication of the chimeric CVB3/28 to the level observed in the wild type CVB3/GA virus. Interestingly, the reciprocal exchange had no effect on the replication efficiency of CVB3/GA. In vivo however, both chimeric viruses experienced slower replication rates in the pancreas compared to the parental strains. The lack of a clear association between the genotype of the viruses and virus replication rates demonstrate the difficulty in modeling the complex interactions between the viruses and the host microenvironment. Work remains to identify the viral genetics that determine replication efficiency in the pancreas as well as in the beta cell itself.

## 5.3 Impact of Microenvironment on Ability of CVB to Initiate T1D in the Host

Given that an understanding of all factors involved in the development of T1D is incomplete, and that there are not HEV-linked outbreaks of T1D in human populations that are genetically more at risk for T1D, it is reasonable to postulate that at any given snapshot in time, the majority of the human population is refractory to T1D development. If correct, then one should be able to identify some of the host characteristics associated with susceptibility or resistance to diabetes development. For example, when young female NOD mice (4-6 weeks old) are inoculated with CVB3, the virus replicates to high titers in the pancreas, but sudden T1D onset does not occur (Tracy et al. 2002); in fact, infection of young NOD mice with CVB provides significant protection from T1D onset later in life. T1D onset is either significantly delayed or prevented in these mice. Similar data has been obtained with several CVB serotypes (Tracy et al. 2002). Conversely, infection of older prediabetic NOD mice (12 weeks of age) with the *same strains* of CVB3 has been shown to induce sudden onset T1D (Tracy et al. 2002). These studies demonstrate that even though the insult is stable (i.e., the virus strain and dose is consistent), the outcome of infection differs depending on the age of the NOD mouse host, indicating that both host and pathogen characteristics are important in determining disease outcome.

What are the key differences between young and old NOD mice, and what relevance does this have to the human condition? A primary difference between young and old NOD mice is the degree of insulitis that is observed in the pancreas. Mice with no insulitis that are exposed to virulent strains of CVB do not develop T1D (Tracy et al. 2002), while older mice with naturally occurring, autoimmune insulitis are susceptible to sudden onset T1D following CVB exposure (Drescher et al. 2004; Kanno et al. 2006). The relevance of these findings for humans is difficult to state specifically, as the extent of insulitis in any given person at any time is unknown. However, one may presume that insulitis increases in individuals who are at risk for T1D development over time and so by analogy to the mouse model, exposure of these individuals to CVB may initiate sudden onset T1D. Considering the human data regarding HEV exposure, we postulate that individuals predisposed to T1D development who are also exposed to CVBs at a young age may be protected from later development of T1D. While the mechanism of this is at present undefined, it likely involves regulation of the adaptive immune responses possibly via the induction of regulatory T cells (Herman et al. 2004; Tang et al. 2004). Further testing of the hygiene hypothesis in the context of T1D development is worth examining.

Some studies have suggested that the diabetogenic potential of CVB is associated with the ability of select viral strains [or even specific serotypes (Filippi and von Herrath 2005)] to enter the pancreatic islet cells and destroy the insulin-producing beta cells. Immunohistochemical staining of pancreata from NOD mice was performed to test if the coxsackie-adenovirus receptor (CAR) is expressed in islets under normal conditions. These studies, which demonstrated that CAR was expressed in both young and old uninfected NOD mice (Drescher et al. 2004), indicate there is no obvious block to CVB infection of the islets at the level of viral receptor expression. Human pancreas was also demonstrated to express CAR within islets in the same study. Extending these data, prediabetic NOD mice (12 weeks old) were inoculated with CVB3 and immunohistochemical staining was performed. Using an antibody specific for CVB capsid proteins with an additional antibody specific for one of the cell types of the islet (glucagon, insulin, pancreatic polypeptide, and somatostatin), it was determined that all of the endocrine cell types of the islet have the capacity to become infected by CVB3, although the majority of the virus was localized to the acinar tissue (Kanno et al. 2006). Further, CVB3/GA (a strain that requires high titers of virus to be inoculated to initiate T1D), was also detected in islets, demonstrating that the ability to induce T1D is linked to the presence of the virus in the islet prior to disease onset.

Are pancreatic islet cells of young NOD mice susceptible to CVB3 infection? As CAR is demonstrable in such islets (Drescher et al. 2004), islets should be able to be infected by CVB3; however, analysis of islets from young NOD mice inoculated with CVB has shown no evidence of viral infection (Tracy et al. 2002). If CVB enters healthy islets through interacting with CAR, productive viral replication is inhibited in some manner, and therefore, T1D does not occur as it does in older prediabetic mice in which productive islet infections are readily observed (Kanno et al. 2006). Islets in young NOD mice are hypothesized to be protected by an islet-specific production of type 1 interferons (Chehadeh et al. 2000). Although wild type CVBs do not productively infect healthy islets in young NOD mice, it has been shown that the susceptibility of such islets to infection by CVB can be altered using a chimeric CVB virus that expresses murine interleukin-4 (IL-4) (Drescher et al. 2004). The CVB3-mIL4 strain (Chapman et al. 2000) apparently alters the microenvironment in pancreatic islets, allowing productive replication of this virus to occur. It is postulated that the expression of IL-4 in the chimeric virus overcomes the innate defenses of the islets of the NOD mice, perhaps by altering the levels of type 1 interferons in the islets. Inoculation of young NOD mice with poly I:C, a potent inducer of the innate immune response, lowers CVB3-mIL4 titers in the pancreas, consistent with this hypothesis (Drescher et al. 2004). These data suggest that although islets may be subject to CVB infection through expression of the CAR protein, the islets themselves can control whether or not the virus infection proceeds toward a pathogenic outcome. These results suggest that a fine cooperation between progress of the host autoimmune disease (insulitis) and the viral infection is required to initiate events whose outcome is T1D. It may be argued that it is the state of the islet that is the final arbiter of the process, all other things being maintained constant.

CVB4/E2 is a strain of CVB4 that has been shown to initiate T1D onset in SJL/J mice (Yap et al. 2003). Studies comparing CVB4/E2 with CVB4/JVB, a strain not associated with T1D onset, suggest that beta cell damage is not the sole determinant of whether an animal will become diabetic following infection with the CVB. These studies hypothesize that CVB4/E2-infected mice are unable to effectively support islet neogenesis, and the inability to trigger and sustain pancreatic islet repair mechanisms are key parameters in determining whether a mouse becomes diabetic following infection with CVB (Yap et al. 2003). Similar to results obtained in NOD mouse studies, both strains of CVB4 were primarily localized to the acinar tissue of the injected SJL/J mouse. The authors further hypothesize that the diabetic potential of the virus is related to the extent of acinar tissue damage, supporting an indirect mechanism of islet cell damage and subsequent onset of T1D (Yap et al. 2003). In this case, one might propose that exceptionally extensive pancreatitis would also result in bystander loss of islet tissue, thereby initiating T1D. How relevant this mechanism may be vis-a-vis direct viral infection and loss of viable islet tissue is not clear, as CVB-associated pancreatitis is well known (Ursing 1973; Kloppel 2004; Huber and Ramsingh 2004; Arnesjo et al. 1976), while T1D associated with acute pancreatitis is rare (Taniguchi et al. 2001; Lernmark 2000; Imagawa et al. 2000). Nonetheless, T1D has been associated with chronic pancreatitis (Larsen 1993), suggesting such a mechanism may exist with another type of pancreatitis.

## 5.4 Final Thoughts

How do we relate inferences from the hygiene hypothesis, human T1D incidences, and data from murine models to a better understanding of T1D etiology? First, it must be remembered that the percentage of T1D cases attributable to any specific viral infection has not been established; further, such data will be difficult to obtain for a variety of reasons. The majority of T1D cases must have an environmental influence, this much is clear. Given at present a lack of a similar body of evidence for the significant involvement of any other infectious agent in T1D onset, the impact of HEV infections must therefore be considered to be primary. Increases in societal development have in general led to increases in levels of hygiene, with concurrent decreases in the frequency of HEV infections. This has also been linked in time to increases in cases of T1D (Viskari et al. 2000). Although only inferential, these observations are not likely to be merely coincidental. The analogy to the increase in hygiene during the twentieth century and the occurrence of annual poliovirus epidemics, events that were quite rare prior to that time, is stark.

How does this take into account altered exposure patterns to HEV in individuals living in industrialized countries? Widespread availability of microbiologically cleaner water supplies in developed countries would reduce the likelihood of exposure to infectious agents commonly found in human excrement from a young age onward. Children are therefore growing older with fewer HEV exposures and are being increasingly exposed to natural HEV infections at older ages, while infants that are exposed have a much lower chance of having HEV immunity acquired passively from their mothers during nursing.

How would one model the above studies to reflect the human condition? We suggest that at least two main factors influence whether a person will develop T1D following infection with HEV. The most important factor is whether a person is genetically predisposed to developing T1D and whether that individual has preexisting insulitis at the time of CVB infection. That is, is the individual immediately prediabetic, similar to older prediabetic NOD mice? If these people are exposed to certain strains and/or doses of CVB (that is a high dose of a poorly replicating virus or a low to moderate dose of a highly virulent strain of virus), then these individuals should have an increased risk of developing T1D. In contrast, individuals in less developed countries with poorer water quality levels have a higher risk of continuous exposure to fecal-oral-transmitted HEV from early life onward, allowing them to naturally develop robust immunity to these viruses. This is coupled with increased levels of maternal antibodies that are transferred to the very young and that bestow partial immunity to the children, thereby reducing the opportunity for extensive pathology due to HEV infection.

Of what practical use is this knowledge? Understanding the infectious initiators of T1D in humans will potentially permit development of vaccine technology to prevent T1D. Our current understanding of this disease strongly suggests that while specific HEV (the CVB and probably other HEV-B members) can initiate T1D under the right conditions, this can be prevented by early exposure (in effect, vaccination) to the same viruses. This leads one to the inevitable conclusion that vaccinating to prevent some, if not many, cases of T1D should be feasible. Combined with recent results, now being tested in humans, that newly diagnosed T1D may be reversed through anti-CD3 antibody therapy (Shoda et al. 2005; Belghith et al. 2003; Bresson et al. 2006), there is the potential not only for preventing T1D but rescuing those from the diabetic abyss.

**Acknowledgements** The laboratory of KMD is supported by grants from the National Institutes of Health, National Multiple Sclerosis Society, and the American Cancer Society. Some of these investigations were conducted in a facility constructed with support from Research Facilities Improvement Program (1 C06 RR17417-01) from the National Center for Research Resources, National Institutes of Health (KMD). The laboratory of SMT is supported by grants from the Juvenile Diabetes Research Foundation and the American Diabetes Association.

# References

Anderson MS, Bluestone JA (2005) The NOD mouse: a model of immune dysregulation. Ann Rev Immunol 23:447-485

Andreoletti L, Hober D, Hober-Vandenberghe C, Belaich S, Vantyghem MC, Lefebvre J, Wattre P (1997) Detection of coxsackie B virus RNA sequences in whole blood samples from adult patients at the onset of type 1 diabetes mellitus. J Med Virol 52:121-127

Arnesjo B, Eden T, Ihse I, Nordenfelt E, Ursing B (1976) Enterovirus infections in acute pancreatitis: a possible etiological connection. Scand J Gastroenterol 11:645-649

Atkinson MA, Leiter EH (1999) The NOD mouse model of type 1 diabetes: as good as it gets? Nat Med 5:601-604

Bach JF (2001) Protective role of infections and vaccinations on autoimmune diseases. Autoimmunity 16:347-353

Barnett AH, Eff C, Leslie RD, Pyke DA (1981) Diabetes in identical twins: a study of 200 pairs. Diabetologia 20:87-93

Belghith M, Bluestone JA, Barriot S, Megret J, Bach JF, Chatenoud L (2003) TGF-beta-dependent mechanisms mediate restoration of self-tolerance induced by antibodies to CD3 in overt autoimmune diabetes. Nat Med 9:1202-1208

Birgisfottir BE, Hill JP, Thorsson AV, Thorsdottir I (2006) Lower consumption of cow milk protein A1 beta-casein at 2 years of age, rather than consumption among 11- to 14-year-old adolescents, may explain the lower incidence of type 1 diabetes in Iceland than in Scandinavia. Ann Nut Metab 50:177-183

Brenden N, Rabbani H, Abedi-Valugerdi M (2001) Analysis of mercury-induced immune activation in nonobese diabetic (NOD) mice. Clin Exp Immunol 202-210

Bresson D, Togher L, Rodrigo E, Chen Y, Bluestone JA, Herold KC, Von Herrath MG (2006) Anti-CD3 and nasal proinsulin combination therapy enhances remission from recent-onset autoimmune diabetes by inducing Tregs. J Clin Invest 116:1371-1381

Bunimovich-Mendrazitsky S, Stone L (2005) Modeling polio as a disease of development. J Theor Biol 237:302-315

Cabrera-Rode E, Sarmiento L, Tiberti C, Molina G, Barrios J, Hernandez D, Diaz-Horta O, Di Mario U (2003) Type 1 diabetes islet associated antibodies in subjects infected by echovirus 16. Diabetologia 46:1348-1353

Cabrera-Rode E, Sarmiento L, Molina G, Perez C, Arranz C, Galvan JA, Prieto M, Barrios J, Palomera R, Fonseca M, Mas P, Diaz-Diaz O, Diaz-Horta O (2005) Islet cell related antibodies and type 1 diabetes associated with echovirus 30 epidemic: a case report. J Med Virol 76:373-377

Centers for Disease Control (2000a) Enterovirus surveillance-United States, 1997-1999. Morb Mort Wkly Rpt 49:913-916

Centers for Disease Control (2000b) Non-polio enterovirus surveillance-United States, 2002-2004. Morb Mort Wkly Rpt 46:748-750

Centers for Disease Control (2006a) Measles. Centers for Disease Control, Atlanta

Centers for Disease Control (2006b) Enterovirus surveillance-United States, 2002-2004. Morbid Mortal Wkly Rpt 55:153-156

Chapman N, Kim KS, Tracy S, Jackson J, Hofling KLJS, Malone J, Kolbeck P (2000) Coxsackievirus expression of the murine secretory protein IL-4 induces increased synthesis of IgG1 in mice. J Virol 74:7952-7962

Chehadeh W, Weill J, Vantyghem MC, Alm G, Lefebvre J, Wattre P, Hober D (2000) Increased level of interferon-$\alpha$ in blood of patients with insulin-dependent diabetes mellitus-relationship with coxsackievirus B infection. J Infect Dis 181:1929-1939

Christen U, Wolfe T, Mohrle U, Hughes AC, Rodrigo E, Green EA, Flavell RA (2001) A dual role for TNF-alpha in type 1 diabetes: islet specific expression abrogates the autoimmune process when induced late but not early during pathogenesis. J Immunol 166:7023-7032

Diaz-Horta O, Bello M, Cabrera-Rode E, Suarez J, Mas P, Garcia I, Abalos I, Jofra R, Molina G, Diaz-Diaz O, Dimario U (2001) Echovirus 4 and type 1 diabetes mellitus. Autoimmunity 34:275-281

Domingo E, Escarmis C, Sevilla N, Moya A, Elena SF, Quer J, Novella IS, Holland JJ (1997) Basic concepts in RNA virus evolution. FASEB J 10:859-864

Domingo E, Martin V, Perales C, Grande-Perez A, Garcia-Arriaza J, Arias A (2006) Viruses as quasispecies: biological implications. Curr Top Microbiol Immunol 299:51-82

Dorman J, Bunker CH (2000) HLA-DQ locus of the human leukocyte antigen complex and type 1 diabetes mellitus: a HuGE review. Epidemiol Rev 22:218-227

Dotta F, Censini S, van Halteren AG, Marselli L, Masini M, Dionisi S, Mosca F, Boggi U, Muda AO, Prato SD, Elliott JF, Covacci A, Rappuoli R, Roep BO, Marchetti P (2007) Coxsackie B4 virus infection of beta cells and natural kill insulitis in recent-onset type 1 diabetic patients. Proc Natl Acad Sci U S A 104:5115-5120

Drescher KM, Kono K, Bopegamage S, Carson SD, Tracy S (2004) Coxsackievirus B3 infection and type 1 diabetes development in NOD mice: insulitis determines susceptibility of pancreatic islets to virus infection. Virology 329:381-394

Dunn G, Bradrick S, Chapman N, Tracy S, Romero J (2003) The stem loop II within the 5' non-translated region of clinical coxsackievirus B3 genome determines cardiovirulence phenotype in a murine model. J Infect Dis 15:1552-1561

Dunn JJ, Chapman NM, Tracy S, Romero JR (2000) Natural genetics of cardiovirulence in coxsackievirus B3 clinical isolates: localization to the 5' non-translated region. J Virol 74:4787-4794

Filippi C, von Herrath M (2005) How viral infection affect the autoimmune process leading to type 1 diabetes. Cell Immunol 233:125-132

Frisk G, Tuvemo T (2004) Enterovirus infections with beta-cell tropic strains are frequent in siblings of children diagnosed with type 1 diabetes children and in association with elevated levels of GAD65 antibodies. J Med Virol 73:450-459

Fuchtenbusch M, Irnstetter G, Jager G, Ziegler AG (2001) No evidence for an association of coxsackievirus infections during pregnancy and early childhood with the development of islet autoantibodies in offspring of mothers or fathers with type 1 diabetes. J Autoimmunol 17:333-340

Gamble DR (1976) A possible virus etiology for juvenile diabetes. In: Krutzfeld W, Kobberling J, Neel J (eds) The genetics of diabetes mellitus. Springer Verlag, Berlin, pp 95-105

Goldacre MJ, Wotton CJ, Yeates D, Seagroatt V, Neil A (2005) Hospital admission for selected single virus infections prior to diabetes mellitus. Diabetes Res Clin Pract 69:256-261

Green A, Patterson CC; EURODIAB TIGER Study Group (2001) Trends in the incidence of childhood-onset diabetes in Europe 1989-1998. Diabetologia 44:B3-B8

Haller MJ, Atkinson MA, Schatz, D (2005) Type 1 diabetes mellitus: etiology, presentation, and management. Pediatr Clin North Am 52:1553-1578

Herman AE, Freeman GJ, Mathis D, Benoist C (2004) CD4+CD25+ T regulatory cells dependent on ICOS promote regulation of effector cells in the prediabetic lesion. J Exp Med 199:1479-1489

Hermitte L, Vialettes B, Naquet P, Atlan C, Payan MJ, Vague P (1990) Paradoxical lessening of autoimmune processes in non-obese diabetic mice after infection with the diabetogenic variant of encephalomyocarditis virus. Eur J Immunol 20:1297-1303

Hierholzer JC, Farris W (1974) Follow-up of children infected in a coxsackievirus B3 and B4 outbreak: no evidence of diabetes mellitus. J Infect Dis 129:741-746

Holz A, Brett K, Oldstone MB (2001) Constitutive beta cell expression of IL-12 does not perturb self-tolerance but intensifies established autoimmune diabetes. J Clin Invest 108:1749-1758

Huber S, Ramsingh AI (2004) Coxsackievirus-induced pancreatitis. Viral Immunol 17:358-369

Hyoty H (2002) Enterovirus infections and type 1 diabetes. Ann Med 34:138-147

Imagawa A, Hanafusa T, Miyagawa J, Matsuzawa Y (2000) A novel subtype of type 1 diabetes mellitus characterized by a rapid onset and an absence of diabetes-related antibodies. N Engl J Med 342:301-307

Iowa Measles Response Team (2005) The cost of containing one case of measles: the economic impact on the public health infrastructure-Iowa, 2004. Pediatrics 116:e1-e4

Jacob CO, Aiso S, Michie SA, McDevitt HO, Acha-Orbea H (1990) Prevention of diabetes in nonobese diabetic mice by tumor necrosis factor (TNF): similarities between TNF-alpha and interleukin-1. Proc Natl Acad Sci U S A 87:968-972

James JA, Harley JB (1998) B-cell epitope spreading in autoimmunity. Immunol Rev 164:185-200

Jenson AB, Rosenberg HS (1984) Multiple viruses in diabetes mellitus. Prog Med Virol 29:197-217

Kanno T, Kim K, Kono K, Drescher KM, Chapman NM, Tracy S (2006) Group B coxsackievirus diabetogenic phenotype correlates with replication efficiency. J Virol 80:5637-5643

Karvonen M, Tuomilehto J, Libman I, LaPorte R (1993) A review of the recent epidemiological data on the worldwide incidence of type 1 (insulin-dependent) diabetes mellitus. Diabetologia 36:883-892

Karvonen M, Rusanen J, Sundberg M, Virtala E, Colpaert A, Naukkarinen A, Tuomilehto J (1997) Regional differences in the incidence of childhood insulin dependent diabetes mellitus among children in Finland between 1987 to 1991. Ann Med 29:297-304

Kloppel G (2004) Acute pancreatitis. Semin Diagn Pathol 21:221-226

Kolb H, Elliott RB (1994) Increasing incidence of IDDM a consequence of improved hygiene? Diabetologia 37:729-731

Kumar A, Murray DL, Havlichek DH (2005) Immunizations for the college student: a campus perspective of an outbreak and national and international considerations. Pediatr Clin North Am 52:229-241

Larsen S (1993) Diabetes mellitus secondary to chronic pancreatitis. Dan Med Bull 40:153-162

Lee CK, Kono K, Haas E, Kim KS, Drescher KM, Chapman NM, Tracy S (2005) Characterization of an infectious cDNA copy of the genome of a naturally-occurring, avirulent coxsackievirus B3 clinical isolate. J Gen Virol 86:197-210

Lemanske RF (2004) Viral infections and asthma inception. J Allergy Clin Immunol 114:1023-1026

Lernmark A (2000) Rapid-onset type 1 diabetes with pancreatic exocrine dysfunction. N Engl J Med 342:344-345

Lo SS, Tun RY, Hawa M, Leslie RD (1991) Studies of diabetic twins. Diabetes Metab Rev 7:223-238

Maria H, Elshebani A, Anders O, Torsten T, Gun F (2005) Simultaneous type 1 diabetes onset in mother and son coincident with an enteroviral infection. J Clin Virol 33:158-167

Mikulecky M, Michalkova D (2001) Secular and seasonal cycling of IA2-ab autoantibody in Slovak diabetic children. Biomed Pharmacother 55:106s-109s

Mikulecky M, Michalkova D, Petrovicova A (2000) Coxsackie infections and births of future diabetic children: year, seasonality, and secularity. J Pediatr Endocrin Metab 13:523-527

Monteyne P, Bureau JF, Brahic M (1998) Viruses and multiple sclerosis. Curr Opin Neurol 11:287-291

Nathanson N, Martin J (1979) The epidemiology of poliomyelitis: enigmas surrounding its appearance, epidemicity, and disappearance. Am J Epidemiol 110:672-692

Nishioka K, Nakajima T, Husunuma T, Sato K (1993) Rheumatic manifestation of human leukemia virus infection. Rheum Dis Clin Am 19:489-503

Oldstone MBA (1988) Prevention of type 1 diabetes in NOD mice by virus infection. Science 239:500-502

Oldstone MBA, Ahmed R, Salvato M (1990) Viruses as therapeutic agents. II. Viral reassortants map prevention of insulin-dependant diabetes mellitus to the small RNA of lymphocytic choriomeningitis virus. J Exp Med 171:2091-2100

Onkamo P, Väänänen S, Karvonen M, Tuomilehto J (1999) Worldwide increase in incidence of type 1 diabetes-the analysis of the data on published incidence trends. Diabetologia 42:1395-1403

Otonkoski T, Roivainen M, Vaarala O, Dinesen B, Leipala JA, Hovi T, Knip M (2000) Neonatal type I diabetes associated with maternal echovirus 6 infection: a case report. Diabetologia 43:1235-1238

Paananen A, Ylipaasto P, Rieder E, Hovi T, Galama J, Roivainen M (2003) Molecular and biochemical analysis of echovirus 9 strain isolated from a diabetic child. J Med Virol 69:529-537

Palacios G, Oberste MS (2005) Enteroviruses as agents of emerging infectious diseases. J Neurovirol 11:424-433

Pallansch M, Roos RP (2001) Enteroviruses: polioviruses, coxsackieviruses, echoviruses, and newer enteroviruses. In: Knipe DM, Howley PM (eds) Fields virology. Lipincott Williams and Wilkins, Philadelphia, pp 723-776

Paul JR (1971) A history of poliomyelitis. Yale University Press, New Haven

Pitkaniemi J, Onkamo P, Tuomilehto J, Arjas E (2004) Increasing incidence of type 1 diabetes-a role for genes. BMC Genet 5:5

Rawson H, Crampin A, Noah N (2001) Deaths from chickenpox in England and Wales 1995-7: analysis of routine mortality data. Br Med J 323:1091-1093

Redondo MJ, Yu L, Hawa M, Mackenzie T, Pyke DA, Eisenbarth GS, Leslie RD (2001) Heterogeneity of type I diabetes: analysis of monozygotic twins in Great Britian and the United States. Diabetologia 44:354-362

Romero J, Price C, Dunn J (1997) Genetic divergence among group B coxsackieviruses. Curr Top Microbiol Immunol 223:97-152

Sanders LJ (2002) From Thebes to Toronto and the 21st century: an incredible journey. Diabetes Spectrum 15:56-60

Shoda LK, Young DL, Ramanujan S, Whiting CC, Atkinson MA, Bluestone JA, Eisenbarth GS, Mathis D, Rossini AA, Campbell SE, Kahn R, Kreuwel HT (2005) A comprehensive review of interventions in the NOD mouse model and implications for translation. Immunity 23:115-126

Smith C, Clements G, Riding M, Collins P, Bottaza G, Taylor K (1998) Simultaneous onset of type 1 diabetes mellitus in identical infant twins with enterovirus infection. Diabet Med 15:515-517

Strachan DP (1989) Hay fever, hygiene, and household size. BMJ 299:1259-1260

Strachan DP (2000) Family size, infection and atopy: the first decade of the 'hygiene hypothesis'. Thorax 55 [Suppl 1]:S2-S10

Szopa TM, Titchener PA, Portwood ND, Taylor KW (1993) Diabetes mellitus due to viruses-some recent developments. Diabetologia 36:687-695

Takei I, Asaba Y, Kasatani T, Maruyama T, Watanabe K, Yanagawa T, Saruta T, Ishii T (1992) Suppression of development of diabetes in NOD mice by lactate dehydrogenase virus infection. J Autoimmun 5:665-673

Tang Q, Henriksen KJ, Bi M, Finger EB, Szot G, Ye J, Masteller EL, McDevitt H, Bonyhadi M, Bluestone JA (2004) In vitro-expanded antigen-specific regulatory T cells suppress autoimmune diabetes. J Exp Med 199:1455-1465

Tani N, Doni Y, Kurumatani N, Yonemasu K (1995) Seasonal distribution of adenoviruses, enteroviruses, and reoviruses in urban river water. Microb Immunol 39:577-580

Taniguchi T, Tanaka J, Seko S, Okazaki K, Okamoto M (2001) Association of rapid-onset type 1 diabetes and clinical acute pancreatitis positive for autoantibodies to the exocrine pancreas. Diabetes Care 24:2156-2157

Tracy S, Holfling K, Pirruccello S, Lane PH, Reyna SM, Gauntt C (2000) Group B coxsackievirus myocarditis and pancreatitis in mice: connection between viral virulence phenotypes. J Med Virol 69:4607-4618

Tracy S, Drescher KM, Chapman NM, Kim KS, Carson SD, Pirruccello S, Lane PH, Romero JR, Leser JS (2002) Toward testing the hypothesis that group B coxsackieviruses (CVB) trigger insulin-dependent diabetes: inoculating nonobese diabetic mice with CVB markedly lowers diabetes incidence. J Virol 76:12097-12111

Ursing B (1973) Acute pancreatitis in coxsackie B infection. BMJ 3:524-525

Viskari H, Ludvigsson J, Uibo R, Salur L, Marciulionyte D, Hermann R, Soltesz G, Fuchtenbusch M, Ziegler AG, Kondrashova A, Romanov A, Kaplan B, Laron Z, Koskela P, Vesikari T, Huhtala H, Knip M, Hyoty H (2005) Relationship between the incidence of type 1 diabetes and maternal enterovirus antibodies: time trends and geographical variation. Diabetologia 48:1280-1287

Viskari HR, Koskela P, Lonnrot M, Luonuansuu S, Reunanen A (2000) Can enterovirus infections explain the increasing incidence of type 1 diabetes? Diabetes Care 23:414-416

Von Herrath MG, Allison J, Oldstone MBA (2005) Focal expression of interleukin-2 does not break unresponsiveness to 'self' (viral) antigen expressed in beta cells but enhances development of autoimmune disease (diabetes) after initiation of an anti-self immune response. J Clin Invest 95:477-485

Wahlberg J, Vaarala O, Ludvigsson J (2006) Dietary risk factors for the emergence of type 1 diabetes-related autoantibodies in 2 1/2 year-old Swedish children. Br J Nutr 95:603-608

Wilberz S, Partke H, Dagnaes-Hansen F, Herberg L (1991) Persistent MHV (mouse hepatitis virus) infection reduces the incidence of diabetes mellitus in non-obese diabetic mice. Diabetologia 34:2-5

Williams CH, Oikarinen S, Taurianinen S, Salminen K, Hyoty H, Stanway G (1900) Molecular analysis of an echovirus 3 strain isolated from an individual concurrently with appearance of islet cell and IA-2 autoantibodies. J Clin Microbiol 44:441-448

Yap IS, Giddings G, Pocock E, Chantler JK (2003) Lack of islet neogenesis plays a key role in beta-cell depletion in mice infected with a diabetogenic variant of coxsackievirus B4. J Gen Virol 84:3051-3068

Yoon JW, Onodera T, Notkins AL (1978) Virus-induced diabetes mellitus. XV. Beta cell damage and insulin-dependent hyperglycemia in mice infected with Coxsackie virus B4. J Exp Med 148:1068-1080

Yoon JW, Austin M, Onodera T, Notkins AL (1979) Isolation of a virus from the pancreas of a child with diabetic ketoacidosis. N Engl J Med 300:1173-1179

# Persistent Coxsackievirus Infection: Enterovirus Persistence in Chronic Myocarditis and Dilated Cardiomyopathy

N. M. Chapman(✉) and K.-S. Kim

| | | |
|---|---|---|
| 1 | Introduction............................................................................................................ | 276 |
| 2 | Persistent Enterovirus Infections in Human Myocarditis and | |
| | Dilated Cardiomyopathy........................................................................................ | 276 |
| | 2.1 Efficiency of Detection of Viral RNA.......................................................... | 276 |
| | 2.2 Evidence of Replicating Enterovirus ............................................................ | 278 |
| | 2.3 Correlation of Enterovirus Infection with Severity of Disease..................... | 279 |
| 3 | Persistence of Enterovirus in Murine Models of Myocarditis................................ | 280 |
| | 3.1 Coxsackievirus B3 Persistence and Inbred Mouse Strains........................... | 280 |
| | 3.2 Enterovirus Persists in the Form of a Replicating Defective Virus ............. | 281 |
| | 3.3 T Cells in Myocarditis and Persisting Coxsackievirus ................................ | 282 |
| | 3.4 Effects of Cytokine Expression on Autoimmunity May Also | |
| | Alter Persistent Enterovirus Replication...................................................... | 283 |
| 4 | Effects of Persistent Enterovirus Replication upon Cardiac Cells......................... | 284 |
| 5 | Interactions Between Acute Virus Infection, Autoimmunity, and | |
| | Persistent Enterovirus Infection: Making It Possible for TD Viruses to Occur..... | 285 |
| References................................................................................................................... | | 286 |

**Abstract** Enteroviral infection of the heart has been noted in a significant proportion of cases of myocarditis and dilated cardiomyopathy. The presence of enterovirus RNA at stages of disease after acute infection and correlation of enterovirus replication with worse clinical outcome suggests continued replication of the virus is involved in the progression of the disease. This finding is mirrored by the murine model of coxsackievirus B3 myocarditis, in which virus persists through the evolution of the virus to a terminally deleted defective form which persists in the myocardium. Studies of the mechanism of induction of myocarditis by coxsackievirus B3 require assessment of the effects of alterations of the immune response upon virus persistence in this form. As expression of viral proteins in the heart have been shown to generate significant impairment of cardiomyocyte function and promote generation of dilated cardiomyopathy, the role of virus persistence is likely to include direct effects of viral replication as well as induction of inflammation in the heart. Factors that control the

N. M. Chapman
University of Nebraska Medical Center, Omaha, NE 68198-6495, USA
nchapman@unmc.edu

S. Tracy et al. (eds.), *Group B Coxsackieviruses. Current Topics in Microbiology and Immunology 323.*
© Springer-Verlag Berlin Heidelberg 2008

extent of cardiac infection with terminally deleted enteroviruses and the relative roles of continued immune response of the virus vs viral modification of cardiac function need to be measured to find effective therapies for the human disease.

# 1 Introduction

Myocarditis is a frequent acute disease (1% of the population; Friman et al. 1995) based on routine autopsy studies. In those in whom symptoms are severe enough to be diagnosed, there is a significant risk of progression to dilated cardiomyopathy (DCM) (Mason 2003; Kawai 1999). Viral infections of the myocardium are often associated with acute myocarditis when assayed and the presence of virus is found more frequently in cardiomyopathic hearts than controls (Baboonian and Treasure 1997; Spotnitz and Lesch 2006). Although the induction of autoimmunity and immunopathologic mechanisms induced by the acute infection have been invoked to explain later stage disease such as chronic myocarditis and dilated cardiomyopathy (Fairweather et al. 2005; Huber and Sartini 2005), the fact remains that the role of enteroviruses in these diseases has been invoked because enteroviral RNA is present in these hearts at a late stage of disease. As human enteroviruses do not have a mode for latency, persistence of these positive-strand RNA viruses implies some level of continued replication. Multiple mechanisms have been proposed for the progression of viral myocarditis to DCM, including an adjuvant activity of virus infection inducing autoimmunity (Fairweather et al. 2005), and activation of a dormant viral infection to induce proteolytic activities inducing DCM or direct cardiotoxic activities of viral infection (Spotnitz and Lesch 2006). The evolution of coxsackieviruses in murine and human myocarditic hearts to a slowly replicating defective form has provided a viral explanation for chronic myocarditis and DCM and the persistence of enteroviral RNA in the absence of infectious virus (as assayed by cytopathic effect) (Kim et al. 2005). As much of the work on the mechanism of the immunopathology of chronic myocarditis has been done in inbred stains of mice in which coxsackievirus B3 persists as a replicating defective virus, the effects of the modification of the immune response have to take into account alterations in the extent of viral replication likely to occur.

# 2 Persistent Enterovirus Infections in Human Myocarditis and Dilated Cardiomyopathy

## 2.1 Efficiency of Detection of Viral RNA

Enteroviral RNA or protein is detected in 6%-70% of myocarditic hearts (Jin et al. 1990; Why et al. 1994; Martin et al. 1994; Pauschinger et al. 1999; Arbustini et al. 2000; Li et al. 2000; Shirali et al. 2001; Chimenti et al. 2001; Alter et al. 2001,

2002, Frustaci et al. 2003; Bowles et al. 2003; Zhang et al. 2004; Angelini et al. 2000; Calabrese et al. 2002). Enterovirus has been detected less frequently in cardiomyopathy ranging from 0% to 70% (Cochrane et al. 1991; Zoll et al. 1992; Jin et al. 1990; Schwaiger et al. 1993; Kammerer et al. 1994; Satoh et al. 1994; Andreoletti et al. 1996;, Arbustini et al. 1997; Archard et al. 1998; Li et al. 2000; Fujioka et al. 2000; Bowles et al. 2002; Alter et al. 2001; Zhang et al. 2004; Grasso et al. 1992; Calabrese et al. 2000; Mahon et al. 2001; Calabrese et al. 2002). The wide variation seen in detection of enteroviruses, particularly in later-stage disease, leads to the question of whether enteroviruses are significant sources of dilated cardiomyopathy (DCM). The answer may lie in the variety of methods used to detect enterovirus RNA over the nearly 20 years of these studies.

Because it is based on amplification of a nucleic acid, the most sensitive method for detection of enterovirus should be nested reverse transcription polymerase chain reaction (RT-PCR), but in two cases, an immunohistochemical method specific for capsid protein detected enterovirus in samples in which no signal was obtained by RT-PCR (Luppi et al. 2003; Li et al. 2000). This might be explained by dependency of the RT-PCR method upon extent of RNA extraction and quality of RNA (which is frequently impaired in fixed tissues) as well as the efficiency of primer design. Nested or semi-nested RT-PCR using primers specific for the highly conserved cloverleaf and internal ribosome entry sequence (IRES) (Kammerer et al. 1994; Pauschinger et al. 1999; Arbustini et al. 1997; Archard et al. 1998; Fujioka et al. 2004) are likely to have the highest rates of enterovirus RNA detection due to conservation of the primer annealing site and higher levels of amplification. Unfortunately, amplification of this sequence does not identify the specific enterovirus serotype due to the high rate of recombination between the 5' nontranslated region and the capsid encoding region in the enterovirus B group of viruses (Oberste et al. 2004a, 2004b), although limited data on isolation and serology (Baboonian and Treasure 1997) as well as the cardiac expression of the coxsackievirus B receptor (Tomko et al. 1997) suggest that the enteroviruses detected in myocarditic hearts belong to the coxsackievirus B (CVB) group. The enterovirus B database of 5' NTR sequences has grown considerably over the past 20 years. As less complete data were used for primer design in many of the studies, efficiency of cDNA amplification is likely to vary from study to study.

Some studies have detected enterovirus by generation of sufficient cDNA to detect with electrophoresis (Shirali et al. 2001; Luppi et al. 2003), but enteroviral cDNA is usually more sensitively and accurately detected by hybridization to an oligonucleotide or larger DNA probe. Although few of the many human studies detecting the presence of viral RNA with RT-PCR have sequenced the amplified product to demonstrate that amplified sequence is from non-laboratory strains (Giacca et al. 1994; Kammerer et al. 1994; Archard et al. 1998; Fujioka et al. 2000, 2004), sequence analysis is the gold standard for detection of contamination with laboratory strains of virus because contamination of samples from laboratory strains has been noted (Giacca et al. 1994). Although the 5' NTR is well conserved, sufficient sequence divergence from the prototype strains has occurred to provide significant sequence differences for this control. For example, a CVB2 viral RNA isolated from a heart tissue of a myocarditis patient from Japan in 2002 differs from the prototypic CVB2 Ohio strain in the 198 nucleotides amplified using the E1 and E2 primers (Chapman et al. 1990) by 24 nucleotides (N.M. Chapman, K.-S. Kim, and S. Tracy, unpublished observations). Detection by hybridization can increase the sensitivity of these studies but sequence analysis gives best verification of enterovirus presence.

Enterovirus-positive cardiomyopathy may be undercounted due to the fact that it is often not possible to use multiple biopsies for determination of enteroviral RNA presence, although areas of enterovirus infection are known to be limited (Li et al. 2000; Zhang et al. 2000). In a study in which both two biopsies and single biopsy detection was employed, detection in double biopsies increased the level of detection by nearly threefold over cases with only a single biopsy (Pauschinger et al. 1999). The percentage of cases in which enterovirus myocarditis has been detected in recent studies is lower than 20%-25% (Kuhl et al. 2005; Mahrholdt et al. 2006). Other pathogens (for which no tests were conducted in earlier studies) have been detected, but these studies still have a large proportion of cases in which no etiologic agent is detected (Kuhl et al. 2005; Mahrholdt et al. 2006). As these studies have used myocardial biopsy tissue for detection of pathogens, an extremely low amount of tissue has been assayed, but the relative scarcity of autopsy and explanted heart tissue limits the predictive value of studies. It is possible that the number of cases of enterovirus-induced myocarditis would be greater if studies routinely used sensitive semi-nested RT-PCR because, during replication in the myocardium, defective terminally deleted enteroviruses are quickly selected, leading to a lower level of viral RNA in infected cells (Kim et al. 2005). It is also possible that enteroviral myocarditis is declining, although enterovirus and even coxsackievirus isolations are still quite common in Europe and in North America (Sadeharju et al. 2007; Antona et al. 2007; Khetsuriani et al. 2006; Witso et al. 2006). However, seropositivity for enterovirus appears to be decreasing at least in some countries in Europe (Viskari et al. 2004, 2005). One possibility is that universal poliovirus vaccination has led to sufficient levels of non-neutralizing but cross-reactive binding immune response (Juhela et al. 1999), which may reduce the overall level of replication of an enterovirus upon exposure so that, while there is no protection against enterovirus infection, there is less severe disease. The variability in proportion of enterovirus-positive patients with myocarditis and DCM may be due in part to variability in sensitivity and accuracy of methodology as well as true differences in the patient populations assayed. Sufficient studies exist to show that enteroviral RNA is present in a significant number of patients with both myocarditis and dilated cardiomyopathy.

## 2.2 *Evidence of Replicating Enterovirus*

Although enterovirus is rarely cultured from cardiomyopathies, there is evidence that enterovirus continues to replicate in later-stage human disease. Detection of negative-strand enteroviral RNA as well as positive-strand in cases of chronic myocarditis and cardiomyopathy by RT-PCR (Why et al. 1994; Pauschinger et al. 1999; Fujioka et al. 2000, 2004; Satoh et al. 2004; Andreoletti et al. 2000) and the detection of enteroviral protein in cardiomyopathic heart muscle (Andreoletti et al. 2000; Li et al. 2000; Satoh et al. 2004; Zhang et al. 2004) has provided evidence of persistent enterovirus replication in the heart. As the negative strand of enteroviruses is an antigenome that is not detected in preparations of encapsidated virus of wild type enteroviruses, the presence of this intracellular RNA is expected to be a sign of replication intermediates. However, as CVBs have been shown to evolve to a terminally deleted defective form (TD virus) that can encapsidate the negative strand (Kim et al. 2005), the detection of negative strand may point to persistence due to replication and encapsidation of TD virus. The enteroviral proteins in cardiomyopathy are detected via a site in the viral capsid conserved in enteroviruses that is also present in encapsidated virus, but the presence of complete virion would argue for a higher rate of isolation of infectious virus than has been observed (Rey et al. 2001). Graft failure due to infection of the transplanted heart has been noted in a case of enterovirus-positive cardiomyopathy (Calabrese et al. 2002), again indicating the persistence of replicating virus (in this case elsewhere in the body). The persistence of enterovirus occurs as a replicating virus in a significant number of cases.

## 2.3 Correlation of Enterovirus Infection with Severity of Disease

There are conflicting reports as to outcome of enterovirus-positive myocarditis. One report finds that a more benign outcome for patients with enterovirus-positive DCM than for patients in which cardiac virus was not detected (Figulla et al. 1995). Even though a sensitive method of in situ hybridization was used for this study, in general RT-PCR will detect low levels of viral RNA more sensitively, raising the question as to whether the virus-negative group contained patients with low-level infection of the heart with enterovirus. Other studies have demonstrated that the presence of enterovirus or other virus in hearts of patients with DCM was associated with a worse outcome (Kuhl et al. 2005; Why et al. 1994). One of these studies used hybridization of slot-blotted RNA from hearts to detect virus (Why et al. 1994), which can be more sensitive than in situ hybridization as more RNA is concentrated and available for detection. The second study to associate enterovirus presence with worse outcome of disease (Kuhl et al. 2005) used amplification of cDNA from virus with efficient primers that anneal to conserved regions of the 5' NTR with sequencing to verify specificity of amplification. It can be argued that increased sensitivity of detection of virus is demonstrating that persistence of enterovirus in the heart can have deleterious effects.

Treatment of enterovirus-positive patients with interferon β (IFNβ) resolved viral infection, reduced inflammation, and either halted worsening of the disease or improved ventricular function (Kuhl et al. 2003), indicating that this antiviral therapy also reduced the inflammatory disease. Patients who do not respond well to immunosuppressive therapy have a high rate of persisting enterovirus in the heart (Frustaci et al. 2003), indicating that persistence of enteroviral RNA can be associated with a worse outcome for myocardial disease and one that cannot be overcome by anti-inflammatory treatment. Overall, a body of evidence on detection of enterovirus infections in patients with late-stage disease suggests that enteroviruses persist in a replicating form in a significant proportion of patients with viral myocarditis and DCM, and this persistence is associated with a more deleterious outcome.

## 3 Persistence of Enterovirus in Murine Models of Myocarditis

### 3.1 Coxsackievirus B3 Persistence and Inbred Mouse Strains

Enteroviral involvement in myocarditis has been well modeled by coxsackievirus B3 infection of inbred mice (Woodruff 1980; Huber and Sartini 2005; Gauntt and Huber 2003; Kim et al. 2001). These studies have predominantly utilized a few highly cardiovirulent CVB3 strains and attenuated strains generated by passage in cell culture (Tracy et al. 1992; Knowlton et al. 1996; Lee et al. 1997; Schmidtke et al. 2000). The pattern of virus replication as well as the extent and type of pathogenesis induced by virus infection vary by the strain of inbred mouse used (Gauntt et al. 1984; Huber 1997; Wolfgram et al. 1986). The study of mechanisms of this disease using the murine model of CVB3-induced myocarditis is complicated by the type of myocarditis studied, the early-stage acute disease occurring within 7-14 days of inoculation, and late-stage chronic disease starting 3-4 weeks after inoculation. Studies of resistant vs susceptible mouse strains inoculated with cardiovirulent virus have demonstrated a persistence of infectious virus in the heart beyond 7 days post-inoculation in the susceptible mouse strains such as BALB/c (Lodge et al. 1987), NMRI (Gluck et al. 2001), C3H/HeJ (Tracy et al. 1992), and SWR (Ouyang et al. 1995) and replication only during the acute stage of disease in hearts of the resistant C57BL/6 (Henke et al. 1995) and DBA/1J (Klingel et al. 1992).

It is known that CVB3 RNA can be detected in hearts of inoculated mice by sensitive RT-PCR assay out to 90 days post inoculation (Kim et al. 2005; Reetoo et al. 2000) and in muscle (Tam and Messner 1999). C57BL/6 mice that have acute disease but do not develop chronic myocarditis upon inoculation with CVB3 do not have late-stage persistence of viral RNA by RT-PCR assay nor by in situ detection of viral RNA and proteins in the heart (Grun et al. 2005). Mice that are susceptible to chronic myocarditis, such as BALB/c (Adachi et al. 1996), NMRI (Gluck et al. 2001), A/J (Kanno et al. 2006; Kim et al. 2005), and SWR (Reetoo et al. 2000) strains, do have persistence of viral RNA, proving that the murine model has a similarity to the human disease.

## 3.2 Enterovirus Persists in the Form of a Replicating Defective Virus

We have recently demonstrated that the form in which this persistence occurs is by evolution of the virus to a 5′ terminally deleted, defective virus (Kim et al. 2005). Examination of HeLa cultures of heart homogenates from A/J and C3H/HeJ mice inoculated 2-8 weeks previously with CVB3 showed no sign of cytopathic effect, although RT-PCR assay for CVB3 RNA was positive. Examination of the cardiac virus replicating in these cultures showed deletions of the 5′ terminal 7-49 nucleotides were responsible for the lack of cytopathic effect. These terminally deleted (TD) viruses replicate with much reduced yield of virus, package negative-strand RNA in virion, and can persist in hearts of TD-inoculated mice as long as 5 months (Kim et al. 2005). Deletion of the 5′ terminal sequence of the virus genome generates a negative strand with a 3′ terminal deletion, a deletion in a region that is critical for positive-strand replication of enteroviruses (Sharma et al. 2005). This region is a site of binding of a host factor, hnRNPC1 (Brunner et al. 2005), a nuclear protein known to be transported to the cytoplasm during the cell cycle (Kim et al. 2003). As CVB3 replication is affected by cell cycle arrest (Feuer et al. 2002), it is possible that in differentiated cells such as the cardiomyocyte, which do not frequently divide (Soonpaa and Field 1997), the production of TD virus is favored due to altered availability of essential host factors for positive-strand replication. The alteration in the positive- to negative-strand ratio in these viruses is probably due to a defect in the preferential replication of the positive strand (Sharma et al. 2005). Lowered rates of synthesis of this strand, which in enteroviruses also serves as the viral mRNA, are likely to slow the intracellular replication of the virus and limit the ability to produce encapsidated virus due to reduced concentration of capsid protein. The inability to detect these viruses with cytopathic assays, despite their ability to replicate in culture, is due to the low yield of virus per infected cell, which allows uninfected cells to rapidly fill in the monolayer.

Infection of susceptible mice with a cardiovirulent virus causes a high rate of infection of the myocardium. As TD viruses evolve in passage of CVB3 in primary cultures of cardiac or pancreatic cells (Kim et al. 2005; K.-S. Kim, N.M. Chapman, S. Tracy, unpublished data), production of TD virus may require only that these cells be infected. The extent of CVB3 replication in the heart during acute disease may be determined by the extent to which the myocardium becomes infected with TD virus. Although no virus can be detected by plaque assay at the point at which chronic myocarditis is considered to exist, slowly replicating TD virus is present in the heart muscle (Kim et al. 2005).

The validity of this mechanism of persistence by defective enterovirus has been confirmed by the detection of 5′ terminally deleted genomes in the heart of a human case of myocarditis, the only case to date in which the 5′ terminus of the persisting enterovirus has been examined (N.M. Chapman, K.-S. Kim, K. Oka, S. Tracy, unpublished data). The low levels of replication of these defective viruses also suggest the necessity for sensitive assays to detect virus-infected cells. In one study, use of a semi-nested RT-PCR allowed detection of CVB3 to day 98 post-inoculation but standard RT-PCR could not detect virus at day 28 (Gluck et al. 2001). In our study of these defective enteroviruses in A/J and C3H/HeJ mice, virus was detected in a greater number of samples when amplified by culture of heart homogenates in HeLa cells prior to detection using RT-PCR (Kim et al. 2005). The low level of viral RNA present in cells infected with TD virus may have reduced enterovirus detection in human cardiomyopathic hearts, especially when assays are limited to the small amount of tissue available from endomyocardial biopsies. Although RNA replication of TD viruses is impaired, this type of virus has been shown to persist for as long as 5 months in mice (Kim et al. 2005). Because the deletion is in a noncoding region, TD viruses can produce all of the enterovirus proteins, allowing for long-term effects upon the heart and immune response to all viral epitopes. The ability of enteroviruses to evolve to this form explains the persistence of viral RNA (detectable by RT-PCR) in the absence of infectious virus (detected by plaque or $TCID_{50}$ assay).

## 3.3 T Cells in Myocarditis and Persisting Coxsackievirus

A large number of studies using mutation and knockouts of portions of the innate and adaptive immune system have implicated immunopathogenic mechanisms in myocarditis and DCM, but the conclusions of many studies utilizing mutants and knockouts of portions of the immune response were obscured by differences in viral replication dictated by the genetic background of the host (reviewed by Tam 2006). Studies both of TLR signaling in response to CVB3 infection (Fairweather et al. 2003) and signal transduction in T cell activation (Liu et al. 2000) show that both the immune response and the ability of the virus to replicate are affected, thus obscuring the immunopathologic mechanism. The requirement for virus-mediated signaling to increase T cell infiltration of the heart may explain the association of persistent replication of CVB3 in the heart with myocarditis. Many alterations of the immune response result in CVB3 persistence, thus leading to questions of which effect of the alteration (change in immune effectors vs the effects of the persistent virus replication) is more important (Klingel et al. 2003).

A series of studies by Huber and colleagues have demonstrated that the relative size of the $V\gamma1^+$ and $V\gamma4^+$ subsets of the CD1d-restricted $\gamma\delta$ T cells induced by virus infection correlates with the extent of myocarditis in the C57BL/6 (Huber et al. 2000) and BALB/c mice (Huber et al. 2001). Studies of a mouse-passaged cardiovirulent CVB3, CVB3/H3 (isolated from the heart of a mouse inoculated with CVB3/Woodruff) and an antibody-escape mutant of this virus, the noncardiovirulent CVB3/H310A1 (Van Houten et al. 1991), demonstrated less induction of CD1d by the noncardiovirulent virus (Huber et al. 2003). The location of mutations found in the antibody-escape variant, CVB3/H310A1, and in two attenuated antibody-escape mutants of another murine-adapted CVB3 (Stadnick et al. 2004) to a putative DAF-binding site on the CVB3 capsid and the reduced myocarditis induced by CVB3/H3 in mice that do not express DAF (Huber et al. 2006) suggest the importance of this interaction to cardiovirulence in mice. Although human CVB isolates do not bind murine DAF (Spiller et al. 2000), a mechanism for enhanced entry of virus into human cells via CAR through tyrosine kinase signaling induced by interaction with human DAF (Coyne and Bergelson 2006) suggest a similar mechanism may enhance infection in human beings. It is possible that while DAF interaction may induce CD1 and enhance myocarditis via recruitment of $\gamma\delta$ T cells, it also is enhancing the level of infection of cells in the myocardium and persistence of CVB3 at the chronic stage due to enhanced entry of CVBs by disruption of tight junctions containing CAR. However, as the virulence of human CVB3 isolates have been determined in mice, the cardiovirulence of some isolates and the relative lack of cardiovirulence of most cannot be explained by DAF-binding differences (Tracy et al. 2000). A study of mice susceptible to CVB3-induced chronic myocarditis demonstrated that there was increased expression of genes associated with MHC class I antigen presentation in these mice, perhaps enhancing the T cell-mediated effects of the persisting coxsackievirus without clearing the virus infection (Szalay et al. 2006). While the role of T cells in CVB3-induced myocarditis may be critical for the murine model, this role may be dependent upon the continued presence of CVB3 during myocarditis.

## *3.4 Effects of Cytokine Expression on Autoimmunity May Also Alter Persistent Enterovirus Replication*

Late-stage chronic myocarditis in the CVB3-inoculated mouse models of dilated cardiomyopathy in human beings and late-stage disease has been assumed to involve the induction of autoimmunity to an assortment of antigens in the heart (Huber et al. 2006). This model has been driven by the ability to induce chronic myocarditis by immunization with heart antigens in the presence of adjuvants using inbred mice also susceptible to CVB3-mediated myocarditis (Neu et al. 1987; Pummerer et al. 1996). The other finding that suggests autoimmune mechanisms predominate in late disease is that viral titers decrease in susceptible mice with time and cannot be detected in late-stage disease by cytopathic assays in cell culture. As CVB3 can persist in these strains of mice without producing virus detectable by plaque assay, it is likely that the disease seen in CVB3-induced chronic myocarditis has a contribution from the replicating virus as well as the detectable autoimmune response. Induction of tolerance to cardiac myosin does not alter the degree of disease induced by CVB3 infection in the NOD mouse (Horwitz et al. 2005) perhaps because of the role of persisting virus. Because the murine model has suggested immunosuppression as a critical method for treatment, the degree to which persisting virus contributes to disease in this model is important to assess.

So how is the model of autoimmune heart disease triggered by low-dose inoculation with cardiovirulent CVB3 in BALB/c and A/J mice (Fairweather et al. 2005b) affected by TD virus persistence? Cytokine production in chronic myocarditis has been suggested to play a role in the development of myocardial disease through autoimmune responses (Fairweather and Rose 2005). Human monocytes have been shown to produce tumor necrosis factor alpha (TNFα), interleukin-1 beta (IL-1β) and IL-6 upon CVB3 infection (Henke et al. 1992a, 1992b) (although one study showed only low levels; Hofmann et al. 2001). Human vascular endothelial cells (Conaldi et al. 1997; Zanone et al. 2003) and myocardial fibroblasts (Heim et al. 2000) have also been shown to produce TNFα, IL-1β, IL-6 and IL-8 upon CVB3 infection. The persistent replication of CVB3 in chronic myocarditis in the murine model is also accompanied by production of proinflammatory cytokines (Gluck et al. 2001), suggesting that the persistent virus contributes to both the cytokine-mediated disease seen in chronic myocarditis and the lymphocyte infiltration of the myocardium. CVB3-induced chronic myocarditis due to a deficiency of IFNγ in the BALB/c mouse (Fairweather et al. 2005) may be explained by the inability to clear the persistent TD infection through the action of IFNγ (Szalay et al. 2006). It is not surprising that transforming growth factor beta (TGFβ) expression decreased chronic myocarditis (Horwitz et al. 2006) as TGFβ decreases expression of CAR (Lacher et al. 2006), thus making persistent replication less likely. In many cases, modulation of the levels of cytokines, receptors, and immune cells of various types is likely to alter signaling to activate tyrosine kinase, which is involved in CVB3 infection of cells (Coyne and Bergelson 2006) as well as the development of autoimmunity. The relative role that autoimmunity plays in the mouse models of chronic myocarditis and DCM will not be well understood until quantitative assays of amounts of virus and localization of the type of cell infected are assessed: present studies rarely quantitate the extent of persistent CVB3 infection as plaque assay cannot detect TD virus.

## 4 Effects of Persistent Enterovirus Replication upon Cardiac Cells

We know that TD viruses can be selected in replication of wild type CVB3 in cardiac and pancreatic primary cell cultures (K.-S. Kim, N.M. Chapman, S. Tracy, unpublished observations). In addition, this process is not limited to specific virus strains but has been shown to occur in cell culture passage of noncardiovirulent CVB3 strains and with a CVB2 strain in a human case of myocarditis (N.M. Chapman, K.-S. Kim, K. Oka, S. Tracy, unpublished observations). This implies that the limiting factor for persistent replication in the heart may be sufficient replication in cell types in which there is a selection for the TD form of the virus. These viruses have a profound defect of positive-strand replication, one which is likely to slow viral replication although not prevent translation of proteins and production of infectious virion. Such replication is likely to affect function of cardiomyocytes, as has been seen in cardiomyocytes transfected with replication-incompetent virus (Wessely et al. 1998a) or transgenic mice expressing coxsackievirus proteins

(Wessely et al. 1998b) or just the viral protease 2A (Xiong et al. 2007). While immunopathogenic activities against uninfected and infected cells have been shown to play a role in cardiomyopathy, the extent of disease is likely to be affected by the virulence of the infecting virus, as this may determine whether the extent of the TD virus infection.

## 5 Interactions Between Acute Virus Infection, Autoimmunity, and Persistent Enterovirus Infection: Making It Possible for TD Viruses to Occur

To generate sufficient TD virus, infection of the heart for chronic myocarditis may require a high level of replication in the heart muscle during the acute infection, as cardiovirulent viruses are more likely to induce persistent virus infection than noncardiovirulent variants in the mouse model, and deficits of the immune system in a resistant mouse strain can render it likely to have a persistent virus infection. Although expression of cytokines such as TGFβ, TNFα, and IFNγ appear to downregulate expression of the CVB receptor (Vincent et al. 2004; Lacher et al. 2006), the receptor is upregulated in myocardial areas of tissue necrosis and remodeling (Fechner et al. 2003). Indeed, CAR expression is upregulated in human hearts with dilated cardiomyopathy and can be induced in hearts by experimental autoimmune myocarditis (Ito et al. 2000). It is possible that the initial insult of an acute enterovirus infection in the heart seeds the myocardium with TD virus capable of inducing alterations of the muscle due to prolonged viral protein expression. This persistent infection may help to induce or may accompany an autoimmune response to the heart. The damage to the heart muscle from the altered infected cardiomyocytes, the infiltration of the myocardium induced by altered signaling and cytokine production of the infected cells, and the autoimmune damage to uninfected cells are likely to make CAR more highly expressed and more available (Coyne and Bergelson 2006), thus making more cardiac cells susceptible to the persisting virus infection. Prior cardiac infection with other viruses or infection after the original enteroviral myocarditis are likely to enhance the environment for TD virus infection of cardiomyocytes, generating a situation in which there is a high probability that DCM patients will have had multiple cardiac infections (Kuhl et al. 2005). Although it has been proposed that microbial infections cause a fertile field for creation of an autoreactive T cell response (von Herrath et al. 2003), it may also be true that the original acute infection with an enterovirus and any subsequent autoimmunity create a fertile field for the generation of a persistent enterovirus infection of the heart due to creation of more accessible and higher levels of receptor for the virus on uninfected cells. This more widespread infection may create cardiomyopathy from the effects of enteroviral proteins on cardiomyocyte function. This would suggest that both antiviral and immunosuppressive therapies would have positive effects upon enterovirus-associated DCM and the degree to which the immunosuppressive therapy works might well depend on the extent of the enterovirus infection of the heart. If the heart has a large population of TD infected cells, antiviral therapy might be required to allow the heart to heal without further virus infection.

# References

Adachi K, Muraishi A, Seki Y, Yamaki K, Yoshizuka M (1996) Coxsackievirus B3 genomes detected by polymerase chain reaction: evidence of latent persistency in the myocardium in experimental murine myocarditis. Histol Histopathol 11:587-596

Alter P, Jobmann M, Meyer E, Pankuweit S, Maisch B (2001) Apoptosis in myocarditis and dilated cardiomyopathy: does enterovirus genome persistence protect from apoptosis? An endomyocardial biopsy study. Cardiovasc Pathol 10:229-234

Andreoletti L, Bourlet T, Moukassa D, Rey L, Hot D, Li Y, Lambert V, Gosselin B, Mosnier JF, Stankowiak C, Wattre P (2000) Enteroviruses can persist with or without active viral replication in cardiac tissue of patients with end-stage ischemic or dilated cardiomyopathy. J Infect Dis 182:1222-1227

Andreoletti L, Hober D, Decoene C, Copin MC, Lobert PE, Dewilde A, Stankowiac C, Wattre P (1996) Detection of enteroviral RNA by polymerase chain reaction in endomyocardial tissue of patients with chronic cardiac diseases. J Med Virol 48:53-59

Angelini A, Calzolari V, Calabrese F, Boffa GM, Maddalena F, Chioin R, Thiene G (2000) Myocarditis mimicking acute myocardial infarction: role of endomyocardial biopsy in the differential diagnosis. Heart 84:245-250

Angelini A, Crosato M, Boffa GM, Calabrese F, Calzolari V, Chioin R, Daliento L, Thiene G (2002) Active versus borderline myocarditis: clinicopathological correlates and prognostic implications. Heart 87:210-215

Antona D, Leveque N, Chomel JJ, Dubrou S, Levy-Bruhl D, Lina B (2007) Surveillance of enteroviruses in France, 2000-2004. Eur J Clin Microbiol Infect Dis 26:403-412

Arbustini E, Grasso M, Porcu E, Bellini O, Diegoli M, Fasani R, Banchieri N, Pilotto A, Morbini P, Dal Bello B, Campana C, Gavazzi A, Vigano M (1997) Enteroviral RNA and virus-like particles in the skeletal muscle of patients with idiopathic dilated cardiomyopathy. Am J Cardiol 80:1188-1193

Arbustini E, Porcu E, Bellini O, Grasso M, Pilotto A, Dal Bello B, Morbini P, Diegoli M, Gavazzi A, Specchia G, Tavazzi L (2000) Enteroviral infection causing fatal myocarditis and subclinical myopathy. Heart 83:86-90

Archard LC, Khan MA, Soteriou BA, Zhang H, Why HJ, Robinson NM, Richardson PJ (1998) Characterization of Coxsackie B virus RNA in myocardium from patients with dilated cardiomyopathy by nucleotide sequencing of reverse transcription-nested polymerase chain reaction products. Hum Pathol 29:578-584

Baboonian C, Treasure T (1997) Meta-analysis of the association of enteroviruses with human heart disease. Heart 78:539-543

Bowles NE, Ni J, Kearney DL, Pauschinger M, Schultheiss HP, McCarthy R, Hare J, Bricker JT, Bowles KR, Towbin JA (2003) Detection of viruses in myocardial tissues by polymerase chain reaction. Evidence of adenovirus as a common cause of myocarditis in children and adults. J Am Coll Cardiol 42:466-472

Bowles NE, Ni J, Marcus F, Towbin JA (2002) The detection of cardiotropic viruses in the myocardium of patients with arrhythmogenic right ventricular dysplasia/cardiomyopathy. J Am Coll Cardiol 39:892-895

Brunner JE, Nguyen JH, Roehl HH, Ho TV, Swiderek KM, Semler BL (2005) Functional interaction of heterogeneous nuclear ribonucleoprotein C with poliovirus RNA synthesis initiation complexes. J Virol 79:3254-3266

Calabrese F, Angelini A, Thiene G, Basso C, Nava A, Valente M (2000) No detection of enteroviral genome in the myocardium of patients with arrhythmogenic right ventricular cardiomyopathy. J Clin Pathol 53:382-387

Calabrese F, Rigo E, Milanesi O, Boffa GM, Angelini A, Valente M, Thiene G (2002) Molecular diagnosis of myocarditis and dilated cardiomyopathy in children: clinicopathologic features and prognostic implications. Diagn Mol Pathol 11:212-221

Chapman NM, Tracy S, Gauntt CJ, Fortmueller U (1990) Molecular detection and identification of enteroviruses using enzymatic amplification and nucleic acid hybridization. J Clin Microbiol 28:843-850

Chimenti C, Calabrese F, Thiene G, Pieroni M, Maseri A, Frustaci A (2001) Inflammatory left ventricular microaneurysms as a cause of apparently idiopathic ventricular tachyarrhythmias. Circulation 104:168-173

Cochrane HR, May FE, Ashcroft T, Dark JH (1991) Enteroviruses and idiopathic dilated cardiomyopathy. J Pathol 163:129-131

Conaldi PG, Serra C, Mossa A, Falcone V, Basolo F, Camussi G, Dolei A, Toniolo A (1997) Persistent infection of human vascular endothelial cells by group B coxsackieviruses. J Infect Dis 175:693-696

Coyne CB, Bergelson JM (2006) Virus-induced Abl and Fyn kinase signals permit coxsackievirus entry through epithelial tight junctions. Cell 124:119-131

Fairweather D, Frisancho-Kiss S, Rose NR (2005a) Viruses as adjuvants for autoimmunity: evidence from Coxsackievirus-induced myocarditis. Rev Med Virol 15:17-27

Fairweather D, Frisancho-Kiss S, Yusung SA, Barrett MA, Davis SE, Steele RA, Gatewood SJ, Rose NR (2005b) IL-12 protects against coxsackievirus B3-induced myocarditis by increasing IFN-gamma and macrophage and neutrophil populations in the heart. J Immunol 174:261-269

Fairweather D, Rose NR (2005) Inflammatory heart disease: a role for cytokines. Lupus 14:646-651

Fairweather D, Yusung S, Frisancho S, Barrett M, Gatewood S, Steele R, Rose NR (2003) IL-12 receptor beta 1 and Toll-like receptor 4 increase IL-1 beta- and IL-18-associated myocarditis and coxsackievirus replication. J Immunol 170:4731-4737

Fechner H, Noutsias M, Tschoepe C, Hinze K, Wang X, Escher F, Pauschinger M, Dekkers D, Vetter R, Paul M, Lamers J, Schultheiss HP, Poller W (2003) Induction of coxsackievirus-adenovirus-receptor expression during myocardial tissue formation and remodeling: identification of a cell-to-cell contact-dependent regulatory mechanism. Circulation 107:876-882

Feuer R, Mena I, Pagarigan R, Slifka MK, Whitton JL (2002) Cell cycle status affects coxsackievirus replication, persistence, and reactivation in vitro. J Virol 76:4430-4440

Figulla HR, Stille-Siegener M, Mall G, Heim A, Kreuzer H (1995) Myocardial enterovirus infection with left ventricular dysfunction: a benign disease compared with idiopathic dilated cardiomyopathy. J Am Coll Cardiol 25:1170-1175

Friman G, Wesslen L, Fohlman J, Karjalainen J, Rolf C (1995) The epidemiology of infectious myocarditis, lymphocytic myocarditis and dilated cardiomyopathy. Eur Heart J 16 [Suppl O]: 36-41

Frustaci A, Chimenti C, Calabrese F, Pieroni M, Thiene G, Maseri A (2003) Immunosuppressive therapy for active lymphocytic myocarditis: virological and immunologic profile of responders versus nonresponders. Circulation 107:857-863

Fujioka S, Kitaura Y, Deguchi H, Shimizu A, Isomura T, Suma H, Sabbah HN (2004) Evidence of viral infection in the myocardium of American and Japanese patients with idiopathic dilated cardiomyopathy. Am J Cardiol 94:602-605

Fujioka S, Kitaura Y, Ukimura A, Deguchi H, Kawamura K, Isomura T, Suma H, Shimizu A (2000) Evaluation of viral infection in the myocardium of patients with idiopathic dilated cardiomyopathy. J Am Coll Cardiol 36:1920-1926

Gauntt C, Huber S (2003) Coxsackievirus experimental heart diseases. Front Biosci 8: e23-35

Gauntt CJ, Gomez PT, Duffey PS, Grant JA, Trent DW, Witherspoon SM, Paque RE (1984) Characterization and myocarditic capabilities of coxsackievirus B3 variants in selected mouse strains. J Virol 52:598-605

Giacca M, Severini GM, Mestroni L, Salvi A, Lardieri G, Falaschi A, Camerini F (1994) Low frequency of detection by nested polymerase chain reaction of enterovirus ribonucleic acid in endomyocardial tissue of patients with idiopathic dilated cardiomyopathy. J Am Coll Cardiol 24:1033-1040

Gluck B, Schmidtke M, Merkle I, Stelzner A, Gemsa D (2001) Persistent expression of cytokines in the chronic stage of CVB3-induced myocarditis in NMRI mice. J Mol Cell Cardiol 33:1615-1626

Grasso M, Arbustini E, Silini E, Diegoli M, Percivalle E, Ratti G, Bramerio M, Gavazzi A, Vigano M, Milanesi G (1992) Search for Coxsackievirus B3 RNA in idiopathic dilated cardiomyopathy using gene amplification by polymerase chain reaction. Am J Cardiol 69:658-664

Grun K, Markova B, Bohmer FD, Berndt A, Kosmehl H, Leipner C (2005) Elevated expression of PDGF-C in coxsackievirus B3-induced chronic myocarditis. Eur Heart J 26:728-739

Heim A, Zeuke S, Weiss S, Ruschewski W, Grumbach IM (2000) Transient induction of cytokine production in human myocardial fibroblasts by coxsackievirus B3. Circ Res 86:753-759

Henke A, Huber S, Stelzner A, Whitton JL (1995) The role of CD8+ T lymphocytes in coxsackievirus B3-induced myocarditis. J Virol 69:6720-6728

Henke A, Mohr C, Sprenger H, Graebner C, Stelzner A, Nain M, Gemsa D (1992a) Coxsackievirus B3-induced production of tumor necrosis factor-alpha IL-1 beta, IL-6 in human monocytes. J Immunol 148:2270-2277

Henke A, Spengler HP, Stelzner A, Nain M, Gemsa D (1992b) Lipopolysaccharide suppresses cytokine release from coxsackie virus-infected human monocytes. Res Immunol 143:65-70

Hofmann P, Schmidtke M, Stelzner A, Gemsa D (2001) Suppression of proinflammatory cytokines and induction of IL-10 in human monocytes after coxsackievirus B3 infection. J Med Virol 64:487-498

Horwitz MS, Ilic A, Fine C, Sarvetnick N (2005) Induction of antigen specific peripheral humoral tolerance to cardiac myosin does not prevent CB3-mediated autoimmune myocarditis. J Autoimmun 25:102-111

Horwitz MS, Knudsen M, Ilic A, Fine C, Sarvetnick N (2006) Transforming growth factor-beta inhibits coxsackievirus-mediated autoimmune myocarditis. Viral Immunol 19:722-733

Huber S, Sartini D, Exley M (2003) Role of CD1d in coxsackievirus B3-induced myocarditis. J Immunol 170:3147-3153

Huber S, Song WC, Sartini D (2006) Decay-accelerating factor (CD55) promotes CD1d expression and Vgamma4+ T-cell activation in coxsackievirus B3-induced myocarditis. Viral Immunol 19:156-166

Huber SA (1997) Coxsackievirus-induced myocarditis is dependent on distinct immunopathogenic responses in different strains of mice. Lab Invest 76:691-701

Huber SA, Graveline D, Born WK, O'Brien RL (2001) Cytokine production by Vgamma(+)-T-cell subsets is an important factor determining CD4(+)-Th-cell phenotype and susceptibility of BALB/c mice to coxsackievirus B3-induced myocarditis. J Virol 75:5860-5869

Huber SA, Graveline D, Newell MK, Born WK, O'Brien RL (2000) V gamma 1+ T cells suppress and V gamma 4+ T cells promote susceptibility to coxsackievirus B3-induced myocarditis in mice. J Immunol 165:4174-4181

Huber SA, Sartini D (2005) Roles of tumor necrosis factor alpha (TNF-alpha) and the p55 TNF receptor in CD1d induction and coxsackievirus B3-induced myocarditis. J Virol 79:2659-2665

Ito M, Kodama M, Masuko M, Yamaura M, Fuse K, Uesugi Y, Hirono S, Okura Y, Kato K, Hotta Y, Honda T, Kuwano R, Aizawa Y (2000) Expression of coxsackievirus and adenovirus receptor in hearts of rats with experimental autoimmune myocarditis. Circ Res 86:275-280

Jin O, Sole MJ, Butany JW, Chia WK, McLaughlin PR, Liu P, Liew CC (1990) Detection of enterovirus RNA in myocardial biopsies from patients with myocarditis and cardiomyopathy using gene amplification by polymerase chain reaction. Circulation 82:8-16

Juhela S, Hyoty H, Uibo R, Meriste SH, Uibo O, Lonnrot M, Halminen M, Simell O, Ilonen J (1999) Comparison of enterovirus-specific cellular immunity in two populations of young children vaccinated with inactivated or live poliovirus vaccines. Clin Exp Immunol 117:100-105

Kammerer U, Kunkel B, Korn K (1994) Nested PCR for specific detection and rapid identification of human picornaviruses. J Clin Microbiol 32:285-291

Kanno T, Kim K, Kono K, Drescher KM, Chapman NM, Tracy S (2006) Group B coxsackievirus diabetogenic phenotype correlates with replication efficiency. J Virol 80:5637-5643

Kawai C (1999) From myocarditis to cardiomyopathy: mechanisms of inflammation and cell death: learning from the past for the future. Circulation 99:1091-1100

Khetsuriani N, Lamonte-Fowlkes A, Oberst S, Pallansch MA (2006) Enterovirus surveillance - United States, 1970-2005. MMWR Surveill Summ 55:1-20

Kim JH, Paek KY, Choi K, Kim TD, Hahm B, Kim KT, Jang SK (2003) Heterogeneous nuclear ribonucleoprotein C modulates translation of c-myc mRNA in a cell cycle phase-dependent manner. Mol Cell Biol 23:708-720

Kim KS, Hufnagel G, Chapman NM, Tracy S (2001) The group B coxsackieviruses and myocarditis. Rev Med Virol 11:355-368

Kim KS, Tracy S, Tapprich W, Bailey J, Lee CK, Kim K, Barry WH, Chapman NM (2005) 5′-Terminal deletions occur in coxsackievirus B3 during replication in murine hearts and cardiac myocyte cultures and correlate with encapsidation of negative-strand viral RNA. J Virol 79:7024-7041

Klingel K, Hohenadl C, Canu A, Albrecht M, Seemann M, Mall G, Kandolf R (1992) Ongoing enterovirus-induced myocarditis is associated with persistent heart muscle infection: quantitative analysis of virus replication, tissue damage, and inflammation. Proc Natl Acad Sci USA 89:314-318

Klingel K, Schnorr JJ, Sauter M, Szalay G, Kandolf R (2003) beta2-microglobulin-associated regulation of interferon-gamma and virus-specific immunoglobulin G confer resistance against the development of chronic coxsackievirus myocarditis. Am J Pathol 162:1709-1720

Knowlton KU, Jeon ES, Berkley N, Wessely R, Huber S (1996) A mutation in the puff region of VP2 attenuates the myocarditic phenotype of an infectious cDNA of the Woodruff variant of coxsackievirus B3. J Virol 70:7811-7818

Kuhl U, Pauschinger M, Schwimmbeck PL, Seeberg B, Lober C, Noutsias M, Poller W, Schultheiss HP (2003) Interferon-beta treatment eliminates cardiotropic viruses and improves left ventricular function in patients with myocardial persistence of viral genomes and left ventricular dysfunction. Circulation 107:2793-2798

Kuhl U, Pauschinger M, Seeberg B, Lassner D, Noutsias M, Poller W, Schultheiss HP (2005) Viral persistence in the myocardium is associated with progressive cardiac dysfunction. Circulation 112:1965-1970

Lacher MD, Tiirikainen MI, Saunier EF, Christian C, Anders M, Oft M, Balmain A, Akhurst RJ, Korn WM (2006) Transforming growth factor-beta receptor inhibition enhances adenoviral infectability of carcinoma cells via up-regulation of coxsackie and adenovirus receptor in conjunction with reversal of epithelial-mesenchymal transition. Cancer Res 66:1648-1657

Lee C, Maull E, Chapman N, Tracy S, Wood J, Gauntt C (1997) Generation of an infectious cDNA of a highly cardiovirulent coxsackievirus B3(CVB3m) and comparison to other infectious CVB3 cDNAs. Virus Res 50:225-235

Li Y, Bourlet T, Andreoletti L, Mosnier JF, Peng T, Yang Y, Archard LC, Pozzetto B, Zhang H (2000) Enteroviral capsid protein VP1 is present in myocardial tissues from some patients with myocarditis or dilated cardiomyopathy. Circulation 101:231-234

Liu P, Aitken K, Kong YY, Opavsky MA, Martino T, Dawood F, Wen WH, Kozieradzki I, Bachmaier K, Straus D, Mak TW, Penninger JM (2000) The tyrosine kinase p56lck is essential in coxsackievirus B3-mediated heart disease. Nat Med 6:429-434

Lodge PA, Herzum M, Olszewski J, Huber SA (1987) Coxsackievirus B-3 myocarditis. Acute and chronic forms of the disease caused by different immunopathogenic mechanisms. Am J Pathol 128:455-463

Luppi P, Rudert W, Licata A, Riboni S, Betters D, Cotrufo M, Frati G, Condorelli G, Trucco M (2003) Expansion of specific alphabeta+ T-cell subsets in the myocardium of patients with myocarditis and idiopathic dilated cardiomyopathy associated with Coxsackievirus B infection. Hum Immunol 64:194-210

Mahon NG, Zal B, Arno G, Risley P, Pinto-Basto J, McKenna WJ, Davies MJ, Baboonian C (2001) Absence of viral nucleic acids in early and late dilated cardiomyopathy. Heart 86:687-692

Mahrholdt H, Wagner A, Deluigi CC, Kispert E, Hager S, Meinhardt G, Vogelsberg H, Fritz P, Dippon J, Bock CT, Klingel K, Kandolf R, Sechtem U (2006) Presentation, patterns of myocardial damage, and clinical course of viral myocarditis. Circulation 114:1581-1590

Martin AB, Webber S, Fricker FJ, Jaffe R, Demmler G, Kearney D, Zhang YH, Bodurtha J, Gelb B, Ni J et al. (1994) Acute myocarditis. Rapid diagnosis by PCR in children. Circulation 90:330-339

Mason JW (2003) Myocarditis and dilated cardiomyopathy: an inflammatory link. Cardiovasc Res 60:5-10

Neu N, Rose NR, Beisel KW, Herskowitz A, Gurri-Glass G, Craig SW (1987) Cardiac myosin induces myocarditis in genetically predisposed mice. J Immunol 139:3630-3636

Oberste MS, Maher K, Pallansch MA (2004a) Evidence for frequent recombination within species human enterovirus B based on complete genomic sequences of all thirty-seven serotypes. J Virol 78:855-867

Oberste MS, Penaranda S, Pallansch MA (2004b) RNA recombination plays a major role in genomic change during circulation of coxsackie B viruses. J Virol 78:2948-2955

Ouyang X, Zhang H, Bayston TA, Archard LC (1995) Detection of Coxsackievirus B3 RNA in mouse myocarditis by nested polymerase chain reaction. Clin Diagn Virol 3:233-245

Pauschinger M, Doerner A, Kuehl U, Schwimmbeck PL, Poller W, Kandolf R, Schultheiss HP (1999) Enteroviral RNA replication in the myocardium of patients with left ventricular dysfunction and clinically suspected myocarditis. Circulation 99:889-895

Pummerer CL, Grassl G, Sailer M, Bachmaier KW, Penninger JM, Neu N (1996) Cardiac myosin-induced myocarditis: target recognition by autoreactive T cells requires prior activation of cardiac interstitial cells. Lab Invest 74:845-852

Reetoo KN, Osman SA, Illavia SJ, Cameron-Wilson CL, Banatvala JE, Muir P (2000) Quantitative analysis of viral RNA kinetics in coxsackievirus B3-induced murine myocarditis: biphasic pattern of clearance following acute infection, with persistence of residual viral RNA throughout and beyond the inflammatory phase of disease. J Gen Virol 81:2755-2762

Rey L, Lambert V, Wattre P, Andreoletti L (2001) Detection of enteroviruses ribonucleic acid sequences in endomyocardial tissue from adult patients with chronic dilated cardiomyopathy by a rapid RT-PCR and hybridization assay. J Med Virol 64:133-140

Sadeharju K, Knip M, Virtanen SM, Savilahti E, Tauriainen S, Koskela P, Akerblom HK, Hyoty H (2007) Maternal antibodies in breast milk protect the child from enterovirus infections. Pediatrics 119:941-946

Satoh M, Nakamura M, Akatsu T, Shimoda Y, Segawa I, Hiramori K (2004) Toll-like receptor 4 is expressed with enteroviral replication in myocardium from patients with dilated cardiomyopathy. Lab Invest 84:173-181

Satoh M, Tamura G, Segawa I, Hiramori K, Satodate R (1994) Enteroviral RNA in dilated cardiomyopathy. Eur Heart J 15:934-939

Schmidtke M, Selinka HC, Heim A, Jahn B, Tonew M, Kandolf R, Stelzner A, Zell R (2000) Attachment of coxsackievirus B3 variants to various cell lines: mapping of phenotypic differences to capsid protein VP1. Virology 275:77-88

Schwaiger A, Umlauft F, Weyrer K, Larcher C, Lyons J, Muhlberger V, Dietze O, Grunewald K (1993) Detection of enteroviral ribonucleic acid in myocardial biopsies from patients with idiopathic dilated cardiomyopathy by polymerase chain reaction. Am Heart J 126:406-410

Sharma N, O'Donnell BJ, Flanegan JB (2005) 3'-Terminal sequence in poliovirus negative-strand templates is the primary cis-acting element required for VPgpUpU-primed positive-strand initiation. J Virol 79:3565-3577

Shirali GS, Ni J, Chinnock RE, Johnston JK, Rosenthal GL, Bowles NE, Towbin JA (2001) Association of viral genome with graft loss in children after cardiac transplantation. N Engl J Med 344:1498-1503

Soonpaa MH, Field LJ (1997) Assessment of cardiomyocyte DNA synthesis in normal and injured adult mouse hearts. Am J Physiol 272: H220-H226

Spiller OB, Goodfellow IG, Evans DJ, Almond JW, Morgan BP (2000) Echoviruses and coxsackie B viruses that use human decay-accelerating factor (DAF) as a receptor do not bind the rodent analogues of DAF. J Infect Dis 181:340–343

Spotnitz MD, Lesch M (2006) Idiopathic dilated cardiomyopathy as a late complication of healed viral (Coxsackie B virus) myocarditis: historical analysis, review of the literature, and a postulated unifying hypothesis. Prog Cardiovasc Dis 49:42-57

Stadnick E, Dan M, Sadeghi A, Chantler JK (2004) Attenuating mutations in coxsackievirus B3 map to a conformational epitope that comprises the puff region of VP2 and the knob of VP3. J Virol 78:13987-4002

Szalay G, Meiners S, Voigt A, Lauber J, Spieth C, Speer N, Sauter M, Kuckelhorn U, Zell A, Klingel K, Stangl K, Kandolf R (2006) Ongoing coxsackievirus myocarditis is associated with increased formation and activity of myocordial immunoproteasomes. Am J Pathol 168:1542–1552

Tam PE (2006) Coxsackievirus myocarditis: interplay between virus and host in the pathogenesis of heart disease. Viral Immunol 19:133-146

Tam PE, Messner RP (1999) Molecular mechanisms of coxsackievirus persistence in chronic inflammatory myopathy: viral RNA persists through formation of a double-stranded complex without associated genomic mutations or evolution. J Virol 73:10113-10121

Tomko RP, Xu R, Philipson L (1997) HCAR and MCAR: the human and mouse cellular receptors for subgroup C adenoviruses and group B coxsackieviruses. Proc Natl Acad Sci USA 94:3352-3356

Tracy S, Chapman NM, Tu Z (1992) Coxsackievirus B3 from an infectious cDNA copy of the genome is cardiovirulent in mice. Arch Virol 122:399-409

Tracy S, Hofling K, Pirruccello S, Lane PH, Reyna SM, Gauntt CJ (2000) Group B coxsackievirus myocarditis and pancreatitis: connection between viral virulence phenotypes in mice. J Med Virol 62:70-81

Van Houten N, Bouchard PE, Moraska A, Huber SA (1991) Selection of an attenuated Coxsackievirus B3 variant, using a monoclonal antibody reactive to myocyte antigen. J Virol 65:1286-1290

Vincent T, Pettersson RF, Crystal RG, Leopold PL (2004) Cytokine-mediated downregulation of coxsackievirus-adenovirus receptor in endothelial cells. J Virol 78:8047-8058

Viskari H, Ludvigsson J, Uibo R, Salur L, Marciulionyte D, Hermann R, Soltesz G, Fuchtenbusch M, Ziegler AG, Kondrashova A, Romanov A, Kaplan B, Laron Z, Koskela P, Vesikari T, Huhtala H, Knip M, Hyoty H (2005) Relationship between the incidence of type 1 diabetes and maternal enterovirus antibodies: time trends and geographical variation. Diabetologia 48:1280-1287

Viskari H, Ludvigsson J, Uibo R, Salur L, Marciulionyte D, Hermann R, Soltesz G, Fuchtenbusch M, Ziegler AG, Kondrashova A, Romanov A, Knip M, Hyoty H (2004) Relationship between the incidence of type 1 diabetes and enterovirus infections in different European populations: results from the EPIVIR project. J Med Virol 72:610-617

von Herrath MG, Fujinami RS, Whitton JL (2003) Microorganisms and autoimmunity: making the barren field fertile? Nat Rev Microbiol 1:151-157

Wessely R, Henke A, Zell R, Kandolf R, Knowlton KU (1998a) Low-level expression of a mutant coxsackieviral cDNA induces a myocytopathic effect in culture: an approach to the study of enteroviral persistence in cardiac myocytes. Circulation 98:450-457

Wessely R, Klingel K, Santana LF, Dalton N, Hongo M, Jonathan Lederer W, Kandolf R, Knowlton KU (1998b) Transgenic expression of replication-restricted enteroviral genomes in heart muscle induces defective excitation-contraction coupling and dilated cardiomyopathy. J Clin Invest 102:1444-1453

Why HJ, Meany BT, Richardson PJ, Olsen EG, Bowles NE, Cunningham L, Freeke CA, Archard LC (1994) Clinical and prognostic significance of detection of enteroviral RNA in the myocardium of patients with myocarditis or dilated cardiomyopathy. Circulation 89:2582-2589

Witso E, Palacios G, Cinek O, Stene LC, Grinde B, Janowitz D, Lipkin WI, Ronningen KS (2006) High prevalence of human enterovirus a infections in natural circulation of human enteroviruses. J Clin Microbiol 44:4095-4100

Wolfgram LJ, Beisel KW, Herskowitz A, Rose NR (1986) Variations in the susceptibility to Coxsackievirus B3-induced myocarditis among different strains of mice. J Immunol 136:1846-1852

Woodruff JF (1980) Viral myocarditis. A review. Am J Pathol 101:425-484

Xiong D, Yajima T, Lim BK, Stenbit A, Dublin A, Dalton ND, Summers-Torres D, Molkentin JD, Duplain H, Wessely R, Chen J, Knowlton KU (2007) Inducible cardiac-restricted expression of enteroviral protease 2A is sufficient to induce dilated cardiomyopathy. Circulation 115:94-102

Zanone MM, Favaro E, Conaldi PG, Greening J, Bottelli A, Perin PC, Klein NJ, Peakman M, Camussi G (2003) Persistent infection of human microvascular endothelial cells by coxsackie B viruses induces increased expression of adhesion molecules. J Immunol 171:438-446

Zhang H, Li Y, McClean DR, Richardson PJ, Florio R, Sheppard M, Morrison K, Latif N, Dunn MJ, Archard LC (2004) Detection of enterovirus capsid protein VP1 in myocardium from cases of myocarditis or dilated cardiomyopathy by immunohistochemistry: further evidence of enterovirus persistence in myocytes. Med Microbiol Immunol (Berl) 193:109-114

Zhang H, Li Y, Peng T, Aasa M, Zhang L, Yang Y, Archard LC (2000) Localization of enteroviral antigen in myocardium and other tissues from patients with heart muscle disease by an improved immunohistochemical technique. J Histochem Cytochem 48:579-584

Zoll GJ, Melchers WJ, Kopecka H, Jambroes G, van der Poel HJ, Galama JM (1992) General primer-mediated polymerase chain reaction for detection of enteroviruses: application for diagnostic routine and persistent infections. J Clin Microbiol 30:160-165

# Autoimmunity in Coxsackievirus Infection

N. R. Rose

| | | |
|---|---|---|
| 1 | Introduction | 294 |
| | 1.1 Some Basic Definitions | 294 |
| | 1.2 Establishing an Autoimmune Disease | 295 |
| | 1.3 Virus-Induced Autoimmunity | 296 |
| | 1.4 Immunopathology of Viral Infection | 296 |
| 2 | Coxsackievirus B3-Induced Myocarditis | 297 |
| | 2.1 Spectrum of Responses in Humans | 297 |
| | 2.2 Spectrum of Responses in Mice | 297 |
| 3 | Autoimmune Myocarditis | 299 |
| | 3.1 Two Phases of Coxsackievirus-Induced Myocarditis | 299 |
| | 3.2 Cardiac Myosin Antibodies in Late-Phase Disease | 299 |
| | 3.3 Cardiac Myosin-Induced Myocarditis | 300 |
| 4 | The Basis of Susceptibility to Autoimmune Myocarditis | 301 |
| | 4.1 Innate Immune Response | 301 |
| | 4.2 The Adjuvant Effect | 302 |
| | 4.3 Complement | 303 |
| | 4.4 Nitric Oxide | 303 |
| | 4.5 The $Th_1/Th_2$ Paradigm | 304 |
| | 4.6 Genetic Regulation of Autoimmune Myocarditis | 305 |
| 5 | Pathogenetic Mechanisms | 306 |
| | 5.1 T Cells | 306 |
| | 5.2 Macrophage | 306 |
| | 5.3 Antibody | 307 |
| 6 | Autoimmunity in Human Myocarditis | 307 |
| References | | 309 |

**Abstract** Viral infections frequently result in the production of autoantibodies. In most cases, these autoantibodies are low-affinity IgMs that exhibit extensive cross-reactions. Sometimes these virus-triggered immune responses progress to a

---

N. R. Rose
Johns Hopkins Center for Autoimmune Disease Research, Johns Hopkins University,
615 N. Wolfe St., E5009, Baltimore, MD 21205, USA
nrrose@jhmi.edu

pathogenic autoimmunity to form autoimmune disease. To delineate the mechanisms determining induction of autoimmune disease, we have investigated in detail a model of autoimmune myocarditis induced in genetically susceptible mice by infection with a cardiotropic strain of coxsackievirus B3. We found that the autoimmune sequelae of the viral infection can be simulated by immunization of the susceptible mice with murine cardiac myosin. In both models of the disease, the determination of whether to progress from a contained viral myocarditis to a pathogenic autoimmune response is made within hours after induction of infection and is characterized by production of a few key cytokines, including IL-1β and TNFα. Many of the lessons learned from study of these models are applicable to human myocarditis and dilated cardiomyopathy.

# 1 Introduction

The immune system is designed to recognize the universe of antigens. To limit those immune responses directed to antigen of the host that might cause harm, mechanisms of central and peripheral tolerance have evolved. In reality, however, many host antigens are also found on existing or emerging pathogenic microorganisms. Eliminating all such reactions entirely might leave the host vulnerable to uncontrolled infection. A balance must be struck; the immune response repertoire must remain sufficiently diverse to recognize and eliminate present and future pathogens without attacking the host. Thus, autoreactive responses are constantly being generated, but are normally kept under control by homeostatic mechanisms. When these mechanisms fail or are overwhelmed, autoimmune disease may ensue.

It is not surprising, therefore, that autoimmunity in the form of autoantibodies commonly results from infection. The issue discussed in this chapter is when and how benign autoimmunity, induced by an infecting virus, sometimes gives rise to autoimmune disease by providing an unusually strong immunogenic stimulus, by overcoming homeostatic regulation, or by a combination of both.

## 1.1 *Some Basic Definitions*

Autoimmune disease represents a large and growing problem in the United States and other industrialized societies. In a recent report, the U.S. National Institutes of Health estimated that some 5-8% of Americans suffer from one of the 80 recognized autoimmune disorders. The dimensions of the problem, however, depend on the definition of autoimmune disease.

In decades past, it was believed that autoimmunity was a rare event that can occur only when a few isolated or sequestered antigens are exposed to the host's immune system. The lens, testis and, to some extent, brain were regarded as privileged sites in which the normal mechanisms of self-tolerance and immune regulation do not apply. In more recent years, we have learned that immune responses to many autologous antigens occur frequently as part of normal physiology, taking the form of naturally occurring circulating autoantibodies or self-reactive B and T cells. Natural autoimmunity is normally held in check by homeostatic regulatory mechanisms. Only when these regulatory mechanisms are rendered ineffective or breached does limited autoimmune response progress to autoimmune disease. Within the context described above, autoimmune disease can be defined as the pathologic consequence of an unchecked autoimmune response caused by a potent antigenic stimulus and loss of normal homeostasis.

## 1.2 Establishing an Autoimmune Disease

In practical terms, establishing that autoimmunity is the cause of a human disease, rather than its consequence or an innocent bystander, is a formidable task. The examples of human diseases that are clearly caused by autoimmunity are generally confined to those instances where the damage is attributable to circulating autoantibody. In some cases, the antibody is transferred to an infant as an experiment of nature during pregnancy, as exemplified by neonatal hyperthyroidism, myasthenia gravis, pemphigus, and lupus-associated congenital heart block. Sometimes antibody transfer can reproduce the disease in a vulnerable experimental animal as with pemphigus and pemphigoid. When blood cells are the target of an autoimmune response, the disease can be approximated in the test tube as represented by some forms of acquired hemolytic anemia, idiopathic thrombocytopenia, and leukopenia.

Putting aside these few examples of definite autoimmune disease, investigations to establish the autoimmune basis of human disease generally depend upon the availability or production of animal models in which transfer of the disease by antibody or T cells is feasible. A model can sometimes be developed by experimental immunization of rats or mice if the requisite antigen has been identified. Usually a powerful adjuvant is required to induce disease, and only certain strains of animals are susceptible. Well-known examples include experimental autoimmune thyroiditis, orchitis, uveitis, and encephalomyelitis. It is uncertain how closely these models reproduce human disease. While they may share many characteristics of pathogenesis, they clearly differ in etiology.

An alternative strategy is to identify in an experimental animal a spontaneous disease that closely resembles the human equivalent. A number of genetically determined models of lupus, hemolytic anemia, thyroiditis, and diabetes have been discovered and studied in detail. Transfer of the disease is then possible using antibody, lymphocytes, or a combination of the two. The transfer may definitely assign the cause of the disease to an autoimmune response. More recently, genetically engineered mice have proved to be of great value in replicating human diseases in rodents. Manipulations may involve interrupting normal production of major regulatory cytokines by genetic homologous recombination or transduction, or by inserting in mice human major histocompatibility complexes (MHC) that direct susceptibility. In animal experiments, these molecular genetic approaches have greatly amplified the opportunities for developing animal models to analyze human diseases. But they are, in the last instance, limited models and not necessarily the full picture. Useful though they have proven to be for studies of pathogenetic mechanisms, experimental models of human diseases in animals have intrinsic limitations in investigations of causation.

## 1.3 Virus-Induced Autoimmunity

The association of viral infection and autoimmunity has been remarked upon frequently, but the mechanisms underlying the association have not been well understood (Rose and Griffin 1991). Viral infections regularly initiate the production of autoimmune responses, but only rarely are they pathogenic. Several different mechanisms have been proposed to explain virus-induced autoimmunity. Sometimes the responses result from cross-reactions between antigens of the virus and those of the host, so-called molecular (or, more properly, epitope) mimicry (Rose and Mackay 2000; Cunningham 2004). Recent investigations have emphasized the frequent occurrence of molecular mimicry in nature, both at the level of antibodies and T cells. Virus infection may amplify preexisting, naturally occurring autoantibodies and self-reactive T cells. This amplification of our innate autoimmune repertoire may result from the adjuvant effect of the viral infection itself (Rose and Afanasyeva 2003; Fairweather et al. 2005). Viral infection can also release and alter normally isolated, inaccessible self-antigens. Thus, we are not surprised when many viral infections, or even immunization using live vaccines, increases the levels of preexisting autoantibodies. The problem facing the investigator is to determine whether these common autoimmune consequences of viral infection are harmless or pathogenetic; that is, whether they contribute in some meaningful way to a disease process. If such evidence can be mustered, the disease can be regarded as autoimmune within the definition developed above. There are still very few well-established instances of a human autoimmune disease resulting from a particular viral infection.

## 1.4 Immunopathology of Viral Infection

The host has a number of weapons protect against viral infection, although they sometimes result in host cell death by apoptosis or necrosis. Initially, the host defends itself by producing neutralizing antibodies to prevent attachment and entry of the virus into the cell. Following viral entry into cells, viral infections instigate immunopathologic responses resulting from the mechanisms by which the host attempts to clear virus. For instance, the host may respond by mounting a T-cell-mediated immune reaction to virus-infected cells that inflicts damage and leads to elimination of the infected cells. Such host responses are essential for the arrest of viral infection, and an impaired cellular response on the part of the host can result in overwhelming infection. Whether an immune response on balance benefits to the host depends, among other factors, upon the strength and promptness of the initial immune response, the extent of host-cell damage caused by T-cell-mediated immunity, and the possibilities for replacing essential cells that have sustained injury. Overcoming viral infection depends upon eliminating or containing the virus while maintaining, or even enhancing, the host's immune response. Treatment of the immunopathic consequences, in contrast, may involve suppressing the host's immune response. Distinguishing the damage caused by the virus from that due to immunopathic defense mechanisms remains a clinical priority.

Based on these considerations, investigation of the pathogenic effects of the viral infection must distinguish three categories of cellular damage: (1) direct effects due to viral infection of the cell, (2) immunopathic effects related to the host response to active or persistent viral infection, and (3) autoimmune responses initiated by the viral infection resulting from molecular mimicry, alteration or liberation of host products, or immunopotentiating effects of viral infection. To better define these mechanisms, my research group undertook a systematic study of coxsackievirus-induced myocarditis in the early 1980s.

## 2 Coxsackievirus B3-Induced Myocarditis

### 2.1 Spectrum of Responses in Humans

Group B coxsackievirus infections of humans can take many forms, but frequently produce a flulike syndrome with fever, arthralgias, and malaise (Tam 2006). Myocarditis and pericarditis are major causes of sudden death and have been estimated to occur in about 5% of cases of coxsackievirus B3 (CB3) infection. The incidence, however, is highly age-related because the occurrence of myocarditis is seen predominantly in individuals younger than 40 years of age (Feldman and McNamara 2000). In fact, newborns are particularly susceptible to coxsackievirus B infections, which can lead to fulminating myocarditis, encephalitis, meningitis, or other life-threatening diseases. Although the basis of this marked age dependency on disease outcome is not understood, it underlines the critical role of the host response to the viral infection and deserves much more investigation.

### 2.2 Spectrum of Responses in Mice

The great majority of patients deal with CB3 infections well and recover completely. We hypothesized, therefore, that if a pathogenic autoimmune response follows CB3 infection, it must signify an infrequent, genetically determined susceptibility. Since the response of humans to this viral infection is genetically diverse, we began by studying genetic differences in susceptibility to viral disease using 22 genetically well-defined inbred mouse strains, including congenic combinations differing only at the murine MHC, H-2.

In these 4- to 6-week-old mice, the investigation revealed strain-related differences in content of infectious CB3 in the heart and other tissues as well as in levels of virus-specific neutralizing antibody in the serum. Briefly, we found that infectious CB3 was present in the bloodstream of all of the mice examined at day 2 and was undetectable by day 5 (Herskowitz et al. 1985; Wolfgram et al. 1986.) Strains differed significantly in the degree of viremia on days 2 and 3. For example, strain A.SW had higher mean levels of virus on days 2 and 3 than B10.S mice, even though the two strains share the same H-2 haplotype, implicating non-MHC background genes in regulating the ability of the mouse to clear virus. On the other hand, A.BY mice had significantly more virus in the blood than did mice of the A.CA strain, which differed only at the H-2 locus. Thus, both MHC and non-MHC genes are involved in providing protective immunity for the host.

When the virus content of the heart was examined, all strains showed significant levels of viable virus on day 2 with a peak on day 5. The last day in which viable virus could be recovered from the hearts of any of the mice was day 9 after infection. There were, however, marked differences among the mouse strains in the incidence and severity of virus-induced myocarditis, so that, for example, the A.BY and A.SW strains, which had the highest levels of viremia, showed the greatest prevalence and severity of myocarditis.

Significant differences were found in the time of appearance of neutralizing antibody in the serum. The more resistant strains all exhibited neutralizing antibodies on day 3, whereas the more susceptible strains had demonstrable antibody only on day 5. Initially, then, the prompt production of neutralizing antibody controls the later course of viral infection.

On day 15, dramatic changes were seen in the histologic appearance of myocarditis. Strains sharing the B10 background, such as B10.S and B10.PL, showed evidence of healing, represented by focal necrotic lesions with minor residual cellular infiltration. By day 21, evidence of the earlier inflammation had virtually disappeared. On the other hand, strains on the A background, A/J A.BY, A.SW, A.CA still showed an active diffuse interstitial infiltrate. Despite the considerable differences in the severity of the earlier viral disease among the A strains, they were indistinguishable in both incidence and severity of myocardial inflammation on day 15. For example, the A.CA strain, which had a more modest level of viremia, had an inflammatory response on day 15 that matched the A.SW strain. Thus the extent of preliminary viral infection did not determine the severity of later disease.

Careful study brought out differences in the histologic pattern of early (day 9) and late (day 21) myocarditis in A strain mice. The early disease was characterized by focal myocyte death and contraction band necrosis, together with infiltrates of polymorphonuclear neutrophils and macrophages, suggestive of direct viral damage. The strains with continuing myocarditis showed a more generalized mononuclear infiltrate that persisted for at least 45 days, accompanied by increasing presence of fibrous connective tissue. Clearly, we were dealing with two different diseases, an early viral infection and a later pathologic response in the absence of infectious virus (Wolfgram et al. 1986, 1989). These experiments, of course, could not exclude the continuing presence of latent or persistent virus. Yet isolated T cells from mice with CB3-induced myocarditis showed cytolytic activity against both virus infected and noninfected cardiomyocyte in vitro (Hill et al. 2002). Later, Fairweather et al. (2001) reported that viral RNA persisted in strains of mice that resolved disease on day 21 as well as those with continuing myocarditis. We could infer, therefore, that persistent virus infection did not account for the late phase of disease, but that autoimmunologic mechanisms were responsible.

## 3 Autoimmune Myocarditis

### 3.1 Two Phases of Coxsackievirus-Induced Myocarditis

All of the mice were examined for the production of heart-reactive autoantibodies by indirect immunofluorescence (Wolfgram et al. 1985). Most of the sera taken on days 5-7 after CB3 infection reacted with both cardiac and skeletal muscle. On day 15, however, all of the A strain mice developed autoantibodies that reacted mainly with cardiac muscle. These antibodies persisted through day 45 in most of the animals.

Several patterns of immunofluorescent localization were distinguished with sera from CB3-infected animals (Alvarez et al. 1987). One pattern, termed antisarcolemal, represented the reaction with the myocyte membrane and resembled to the pattern described with many human sera from patients with postviral myocarditis, rheumatic carditis, postpericardiotomy syndrome and postmyocardial infarction syndrome. A second reaction pattern suggested the localization on the intracellular contractual proteins and resembled the antifibrillary antibodies described in many human cases of postviral myocarditis.

To determine more conclusively the specificity of the heart-reactive autoantibodies, absorption experiments were carried out with insoluble fractions of various mouse tissues. Heart homogenate readily absorbed all of the reactivity from the sera with heart, whereas skeletal muscle homogenate had only a partial effect on the reaction with heart, although it removed all of the antibody reactive with skeletal muscle. Other tissue fractions (such as kidney and liver) were unable to absorb any of the reaction with heart. We concluded that production of heart-specific autoantibodies marked the development of late-phase disease in the susceptible strains.

### 3.2 Cardiac Myosin Antibodies in Late-Phase Disease

The presence of heart-specific antibodies in the late phase of disease suggested that this continuing form of myocarditis may be related to autoimmunity. We therefore embarked on a program to identify the responsible antigen and to reproduce the disease by experimental immunization. For that purpose, we used sera from those CB3-infected mice that developed the ongoing myocarditis (Neu et al. 1987a). Samples were collected early (5-7 days) and later (15-21 days) after infection. In addition, serum samples from mice that failed to develop the ongoing myocarditis were included in the investigation. The sera as well as normal control sera were tested with heart and skeletal muscle preparations using Western immunoblots. Serum samples taken from the mice susceptible to the late phase of myocarditis (such as A strain mice) reacted strongly with a 200-kDa protein band present in the extracts from both heart and skeletal muscle. In contrast, samples from the early time point or from mice resistant to late-phase myocarditis showed only weak reactions with this band, as did a normal mouse serum. Further studies using monoclonal antibodies identified this band as the $\alpha$ heavy chain of myosin.

Absorptions were then performed using purified cardiac myosin bound to Sepharose beads. The antibody eluted from the beads reacted with the heart sections showing the fibrillary band staining pattern. Thus, myosin-specific antibodies dominate the post-CB3 infection response of mice that developed late-phase myocarditis.

Further experiments used more quantitative ELISA and competitive inhibition assays. Serum samples were taken days 15-21 after CB3 infection from mice susceptible or resistant to the late-phase of myocarditis. Serum from both susceptible and resistant mice had antibodies, mainly of the IgM class, reactive with highly purified mouse skeletal and cardiac myosins, as did some of the normal mouse sera tested. In contrast, only serum samples from the susceptible strains (such as the A strains) contained high levels of IgG antibody reactive only with cardiac myosin. Thus, IgM antibodies reacted equally well with myosin prepared from heart, skeletal muscle, and, to a slightly lesser degree, brain, whereas the IgG antibody found in susceptible mice reacted predominantly with heart myosin. Finally, quantitative inhibition studies revealed that two specificities of IgG antibodies are present in the sera of susceptible strains of mice, one reacting specifically with cardiac myosin, the other reacting with cardiac, skeletal, and brain myosin. The results showed that the IgM antibodies to myosin frequently present in serum of virus-infected (and even some uninfected) mice did not distinguish cardiac from other forms of myosin. In contrast, sera from susceptible, but not resistant, mice contained an IgG antibody that specifically recognized the cardiac isoform of myosin. Further detailed studies have shown that the myosin-specific antibody found in susceptible A/J strain mice is predominantly of the IgG1 subclass. For certain strains of mice, the precise amino acid sequences recognized by the cardiac myosin-specific antibodies have been identified (Donermeyer et al. 1995; Pummerer et al. 1996).

## 3.3 Cardiac Myosin-Induced Myocarditis

The next step in establishing the autoimmune origin of the late phase of myocarditis required reproducing the disease by experimental immunization with the candidate antigen, cardiac myosin (Neu et al. 1987b; Rose et al. 1987). For these experiments, we selected a number of strains that were known to be susceptible to the late phase of CB3-induced myocarditis, all of which had the same A background, but differed at H-2. For comparison, mice classified as resistant to late-phase disease were included, such as C57BL/10, C57BL/6 and B10.A. The purified myosin was first emulsified in complete Freund adjuvant (CFA). Six-week-old mice were injected twice with varying doses of purified cardiac or skeletal muscle myosin at 7-day intervals. Control groups received injections of CFA alone. Serum samples were taken periodically and tested for antibodies to heart and skeletal muscle myosin by ELISA. At the end of the experiment, hearts were removed and evaluated histologically. The findings clearly showed that all of the A strains susceptible to late-phase myocarditis developed strong IgG antibodies specific for cardiac myosin. Most of the animals also developed a histologic picture typical of the late-phase myocarditis. Although all of the A strain congenics were reactive, the incidence and severity of disease varied markedly with the H-2 haplotype. For example, A.SW and A.CA mice developed severe myocarditis with a high incidence, whereas A.BY animals showed only minimal or no disease. We concluded that the A background provided genes important for general susceptibility to cardiac myosin-induced myocarditis and that genes within the H-2 segment were important in directing and regulating the response.

In contrast to the myocarditis and cardiac myosin-specific IgG antibodies evident in all of the A strain mice, no cardiac myosin-specific IgG antibodies were produced by the C57BL/10 or C57BL/6 mice resistant to late-phase myocarditis. Of importance, none of the mice administered skeletal muscle myosin developed myocarditis whether they were susceptible or resistant to the cardiac myosin-induced disease.

These studies proved that mouse strains that are susceptible to CB3-induced late-phase myocarditis are also susceptible to autoimmune myocarditis induced by immunization with purified cardiac myosin in CFA. Thus, the genetic predisposition to develop an immune response to cardiac myosin is the critical factor in determining susceptibility to the late phase of the viral disease. These experiments further confirmed that the severity of the early phase of disease associated with direct viral injury is not predictive of late-phase autoimmune myocarditis. We later found that porcine myosin can induce the same disease in the susceptible A strains of mice (Wang et al. 1999). Genetic susceptibility to myosin-induced myocarditis coincided precisely with susceptibility to the virus-induced autoimmune disease. As firm evidence of its autoimmune etiology, the late-phase disease was reproduced by adoptive transfer of lymphocytes stimulated with cardiac myosin. These experiments established for the first time that the virus-induced autoimmune disease can be reproduced in the absence of viral infection by direct immunization with a defined antigen.

## 4 The Basis of Susceptibility to Autoimmune Myocarditis

### *4.1 Innate Immune Response*

We next asked what determines whether the mouse proceeds to an autoimmune disease or recovers completely from the initial viral infection. Similar questions arise with the myosin-induced model of disease. Does the early or innate immune response dictate the later events in the later, adaptive immunity (Kaya and Rose 2005)? To investigate the issue, we undertook a series of experiments using bacterial lipopolysaccharide (LPS), a potent agent known to augment and even redirect the innate immune response. We found that infecting mice with CB3 and simultaneously co-treating with LPS greatly increased the mortality beyond that seen with viral infection alone (Lane et al. 1991). A most striking finding of the study was the unexpected observation that even strains of mice that normally failed to develop the late-phase myocarditis, C57BL/10 and C57BL/6, now produced the typical autoimmune heart disease. Hence, a potent stimulus of the innate immune response overcame the genetic resistance of these mice.

The discovery that LPS treatment of CB3-infected mice that are normally resistant to autoimmune myocarditis raised the question of the mechanism by which this agent works. Serum samples of the LPS-treated animals showed heightened levels of the number of cytokines, but the most prominent ones were IL1$\beta$ and TNF$\alpha$ (Lane et al. 1992). In addition, large numbers of IL1$\beta$- and TNF$\alpha$-containing leukocytes were found within the inflammatory lesions in the heart. Direct evidence for the critical role of these two proinflammatory cytokines in promoting myocarditis even in genetically resistant strains of mice was produced by treating these resistant mice with recombinant cytokines at the time of CB3 infection. Finally, we found that immunizing resistant mice with the cardiac antigen and co-treating with LPS or administering recombinant IL1$\beta$ or TNF$\alpha$ also resulted in severe myocarditis. Subsequent experiments have shown that the differential production of these key cytokines, IL1$\beta$ and TNF$\alpha$, along with IL18, occurs as early as 6 h after infection with CB3 (Fairweather et al. 2004b). The very earliest innate immune response to the virus (or to antigen administered with adjuvant) is a powerful determinant of whether the animal will subsequently develop a pathogenic autoimmune response.

## 4.2 The Adjuvant Effect

LPS provides valuable insights into mechanisms by which the adjuvant effect of the infection may promote autoimmune disease (Fairweather et al. 2005). Typically, adjuvants interact with cells of the innate immune response to enhance antigen processing and presentation, upregulate essential cell surface molecules such as MHC class II and B-7 1/2 involved in antigen prevention and stimulate production of proinflammatory cytokines. Striking, these basic mechanisms are largely parallel in virus-induced and myosin-induced myocarditis. LPS mediates its effects, including upregulating production of the critical cytokines TNF$\alpha$, IL1$\beta$, and IL-18, signaling through TLR4. Viral infection as well as the mycobacterial component of CFA also activate TLR4. Accordingly, TLR4-deficient mice develop less severe disease and reduced viral replication in the heart on day 12 after CB3 infection despite equivalent virus in the heart on day 2 (Fairweather et al. 2003). The animals produced reduced levels of IL-1$\beta$ and IL-18 in their hearts on day 12, showing that signaling through

TLR4 contributes to the inflammatory autoimmune disease (Afanasyeva and Rose 2004).

## 4.3 Complement

Complement is another major component of the initial innate immune response that effects the later susceptibility of mice to autoimmune myocarditis induced either by viral infection or myosin immunization. Administration to mice of cobra venom factor to deplete complement component C3 at the time of immunization with cardiac myosin reduced the myosin-specific IgG antibody response and prevented the development of myocarditis (Kaya et al. 2001). Depletion of C3 at a later time during progression to disease was ineffective. The incidence and severity of disease was also reduced in mice genetically deficient in the C3 receptors, CR1 and CR2. Blockade of CR1 and CR2 at the time of myosin immunization with a monoclonal antibody that binds to the extracellular domain shared by these two receptors abrogated disease and reduced the production of myosin-specific antibodies. These effects were associated with decreased production of the two key cytokines, IL1$\beta$ and TNF$\alpha$, by splenocytes stimulated with cardiac myosin. Further evidence that the effect of complement occurs during the early, inductive stage of the immune response rather than the effector phase comes from these experiments conducted in A/J mice that are congenically deficient in C5 and, therefore, unable to develop the cell-damaging membrane attack complex. The precise role of complement in the induction of the disease requires further investigation. The complement receptors CR1 and CR2 are expressed on mouse B cells, but also on a subpopulation of activated mouse T cells, suggesting that complement may act directly on T cells and modify the course of the later, adaptive immune response (Kaya et al. 2005). CR1/CR2 deficiency also increases virus-mediated myocarditis, emphasizing that an agent may have opposite effects on viral disease vs autoimmune disease, demonstrating its differing effect on protective immunity and autoimmunity.

## 4.4 Nitric Oxide

In other studies, we found additional differences between the effect of a treatment during early viral myocarditis and later autoimmune disease. Nitric oxide (NO) exerts strong antiviral activity in CB3 infection (Lowenstein et al. 1996). Aminoguanidine (AG) blockage of inducible NO synthase soon after virus infection increased the severity of cardiac damage. Later drug treatment (10-20 days after infection) reduced the severity of the interstitial cardiac lesions in B10.M female mice, which are moderately susceptible to autoimmune disease (Hill and Rose 2001). We concluded that NO plays a protective role in the viral disease, but it enhances the later autoimmune myocarditis, perhaps by impairing cardiac contractibility (Hill et al. 2002).

## 4.5 The $Th_1/Th_2$ Paradigm

The critical role of CD4+ T cells in the induction of autoimmune myocarditis was first shown by Smith and Allen (1991) and confirmed in our laboratory (Afanasyeva et al. 2001a). In 1980, Mosmann and colleagues distinguished two subsets of CD4+ T cells based upon their production of certain key cytokines. $Th_1$ cells were characterized by secretion of IL-2 and IFN-γ, whereas $Th_2$ cells secreted IL-4 and IL-5. $TH_1$ cells were primarily associated with cell-mediated immunity and $Th_2$ cells with antibody production. Furthermore, the two sets of cytokines seem to be mutually inhibitory. While this paradigm has proven to be of great value in resolving many questions, it has not fit well with the findings in studies of autoimmune diseases such as myocarditis (Gor et al. 2003).

IL-12 is produced mainly by antigen-presenting cells and induces $Th_1$ responses. We found that IL-12 signaling through IL-12Rβ1 and STAT4 promotes autoimmune myocarditis (Afanasyeva et al. 2001b). IL-12β1 receptor knock-out and STAT4 knock-out mice were completely resistant to the myosin-induced disease. Recombinant IL-12 exacerbated disease and suppressed the production of IL-4, IL-5, and IL-13.

Because IL-12 leads to the production of IFN-γ, we tested the role of IFN-γ in myocarditis. Administration of anti-IFN-γ or use of IFN-γ knock-out mice exacerbated disease, leading to a classical picture of dilated cardiomyopathy and, in many animals, heart failure (Afanasyeva et al. 2005).

IFN-γ has been reported to exert negative inotropic effects on cardiac myocytes, possibly through activation of inducible NO (Hill et al. 2002). In addition, IFN-γ deficiency led to reduced apoptosis and accumulation of activated CD4+ T cells, suggestive of a reduction in activation-induced cell death.

IL-4 is the prototypic $Th_2$ cytokine. We found that anti-IL-4 administration reduced the severity of autoimmune myocarditis in highly susceptible A/J mice (Afanasyeva et al. 2001a). Moreover, careful study of the disease in this strain revealed many characteristics of a $Th_2$ immune response, including the presence of eosinophils and giant cells in the cardiac infiltrate, raised levels of cardiac myosin-specific IgG1 and elevated IgE.

The classical $Th_1$-$Th_2$ paradigm does not fit the findings in autoimmune myocarditis. To pursue the issue, we tested a prototypic $Th_2$ cytokine, IL-13, which shares a receptor with IL-4. IL-13 knock-out mice developed a severe form of myocardial inflammation and impaired cardiac function, recalling in some measure the IFN-γ-deprived animals (D. Cihakova, submitted). IL-4 and IL-13 have opposite and opposing actions. In a somewhat similar manner, IL-12 and IFN-γ exert opposite effects. A good candidate to explain the action of IL-12 is IL-23, which shares the IL-12Rβ1 receptor. This heterochimeric cytokine is composed of a p40 subunit shared with IL-12 and a unique p19. Unlike IL-12, it does not promote development of IFN-γ, but a cytokine implicated in autoimmune inflammation (Fairweather et al. 2004a). It was shown that severe myocarditis in IFN-γ deficient mice is driven by IL-17 (Rangachari et al. 2006). Additionally blocking of IL-17 significantly reduced the EAM severity, although it did not diminish it completely (Sonderegger et al. 2006; Cihakova, Rose unpublished data). Thus, it appears that $Th_1$, $Th_2$ as well as $Th_{17}$ immunopathic mechanisms contribute to the various forms of myocarditis.

## 4.6 Genetic Regulation of Autoimmune Myocarditis

Although induced by quite different mechanisms, virus-induced myocarditis and myosin-induced myocarditis are subject to similar genetic controls (Guler et al. 2005a). For example, most A background mice such as A/J (H-2a) and A.SW (H-2s), differing only at the MHC, develop moderate to severe inflammation following infection with CB3 virus on immunization with cardiac myosin. In contrast, B strains of mice with the same H-2 haplotype on a C57 BL/10 background, B10.A (H-2a) and B10.S (H-2s) resist the induction of autoimmune myocarditis induced by viral infection or myosin immunization. These results emphasize the importance of non-MHC genes in determining the susceptibility to cardiac autoimmunity. Nevertheless, MHC genes still modulate the severity of disease since A.BY mice on a susceptible A background develop only mild disease, whereas female (but not male) B10.M mice, bearing the haplotype to resistant B10 background, are moderately susceptible (Hill and Rose 2001).

Elliott et al. (2003) described a spontaneous form of autoimmune myocarditis in NOD mice in which the murine class II MHC I-Aβ chain was deleted and substituted with human HLA-DQ8, a haplotype sometimes associated with susceptibility to cardiomyopathy in patients. Other strains of mice deficient in murine class II MHC but bearing the human DQ8 haplotype did not develop the autoimmune myocarditis, nor did disease appear in NOD mice bearing other HLA haplotypes. In contrast, transgenic NOD mice bearing the HLA-DQ6 haplotype spontaneously developed a different autoimmune disease, thyroiditis. These experiments suggest that non-MHC background genes from the NOD parent collectively provide a heightened susceptibility to a range of autoimmune diseases, while the specific MHC directs the response to a particular disorder.

In order to investigate further the role of non-MHC background genes in myocarditis, we performed F1 and F2 crosses between A.SW and B10.S mice that share the same H-2s haplotype. F1 mice showed intermediate susceptibility to autoimmune myocarditis with wide variance even though they are genetically identical. The finding probably illustrates the stochastic randomness of the multiple genes involved in the inheritance of these autoimmune characteristics as well as the input of subtle environmental factors. To study genetic segregation, F2 animals were produced and subjected to full genome analysis using single-strand-length polymorphic markers (Guler et al. 2005b). The analysis localized two critical loci on murine chromosome 6 and chromosome 1. Interestingly, both of these loci had previously been identified in determining susceptibility to other autoimmune diseases in mice such as diabetes in NOD mice. Since both loci contain genes that influence T cell apoptosis, they may well represent important common immune regulatory genes.

## 5 Pathogenetic Mechanisms

### 5.1 T Cells

The ability to reproduce the late phase of virus-induced myocarditis by immunization with murine cardiac myosin established the autoimmune basis of the disease. The immunopathic mechanisms of cardiac injury and heart failure, however, have not been clearly defined. As mentioned above, we found that IFNγ deficiency promotes severe inflammation and markedly impairs cardiac function (Afanasyeva et al. 2005). Cardiac dysfunction in these IFNγ-deficient animals is associated with expansion of activated T cells and reduced apoptosis of $CD4^+$ (but not $CD8^+$) T cells. T cells of the deficient mice showed expansion of $CD4^+$ $CD44^+$ $CD25^-$ subset of T cells. The results suggest that severe autoimmune myocarditis is associated with the preferential expansion of $CD25^-$ effector T cells, perhaps related to diminished activation-induced cell death. It is also possible that a subpopulation of $CD4^+$ $CD25^+$ T cells with regulatory function retard the development of the disease.

Adoptive transfer experiments further support the central role of $CD4^+$ T cells in initiating myosin-induced or experimental myocarditis (Hayward et al. 2006). Depletion of this T cell subpopulation with monoclonal antibody reduces disease, whereas transfer of $CD4^+$ T cells from cardiac myosin-immunized mice to naïve syngeneic recipients successfully transfers inflammatory myocarditis. In quantitative experiments, $CD4^+$ T cells were markers of disease progression and cardiac dysfunction (Afanasyeva et al. 2004a).

Experiments depleting $CD8^+$ T cells have given ambiguous results. In some investigations, reducing $CD8^+$ T cells by administration of monoclonal antibody has reduced the severity of myocarditis, whereas genetically deficient mice develop more severe disease following myosin immunization (Smith and Allen 1993; Penninger et al. 1993). In a model of spontaneous myocarditis, Hayward et al. (2006) found by adaptive transfer that mice lacking $CD8^+$ T cells failed to develop cardiac disease. Thus, different subpopulations of $CD8^+$ T cells may exert effector or regulatory functions (Afanasyeva et al. 2004a).

### 5.2 Macrophage

The inflammatory infiltrate observed in the myocardial lesions of myocarditis consists of more than 70% mononuclear cells (Afanasyeva et al. 2004b). In order to determine their critical role in the pathogenesis of myocarditis, we inhibited mononuclear cell migration by administration of monoclonal antibodies to monocyte chemoattractant protein 1 (MCP-1) and macrophage inflammatory protein 1α (MIP-1α) (Goser et al. 2005). The severity and prevalence of myocarditis were reduced in the antibody-treated animals. In addition, mice genetically deficient in the respective chemokine receptors also showed diminished disease. The studies confirm an important role of macrophages in initiating and dictating severity of autoimmune myocarditis.

## 5.3 Antibody

In addition to these cellular components, the role of antibody needs to be considered. Mice immunized with cardiac myosin show localization of IgG1 antibody on cardiac myocyte surfaces (Neumann et al. 1992; Afanasyeva and Rose. 2004). In preliminary experiments carried out in our laboratory using BALB/c mice, we were unable to transfer myocarditis even by repeated injections of serum containing high levels of myosin antibody. On the other hand, co-transfer of antibody with splenic cells from immunized donors produced more cardiac inflammation than the cell transfer alone. Thus antibody and T cells may work together.

Other investigators have found that transfer of antibody alone in selected strains of mice can induce inflammatory myocarditis. Monoclonal antibody to cardiac myosin induced myocarditis in DBA/2 recipients, but not in BALB/c of mice (Liao et al. 1995). Similarly, cardiac myosin-induced antibody produced in other strains of mice can induce disease in DBA/2 recipients, but not in other strains. These findings accord with the observation that DBA/2 mice express cardiac myosin-like epitopes on the surface of its myocytes. In most strains of mice, antigenic determinants of cardiac myosin are expressed only after myocardial damage has been induced.

Another important characteristic of severe myocarditis (as well as other forms of organ-specific autoimmunity) is the production of multiple autoantibodies. Animals developing autoimmune myocarditis following CB3 infection or cardiac myosin immunization subsequently develop antibodies to additional cardiac antigens, including the adenine nucleotide translocator (ANT) protein, branched chain keto dehydrogenase (BCKD), and β-1 adrenergic receptors (Neumann et al. 1994). In addition, animals immunized with another cardiac autoantigen, troponin I, develop inflammatory heart disease and, after 90 days, show antibody to cardiac myosin (Goser et al. 2006). These findings of multiple autoantibodies to the target organ provide important clues when the results on studies of animal models are translated to human investigations. The presence of several antibodies to cardiac antigens may indicate the presence of significant cardiac damage.

## 6 Autoimmunity in Human Myocarditis

There is a large body of evidence supporting the view that some forms of myocarditis and dilated cardiomyopathy in humans result from a pathogenic autoimmune response. This evidence has been reviewed in a recent publication and will not be considered in detail here (Rose 2006).

Many patients with DCM have antibodies directed to multiple cardiac antigens, especially cardiac myosin (Caforio et al. 1992; Neumann et al. 1990). The antibodies may produce the fibrillary or sarcolemal patterns described in the mouse models. Some studies have shown the presence of antibody specific for the α and β isoforms of myosin heavy chain, especially in the IgG3 subclass (Warraich et al. 1999). Similar antibodies to cardiac myosin, but with differing specificities, have been found in patients with other forms of heart disease, including ischemic heart disease, Chagas disease, and rheumatic heart disease. These antibodies are relatively infrequent in individuals without any demonstrable heart disease. Among the other antibodies that have been described in human cases of myocarditis and DCM are antibodies to ANT, BCKD, and the β-1 adrenergic receptor (Feldman and McNamara 2000; Sundstrom et al. 2006). Elevated titers of autoantibodies to troponin I have recently been described in patients with DCM (Goser et al. 2006).

The presence of antibodies may actually precede the development of disease. For example, Caforio et al. (2001) reported that cardiac antibodies were detected in symptom-free relatives of patients with DCM and associated with echocardiographic changes suggestive of early disease.

In recent years, the antibodies receiving the greatest attention in patients with dilated cardiomyopathy are those directed to the cardiac β-1 adrenergic receptors. Their possible role in the pathogenesis of DCM was highlighted by the observation that the number of receptors, but not their affinity, decreased in DCM patients but not in patients with other cardiac diseases. Limas et al. (1989) presented strong evidence that the decrement in receptors was due to the presence of an autoantibody in the serum. The antibodies were relatively specific since they were ineffective against cardiac α1 adrenergic receptors and β2 adrenergic receptors in the lung. Antibodies to the β1 adrenergic receptor are also found in sera of patients with Chagas disease and ischemic cardiomyopathy. There is a great deal of evidence that the receptors are functionally important since improvements in myocardial contractility, exercise tolerance, and survival have been described in patients with chronic heart failure treated chronically with receptor antagonists (Freedman and Lefkowitz 2004). The functional autoimmune epitope on the receptor has been localized to the 169-193 sequence of the second extracellular loop (Fu et al. 1993; Magnusson et al. 1990). The rationale for a therapeutic approach based on these receptors comes from experiments using the rat model of myosin-induced myocarditis. It was shown that β2 adrenergic receptor agonists suppressed the development of the disease, perhaps by inhibiting the activation of cardiac-myosin-specific T cells and shifting the balance in the $Th_1:Th_2$ ratio towards $Th_2$ cytokines (Nishii et al. 2006).

The existence of functionally active autoantibodies against the β1 adrenergic receptor prompted the idea that removal of the circulating autoantibodies in DCM patients may lead to improvements in cardiac function. Felix et al. (2006) prepared absorption columns that contained a polyclonal antihuman immunoglobulin antibody produced in sheep. They found that the antireceptor antibodies in DCM patients were largely eliminated and remained low for as long as 1 year. The immunoglobulin levels were restored by administration of intravenous immunoglobulin, which may by itself have some beneficial therapeutic effect. Similar results were obtained with affinity absorption columns prepared by protein A from *Staphylococcus aureus* immobilized on sepharose columns (Staudt et al. 2002). A further refinement was the use of an immunoabsorption system charged with the functional peptide of the receptor (Schimke et al. 2005; Wallukat et al. 2006). Recent pilot studies show that removal of these autoantibodies by immunoabsorption improves in cardiac function in DCM patients. The antibodies of functional significance appear to belong primarily to IgG3 subclass in the instance of both anti β1 adrenergic receptor antibodies and antimyosin antibodies (Staudt et al. 2005; Warraich et al. 2002).

It is still uncertain whether these studies have long-range implications for therapy of myocarditis or DCM. Large-scale clinical trials of general immunosuppression have failed to show significant benefits. These results, however, are difficult to interpret because of the relatively high spontaneous remission rate in patients receiving only supportive treatment (Oakley 2000). In addition, patients were not sorted for evidence of immune-mediated disease on a basis of heart-specific antibodies or expression of MHC class II on myocytes. Wojnicz et al. (2001) assessed MHC class II expression on endomyocardial biopsies in order to identify inflammatory cardiomyopathy patients and found that they were more likely to respond to immunosuppression than patients who lack this evidence of immune activation. A recent report by Frustaci et al. (2006) determined viral genomic expression as well as circulating cardiac autoantibodies in a cohort of biopsied patients who had failed to respond to conventional supportive therapy. All patients were treated by immunosuppression and classified as responders or nonresponders after a 1-year follow-up. Patients with evidence of viral persistence were unlikely to show a beneficial response to treatment with prednisone and cyclosporin, whereas over 90% of the antibody-positive patients had a favorable response. These studies support the view that a portion of patients with myocarditis and dilated cardiomyopathy whose disease is primarily immune-mediated and who might, therefore, benefit from immune-based therapies. A task for the future is to develop reliable, noninvasive tests for this subpopulation.

**Acknowledgement** I thank Dr. Daniela Cihakova and Jobert G. Barin for carefully reviewing the manuscript. My research is supported by NIH grants HL067290, HL077611 and HL070729.

# References

Afanasyeva M, Rose NR (2004) Viral infection and heart disease: autoimmune mechanisms. In: Schoenfeld Y, Rose NR (eds) Infection and autoimmunity. Elsevier, Amsterdam, pp 299–318
Afanasyeva M, Wang Y, Kaya Z, Park S, Zilliox MJ, Schofield BH, Hill SL, Rose NR (2001a) Experimental autoimmune myocarditis in A/J mice is an interleukin-4-dependent disease with a Th2 phenotype. Am J Pathol 159:193–203

Afanasyeva M, Wang Y, Kaya Z, Stafford EA, Dohmen KM, Sadighi Akha AA, Rose NR (2001b) Interleukin-12 recepter/STAT4 signaling is required for the development of autoimmune myocarditis in mice by an interferon-γ-independent pathway. Circulation 104:3145-3151

Afanasyeva M, Georgakopoulos D, Belardi DF, Ramsundar AC, Barin JG, Kass DA, Rose NR (2004a) Quantitative analysis of myocardial inflammation by flow cytometry in murine autoimmune myocarditis. Am J Pathol 164:807-815

Afanasyeva M, Georgakopoulos D, Rose NR (2004b) Autoimmune myocarditis: cellular mediators of cardiac dysfunction. Autoimmun Rev 3:476-486

Afanasyeva M, Georgakopoulos D, Belardi DF, Bedja D, Fairweather D, Wang Y, Kaya Z, Gabrielson KL, Rodriguez ER, Caturegli P, Kass DA and Rose NR (2005) Impaired upregulation of CD25 on CD4+ T cells in IFN-γ knockout mice is associated with progression of myocarditis to heart failure. Proc Natl Acad Sci U S A 102:180-185

Alvarez FL, Neu N, Rose NR, Craig SW, Beisel KW (1987) Heart-specific autoantibodies induced by Coxsackievirus $B_3$: identification of heart autoantigens. Clin Immunol Immunopathol 43:129-139

Barin JG, Talor MV, Sharma RB, Rose NR, Burek CL (2005) Iodination of murine thyroglobulin enhances autoimmune reactivity in the NODH2h4 mouse. Clin Exp Immunol 142:251-259

Caforio AL, Grazzini M, Mann JM, Keeling PJ, Bottazzo GF, McKenna WJ, Schiaffino S (1992) Identification of alpha- and beta-cardiac myosin heavy chain isoforms as major autoantigens in dilated cardiomyopathy. Circulation 85:1734-1742

Caforio AL, Mahon NJ, McKenna WJ (2001) Cardiac autoantibodies to myosin and other heart-specific autoantigens in myocarditis and dilated cardiomyopathy. Autoimmunity 34:199-204

Caforio AL, Mahon NJ, Tona F, McKenna WJ (2002) Circulating cardiac autoantibodies in dilated cardiomyopathy and myocarditis: pathogenetic and clinical significance. Eur J Heart Fail 4:411-417

Camargo PR, Snitcowsky R, da Luz PL, Mazzieri R, Higuchi ML, Rati M, Stolf N, Ebaid M, Pileggi F (1995) Favorable effects of immunosuppressive therapy in children with dilated cardiomyopathy and active myocarditis. Pediatr Cardiol 16:61-68

Cihakova D, Jobert G. Barin, Marina Afanasyeva, Miho Kimura, DeLisa Fairweather, Michael Berg, Monica V Talor, G. Christian Baldeviano, Sylvia Frisancho, Kathleen Gabrielson, Djahida Bedja, Noel R. Rose (2008) IL 13 protects against experimental autoimmune myocarditis by regulating macrophage differentiation. Amer J Pathol Accepted.

Cunningham MW (2004) T cell mimicry in inflammatory heart disease. Mol Immunol 40:1121-1127

Dec GW Jr, Palacios IF, Fallon JT, Aretz HT, Mills J, Lee DC, Johnson RA (1985) Active myocarditis in the spectrum of acute dilated cardiomyopathies: clinical features, histologic correlates, and clinical outcome. N Engl J Med 312:885-890

Donermeyer DL, Beisel KW, Allen PM, Smith SC (1995) Myocarditis-inducing epitope of myosin binds constitutively and stably to I-Ak on antigen-presenting cells in the heart. J Exp Med 182:1291-1300

Dorffel WV, Wallukat G, Dorffel Y, Felix SB, Baumann G (2004) Immunoadsorption in idiopathic dilated cardiomyopathy, a 3-year follow-up. Int J Cardiol 97:529-534

Elliott JF, Liu J, Yuan ZN, Bautista-Lopez N, Wallbank SL, Suzuki K, Rayner D, Nation P, Robertson MA, Liu G, Kavanagh KM (2003) Autoimmune cardiomyopathy and heart block develop spontaneously in HLA-DQ8 transgenic IAbeta knockout NOD mice. Proc Natl Acad Sci U S A 100:13447-13452

Fairweather D, Kaya Z, Shellam GR, Lawson CM, Rose NR (2001) From infection to autoimmunity. J Autoimmun 16:175-186

Fairweather D, Yusung S, Frisancho S, Barrett M, Gatewood S, Steele R, Rose NR (2003) IL-12 receptor beta 1 and Toll-like receptor 4 increase IL-1 beta- and IL-18-associated myocarditis and coxsackievirus replication. J Immunol 170:4731-4737

Fairweather D, Afanasyeva M, Rose NR (2004a) Cellular immunity: a role for cytokines. In: Doria A, Pauletto P (eds) Handbook of systemic autoimmune diseases. Vol 1: The heart in systemic autoimmune diseases. Elsevier, Amsterdam, pp 3-17

Fairweather D, Frisancho-Kiss S, Gatewood S, Njoku D, Steele R, Barrett M, Rose NR (2004b) Mast cells and innate cytokines are associated with susceptibility to autoimmune heart disease following Coxsackievirus B3 infection. Autoimmunity 37:131-145

Fairweather D, Frisancho-Kiss S, Rose NR (2005) Viruses as adjuvants for autoimmunity: evidence from Coxsackievirus-induced myocarditis. Rev Med Virol 15:17-27

Fairweather D, Frisancho-Kiss S, Njoku DB, Nyland JF, Kaya Z, Yusung SA, Davis SE, Frisancho JA, Barrett MA, Rose NR (2006) Complement receptor 1 and 2 deficiency increases Coxsackievirus B3-induced myocarditis, dilated cardiomyopathy, and heart failure by increasing macrophages IL-1β, and immune complex deposition in the heart. J of Immunol 176:3516-3524

Feldman AM, McNamara D (2000) Myocarditis. N Engl J Med 343:1388-1398

Felix SB, Staudt A (2006) Non-specific immunoadsorption in patients with dilated cardiomyopathy: mechanisms and clinical effects. Int J Cardiol 112:30-33

Felix SB, Staudt A, Friedrich GB (2001) Improvement of cardiac function after immunoadsorption in patients with dilated cardiomyopathy. Autoimmunity 34:211-215

Freedman NJ, Lefkowitz RJ (2004) Anti-beta(1)-adrenergic receptor antibodies and heart failure: causation, not just correlation. J Clin Invest 113:1379-1382

Frisancho-Kiss S, Nyland JF, Davis SE, Barrett MA, Gatewood SJL, Njoku DB, Cihakova D, Silbergeld EK, Rose NR, Fairweather D (2006) Cutting edge: T cell Ig mucin-3 reduces inflammatory heart disease by increasing CTLA-4 during innate immunity. J Immunol 176:6411-6415

Frustaci A, Pieroni M, Chimenti C (2006) Immunosuppressive treatment of chronic non-viral myocarditis. Ernst Schering Res Found Workshop 55:343-351

Fu LX, Magnusson Y, Bergh CH, Liljeqvist JA, Waagstein F, Hjalmarson A, Hoebeke J (1993) Localization of a functional autoimmune epitope on the muscarinic acetylcholine receptor-2 in patients with idiopathic dilated cardiomyopathy. J Clin Invest 91:1964-1968

Gagliardi MG (2006) Dilated cardiomyopathy in children. Acta Paediatr Suppl 95:14-16

Gagliardi MG, Bevilacqua M, Bassano C, Leonardi B, Boldrini R, Camassei FD, Fierabracci A, Ugazio AG, Bottazzo GF (2004) Long term follow up of children with myocarditis treated by immunosuppression and of children with dilated cardiomyopathy. Heart 90:1167-1171

Gor DE, Rose NR, Greenspan NS (2003) Commentary: $T_H1$-$T_H2$: a procrustean paradigm. Nat Immunol 4:503-505

Goser S, Ottl R, Brodner A, Dengler TJ, Torzewski J, Egashira K, Rose NR, Katus HA, Kaya Z (2005) Critical role for monocyte chemoattractant protein-1 and macrophage inflammatory protein-1α in induction of experimental autoimmune myocarditis and effective anti-monocyte chemoattractant protein-1 gene therapy. Circulation 112:3400-3407

Goser S, Andrassy M, Buss SJ, Leuschner F, Volz CH, Ottl R, Zittrich S, Blaudeck N, Hardt SE, Pfitzer G, Rose NR, Katus HA, Kaya Z (2006) Cardiac troponin I but not cardiac troponin T induces severe autoimmune inflammation in the myocardium. Circulation 114:1693-1702

Guler ML, Ligons D, Rose NR (2005a) Genetics of autoimmune myocarditis. In: Oksenberg J, Brassat D (eds) Immunogenetics of autoimmune disease. Landis Bioscience/ Springer, Berlin New York Heidelberg, pp 144-151

Guler ML, Ligons DL, Wang Y, Bianco M, Broman KW, Rose NR (2005b) Two autoimmune diabetes loci influencing T cell apoptosis control susceptibility to experimental autoimmune myocarditis. J Immunol 174:2167-2173

Hayward SL, Bautista-Lopez N, Suzuki K, Atrazhev A, Dickie P, Elliott JF (2006) CD4 T cells play major effector role and CD8 T cells initiating role in spontaneous autoimmune myocarditis of HLA-DQ8 transgenic IAb knockout nonobese diabetic mice. J Immunol 176:7715-7725

Herskowitz A, Beisel KW, Wolfgram LJ, Rose NR (1985) Coxsackievirus $B_3$ murine myocarditis: wide pathologic spectrum in genetically defined inbred strains. Hum Pathol 16:671-673

Hill SL, Rose NR (2001) The transition from viral to autoimmune myocarditis. Autoimmunity 34:169-176

Hill SL, Afanasyeva M, Rose NR (2002) Autoimmune myocarditis. In: Theofilopoulos AN, Bona CA (eds) The molecular pathology of autoimmune diseases, 2$^{nd}$ edn. Taylor and Francis, New York, pp 951-964

Jane-wit D, Tuohy VK (2006) Autoimmune cardiac-specific T cell responses in dilated cardiomyopathy. Int J Cardiol 112:2-6

Kaya Z, Rose NR (2005) Innate immunity in experimental autoimmune myocarditis. In: Zouali M (ed) Molecular autoimmunity. Springer Science/Business Media, New York, pp 1-12

Kaya Z, Afanasyeva M, Wang Y, Dohmen KM, Schlichting J, Tretter T, Fairweather D, Holers VM, Rose NR (2001) Contribution of the innate immune system to autoimmune myocarditis: a role for complement. Nat Immunol 2:739-745

Kaya Z, Tretter T, Schlichting J, Leuschner F, Afanasyeva M, Katus HA, Rose NR (2005) Complement receptors regulate lipopolysaccharide-induced T-cell stimulation. Immunol 114:493-497

Lane JR, Neumann DA, Lafond-Walker A, Herskowitz A, Rose NR (1991) LPS promotes CB3-induced myocarditis in resistant B10.A mice. Cell Immunol 136:219-233

Lane JR, Neumann DA, Lafond-Walker A, Herskowitz A, Rose NR (1992) Interleukin 1 or tumor necrosis factor can promote Coxsackie B3-induced myocarditis in resistant B10.A mice. J Exp Med 175:1123-1129

Liao Li Sindhwani R, Rojkind M, Factor S, Leinwand L, Diamond B (1995) Antibody-mediated autoimmune myocarditis depends on genetically determined target organ sensitivity. J Exp Med 181:1123-1131

Limas CJ, Goldenberg IF, Limas C (1989) Autoantibodies against beta-adrenoceptors in human idiopathic dilated cardiomyopathy. Circ Res 64:97-103

Lowenstein CJ, Hill SL, Lafond-Walker A, Wu J, Allen G, Landavere M, Rose NR, Herskowitz A (1996) Nitric oxide inhibits viral replication in murine myocarditis. J Clin Invest 97: 1837-1843

Magnani JW, Dec GW (2006) Myocarditis: current trends in diagnosis and treatment. Circulation 113:876-890

Magnusson Y, Marullo S, Hoyer S, Waagstein F, Andersson B, Vahlne A, Guillet JG, Strosberg AD, Hjalmarson A, Hoebeke J (1990) Mapping of a functional autoimmune epitope on the beta 1-adrenergic receptor in patients with idiopathic dilated cardiomyopathy. J Clin Invest 86:1658-1663

Mobini R, Maschke H, Waagstein F (2004) New insights into the pathogenesis of dilated cardiomyopathy: possible underlying autoimmune mechanisms and therapy. Autoimmun Rev 3:277-284

Neu N, Beisel KW, Traystman MD, Rose NW, Craig SW (1987a) Autoantibodies specific for the cardiac myosin isoform are found in mice susceptible to Coxsackievirus B$_3$-induced myocarditis. J Immunol 138:2488-2492

Neu N, Rose NR, Beisel KW, Herskowitz A, Gurri-Glass G, Craig SW (1987b) Cardiac myosin induces myocarditis in genetically predisposed mice. J Immunol 139:3630-3636

Neumann DA, Burek CL, Baughman KL, Rose NL, Herskowitz A (1990) Circulating heart-reactive antibodies in patients with myocarditis or cardiomyopathy. J Am Coll. Cardiol 16:839-846

Neumann DA, Lane JR, Wulff SM, Allen GS, LaFond-Walker A, Herskowitz A, Rose NR (1992) In vivo deposition of myosin-specific autoantibodies in the hearts of mice with experimental autoimmune myocarditis. J Immunol 148:3806-3813

Neumann DA, Rose NR, Ansari AA, Herskowitz A (1994) Induction of multiple heart autoantibodies in mice with Coxsackievirus B3- and cardiac myosin-induced autoimmune myocarditis. Immunology 152:343-350

Nishii M, Inomata T, Niwano H, Takehana H, Takeuchi I, Nakano H, Shinagawa H, Naruke T, Koitabashi T, Nakahata J, Izumi T (2006) Beta2-Adrenergic agonists suppress rat autoimmune myocarditis: potential role of beta2-adrenergic stimulants as new therapeutic agents for myocarditis. Circulation 114:936-944

Oakley CM (2000) Spontaneous early clinical remission of myocarditis without immunosuppressive treatment. Ital Heart J 1:762-763
Penninger JM, Neu N, Timms E, Wallace VA, Koh D-R, Kishihara K, Pummerer C, Mak TW (1993) Induction of experimental autoimmune myocarditis in mice lacking CD3 or CD8 molecules. J Exp Med 178:1837-1842
Pummerer CL, Luze K, Grassi G, Bachmaier K, Offner F, Burrell SK, Lenz DM, Zamborelli TJ, Penninger JM, Neu NJ (1996) Identification of cardiac myosin peptides capable of inducing autoimmune myocarditis in BALB/c Mice. Clin Invest 97:2057-2062
Rangachari M, Mauermann N, Marty RR, Dirnhofer S, Kurrer MO, Komnenovic V, Penninger JM, Eriksson U (2006) T-beta negatively regulates autoimmune myocarditis by suppressing local production of interleukin 17. J Exp Med 203:2009-2019
Rose NR (2005) The significance of autoimmunity in myocarditis. In: Schultheiss HP, Kapp JF, Grotzbach G (eds) Ernst Schering Res Found Workshop 55: Chronic viral and inflammatory cardiomyopathy. Springer, Berlin New York Heidelberg, pp 141-154
Rose NR (2006) Life amidst the contrivances. Nat Immunol 7:1009-1011
Rose NR, Afanasyeva M (2003) From infection to autoimmunity: the adjuvant effect. ASM News 69:132-137
Rose NR, Griffin DE (1991) Virus-induced autoimmunity. In: Talal N (ed) Molecular autoimmunity. Academic, San Diego, pp 247-272
Rose NR, Mackay IR (2000) Molecular mimicry: a critical look at exemplary instances in human diseases. Cell Mol Life Sci 57:542-551
Rose NR, Wolfgram LJ, Herskowitz A, and Beisel KW (1986) Postinfectious autoimmunity: two distinct phases of Coxsackievirus $B_3$-induced myocarditis. Ann N Y Acad Sci 475:146-156
Rose NR, Beisel KW, Herskowitz A, Neu N, Wolfgram LJ, Alvarez FL, Traystman MD, Craig SW (1987) Cardiac myosin and autoimmune myocarditis. In:Evered D, Whelan J (eds) Autoimmunity and autoimmune disease. Ciba Foundation Symposium 129. John Wile, Chichester UK, pp 3-24
Schimke I, Muller J, Dandel M, Gremmels H-D, Bayer W, Wallukat B, Wallukat G, Hetzer R (2005) Reduced oxidative stress in parallel to improved cardiac performance one year after selective removal of anti-beta 1-adrenoreceptor autoantibodies in patients with idiopathic dilated cardiomyopathy: data of a preliminary study. J Clin Apheresis 20:137-142
Smith SC, Allen PM (1991) Myosin-induced acute myocarditis is a T cell-mediated disease. J Immunol 147:2141-2147
Smith SC, Allen PJ (1993) The role of T cells in myosin-induced autoimmune myocarditis. 68:100-106
Sonderegger I, Rohn TA, Kurrer MO, Iezzi G, Zou Y, Kastelein RA, Bachmann MF, Kopf M (2006) Neutralization of IL-17 by active vaccination inhibits IL-23-dependent autoimmune myocarditis. Eur J Immunol 11:2849–2856
Staudt A, Bohm M, Knebel F, Grosse Y, Bischoff C, Hummel A, Dahm JB, Borges A, Jochmann N, Wernecke KD, Wallukat G, Baumann G, Felix SB (2002) Potential role of autoantibodies belonging to the immunoglobulin G-3 subclass in cardiac dysfunction among patients with dilated cardiomyopathy. Circulation 106:2448-2453
Staudt A, Dorr M, Staudt Y, Bohm M, Probst M, Empen K, Plotz S, Maschke HE, Hummel A, Baumann G, Felix SB (2005) Role of immunoglobulin G3 subclass in dilated cardiomyopathy: results from protein A immunoadsorption. Am Heart J 150:729-736
Sundstrom JB, Burek CL, Rose NR, Ansari AA (2006) Cardiovascular diseases. In: Detrick B, Hamilton RG, Folds JD (eds) Manual of molecular and clinical laboratory immunology, 7th edn. ASM Press, Washington DC, pp 1101-1109
Tam PE (2006) Coxsackievirus myocarditis: interplay between virus and host in the pathogenesis of heart disease. Viral Immunol 19:133-146
Wallukat G, Nissen E, Neichel D, Harris J (2002) Spontaneously beating neonatal rat heart myocyte culture-a model to characterize angiotensin II at(1) receptor autoantibodies in patients with preeclampsia. In Vitro Cell Dev Biol Anim 38:376-377

Wallukat G, Podlowski S, Nissen E, Morwinski R, Csonka C, Tosaki A, Blasig IE (2003) Functional and structural characterization of anti-beta1-adrenoceptor autoantibodies of spontaneously hypertensive rats. Mol Cell Biochem 251:67-75

Wallukat G, Englert A, Muller J, Dandel M, Lehmkuhl H, Wallukat B, Hetzer R (2006) Immunoadsorption - rationale and clinical results. Autoimmun Rev 5:225-226

Wang Y, Afanasyeva M, Hill SL, Rose NR (1999) Characterization of murine autoimmune myocarditis induced by self and foreign cardiac myosin. Autoimmunity 31:151-162

Warraich RS, Dunn MJ, Yacoub MH (1999) Subclass specificity of autoantibodies against myosin in patients with idiopathic dilated cardiomyopathy: pro-inflammatory antibodies in DCM patients. Biochem Biophys Res Commun 259:255-261

Warraich RS, Noutsias M, Kazak I, Seeberg B, Dunn MJ, Schultheiss HP, Yacoub MH, Kuhl U (2002) Immunoglobulin G3 cardiac myosin autoantibodies correlate with left ventricular dysfunction in patients with dilated cardiomyopathy: immunoglobulin G3 and clinical correlates. Am Heart J 143:1076-1084

Wojnicz R, Nowalany-Kozielska E, Wojciechowska C, Glanowska G, Wilczewski P, Niklewski T, Zembala M, Polonski L, Rozek MM, Wodniecki J (2001) Randomized placebo-controlled study for immunosuppressive treatment of inflammatory dilated cardiomyopathy: two-year follow-up results. Circulation 104:39-45

Wolfgram LJ, Beisel KW, Rose NR (1985) Heart-specific autoantibodies following murine Coxsackievirus $B_3$ myocarditis. J Exp Med 161:1112-1121

Wolfgram LJ, Beisel KW, Herskowitz A, Rose NR (1986) Variations in the susceptibility to Coxsackievirus $B_3$-induced myocarditis among different strains of mice. J Immunol 136:1846-1852

Wolfgram LJ, Rose NR (1989) Coxsackievirus infection as a trigger of cardiac autoimmunity. Immunol Res 8:61-80

# CVB Infection and Mechanisms of Viral Cardiomyopathy

K. U. Knowlton

| | | |
|---|---|---|
| 1 | Evidence for CVB Infection in Myocarditis | 316 |
| | 1.1 Coxsackievirus Infection is Associated with Myocarditis and Dilated Cardiomyopathy in Humans | 316 |
| | 1.2 Mouse Models of CVB-Mediated Myocarditis | 317 |
| 2 | Overview of Mechanisms by Which CVB3 Can Cause Myocarditis | 317 |
| | 2.1 Immune Mechanisms of Myocarditis | 317 |
| | 2.2 Interactions Between the Cardiac Myocyte and Virus in Myocarditis | 317 |
| 3 | Evidence that Coxsackievirus Can Cause a Direct Myocytopathic Effect | 318 |
| | 3.1 Coxsackievirus Infects the Heart | 318 |
| 4 | There Is a Cause-Effect Relationship Between Viral Infection in the Heart and Myocyte Injury | 319 |
| | 4.1 Evidence of Direct Virus-Induced Myocyte Damage in Viral-Mediated Cardiomyopathy | 319 |
| | 4.2 Evidence of Persistence of the Enteroviral Genome | 320 |
| | 4.3 Low-Level Expression of Coxsackieviral Genome in Cardiac Myocytes Causes Dilated Cardiomyopathy | 320 |
| 5 | Significant Steps in Viral Replication and Pathogenesis in the Cardiac Myocyte | 321 |
| | 5.1 Binding and Entry of Virus into the Cardiac Myocyte | 321 |
| | 5.2 Role of Viral Proteases in Viral Replication and Pathogenesis | 322 |
| | 5.3 Dystrophin-Glycoprotein Complex, Protease 2A, and Sarcolemmal Membrane Stability | 323 |
| 6 | Innate Immune Response Within the Heart and Cardiac Myocyte | 327 |
| | 6.1 Endogenous Interferon Has Little Direct Effect on Early Viral Infection of the Heart | 327 |
| | 6.2 JAK-STAT and SOCS Signaling Within the Cardiac Myocyte and Their Role in Susceptibility to Coxsackieviral Infection | 328 |
| | 6.3 gp130 Signaling in the Heart | 329 |
| | 6.4 Role of SOCS in Viral Myocarditis | 330 |
| References | | 331 |

K. U. Knowlton
Department of Medicine and Institute of Molecular Medicine, University of California at San Diego, 9500 Gilman Drive, La Jolla, CA 92093-0613K, USA
kknowlton@ucsd.edu

**Abstract** Coxsackievirus infection has been demonstrated to be a cause of acute and fulminant viral myocarditis and has been associated with dilated cardiomyopathy. While considerable attention has focused on the role of the cellular and humoral, antigen-specific immune system in viral myocarditis, the interaction between the virus and the infected host myocyte is also important. Coxsackievirus has a relative tropism for the heart that is in part mediated by relatively high levels of the coxsackievirus and adenovirus receptor (CAR) on the cardiac myocyte. Once within the myocyte, coxsackievirus produces proteases, such as protease 2A, that have an important role in viral replication, but can also affect host cell proteins such as dystrophin. Cleavage of dystrophin may have a role in release of the virus from the myocyte since viral infection is increased in the absence of dystrophin. In addition to the direct effect of viral proteins on cardiac myocytes, there is now evidence that the cardiac myocyte has a potent innate immune defense against coxsackieviral infection. Suppressors of cytokine signaling (SOCS) can inhibit an interferon-independent mechanism within the cardiac myocyte. In summary, the interaction between coxsackievirus and the infected myocyte has a significant role in the pathogenesis of viral myocarditis and the susceptibility to viral infection.

# 1 Evidence for CVB Infection in Myocarditis

## 1.1 Coxsackievirus Infection is Associated with Myocarditis and Dilated Cardiomyopathy in Humans

The association of acute myocarditis with coxsackievirus infection in humans was identified as early as the mid-1950s (Dalldorf 1955; Fechner et al. 1963; Kibrick and Benirschke 1958). Reports of isolation of coxsackievirus from the heart or pericardial fluid of patients with acute myocarditis date back to the mid-1960s (Sun 1966; Sun and Smith 1966), with numerous reports since then that have isolated virus from the heart or pericardial fluid or demonstrated the presence of viral proteins in diseased heart tissue (Grist and Bell 1969; Lerner and Wilson 1973; Li et al. 2000; Sainani et al. 1975; Sutinen et al. 1971; Windorfer and Sitzmann 1971; Burch et al. 1968). According to World Health Organization surveys from many different countries, 34.6 per 1,000 of all coxsackie B virus infections are associated with cardiovascular disease (Gerzen et al. 1972).

In addition to the clear association between enteroviral infection and acute myocarditis, it has been shown that dilated cardiomyopathy can also be a sequela of viral myocarditis (For review, see Martino et al. 1994). Attempts to isolate virus from the myocardium of patients with chronic forms of dilated cardiomyopathy have been unsuccessful.

The studies described above indicate that in cases of acute myocarditis, Koch's first postulate (the organism must be regularly found in the lesions of the disease) and second postulate (the organism must be isolated in pure culture) are met. It is also clear that in a subset of patients with dilated cardiomyopathy, the first postulate has been met, but the second is lacking. These reports establish associations between viral infection and heart disease, but, of themselves, do not establish a clear cause-effect relationship.

## 1.2 Mouse Models of CVB-Mediated Myocarditis

In 1969, Wilson et al. (Wilson et al. 1969) found that acute infection with CVB3 in weanling Swiss mice was followed by marked fibrosis, and dystrophic mineralization in the heart, and microscopic myocardial hypertrophy, which persisted for at least 6 months. Coxsackieviral infection of mice has, subsequently, been used widely to study the acute effects of viral infection on the myocardium. Following inoculation of mice with coxsackievirus, the virus can be consistently isolated from the heart. Therefore, the murine model of viral myocarditis fulfils Koch's postulates three (inoculation of such a pure culture of organisms into a host should initiate the disease) and four (the organisms must be recovered once again from the lesions of the host) for the acute phase of viral heart disease.

# 2 Overview of Mechanisms by Which CVB3 Can Cause Myocarditis

## 2.1 Immune Mechanisms of Myocarditis

One of the more prominent features of coxsackieviral infection of the heart is the marked cellular inflammation that is present at 6-7 days after infection and onward. This has appropriately led to a large number of studies on activation of the humoral and cellular immune response in viral myocarditis. Both antigen-specific and autoimmune mechanisms have been implicated in the immune-mediated effects on the heart. The immune response can be beneficial by limiting viral replication and can also be detrimental if directed against heart antigens or by killing potentially viable cells that are near infected cells. Immune mechanisms involved in viral myocarditis are discussed in considerable detail in the chapter by N.R. Rose, this volume.

## 2.2 Interactions Between the Cardiac Myocyte and Virus in Myocarditis

In addition to the important host cellular and humoral immune response, attention has been directed to the interaction between the infected cell and the virus. There are two important considerations that will be emphasized below. First, since the virus is able to infect the cardiac myocyte, it is important to understand mechanisms by which coxsackievirus can directly affect the cardiac myocyte. Second, there are innate, antigen-independent immune defense mechanisms within the cardiac myocyte that can inhibit the pathogenesis of viral myocarditis. When these mechanisms are altered, they can affect susceptibility to viral infection in the heart. The major focus of the mechanisms that will be described in this chapter will be on the earlier stages of viral infection, days 2-5 after inoculation, prior to the onset of a significant cellular immune response.

## 3 Evidence that Coxsackievirus Can Cause a Direct Myocytopathic Effect

### 3.1 Coxsackievirus Infects the Heart

#### 3.1.1 Coxsackievirus Infects Cardiac Cells in Culture and in the Infected Rodent Heart

Cell culture experiments have demonstrated that CVB3 can infect human and rodent cardiac cells. For example, coxsackievirus is able to infect human embryonic fibroblasts and myocytes and can establish a carrier-state infection in culture (Kandolf et al. 1985). In addition, coxsackievirus can infect isolated neonatal rat and mouse cardiac myocytes and isolated adult mouse myocytes (Yajima et al. 2006; Yasukawa et al. 2001). In addition to the evidence that coxsackievirus can infect cardiac myocytes and fibroblasts, there is evidence that enteroviruses can also infect endothelial cells (Tracy et al. 2000) and smooth muscle cells in culture (Godeny et al. 1986). These cultured cell data indicate that several cardiac cell types have the potential to be infected by coxsackievirus.

Supporting the infection of myocytes in culture is the identification of CVB3 genome and capsid proteins in cardiac myocytes in the murine model of myocarditis (Badorff et al. 1999, 2000b; Kandolf et al. 1987). It is likely that there are other cells such as fibroblasts or endothelial cells are infected in the intact heart as well (Klingel et al. 1998).

#### 3.1.2 Coxsackievirus Can Infect Cardiac Myocytes in the Human Heart

In addition to these experimental models that demonstrate direct infection of cardiac cells with coxsackievirus, there is considerable evidence that coxsackievirus can directly infect the intact human heart as well as be determined by isolation of virus from the heart and identification of viral genomes in the heart, as described above. However, these experiments do not identify which cell types are infected.

Demonstration of infection of cardiac myocytes in humans has been accomplished using immunohistochemical staining for enteroviral-specific capsid proteins that could be identified within the cardiac myocyte in patients with fulminant and acute myocarditis and occasionally in patients with chronic cardiomyopathy (Kandolf and Hofschneider 1989; Lee et al. 2006; Li et al. 2000). In addition, in situ hybridization for viral RNA in patients with acute myocarditis and in a subset of patients with dilated cardiomyopathy is consistent with this finding since viral genomes can be identified in the cardiac myocyte (Hohenadl et al. 1991; Kandolf et al. 1987). While these data demonstrate that coxsackievirus can infect cardiac myocytes, the exact incidence of enteroviral infection in myocarditis and dilated cardiomyopathy is not clear and likely varies with the overall incidence of coxsackieviral infection in the population.

The capability of coxsackievirus to directly infect cells in the heart suggests the possibility that coxsackievirus could cause a direct cytopathic effect on the cells acutely and perhaps in the setting of a chronic, persistent infection and that this may have a role in the pathogenesis of viral myocarditis.

## 4 There Is a Cause-Effect Relationship Between Viral Infection in the Heart and Myocyte Injury

### 4.1 Evidence of Direct Virus-Induced Myocyte Damage in Viral-Mediated Cardiomyopathy

While it is generally accepted that there is myocardial damage in viral myocarditis, it is less clear whether the damage is from activation of the immune response or the direct effects of viral infection. While it is likely that activation of the immune response contributes to the myocyte damage in viral myocarditis, there is also considerable evidence that virus can directly injure infected myocardial cells. For example, extensive cardiac damage has been demonstrated following infection of mice with severe combined immunodeficiency (Chow et al. 1992) and athymic mice (Hashimoto and Komatsu 1978; Hashimoto et al. 1983), where the usual cellular immune response is genetically reduced or absent. In addition, evidence of direct myocyte-mediated myocardial injury was identified histologically in infected cardiac myocytes of immunocompetent mice (McManus et al. 1993). Using Evans blue dye, it was subsequently demonstrated that during early time points after infection, there is clear evidence of myocyte membrane disruption in the infected cells (Badorff et al. 1999; Lee et al. 2000; Yajima et al. 2006). There is a high correlation between the cells that are positive for Evans blue dye and those that are infected with coxsackievirus as determined by immunofluorescent staining for coxsackievirus (Yajima et al. 2006). Disruption of membrane proteins is not limited to infected murine cells, but has also been demonstrated in humans in the setting of fulminant myocarditis (Lee et al. 2006).

## 4.2 Evidence of Persistence of the Enteroviral Genome

In addition to the experiments described above in acute viral heart disease, evidence suggests that persistence of the viral genome can contribute to the evolution of ongoing heart disease, though there is less direct evidence that low-level expression of viral genomes in the cardiac myocyte can induce a direct myocytopathic effect and cardiomyopathy (Klingel et al. 1992; Kyu et al. 1992). The data obtained from coxsackievirus-infected murine models imply that myocardial damage in CVB3-infected mice can occur in two phases: an acute phase with prominent virus replication and cellular infiltration. Second, there can be a chronic phase characterized by progressive myocardial disease that may be associated with low-level persistence of viral genomes and progressive cardiomyopathy (Klingel et al. 1992; Kyu et al. 1992).

## 4.3 Low-Level Expression of Coxsackieviral Genome in Cardiac Myocytes Causes Dilated Cardiomyopathy

While there is an association between the presence of enteroviral genomes and cardiomyopathy, it is more difficult to establish a cause-effect relationship between low-level expression of viral genomes and the induction of cardiomyopathy. To accomplish this goal, infectious recombinant coxsackievirus B3 (CVB3) cDNA was mutated at the autocatalytic cleavage site in the VP0 capsid protein from the amino acids asparagine and serine to lysine and alanine, thus preventing formation of infectious virus progeny (Wessely et al. 1998a). This was initially tested in cultured neonatal ventricular myocytes transfected with the CVB3-mutated cDNA copy of the viral genome, and found to induce a cytopathic effect in transfected myocytes (Wessely et al. 1998a).

To determine whether a similar pathogenic effect could also be observed in the intact heart, transgenic mice were generated that expressed the replication-restricted CVB3 cDNA mutant exclusively in the heart, using the cardiac myocyte-specific myosin light chain-2v (MLC-2v) promoter. This allowed for low-level expression of coxsackieviral genomes in the cardiac myocyte without formation of infectious virions, thus preventing a productive viral replication cycle. In addition, the MLC-2v promoter directs expression in the heart at day 8.5 of embryogenesis (Lee et al. 1992), thus preventing activation of a potent immune response against viral antigens (Wessely et al. 1998b). As expected, heart muscle-specific expression of the CVB3 mutant resulted in the synthesis of viral plus- and minus-strand RNA without formation of infectious viral progeny. Histopathologic analysis of transgenic hearts revealed typical morphologic features of myocardial interstitial fibrosis, hypertrophy, and degeneration of myocytes, thus resembling dilated cardiomyopathy in humans. This occurred in the absence of viral neutralizing antibodies. Analysis of isolated myocytes from transgenic mice demonstrated that there is defective excitation-contraction coupling and a decrease in the magnitude of isolated cell shortening. These findings demonstrate that restricted replication of enteroviral genomes in the heart can induce cardiomyopathy with characteristics that are typical of dilated cardiomyopathy in humans (Wessely et al. 1998b).

## 5 Significant Steps in Viral Replication and Pathogenesis in the Cardiac Myocyte

The data above demonstrate that coxsackievirus can infect cardiac myocytes and can induce a cytopathic effect. Much of the knowledge about the steps involved in viral replication come from experiments in non-myocytes and is discussed elsewhere in this volume. However, some of the steps are of particular relevance to the heart. These include the following points.

1. The virus attaches to the target cell by binding to its receptor, the coxsackievirus and adenovirus receptor (CAR). CAR is expressed at high levels in the cardiac myocyte.
2. Once the virus enters the cell and the viral RNA is translated, the viral monocistronic polyprotein is cleaved by the coxsackieviral proteases creating the nonstructural proteins required for viral RNA replication and efficient translation of the viral RNA. The viral proteases can also cleave a small subset of host proteins, a process that facilitates viral replication and can cause a direct cytopathic effect.
3. Replication of the virus positive-strand RNA through a negative-strand intermediate is accompanied by formation of double-stranded RNA that is thought to have a role in induction of the host-cell innate immune response.
4. Once the replicated, viral positive-strand viral genome is encapsidated and released from the cell during a process that is associated with disruption of the cell membrane and cleavage of dystrophin by a viral protease, a myocyte-specific protein, the replicated virus is released from the infected cell to allow secondary infection of other cells (Racaniello 2007). Alterations in these processes could affect how efficiently a virus can replicate in the heart and how susceptible the host is to the induction of viral myocarditis.

### 5.1 Binding and Entry of Virus into the Cardiac Myocyte

The first step of viral infection and replication is attachment of the virus to the target cell. The CAR has been shown to be the major receptor for coxsackievirus. Elegant studies have defined the role for CAR in cells, but less is known regarding its role in the heart (Bergelson et al. 1997, 1998; Carson 2001; Coyne and Bergelson 2006).

One of the likely explanations for the tropism of coxsackievirus for the heart is the relatively high level of expression of the CAR on the cardiac myocyte (Bergelson et al. 1997, 1998; Kashimura et al. 2004). CAR is a membrane adhesion molecule that has two immunoglobulin domains in the extracellular region: a transmembrane domain and an intracellular domain that includes a PDZ-binding motif. CAR is expressed throughout the sarcolemma of the cardiac myocyte during embryonic growth (Kashimura et al. 2004) and is required for normal embryogenesis. This is demonstrated by embryonic lethality in the global cardiac knockout of the CAR gene (Asher et al. 2005; Chen et al. 2006; Dorner et al. 2005). As the heart reaches its fully developed state, CAR localizes predominantly to the intercalated disc where its function is not clearly known. In addition, as the heart ages the level of CAR expression in the heart decreases, but continues to be expressed (Kashimura et al. 2004). CAR expression is increased in heart failure and autoimmune myocarditis (Fechner et al. 2003; Poller et al. 2002). The susceptibility of the heart to viral infection is thought to correlate with the level of CAR expression (Poller et al. 2002).

## 5.2 Role of Viral Proteases in Viral Replication and Pathogenesis

As described previously, viral proteases have an essential role in cis-cleavage of the monocistronic viral polyprotein. In addition to cleavage of the viral polypeptide, enteroviral proteases can cleave host-cell proteins in trans at highly specific proteolytic sites. The enteroviroviral protease 3C can cleave the TATA-binding protein (Clark et al. 1993), the poly(A)-binding protein (Kuyumcu-Martinez et al. 2002), and the poly(ADP-ribose) polymerase (PARP), a nuclear protein involved in DNA repair and apoptosis (Barco et al. 2000). Enteroviral protease 2A has been shown to cleave host proteins involved in translation, such as eukaryotic initiation factors 4G (eIF4G) -1 (Lamphear et al. 1993) and −2 (Gradi et al. 1998), and the polyadenylate binding protein (PABP) (Kerekatte et al. 1999). In addition, it was discovered that protease 2A could cleave the cytoskeletal proteins dystrophin (Badorff et al. 1999) and cytokeratin-8 (Seipelt et al. 2000), a process that facilitates release of virus from the cell (Xiong et al. 2002) (see below). Both eIF4G-1 and −2 are part of the translation initiation complex that is required for efficient translation of host-cell-capped mRNA, which includes the majority of eukaryotic mRNA. Enteroviral-mediated cleavage of eIF4G results in inhibition of host-cell, cap-dependent translation in favor of viral, IRES-mediated protein synthesis (Lamphear et al. 1995). It is possible that cleavage of eIF4G has a cytopathic effect on the cardiac myocyte since inhibition of its cleavage in HeLa cells inhibits the coxsackievirus-mediated cytopathic effect (Zhao et al. 2003). However, the importance of cleavage of eIF4G in cardiac myocytes has not yet been thoroughly evaluated. The role of protease 2A in the cardiac myocyte and the importance of the dystrophin-glycoprotein complex in viral heart disease will be discussed in more detail in the following section.

## 5.3 Dystrophin-Glycoprotein Complex, Protease 2A, and Sarcolemmal Membrane Stability

### 5.3.1 Dystrophin-Glycoprotein Complex in Cardiomyopathy

The dystrophin-glycoprotein complex is important for maintaining myocyte sarcolemmal membrane integrity. Dystrophin is a subsarcolemmal rod-shaped protein that stabilizes the sarcolemma by attaching the actin-cytoskeleton to the extracellular matrix through the dystrophin-associated glycoprotein complex (Durbeej and Campbell 2002). Abnormalities in dystrophin cause Duchenne and Becker muscular dystrophy. While the most visible phenotype of these muscular dystrophies is a severe skeletal myopathy, most patients develop a dilated cardiomyopathy by the early to mid teens that is thought to contribute to their short life span (Towbin 1998). If dystrophin is expressed at relatively normal levels in the skeletal muscle but is decreased in the cardiac muscle, the patients develop a cardiomyopathy without a significant skeletal myopathy. This occurs in patients with X-linked dilated cardiomyopathy that is associated with an inherited abnormality in dystrophin (Muntoni et al. 1993; Towbin et al. 1993). Abnormalities in other members of the dystrophin-glycoprotein complex can also cause dilated cardiomyopathy (Towbin 1998; Tsubata et al. 2000; van der Kooi et al. 1998).

### 5.3.2 Coxsackievirus B3-Derived Protease 2A Directly Cleaves Dystrophin

Given the importance of the dystrophin-glycoprotein complex in hereditary cardiomyopathy, the effect of coxsackievirus infection on the dystrophin-glycoprotein complex was assessed. A neural network algorithm that allowed identification of potential enteroviral protease 2A cleavage sites identified two putative cleavage sites within the dystrophin molecule (see http://www.cbs.dtu.dk/services/NetPicoRNA/). One of the putative protease 2A cleavage sites is in the hinge-3 region of the murine dystrophin molecule, a region previously shown to be accessible to proteases (Blom et al. 1996; Koenig and Kunkel 1990). There is evidence of a similar putative cleavage site in human dystrophin.

The hinge-3 region of dystrophin lies between the actin-binding sites and the β-dystroglycan anchoring motif of dystrophin (Koenig and Kunkel 1990). Cleavage at this site would be predicted to disconnect the (actin-binding) N-terminal region from the sarcolemma where the C-terminal fragment binds β-dystroglycan.

When purified coxsackieviral protease 2A was added to protein extracts from neonatal rat cardiac myocytes or adult human heart, there was time-dependent cleavage of dystrophin of both rat and human dystrophin, but the cleavage of human dystrophin appeared to be less efficient than that of rodent dystrophin (Badorff et al. 1999). It was later demonstrated that the cleavage of dystrophin occurred through a direct cleavage of the dystrophin protein by enteroviral protease

2A rather than through secondary activation of another protease and that cleavage occurred at the predicted protease 2A-cleavage site in the hinge-3 region of dystrophin (Badorff et al. 2000a).

### 5.3.3 Disruption of Dystrophin, the Dystrophin-Glycoprotein Complex, and Sarcolemmal Membrane Integrity in the CVB-Infected Heart

In addition to the ability of protease 2A to cleave dystrophin in vitro, dystrophin is also proteolytically cleaved and its sarcolemmal localization is disrupted in the intact heart of mice that are infected with CVB3. Additionally, the sarcolemmal integrity in these cells is impaired, as determined by Evans blue dye tracer uptake, in a manner similar to that observed in muscular dystrophy (Badorff et al. 1999). Since genetic dystrophin deficiency leads to a loss of dystrophin-associated glycoproteins such as the sarcoglycans (Matsumura et al. 1993), it was shown that dystrophin cleavage during coxsackievirus B3 infection similarly affects the sarcoglycans (Lee et al. 2000).

In analogy to findings from patients with dilated cardiomyopathy due to Duchenne muscular dystrophy, it is possible that the cleavage of dystrophin with its subsequent biochemical, morphological, and functional disruption of the dystrophin-glycoprotein complex during coxsackievirus B3 infection initiates a cascade of events that contributes to the pathogenesis of dilated cardiomyopathy (Badorff et al. 1999) (Fig. 1). In addition, it is also possible that cleavage of dystrophin is an important step in viral replication, facilitating viral release from the infected cell by weakening the cell membrane and allowing infection of the adjacent cardiac myocytes.

### 5.3.4 Dystrophin Deficiency Increases Susceptibility to Virus-Mediated Cardiomyopathy Via an Increase in Membrane Permeability

In the experiments described earlier, it was demonstrated that CVB protease 2A increases cell membrane permeability as assessed by Evans blue dye staining. This is associated with cleavage of dystrophin. Therefore, it was hypothesized that dystrophin deficiency might increase susceptibility to viral infection by decreasing myocyte membrane stability, allowing the virus to exit from the cell more efficiently and infecting adjacent cells. In CVB3-infected, dystrophin-deficient, mdx mice there was significantly more Evans blue dye uptake when compared to the infected, dystrophin-competent mice. In addition, there was also a nearly 100-fold increase of virus titer in the infected, dystrophin-deficient hearts when compared to the infected, dystrophin-competent, wild type mice (Xiong et al. 2002) (Fig. 2).

When cardiac sections were stained for coxsackievirus, in the wild type mice there was bright staining for coxsackievirus in the rare Evans blue dye-positive cells. In the hearts of dystrophin-deficient mice infected with coxsackievirus, however, the majority of Evans blue dye-positive cells contained only remnants of viral

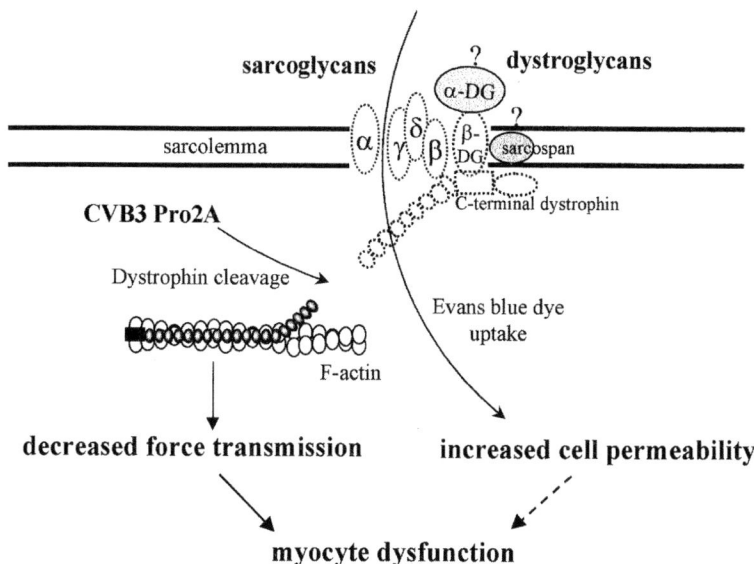

**Fig. 1** Schematic of the pathogenic role of the sarcoglycan complex in the acquired enteroviral cardiomyopathy that demonstrates the loss of the carboxyl-terminal dystrophin, β-dystroglycan (DG), and α-, β-, γ-, and δ-sarcoglycans in the sarcolemmal membrane (From Lee et al. 2000)

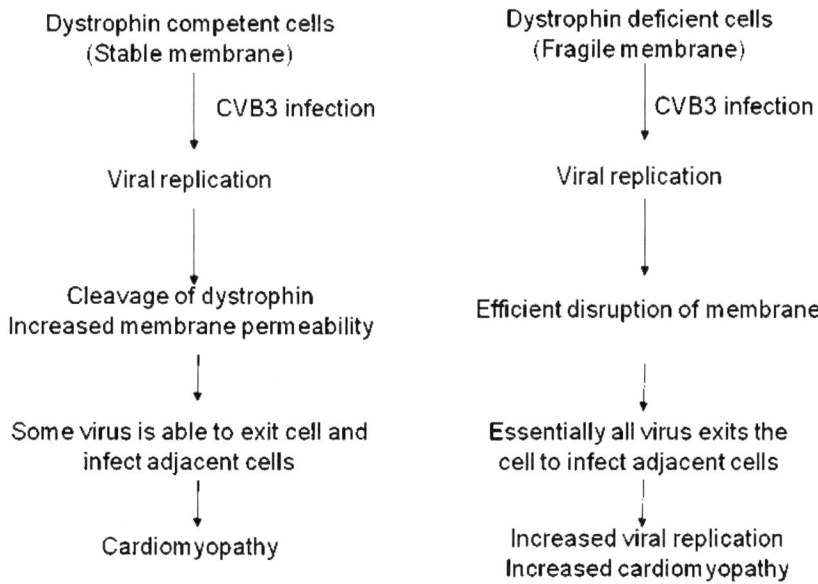

**Fig. 2** Summary of mechanisms by which dystrophin deficiency induces more cardiomyopathy compared to dystrophin-competent myocytes

proteins. Membrane disruption was rarely observed in uninfected mdx mice. This suggested that the absence of dystrophin increases membrane permeability and release of virus from the myocyte. A similar phenomenon was observed in cultured cells where dystrophin expression decreased release of virus from the cells (Xiong et al. 2002).

Enteroviruses are typically released from the cell by disruption of the cell membrane or cell lysis. Because dystrophin provides mechanical support to the sarcolemma, and the absence of dystrophin results in increased susceptibility to sarcomere rupture, the virus exits from the infected cell more efficiently in the absence of dystrophin. This allows more rapid propagation of the virus to adjacent myocytes, resulting in higher viral titers and greater cardiomyopathy. In the presence of dystrophin, the rate and extent of viral release may be slowed by the requirement that protease 2A cleave dystrophin or that there are other viral-mediated mechanisms that lead to membrane disruption. This would result in lower viral titers than are seen in the absence of dystrophin. While there are undoubtedly several mechanisms by which dystrophin deficiency causes cardiomyopathy in individuals with Duchenne muscular dystrophy or X-linked dilated cardiomyopathy, the data in the dystrophin-deficient mouse hearts support the notion that the absence of dystrophin in patients with muscular dystrophies involving the dystrophin-glycoprotein complex increases the susceptibility of such individuals to virally mediated cardiomyopathy and that this could contribute to the development of cardiomyopathy in these patients. The increased susceptibility may be broadly applicable to viruses that are capable of infecting cardiac myocytes and are released from the cell by disruption of the cell membrane. It remains to be seen whether people that develop myocarditis with coxsackieviral infection have genetic mutations in the dystrophin-glycoprotein complex proteins that contribute to an increase in cell membrane permeability and thus an increase in susceptibility to viral infection.

### 5.3.5 Viral Protease 2A Alone in the Heart is Sufficient to Induce Cardiomyopathy

The data described in the preceding section demonstrate that coxsackieviral infection of cardiac myocytes can cause a direct myocytopathic effect. Furthermore, expression of a full-length, replication-defective coxsackieviral genome in the cardiac myocyte is able to cause a cardiomyopathy that appears to be largely independent of the cellular immune response. However, until recently, it was not clear whether expression of coxsackieviral protease 2A alone was sufficient to induce a dilated cardiomyopathy independent of the other viral proteins and viral RNA. Therefore, a transgenic mouse was generated that expressed protease 2A only in the adult cardiac myocyte. Furthermore, the transgenic expression construct was designed to only express protease 2A in an inducible manner following administration of tamoxifen (Xiong et al. 2007).

Within 22 days after induction of protease 2A in the adult heart, the mice developed severe cardiomyopathy. In addition, there was an increase in the number of

Evans blue dye-positive myocytes in the hearts of the mice that expressed protease 2A. This was associated with disruption of dystrophin localization to the sarcolemma in a subset of myocytes. A similar Evans blue dye staining was not observed in mice that had cardiomyopathy due to absence of the muscle LIM protein (MLP), a different model of cardiomyopathy (Xiong et al. 2007). This study demonstrated that the presence of coxsackieviral protease 2A in the cardiac myocyte in the absence of other viral proteins or RNA is sufficient to induce cardiomyopathy with disruption of the sarcolemmal membrane and loss of localization of dystrophin in the intact heart. These findings, however, do not exclude the possibility that protease 2A may also affect other myocyte proteins in addition to dystrophin and may also contribute to the development of cardiomyopathy. Nevertheless, the findings make a strong case for the potential for protease inhibitors in the treatment of coxsackieviral-mediated cardiomyopathy. Protease inhibitors would likely inhibit protease-mediated cytopathic effects in the cardiac myocyte and also inhibit viral replication. A peptide based on the protease 2A dystrophin-cleavage site has been shown to be able to inhibit protease 2A (Badorff et al. 2000a).

## 6 Innate Immune Response Within the Heart and Cardiac Myocyte

Host defense mechanisms can be divided into two broad categories: adaptive and innate. Adaptive immune responses to coxsackievirus infection are discussed in the chapter by N.R. Rose, this volume. Although more effective, the onset of the adaptive immune response is slow (several days). The innate immune response includes the activation of natural killer cells and the production of cytokines. Interruption of the innate immune response in all cells, including circulating cells, can affect susceptibility to coxsackieviral infection. For example, global knockout of myeloid differentiation factor-88 (MyD88) decreases susceptibility to coxsackieviral myocarditis (Fuse et al. 2005). Recently, it has been demonstrated that the cardiac myocyte has innate defense mechanisms and that there are some unique characteristics compared to circulating immune cells and other tissues such as the liver. Alteration of these cardiac myocyte innate defense mechanisms can have a profound effect on the host susceptibility to coxsackieviral infection.

### *6.1 Endogenous Interferon Has Little Direct Effect on Early Viral Infection of the Heart*

Interferons were the first identified cytokines that play a central role in host defense against invasive viruses (Isaacs and Lindemann 1957). Interferons of $\alpha$ and $\beta$ subtype are referred to as type I interferons. The only type II interferon is interferon-$\gamma$.

Interferons exert their effect by binding to specific receptors in the cell membrane that subsequently activate intracellular signaling through Janus kinase (JAK) and signal transducers and activators of transcription (STAT) signaling. It has been shown that both type I and type II interferons can inhibit CVB replication in cultured cells and administration of interferon-$\alpha/\beta$ can ameliorate CVB-induced myocarditis in mice (Fairweather et al. 2004; Karupiah et al. 1993; Matsumori et al. 1988; Wang et al. 2007). In order to determine the effect of interferon receptor-mediated signaling on CVB replication in the heart, mice lacking either type I or type II receptor were infected with CVB3. CVB3 infection in type I receptor-deficient mice lead to a marked increase in viral replication in the liver and a marked increase in mortality in CVB3-infected mice. However, there was no significant increase of viral RNA in the heart of the type I interferon receptor-deficient mice (Fig. 3). These findings show that the presence of type I interferon receptor signaling is required to prevent high-level viral replication in noncardiac organs such as the liver, but that there is no significant effect on early viral replication in the heart of mice infected with CVB3. In contrast, the absence of the type II interferon signaling did not have a significant effect on mortality and resulted in only a mild increase of the viral titers in heart and liver (Wessely et al. 2001). These results demonstrate that endogenous type I interferons have little effect on viral replication in the heart, but it does not exclude the possibility that exogenous administration of interferon could have a beneficial effect on the heart and other organs (Wang et al. 2007).

## 6.2 JAK-STAT and SOCS Signaling Within the Cardiac Myocyte and Their Role in Susceptibility to Coxsackieviral Infection

A family of transmembrane receptors have been identified that transduce their signals to the nucleus via activation of the JAK/STAT pathway. These receptors include the interferon (interferon) receptors, gp130-associated receptors such as the receptors for leukemia inhibitory factor (LIF), cardiotrophin-1 (CT-1), interleukin-6 (IL-6), and the erythropoietin receptor. Ligand binding leads to phosphorylation of JAK1 and JAK2 (Muller et al. 1994). Activated JAKs then phosphorylate the intracellular region of the receptor molecules, which then allow phosphorylation of STAT1 or STAT3 (Greenlund et al. 1994; Heim et al. 1995). Activated STAT1 or STAT3 dimers translocate to the nucleus, where they stimulate expression of a panel of genes. The suppressor of cytokine signaling (SOCS) proteins, also known as JAB (JAK binding protein) and SSI (STAT inducible STAT inhibitor) are upregulated when JAK-signaling is activated.

Upregulation of the SOCS family of proteins has been shown to be a counter-regulatory mechanism (Nicholson et al. 1999). SOCS-mediated counter-regulatory mechanisms can inhibit overactivation of the cytokine signaling cascade (Alexander et al. 1999; Endo et al. 1997; Marine et al. 1999; Naka et al. 1997). SOCS1 is essential for inhibition of interferon signaling, but it can also inhibit most JAK-mediated signaling cascades. Alternatively, endogenous SOCS3 was originally

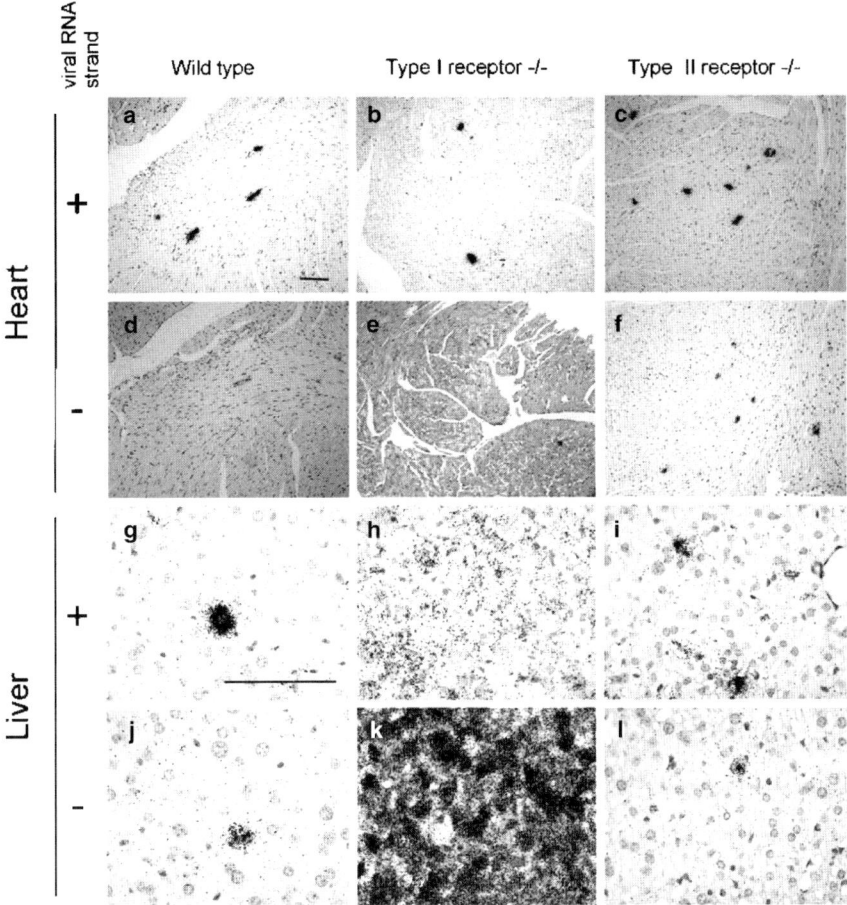

**Fig. 3** In situ hybridization in the heart and liver for coxsackievirus-infected wild type, type I interferon receptor deficient, and type II interferon receptor deficient mice 3 days after inoculation. RNA was probed for both positive-strand (+) and negative-strand (-) viral RNA

thought to act primarily as a negative regulator of gp130 signaling. While it has been implicated in other JAK-STAT signaling cascades, it has little or no effect on type I or type II interferon signaling.

## 6.3 gp130 Signaling in the Heart

gp130 signaling has been shown to have a role in cardiomyocyte survival and induction of ventricular hypertrophy with pressure overload (Hirota et al. 1999). Its stimulation activates STAT3 and SOCS3 (Hirota et al. 1999; Yasukawa et al. 2003).

## 6.4 Role of SOCS in Viral Myocarditis

Given the potential importance of JAK-STAT signaling in the pathogenesis of coxsackieviral myocarditis, transgenic mice that expressed SOCS1 only in the heart were infected with CVB3. Consistent with the fact that SOCS1 inhibits JAK signaling stimulated by a variety of cytokines (Naka et al. 1997; Yasukawa et al. 2000), both STAT1 and STAT3 activation and induction of interferon-responsive genes by CVB3 infection were totally inhibited in the SOCS1 transgenic mouse hearts, indicating that SOCS1 transgenic mice are resistant to stimulation by interferons and gp130-activating cytokines. In addition, CVB3-infected SOCS1 transgenic mice had significantly earlier mortality when compared with their wild type littermates. This was associated with a marked increase in myocyte damage as assessed by an increase in the percent area of myocardial injury than that observed in wild type littermates (Fig. 4). The virus titer in the heart in SOCS1 transgenic mice on the 4th and 5th days after CVB3 infection was also nearly 100-fold greater when compared to the wild type littermates at day 7 after infection. The viral titer in the liver was not elevated in the transgenic mice, showing the cardiac specificity for the effect (Yasukawa et al. 2003). These results demonstrate that there is a potent innate defense mechanism that is within the cardiac myocyte that is inhibited by SOCS expression.

Since SOCS3 inhibits gp130 signaling but has no significant effect on interferon signaling and interferon antiviral effects in the cardiac myocyte, transgenic mice that expressed SOCS3 only in the cardiac myocyte were generated and infected

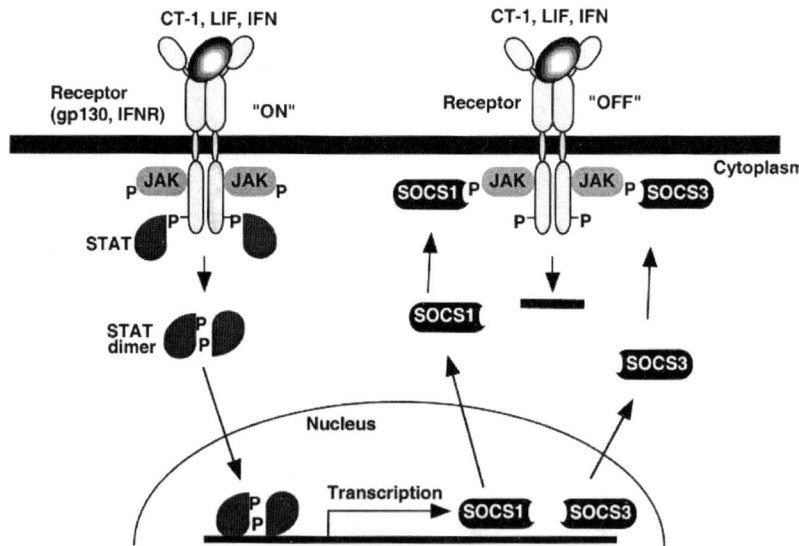

**Fig. 4** Schematic of JAK-STAT signaling and negative feedback regulation by suppressors of cytokine signaling (SOCS)-1 and SOCS3

with CVB3 and found to have essentially the same marked increase in CVB3-mediated cytopathic effects that were observed with SOCS1 expression (Yajima et al. 2006). Consistent with the fact that SOCS3 did not inhibit interferon effects in the myocyte but inhibited gp130-signaling and inhibited a potent cardiotrophin-1-mediated cytoprotective effect against CVB3 infection, it was subsequently shown that at least part of the increased susceptibility to coxsackieviral infection seen with SOCS expression could be replicated by cardiac-specific knockout of gp130, indicating that at least part of the increase in susceptibility in the SOCS3 transgenic mice could be accounted for by inhibition of gp130.

Knockout of interferon receptors in the CVB3 infected mouse had little or no effect on replication of virus in the heart (Wessely et al. 2001), and SOCS3 expression could not inhibit interferon signaling and antiviral effects but markedly increased susceptibility to coxsackieviral infection, both of which strengthen the argument that endogenously produced interferon has little or no antiviral effect in the cardiac myocyte during the early phases of viral replication. Nevertheless, there is a potent innate defense mechanism within the adult cardiac myocyte that is inhibited by SOCS1 and SOCS3. Additional work is ongoing to identify the predominant pathway or pathways that are inhibited by SOCS3. Immunoblots demonstrate that inhibition of gp130 signaling by knockout of the gp130 molecule or the expression of SOCS3 leads to a decrease in the amount of dystrophin-glycoprotein complex proteins that are expressed in the heart (Yajima et al. 2006), thus associating the gp130 signaling cascade with the importance of the dystrophin-glycoprotein complex in the pathogenesis of coxsackieviral mediated myocarditis and cardiomyopathy.

## References

Alexander WS, Starr R, Fenner JE, Scott CL, Handman E, Sprigg NS, Corbin JE, Cornish AL, Darwiche R, Owczarek CM, et al (1999) SOCS1 is a critical inhibitor of interferon gamma signaling and prevents the potentially fatal neonatal actions of this cytokine. Cell 98:597-608

Asher DR, Cerny AM, Weiler SR, Horner JW, Keeler ML, Neptune MA, Jones SN, Bronson RT, Depinho RA, Finberg RW (2005) Coxsackievirus and adenovirus receptor is essential for cardiomyocyte development. Genesis 42:77-85

Badorff C, Lee GH, Lamphear BJ, Martone ME, Campbell KP, Rhoads RE, Knowlton KU (1999) Enteroviral protease 2A cleaves dystrophin: evidence of cytoskeletal disruption in an acquired cardiomyopathy. Nat Med 5:320-326

Badorff C, Berkley N, Mehrotra S, Rhoads RE, Knowlton KU (2000a) Enteroviral protease 2A directly cleaves dystrophin and is inhibited by a dystrophin-based substrate analogue. J Biol Chem 275:1191-1197

Badorff C, Lee SH, Xiong DD, Lee GH, Thistlethwaite PA, Knowlton KU (2000b) Disruption of the dystrophin-glycoprotein complex in human enteroviral cardiomyopathy. Circulation 102, II

Barco A, Feduchi E, Carrasco L (2000) Poliovirus protease 3C(pro) kills cells by apoptosis. Virology 266:352-360

Bergelson JM, Cunningham JA, Droguett G, Kurt-Jones EA, Krithivas A, Hong JS, Horwitz MS, Crowell RL, Finberg RW (1997) Isolation of a common receptor for Coxsackie B viruses and adenoviruses 2 and 5. Science 275:1320-1323

Bergelson JM, Krithivas A, Celi L, Droguett G, Horwitz MS, Wickham T, Crowell RL, Finberg RW (1998) The murine CAR homolog is a receptor for coxsackie B viruses and adenoviruses. J Virol 72:415-419

Blom N, Hansen J, Blaas D, Brunak S (1996) Cleavage site analysis in picornaviral polyproteins: discovering cellular targets by neural networks. 5:2203-2216

Burch GE, Sun SC, Chu KC, Sohal RS, Colcolough HL (1968) Interstitial and coxsackievirus B myocarditis in infants and children. A comparative histologic and immunofluorescent study of 50 autopsied hearts. JAMA 203:1-8

Carson SD (2001) Receptor for the group B coxsackieviruses and adenoviruses: CAR. Rev Med Virol 11:219-226

Chen JW, Zhou B, Yu QC, Shin SJ, Jiao K, Schneider MD, Baldwin HS, Bergelson JM (2006) Cardiomyocyte-specific deletion of the coxsackievirus and adenovirus receptor results in hyperplasia of the embryonic left ventricle and abnormalities of sinuatrial valves. Circ Res 98:923-930

Chow LH, Beisel KW, McManus BM (1992) Enteroviral infection of mice with severe combined immunodeficiency. Evidence for direct viral pathogenesis of myocardial injury. 66:24-31

Clark ME, Lieberman PM, Berk AJ, Dasgupta A (1993) Direct cleavage of human TATA-binding protein by poliovirus protease 3C in vivo and in vitro. Mol Cell Biol 13:1232-1237

Coyne CB, Bergelson JM (2006) Virus-induced Abl and Fyn kinase signals permit coxsackievirus entry through epithelial tight junctions. Cell 124:119-131

Dalldorf G (1955) The Coxsackie viruses. Ann Rev Microbiol 9:277-296

Dorner AA, Wegmann F, Butz S, Wolburg-Buchholz K, Wolburg H, Mack A, Nasdala I, August B, Westermann J, Rathjen FG et al (2005) Coxsackievirus-adenovirus receptor (CAR) is essential for early embryonic cardiac development. J Cell Sci 118:3509-3521

Durbeej M, Campbell KP (2002) Muscular dystrophies involving the dystrophin-glycoprotein complex: an overview of current mouse models. Curr Opin Genet Dev 12:349-361

Endo TA, Masuhara M, Yokouchi M, Suzuki R, Sakamoto H, Mitsui K, Matsumoto A, Tanimura S, Ohtsubo M, Misawa H et al (1997) A new protein containing an SH2 domain that inhibits JAK kinases. Nature 387:921-924

Fairweather D, Frisancho-Kiss S, Yusung SA, Barrett MA, Davis SE, Gatewood SJ, Njoku DB, Rose NR (2004) Interferon-gamma protects against chronic viral myocarditis by reducing mast cell degranulation, fibrosis, and the profibrotic cytokines transforming growth factor-beta 1, interleukin-1 beta, and interleukin-4 in the heart. Am J Pathol 165:1883-1894

Fechner H, Noutsias M, Tschoepe C, Hinze K, Wang X, Escher F, Pauschinger M, Dekkers D, Vetter R, Paul M et al (2003) Induction of coxsackievirus-adenovirus-receptor expression during myocardial tissue formation and remodeling: identification of a cell-to-cell contact-dependent regulatory mechanism. Circulation 107:876-882

Fechner RE, Smith MG, Middlekamp JN (1963) Coxsackie B virus infection of the newborn. Am J Pathol 42:493-505

Fuse K, Chan G, Liu Y, Gudgeon P, Husain M, Chen M, Yeh WC, Akira S, Liu PP (2005) Myeloid differentiation factor-88 plays a crucial role in the pathogenesis of Coxsackievirus B3-induced myocarditis and influences type I interferon production. Circulation 112:2276-2285

Gerzen P, Granath A, Holmgren B, Zetterquist S (1972) Acute myocarditis. A follow-up study. Br Heart J 34:575-583

Godeny EK, Sprague EA, Schwartz CJ, Gauntt CJ (1986) Coxsackievirus group B replication in cultured fetal baboon aortic smooth muscle cells. J Med Virol 20:135-149

Gradi A, Svitkin YV, Imataka H, Sonenberg N (1998) Proteolysis of human eukaryotic translation initiation factor eIF4GII, but not eIF4GI, coincides with the shutoff of host protein synthesis after poliovirus infection. Proc Natl Acad Sci U S A 95:11089-11094

Greenlund AC, Farrar MA, Viviano BL, Schreiber RD (1994) Ligand-induced IFN gamma receptor tyrosine phosphorylation couples the receptor to its signal transduction system (p91) EMBO J 13:1591-1600

Grist NR, Bell EJ (1969) Coxsackie viruses and the heart. Am Heart J 77:295-300

Hashimoto I, Komatsu T (1978) Myocardial changes after infection with Coxsackie virus B3 in nude mice. Br J Exp Pathol 59:13-20

Hashimoto I, Tatsumi M, Nakagawa M (1983) The role of T lymphocytes in the pathogenesis of Coxsackie virus B3 heart disease. Br J Exp Pathol 64:497-504

Heim MH, Kerr IM, Stark GR, Darnell JE Jr (1995) Contribution of STAT SH2 groups to specific interferon signaling by the Jak-STAT pathway. Science 267:1347-1349

Hirota H, Chen J, Betz UA, Rajewsky K, Gu Y, Ross J Jr, Muller W, Chien KR (1999) Loss of a gp130 cardiac muscle cell survival pathway is a critical event in the onset of heart failure during biomechanical stress. Cell 97:189-198

Hohenadl C, Klingel K, Mertsching J, Hofschneider PH, Kandolf R (1991) Strand-specific detection of enteroviral RNA in myocardial tissue by in situ hybridization. Mol Cell Probes 5:11-20

Isaacs A, Lindemann J (1957) Proc R Soc Lond 147:258

Kandolf R, Hofschneider PH (1989) Viral heart disease. Springer Semin Immunopathol 11:1-13

Kandolf R, Canu A, Hofschneider PH (1985) Coxsackie B3 virus can replicate in cultured human foetal heart cells and is inhibited by interferon. J Mol Cell Cardiol 17:167-181

Kandolf R, Ameis D, Kirschner P, Canu A, Hofschneider PH (1987) In situ detection of enteroviral genomes in myocardial cells by nucleic acid hybridization: an approach to the diagnosis of viral heart disease. Proc Natl Acad Sci U S A 84:6272-6276

Karupiah G, Xie QW, Buller RM, Nathan C, Duarte C, MacMicking JD (1993) Inhibition of viral replication by interferon-gamma-induced nitric oxide synthase. Science 261:1445-1448

Kashimura T, Kodama M, Hotta Y, Hosoya J, Yoshida K, Ozawa T, Watanabe R, Okura Y, Kato K, Hanawa H et al (2004) Spatiotemporal changes of coxsackievirus and adenovirus receptor in rat hearts during postnatal development and in cultured cardiomyocytes of neonatal rat. Virchows Arch 444:283-292

Kerekatte V, Keiper BD, Badorff C, Cai A, Knowlton KU, Rhoads RE (1999) Cleavage of Poly(A)-binding protein by coxsackievirus 2A protease in vitro and in vivo: another mechanism for host protein synthesis shutoff? J Virol 73:709-717

Kibrick S, Benirschke K (1958) Severe generalized disease (encephalohepatomyocarditis) occurring in the newborn period and due to infection with Coxsackie virus, group B: evidence of intrauterine infection with this agent. Pediatrics 22:857-875

Klingel K, Hohenadl C, Canu A, Albrecht M, Seemann M, Mall G, Kandolf R (1992) Ongoing enterovirus-induced myocarditis is associated with persistent heart muscle infection: quantitative analysis of virus replication, tissue damage, and inflammation. Proc Natl Acad Sci U S A 89:314-318

Klingel K, Rieger P, Mall G, Selinka HC, Huber M, Kandolf R (1998) Visualization of enteroviral replication in myocardial tissue by ultrastructural in situ hybridization: identification of target cells and cytopathic effects. Lab Invest 78:1227-1237

Koenig M, Kunkel LM (1990) Detailed analysis of the repeat domain of dystrophin reveals four potential hinge segments that may confer flexibility. J Biol Chem 265:4560-4566

Kuyumcu-Martinez NM, Joachims M, Lloyd RE (2002) Efficient cleavage of ribosome-associated poly(A)-binding protein by enterovirus 3C protease. J Virol 76:2062-2074

Kyu B, Matsumori A, Sato Y, Okada I, Chapman NM, Tracy S (1992) Cardiac persistence of cardioviral RNA detected by polymerase chain reaction in a murine model of dilated cardiomyopathy. Circulation 86:522-530

Lamphear BJ, Yan R, Yang F, Waters D, Liebig HD, Klump H, Kuechler E, Skern T, Rhoads RE (1993) Mapping the cleavage site in protein synthesis initiation factor eIF-4 gamma of the 2A proteases from human Coxsackievirus and rhinovirus. J Biol Chem 268:19200-19203

Lamphear BJ, Kirchweger R, Skern T, Rhoads RE (1995) Mapping of functional domains in eukaryotic protein synthesis initiation factor 4G (eIF4G) with picornaviral proteases. Implications for cap-dependent and cap-independent translational initiation. 270:21975-21983

Lee GH, Badorff C, Knowlton KU (2000) Dissociation of sarcoglycans and the dystrophin carboxyl terminus from the sarcolemma in enteroviral cardiomyopathy. Circ Res 87:489-495

Lee KJ, Ross RS, Rockman HA, Harris AN, O'Brien TX, Van Bilsen M, Shubeita HE, Kandolf R, Brem G, Price J (1992) Myosin light chain-2 luciferase transgenic mice reveal distinct regulatory programs for cardiac and skeletal muscle-specific expression of a single contractile protein gene. J Biol Chem 267:15875-15885

Lee YT, Sung K, Shin JO, Jeon ES (2006) Images in cardiovascular medicine. Disruption of dystrophin in acute fulminant coxsackieviral B4 infection. Circulation 113:e76-e77

Lerner AM, Wilson FM (1973) Virus myocardiopathy. Prog Med Virol 15:63-91

Li Y, Bourlet T, Andreoletti L, Mosnier JF, Peng T, Yang Y, Archard LC, Pozzetto B, Zhang H (2000) Enteroviral capsid protein VP1 is present in myocardial tissues from some patients with myocarditis or dilated cardiomyopathy. Circulation 101:231-234

Marine JC, Topham DJ, McKay C, Wang D, Parganas E, Stravopodis D, Yoshimura A, Ihle JN (1999) SOCS1 deficiency causes a lymphocyte-dependent perinatal lethality. Cell 98:609-616

Martino TA, Liu P, Sole MJ (1994) Viral infection and the pathogenesis of dilated cardiomyopathy. 74:182-188

Matsumori A, Tomioka N, Kawai C (1988) Protective effect of recombinant alpha interferon on coxsackievirus B3 myocarditis in mice. Am Heart J 115:1229-1232

Matsumura K, Tomae FM, Ionasescu V, Ervasti JM, Anderson RD, Romero NB, Simon D, Raecan D, Kaplan JC, Fardeau M et al (1993) Deficiency of dystrophin-associated proteins in Duchenne muscular dystrophy patients lacking COOH-terminal domains of dystrophin. J Clin Invest 92:866-871

McManus BM, Chow LH, Wilson JE, Anderson DR, Gulizia JM, Gauntt CJ, Klingel KE, Beisel KW, Kandolf R (1993) Direct myocardial injury by enterovirus: a central role in the evolution of murine myocarditis. Clin Immunol Immunopathol 68:159-169

Muller U, Steinhoff U, Reis LF, Hemmi S, Pavlovic J, Zinkernagel RM, Aguet M (1994) Functional role of type I and type II interferons in antiviral defense. Science 264:1918-1921

Muntoni F, Cau M, Ganau A, Congiu R, Arvedi G, Mateddu A, Marrosu MG, Cianchetti C, Realdi G, Cao A (1993) Brief report: deletion of the dystrophin muscle-promoter region associated with X-linked dilated cardiomyopathy. N Engl J Med 329:921-925

Naka T, Narazaki M, Hirata M, Matsumoto T, Minamoto S, Aono A, Nishimoto N, Kajita T, Taga T, Yoshizaki K et al (1997) Structure and function of a new STAT-induced STAT inhibitor. Nature 387:924-929

Nicholson SE, Willson TA, Farley A, Starr R, Zhang JG, Baca M, Alexander WS, Metcalf D, Hilton DJ, Nicola NA (1999) Mutational analyses of the SOCS proteins suggest a dual domain requirement but distinct mechanisms for inhibition of LIF and IL-6 signal transduction. EMBO J 18:375-385

Poller W, Fechner H, Noutsias M, Tschoepe C, Schultheiss HP (2002) Highly variable expression of virus receptors in the human cardiovascular system. Implications for cardiotropic viral infections and gene therapy. Z Kardiol 91:978-991

Racaniello VR (2007) Picornaviridae: the viruses and their replication. In: Knipe DM, Howley PM (eds) Fields' virology. Lippincott-Raven, Philadelphia, pp

Sainani GS, Dekate MP, Rao CP (1975) Heart disease caused by Coxsackie virus B infection. Br Heart J 37:819-823

Seipelt J, Liebig HD, Sommergruber W, Gerner C, Kuechler E (2000) 2A proteinase of human rhinovirus cleaves cytokeratin 8 in infected HeLa cells. J Biol Chem 275:20084-20089

Sun MC, Smith VM (1966) Hepatitis associated with myocarditis. Unusual manifestation of infection with Coxsackie virus group B, type 3. N Engl J Med 274:190-193

Sutinen S, Kalliomaki JL, Pohjonen R, Vastamaki R (1971) Fatal generalized coxsackie B3 virus infection in an adolescent with successful isolation of the virus from pericardial fluid. Ann Clin Res 3:241-246

Towbin JA (1998) The role of cytoskeletal proteins in cardiomyopathies. Curr Opin Cell Biol 10:131-139

Towbin JA, Hejtmancik JF, Brink P, Gelb B, Zhu XM, Chamberlain JS, McCabe ER, Swift M (1993) X-linked dilated cardiomyopathy. Molecular genetic evidence of linkage to the Duchenne muscular dystrophy dystrophin gene at the Xp21 locus. Circulation 87:1854-1865

Tracy S, Hofling K, Pirruccello S, Lane PH, Reyna SM, Gauntt CJ (2000) Group B coxsackievirus myocarditis and pancreatitis: connection between viral virulence phenotypes in mice. J Med Virol 62:70-81

Tsubata S, Bowles KR, Vatta M, Zintz C, Titus J, Muhonen L, Bowles NE, Towbin JA (2000) Mutations in the human delta-sarcoglycan gene in familial and sporadic dilated cardiomyopathy. J Clin Invest 106:655-662

Van der Kooi AJ, de Voogt WG, Barth PG, Busch HF, Jennekens FG, Jongen PJ, de Visser M (1998) The heart in limb girdle muscular dystrophy. Heart 79:73-77

Wang YX, da Cunha V, Vincelette J, White K, Velichko S, Xu Y, Gross C, Fitch RM, Halks-Miller M, Larsen BR et al (2007) Antiviral and myocyte protective effects of murine interferon beta and alpha2 in coxsackievirus B3-induced myocarditis and epicarditis in Balb/c mice. Am J Physiol Heart Circ Physiol 293:H69-H76

Wessely R, Henke A, Zell R, Kandolf R, Knowlton KU (1998a) Low-level expression of a mutant coxsackieviral cDNA induces a myocytopathic effect in culture: an approach to the study of enteroviral persistence in cardiac myocytes. Circulation 98:450-457

Wessely R, Klingel K, Santana LF, Dalton N, Minoru H, Lederer WJ, Kandolf R, Knowlton KU (1998b) Transgenic expression of replication-restricted enteroviral genomes in heart muscle induces defective excitation-contraction coupling and dilated cardiomyopathy. J Clin Invest 102:1444-1453

Wessely R, Klingel K, Knowlton KU, Kandolf R (2001) Cardioselective infection with coxsackievirus B3 requires intact type I interferon signaling : implications for mortality and early viral replication. Circulation 103:756-761

Wilson FM, Miranda QR, Chason JL, Lerner AM (1969) Residual pathologic changes following murine coxsackie A and B myocarditis. 55:253-265

Windorfer A, Sitzmann FC (1971) Acute virus myocarditis in infants and children. Dtsch Med Wochenschr 96:1177-1184

Xiong D, Lee GH, Badorff C, Dorner A, Lee S, Wolf P, Knowlton KU (2002) Dystrophin deficiency markedly increases enterovirus-induced cardiomyopathy: a genetic predisposition to viral heart disease. Nat Med 8:872-877

Xiong D, Yajima T, Lim BK, Stenbit A, Dublin A, Dalton ND, Summers-Torres D, Molkentin JD, Duplain H, Wessely R et al (2007) Inducible cardiac-restricted expression of enteroviral protease 2A is sufficient to induce dilated cardiomyopathy. Circulation 115:94-102

Yajima T, Yasukawa H, Jeon ES, Xiong D, Dorner A, Iwatate M, Nara M, Zhou H, Summers-Torres D, Hoshijima M et al (2006) Innate defense mechanism against virus infection within the cardiac myocyte requiring gp130-STAT3 signaling. Circulation 114:2364-2373

Yasukawa H, Sasaki A, Yoshimura A (2000) Negative regulation of cytokine signaling pathways. Ann Rev Immunol 18:143-164

Yasukawa H, Hoshijima M, Gu Y, Nakamura T, Pradervand S, Hanada T, Hanakawa Y, Yoshimura A, Ross J Jr, Chien KR (2001) Suppressor of cytokine signaling-3 is a biomechanical stress-inducible gene that suppresses gp130-mediated cardiac myocyte hypertrophy and survival pathways. J Clin Invest 108:1459-1467

Yasukawa H, Yajima T, Duplain H, Iwatate M, Kido M, Hoshijima M, Weitzman MD, Nakamura T, Woodard S, Xiong D et al (2003) The suppressor of cytokine signaling-1 (SOCS1) is a novel therapeutic target for enterovirus-induced cardiac injury. J Clin Invest 111:469-478

Zhao X, Lamphear BJ, Xiong D, Knowlton K, Rhoads RE (2003) Protection of cap-dependent protein synthesis in vivo and in vitro with an eIF4G-1 variant highly resistant to cleavage by Coxsackievirus 2A protease. J Biol Chem 278:4449-4457

# Index

**A**
A33 protein, 75
Acinar cell, 243–249
Acinoductular metaplasia, 245–248, 252, 254
Acute pancreatitis, 241–251, 254, 255
Adjuvant, 295, 296, 301, 302
Akt, 187–189
Antisense oligonucleotides
    therapeutics, 192–193
Apoptosis, 245, 246, 248, 252
Aseptic meningitis, 50, 56
Attenuation, 50, 52, 53
Autoantibody, 295, 308
Autoimmune diseases, 294–296, 304–306
Autoimmunity, 199, 208–210, 212–213,
    260–261, 294–296, 299, 303,
    305, 307–309
    and hygiene, 260–261

**C**
Caveolin-1, phosphorylation, 80. *See also*
    Virus internalization
Cerebrospinal fluid, 228
    hypoglycorrhachia, 229
Chemokines, 249, 250, 253
Chronic pancreatitis, 241, 242, 244, 246–248,
    251–255
Circularization, 98, 103, 104, 109, 110, 113
*cis*-acting replication element (CRE), 41–42
Clathrin
    CVB4 internalization and clathrin-coated
        pits, 79, 81–82
    CVB3 particles colocalization with, 79
Clinical syndromes
    asymptomatic infection, 225
    bornholm disease, 230
    chronic meningoencephalitis
        in agammaglobulinemia, 229

dermatologic, 226
encephalitis, 229, 233
febrile illness, 225, 231
hepatitis, 230–232
herpangina, 227, 228
lower respiratory tract, 227
meningitis, 228, 229, 231
meningoencephalitis, 229, 231
neonatal infection, 230–232
neonatal myocarditis, 231
paralysis, 228, 229
perinatal maternal infection, 231
pharyngitis, 227
pleurodynia, 228, 230
respiratory, 227–228
summer cold, 227
Complement, 303
Coxsackievirus, 316–324, 327, 329
Coxsackievirus and adenovirus receptor
    (CAR), 68
    activation of ERK, 187
    D1 in *Escherichia coli*, 70. *See also protein*
        CAR protein
    carbohydrate modification, 73
    for cell-adhesion, 71
    dimeric and monomeric forms, 72
    to fiber knob (head) domain of
        adenovirus-12, 73
    forming dimers, 72
    PDB-based structure of, 71
    D2, Ig constant (C2) type domain, 71
    entry, 178–180
    mRNA, 69, 74, 77
    protein, 68, 69
        accessibility, 78
        antigens, CVB-infected HeLa cells, 74
        cytoplasmic tail of, 74
        expression, in cells and tissues
            in the adult heart, 77

Coxsackievirus and adenovirus receptor
(CAR) *(cont.)*
   antibody specificity, 77
   in cultures of polarized epithelial
      cells, 78
      Rmcb binding native first Ig motif
         (D1), 76
   extracellular domain of, 74
   homology of, 75
   icosahedral two-fold axes, 73
   inhibition, 72
   interaction with CVB, 78–81
      viral A-particles, formation of, 79
   and occludin in tight junctions of
      choroids plexus epithelium, 75
   production, of inflammatory products,
      74–75
   pseudogenes, 69
   therapeutics, 191–192
Coxsackievirus B3 strain
   cardiovirulence, 283
   CVB3/H3, 282
   CVB3/H310A1, 282
   CVB3/Woodruff, 282
Coxsackievirus, CryoEM structure of, 72
Coxsackieviruses B group (CVB), 68
   cell penetration, 78–81
   generalized DAF-binding phenotypes, 80
CTX protein, 75–77
CVB3 capsid
   the CAR D1 and D1+D2 fragment,
      detection, 73, 74
   CAR dimers with two binding sites, 74
CVB3-CAR complexes, CryoEM
   analysis, 73
Cytokines, 244, 247–251, 253, 255, 294, 295,
   302–305, 308
CXADR gene, 68

**D**
DAF, 282, 283
   activation of ERK, 187
   as alternative receptors, 80
   entry, 178–180
   therapeutics, 191–192
Detection of enteroviral RNA
   frequency
      immunohistochemical, 277
   isolation, 318
   negative strand, 278, 279
   RT-PCR, 277–279, 282
Diabetes, 50, 52, 57–58, 60
   and CVB, 266–268

Diabetogenic CVB, 264–267
Diagnosis
   cell culture, 232, 233
   identification, 232, 233
   interpretation, 233
   nucleic acid sequence-based
      amplification, 232
   reverse transcription-polymerase chain
      reaction, 232
   sensitivity, 232, 233
   serologic, 233
Digestive enzymes, 243, 245, 246, 250
Dilated cardiomyopathy
   outcome, 279, 280
Dynamin protein, 79
Dystrophin, 316, 321–327, 331

**E**
Echovirus, 34, 35
Encephalomyocarditis, 56–57, 59
Endocrine pancreas, 242, 244
Epidemiology, 223–224, 230
   neonatal, 230–231
   seasonality, 224
   shedding, 233
   transmission, 224
Epitope, 296, 307, 308
ERK, 186–187, 189
Eukaryotic translation initiation factor 4G
   (eIF4G)
   cleavage during viral infection, 133, 136
   interaction with viral IRES, 129, 136
Exocrine pancreas, 241, 243, 245–249, 251,
   252, 255

**F**
Fiber knob proteins, 69
Fibrosis, 245–249, 252, 254, 255
Fitness, 52

**G**
GenBank EST database, 69
Genetic regulation, 305–306
Genome sequence or Genome sequencing, 34,
   35, 38, 43

**H**
hCAR1 protein, structure of, 70
Helicase, 93, 97, 112
HLA, 305

Host response, 199–205, 208, 213, 214
Human CAR gene, localization, 69
Hygiene, 51
Hygiene hypothesis, 261, 267, 268

**I**

IFN-gamma, 304–306
IL-4, 304
IL-12, 304
IL-13, 304
IL-17, 305
IL-23, 304
Immune response innate adaptive, 301–303
Immunopathology
   autoimmunity, 276
Inflammation, 242–248, 250–255
Innate immunity, 200–210, 316, 321, 327–331
Insulitis and CVB infection, 267–269
Internal ribosome entry sequence (IRES)
   attenuation mutants in, 139
   classification, 124
   encephalomyocarditis virus, 125
   function, 124–127, 129, 130, 133, 137–139
   IRES trans-activating factors and, 130
   structure, 123–127, 130, 138
Internal ribosome entry site (IRES), 36, 91, 93, 107, 108
Islets of Langerhans, 242, 244

**J**

JAM-A protein, 76

**L**

Latency, 151, 153, 156, 162, 165, 166
Lupus autoantigen (La)
   cleavage during virus infection, 137
   role in viral translation, 130, 131, 137
Lymphocytes, 295, 301

**M**

Macrophages, 245, 248–253, 298, 306–307
Major histocompatibility locus (MHC), 295, 298, 302, 305, 309
Matrixmetallo proteinases, 178, 183–186
Mimicry, 296, 297
MMPs. *See* Matrixmetallo proteinases

Mortality, 231
Murine model
   A/J, 280–283
   BALB/c, 280
   C3H/HeJ, 280–282
   C57BL/6, 280, 282, 301
   detection of CVB3 RNA, 280, 281
   NMRI, 280
   NOD 57, 265–267
   persistence of virus, 280, 282
   SWR, 280
Mutant spectrum, 7, 11, 14, 20–22, 24
Mutation, 3–12, 14–17, 19, 20, 22, 23, 25
Myocarditis, 316–319, 321, 322, 326–328, 330–331

**N**

Necrosis, 244–246
Neurovirulence, 54
NF-κB, 250–252
Nitric oxide, 303–304
NOD mice and CVB, 265–267
NOD mouse, 57
Nonstructural region, 39–41
Non-translated region (NTR), 42
   3′ NTR, 35, 36
   5′ NTR, 36

**P**

Pancreatic cancer, 241, 242, 251–255
Pancreatitis, 50, 52, 54–56, 58
Pancreatitis-associated protein (PAP), 250
PDZ-proteins, 69
Phylogeny, 33, 34, 40, 41
Picornavirus, interaction with Ig-type receptors, 73
Poliovirus, 50, 60, 123–125, 128, 130–132, 135–137, 139
Poliovirus receptor (PVR), 73
Poly(A)-binding protein (PABP)
   in host cell shutoff, 136
   in viral translation shutoff, 132, 134, 137
Poly(rC)-binding protein 2 (PCBP2)
   cleavage during virus infection, 137
   role in viral translation, 132
Polymerase
   3D$^{pol}$, 94–97, 107
Polyprotein, 37–39
Polypyrimidine tract binding protein
   cleavage during virus infection, 137
   role in viral translation, 131, 137
Prevention, 223, 234–235

Primary cell cultures, 284
Proliferating cells, preference of viruses for, 151
Proteinase
　2A, 91, 96
　C, 91, 95, 96
　cysteine, 92, 96
Proteins
　cellular, 91, 92, 94–97, 99, 103–106, 109, 110
　nonstructural, 89, 90, 92, 96, 103–105, 107, 108, 111, 113

## Q
Quasispecies, 53–56, 58
Quiescent cells, 150, 153, 156, 160, 161

## R
Recombination, 3, 4, 7–12, 15–19, 23, 33, 34, 39, 41–43
Replication
　viral, 90, 105
Replicative form (RF), 109, 111
Replicative intermediate (RI), 109, 112
Rhinovirus, 73
RNA replication
　cell cycle arrest, 281
　hnRNPC1, 281
RNA virus, 4, 5, 8–12, 16, 19, 22

## S
Secondary structure
　RNA, 89–92, 97, 101
Sequence diversity, 37–39
Src, 186–187, 189
Stem cells, neural, 155–156
Swine vesicular disease virus (SVDV), 35, 38, 40, 41

## T
Taxonomy, 47
T cells, 201, 202, 204, 206, 208–214, 295–297, 299, 303, 304, 306–308
Terminally deleted defective enterovirus, 275, 279, 281

Ternary complex, 96, 98, 103, 110
Toll-Like Receptors, 200, 202–205
Translation
　cap-independent, 91–93, 108
　host cell
　　mRNA circularization, 136
　　shutoff of, 133, 136
　viral
　　canonical initiation factors in, 124, 127, 128
　　cell-specific restriction, 138–140
　　encephalomyocarditis virus, 125, 140
　　genome circularization, 135, 137
　　IRES (*see* Internal ribosome entry sequence)
　　IRES trans-activating factors, 130
　　shutoff of, 133–138
Treatment, 223, 233–234, 279, 280, 283
　immunoglobulin, 234
　pleconaril, 233, 234

## U
Ubiquitin, 184, 188–189, 192
Upstream of n-ras (UNR)
　role in viral translation, 132, 138

## V
Vesicles
　membranous, 90, 91, 93–97, 99, 101, 105, 106, 108, 111–113
Viral cardiomyopathy, 77. *See also* CAR protein
Viral disease, 8, 22–24
Viral persistence, 11, 20
Viral swarm, 52
Virulence, 49–55, 58, 59
Virus dose and T1D, 264–265
Virus internalization, 80, 81
Viruses and TID onset, 265–266
VPg
　uridylylation, 94–101, 103, 110–112

## X
*Xenopus laevis* cortical thymocyte protein. *See* CTX
X-linked agammaglobulinemia, 229, 234

# Current Topics in Microbiology and Immunology
Volumes published since 1989

Vol. 271: **Koehler, Theresa M. (Ed.):** Anthrax. 2002. 14 figs. X, 169 pp. ISBN 3-540-43497-6

Vol. 272: **Doerfler, Walter; Böhm, Petra (Eds.):** Adenoviruses: Model and Vectors in Virus-Host Interactions. Virion and Structure, Viral Replication, Host Cell Interactions. 2003. 63 figs., approx. 280 pp. ISBN 3-540-00154-9

Vol. 273: **Doerfler, Walter; Böhm, Petra (Eds.):** Adenoviruses: Model and Vectors in VirusHost Interactions. Immune System, Oncogenesis, Gene Therapy. 2004. 35 figs., approx. 280 pp. ISBN 3-540-06851-1

Vol. 274: **Workman, Jerry L. (Ed.):** Protein Complexes that Modify Chromatin. 2003. 38 figs., XII, 296 pp. ISBN 3-540-44208-1

Vol. 275: **Fan, Hung (Ed.):** Jaagsiekte Sheep Retrovirus and Lung Cancer. 2003. 63 figs., XII, 252 pp. ISBN 3-540-44096-3

Vol. 276: **Steinkasserer, Alexander (Ed.):** Dendritic Cells and Virus Infection. 2003. 24 figs., X, 296 pp. ISBN 3-540-44290-1

Vol. 277: **Rethwilm, Axel (Ed.):** Foamy Viruses. 2003. 40 figs., X, 214 pp. ISBN 3-540-44388-6

Vol. 278: **Salomon, Daniel R.; Wilson, Carolyn (Eds.):** Xenotransplantation. 2003. 22 figs., IX, 254 pp. ISBN 3-540-00210-3

Vol. 279: **Thomas, George; Sabatini, David; Hall, Michael N. (Eds.):** TOR. 2004. 49 figs., X, 364 pp. ISBN 3-540-00534X

Vol. 280: **Heber-Katz, Ellen (Ed.):** Regeneration: Stem Cells and Beyond. 2004. 42 figs., XII, 194 pp. ISBN 3-540-02238-4

Vol. 281: **Young, John A. T. (Ed.):** Cellular Factors Involved in Early Steps of Retroviral Replication. 2003. 21 figs., IX, 240 pp. ISBN 3-540-00844-6

Vol. 282: **Stenmark, Harald (Ed.):** Phosphoinositides in Subcellular Targeting and Enzyme Activation. 2003. 20 figs., X, 210 pp. ISBN 3-540-00950-7

Vol. 283: **Kawaoka, Yoshihiro (Ed.):** Biology of Negative Strand RNA Viruses: The Power of Reverse Genetics. 2004. 24 figs., IX, 350 pp. ISBN 3-540-40661-1

Vol. 284: **Harris, David (Ed.):** Mad Cow Disease and Related Spongiform Encephalopathies. 2004. 34 figs., IX, 219 pp. ISBN 3-540-20107-6

Vol. 285: **Marsh, Mark (Ed.):** Membrane Trafficking in Viral Replication. 2004. 19 figs., IX, 259 pp. ISBN 3-540-21430-5

Vol. 286: **Madshus, Inger H. (Ed.):** Signalling from Internalized Growth Factor Receptors. 2004. 19 figs., IX, 187 pp. ISBN 3-540-21038-5

Vol. 287: **Enjuanes, Luis (Ed.):** Coronavirus Replication and Reverse Genetics. 2005. 49 figs., XI, 257 pp. ISBN 3-540- 21494-1

Vol. 288: **Mahy, Brain W. J. (Ed.):** Foot-and-Mouth-Disease Virus. 2005. 16 figs., IX, 178 pp. ISBN 3-540-22419X

Vol. 289: **Griffin, Diane E. (Ed.):** Role of Apoptosis in Infection. 2005. 40 figs., IX, 294 pp. ISBN 3-540-23006-8

Vol. 290: **Singh, Harinder; Grosschedl, Rudolf (Eds.):** Molecular Analysis of B Lymphocyte Development and Activation. 2005. 28 figs., XI, 255 pp. ISBN 3-540-23090-4

Vol. 291: **Boquet, Patrice; Lemichez Emmanuel (Eds.):** Bacterial Virulence Factors and Rho GTPases. 2005. 28 figs., IX, 196 pp. ISBN 3-540-23865-4

Vol. 292: **Fu, Zhen F. (Ed.):** The World of Rhabdoviruses. 2005. 27 figs., X, 210 pp. ISBN 3-540-24011-X

Vol. 293: **Kyewski, Bruno; Suri-Payer, Elisabeth (Eds.):** CD4+CD25+ Regulatory T Cells: Origin, Function and Therapeutic Potential. 2005. 22 figs., XII, 332 pp. ISBN 3-540-24444-1

Vol. 294: **Caligaris-Cappio, Federico, Dalla Favera, Ricardo (Eds.):** Chronic Lymphocytic Leukemia. 2005. 25 figs., VIII, 187 pp. ISBN 3-540-25279-7

Vol. 295: **Sullivan, David J.; Krishna Sanjeew (Eds.):** Malaria: Drugs, Disease and Post-genomic Biology. 2005. 40 figs., XI, 446 pp. ISBN 3-540-25363-7

Vol. 296: **Oldstone, Michael B. A. (Ed.):** Molecular Mimicry: Infection Induced Autoimmune Disease. 2005. 28 figs., VIII, 167 pp. ISBN 3-540-25597-4

Vol. 297: **Langhorne, Jean (Ed.):** Immunology and Immunopathogenesis of Malaria. 2005. 8 figs., XII, 236 pp. ISBN 3-540-25718-7

Vol. 298: **Vivier, Eric; Colonna, Marco (Eds.):** Immunobiology of Natural Killer Cell Receptors. 2005. 27 figs., VIII, 286 pp. ISBN 3-540-26083-8

Vol. 299: **Domingo, Esteban (Ed.):** Quasispecies: Concept and Implications. 2006. 44 figs., XII, 401 pp. ISBN 3-540-26395-0

Vol. 300: **Wiertz, Emmanuel J.H.J.; Kikkert, Marjolein (Eds.):** Dislocation and Degradation of Proteins from the Endoplasmic Reticulum. 2006. 19 figs., VIII, 168 pp. ISBN 3-540-28006-5

Vol. 301: **Doerfler, Walter; Böhm, Petra (Eds.):** DNA Methylation: Basic Mechanisms. 2006. 24 figs., VIII, 324 pp. ISBN 3-540-29114-8

Vol. 302: **Robert N. Eisenman (Ed.):** The Myc/Max/Mad Transcription Factor Network. 2006. 28 figs., XII, 278 pp. ISBN 3-540-23968-5

Vol. 303: **Thomas E. Lane (Ed.):** Chemokines and Viral Infection. 2006. 14 figs. XII, 154 pp. ISBN 3-540-29207-1

Vol. 304: **Stanley A. Plotkin (Ed.):** Mass Vaccination: Global Aspects – Progress and Obstacles. 2006. 40 figs. X, 270 pp. ISBN 3-540-29382-5

Vol. 305: **Radbruch, Andreas; Lipsky, Peter E. (Eds.):** Current Concepts in Autoimmunity. 2006. 29 figs. IIX, 276 pp. ISBN 3-540-29713-8

Vol. 306: **William M. Shafer (Ed.):** Antimicrobial Peptides and Human Disease. 2006. 12 figs. XII, 262 pp. ISBN 3-540-29915-7

Vol. 307: **John L. Casey (Ed.):** Hepatitis Delta Virus. 2006. 22 figs. XII, 228 pp. ISBN 3-540-29801-0

Vol. 308: **Honjo, Tasuku; Melchers, Fritz (Eds.):** Gut-Associated Lymphoid Tissues. 2006. 24 figs. XII, 204 pp. ISBN 3-540-30656-0

Vol. 309: **Polly Roy (Ed.):** Reoviruses: Entry, Assembly and Morphogenesis. 2006. 43 figs. XX, 261 pp. ISBN 3-540-30772-9

Vol. 310: **Doerfler, Walter; Böhm, Petra (Eds.):** DNA Methylation: Development, Genetic Disease and Cancer. 2006. 25 figs. X, 284 pp. ISBN 3-540-31180-7

Vol. 311: **Pulendran, Bali; Ahmed, Rafi (Eds.):** From Innate Immunity to Immunological Memory. 2006. 13 figs. X, 177 pp. ISBN 3-540-32635-9

Vol. 312: **Boshoff, Chris; Weiss, Robin A. (Eds.):** Kaposi Sarcoma Herpesvirus: New Perspectives. 2006. 29 figs. XVI, 330 pp. ISBN 3-540-34343-1

Vol. 313: **Pandolfi, Pier P.; Vogt, Peter K. (Eds.):** Acute Promyelocytic Leukemia. 2007. 16 figs. VIII, 273 pp. ISBN 3-540-34592-2

Vol. 314: **Moody, Branch D. (Ed.):** T Cell Activation by CD1 and Lipid Antigens, 2007, 25 figs. VIII, 348 pp. ISBN 978-3-540-69510-3

Vol. 315: **Childs, James, E.; Mackenzie, John S.; Richt, Jürgen A. (Eds.):** Wildlife and Emerging Zoonotic Diseases: The Biology, Circumstances and Consequences of Cross-Species Transmission. 2007. 49 figs. VII, 524 pp. ISBN 978-3-540-70961-9

Vol. 316: **Pitha, Paula M. (Ed.):** Interferon: The 50th Anniversary. 2007. VII, 391 pp. ISBN 978-3-540-71328-9

Vol. 317: **Dessain, Scott K. (Ed.):** Human Antibody Therapeutics for Viral Disease. 2007. XI, 202 pp. ISBN 978-3-540-72144-4

Vol. 318: **Rodriguez, Moses (Ed.):** Advances in Multiple Sclerosis and Experimental Demyelinating Diseases. 2008. XIV, 376. ISBN 978-3-540-73679-9

Vol. 319: **Manser, Tim (Ed.):** Specialization and Complementation of Humoral Immune Responses to Infection. 2008. XII, 174. ISBN 978-3-540-73899-2

Vol. 320: **Paddison, Patrick J.; Vogt, Peter K. (Eds.):** RNA Interference. 2008. VIII, 273. ISBN 978-3-540-75156-4

Vol. 321: **Bruce, Beutler (Ed.):** Immunology, Phenotype First: How Mutations Have Established New Principles and Pathways in Immunology. 2008. ISBN 978-3-540-75202-8

Vol. 322: **Romeo, Tony (Ed.):** Bacterial Biofilms. 2008. XII, 299. ISBN 978-3-540-75417-6

Printing: Krips bv, Meppel, The Netherlands
Binding: Stürtz, Würzburg, Germany